2nd Annual World Congress on Sustainable Engineering 2012

Topical Conference at the 2012 AIChE Annual Meeting

Pittsburgh, Pennsylvania, USA
28 October - 2 November 2012

ISBN: 978-1-62276-748-9

Printed from e-media with permission by:

Curran Associates, Inc.
57 Morehouse Lane
Red Hook, NY 12571

Some format issues inherent in the e-media version may also appear in this print version.

Printed by Curran Associates, Inc. (2013)

For permission requests, please contact AIChE
at the address below.

AIChE
3 Park Avenue
New York, NY 10016-5991

Phone: (203) 702-7660
Fax: (203) 775-5177

www.aiche.org

Additional copies of this publication are available from:

Curran Associates, Inc.
57 Morehouse Lane
Red Hook, NY 12571 USA
Phone: 845-758-0400
Fax: 845-758-2634
Email: curran@proceedings.com
Web: www.proceedings.com

TABLE OF CONTENTS

(44b) Conceptual Net Energy Output for Biofuel Production from Lignocellulosic Biomass Through Biorefining: A Review

J.Y. Zhu, USDA Forest Service, Forest Products Laboratory, Madison, WI

Conceptual net energy output for biofuel production from lignocellulosic biomass through biorefining: A review

J.Y. Zhu[1,2]

[1] USDA Forest Service, Forest Products Laboratory, Madison, WI, USA

[2] Dept. Biological Systems Engineering, University of Wisconsin-Madison, WI, USA

Abstract

The concept of producing lignocellulosic biofuel, bioproducts, and chemicals through a biorefinery using lignocelluloses has been around for 70 years or more. The recent interest in this old concept arises from the promises of mitigating climate change by substituting some biomass energy for petroleum or other fossil-fuel energy; achieving national energy independence and sustainable economic development; and advances in biotechnology in the past 2-3 decades. Geopolitical and national security reasons have greatly contributed to the need for seeking alternative energy, especially using domestic renewable and sustainable sources.

However, there is a lack of comprehensive information in the retrievable literature on pilot scale process and energy data using promising process technologies and commercially scalable and available capital equipment for lignocellulosic biomass biorefining, despite several pilot scale and demonstration plants are under construction. Unfortunately, bioenergy production is not sustainable without net energy production. Recognizing the heated debate on energy production from lignocellulosic biomass in the scientific community and most arguments were not based on actual laboratory data for biorefinery operation, this presentation is intended to provide an objective view on energy production from lignocellulosic biomass. The process data from approximately a dozen studies that represent state-of-the-art in cellulosic biofuel production concepts, along with literature energy input data for agriculture operations, were analyzed to provide estimates of net energy production. Several scenarios were created for proper allocation energy input to biomass and major agriculture co-product, such as corn or lumber.

For biorefinery operations, increasing solids loading in pretreatment to at least 25% is critical to reducing energy input in a biorefinery. Post thermo-chemical pretreatment size reduction approach should be adopted for energy efficient woody biomass processing. When appropriate pretreatment technologies are used, woody biomass can be processed as efficiently as herbaceous biomass and agricultural residues. Net energy output for cellulosic ethanol was estimated to range approximately from -500 to 2000 MJ/ton biomass (HHV base); indicating that the energy input/output ratio is approximately 1:1 for cellulosic ethanol. However, net energy can reach approximately 4000-7000 MJ/ton of biomass when energy from lignin is included.

(44c) The R&D of Cellulosic Ethanol in China

Xinshu Zhuang, Bio-chem conversion lab, Guangzhou institute of energy conversion,Chinese Academy of Sciences, Guangzhou, WI, China

As the biggest developing country, China needs fuel ethanol to meet its demands of its social and economy development. Basing on the "Mid-long term plan for renewable energy" established in 2007, the annual consumption of fuel ethanol will reach 10 million tons in 2020. Cellulosic ethanol, one of the most important second generation bio-fuel, gains more and more attentions for its substitution of the transportation liquid, lessen the air pollution by the vehicles and promote the rural area economy. In China, more than five demonstration projects have been set up since 2008. In this paper, the main cellulosic ethanol processes applied in China are illustrated, simultaneously the R&D of a green and high efficient process pretreated with hot liquid water is presented.

(108a) Optimal Synthesis and Planning of Sustainable Chemical Processes

Ignacio Grossmann, Department of Chemical Engineering, Carnegie Mellon University, Pittsburgh, PA

Sustainability has recently emerged as a key issue in process systems engineering (PSE). Mathematical programming techniques such as mixed-integer programming and global optimization offer a general modeling framework for including environmental concerns in the synthesis and planning of chemical processes. Including environmental issues in the synthesis and planning of chemical processes poses significant challenges that have not yet been fully solved, and hence merit further attention. One major critical issue is how to systematize the search for alternatives leading to reductions in environmental impact. Furthermore, aside from anticipating the effect of uncertainties, which are quite pronounced in this area, there is the issue on how to cope with competing economic and environmental objectives. In the area of process synthesis we address the energy and water optimization of biofuel plants. We present an optimization framework based on superstructure optimization that effectively accounts for energy and water integration, and illustrate its application to the optimization of corn-based ethanol and lignocellulosic bioethanol plants. In the area of supply chain management we address the environmentally conscious design and planning of hydrogen supply chains for vehicle use using a multiobjective optimization approach coupled with the Eco-indicator 99 to assess the effect on human health. We also address the optimal design of chemical supply chain with uncertain emissions using a chance constrained optimization approach to account for the uncertainties associated with the parameters of LCA models (e.g. emissions, potential harm, etc.).

(108b) Sustainable Manufacturing Advances in Research and Technology: Challenges and Opportunities

Yinlun Huang, Dept. of Chemical Engineering & Material Science, Wayne State University, Detroit, MI

The quest for engineering sustainability reflects a crucial paradigm shift in the 21st century, i.e., from the traditional techno-economic paradigm to triple-bottom-lines-based sustainable system design, manufacturing, and management at various length-time scales. However, sustainability is an extremely complex area of research and practice in terms of scope, contents, and spatial/temporal aspects, and sustainability science is far from exact today. Manufacturing Sustainability needs not only evaluation of the status quo, but also prediction and strategic decision-making for the future, all under the situation of information uncertainty and insufficient knowledge. Thus, how to characterize, analyze, design, and restructure manufacturing systems, from the material-product-process level to the large-scale industrial system level, in the context of sustainability, is a new research frontier in engineering that needs much exploration.

In this presentation, the challenges, opportunities, and research frontiers in manufacturing sustainability research and practice will be discussed first. Then, a Multiscale Design-Manufacturing for Sustainability (MDMS) methodology will be introduced, which is developed by resorting to basic sustainability science, multiscale complex systems theory, uncertainty theory, and control theory. The methodology is capable of performing hierarchical system modeling, comprehensive sustainability analysis, and multiscale decision making under uncertainty. The methodological efficacy will be demonstrated by a number of applied studies, ranging from sustainable design of nanomaterials and manufacturing, to collaborative profitable pollution prevention in surface coating systems, and to industrial-zone-based technology network sustainability. Finally, an NSF funded academic-industrial collaboration involving eight countries on sustainable manufacturing research, education, and technology development will be highlighted and future directions will be discussed.

(115a) Comparison of Environmental Releases in Ethanol and Gasoline Production

Debalina Sengupta[1], Raymond L. Smith[2] and Troy R. Hawkins[2], (1)National Risk Management Research Laboratory, ORISE Post Doctoral Fellow at US Environmental Protection Agency, Cincinnati, OH, (2)National Risk Management Research Laboratory, U.S. Environmental Protection Agency, Cincinnati, OH

As the use of biofuels for transportation grows, it is important to scientifically assess the impacts associated with their production. Currently corn ethanol provides the largest volume of transportation biofuel in the United States with more than 13 billion gallons produced in 2010 and 15 billion gallons projected by 2015.

Most prior studies of fuel ethanol plants include greenhouse gas emissions and in some cases criteria air pollutants. In this contribution the focus is on assessing all reported pollutants released from ethanol plants and petroleum refineries. Ethanol is viewed as an alternative to gasoline, and a model is provided to describe the environmental releases at a facility level for ethanol and gasoline production.

Data is collected from the National Emissions Inventory (NEI) database and compiled for emissions for the species in the database. This includes criteria air pollutants and hazardous air pollutants. This research also shows a relative comparison of toxic releases in the United States with data compiled from the Toxics Release Inventory (TRI). Water and land releases are available from the TRI database.

In this work, the focus is on the dry grind ethanol production process rather than the wet grind alternative because the dry grind process is more directly associated with ethanol as the main product, is responsible for over 85% of US ethanol production, and is the primary technology anticipated for future ethanol production. This study addresses suitable methods and criteria for allocating the co-products of ethanol and gasoline production.

The present work attempts to capture these results and present them for further evaluation.

(115b) Effect of Siloxane Present in Renewable Natural Gas On the Operation of Various Natural Gas Equipment

Nitin Narayanan Nair[1], Jorge Gutierrez[2], Jack Chen[2], Alireza Divsalar[3], Arjun Vas[4], Tongyu Zhao[4], May Lew[2], Fokion N. Egolfopoulos[5] and Theodore Tsotsis[1], (1)Mork Family Department of Chemical Engineering and Materials Science, University of Southern California, Los Angeles, CA, (2)Southern California Gas Company, Los Angeles, CA, (3)Chemical Engineering, University of Southern California, CA, (4)Chemical Engineering, University of Southern California, Los Angeles, CA, (5)Department of Aerospace and Mechanical Engineering, University of Southern California, Los Angeles, CA

Effect of Siloxane Present in Renewable Natural Gas on the Operation of Various Natural Gas Equipment

N. N. Nair[a], J. Gutierrez[c], J. Chen[c], A. Divsalar[a], A. Vas[a], T. Zhao[a], M. Lew[c], F. N. Egolfopoulos[b], T. T. Tsotsis[a]

[a]Department of Chemical Engineering and Materials Science, University of Southern California, CA 90089, USA

[b]Department of Aerospace and Mechanical Engineering, University of Southern California, CA 90089, USA

[c] *Engineering Analysis Center, ML SC723B, Southern California Gas Company*

Abstract

The goal of this project is to systematically investigate, under well-controlled laboratory conditions, the operation of different appliances when using simulated renewable natural gas (RNG), particularly as it relates to the impact of siloxanes a key impurity found in such fuels. Two different devices were investigated, namely a small electric generator operating on RNG, and a typical household furnace. Both appliances have been operated for extended periods of time when using siloxane-containing RNG and their operation has been compared with similar appliances operating on conventional natural gas. The impact of the siloxane combustion (which results in solid components that tend to deposit and coat various surfaces) has been investigated in greater detail. It has been shown that the presence of siloxanes has a very detrimental impact on the life of the various sensors that are utilized to control the operation of such devices. The same is true with the impact on the operating life of catalyst beds which are connected to the devices in order to reduce the pollutants emitted. In addition to coating the various internal surfaces, silica microparticulates resulting from siloxane combustion find their way into the condensate (in the case of the furnace) and the spent oil (in the case of the engine). Studies are ongoing to identify the complete failure mode of these devices and the results will be presented at the meeting.

Keywords: Siloxanes, biomethane, biogas, renewable natural gas, SiO_2 particles

ACKNOWLEDGMENT:

The support of the support of the Southern California Gas Company is gratefully acknowledged, as are many helpful discussions with Dr. Siari Sosa, Mr. Scott Kysar, and Mr. Gilbert Ching

(115c) Population Dynamics of Microalgae

Alim Dewan, Siva A. Vanapalli and M. Nazmul Karim, Chemical Engineering, Texas Tech University, Lubbock, TX

One of the distinctive features of cell-division of microalgae is that various single-cells can divide into different numbers although they are from the same species. This has led us to study the population dynamics of the microalgae starting from single and multiple cells. The goal was to see how the population of algae cells changes over time, or if the daughter cells divide into same numbers as do the mother cells, and if it is possible to artificially select a high-growing algae cell colony. We have investigated the population dynamics of *Chlorella vulgaris* in nano-liter scale static droplet array in PDMS microfluidic chip. Each of the droplets in the chip mimicked a batch bioreactor. The change of number of cells was observed until the growth of the cells reached to a stationary phase. We found that a single *Chlorella vulgaris* cell can divide into 3, 4, 5, 6, and 16 cells. There is a significant heterogeneity in the population dynamics even if the initial cell numbers are the same. We also have observed that the daughter cells do not divide into same numbers as do the mother cells. Currently, using a dynamic droplet generation method, we are developing a method of isolating algae cells that divide into higher number of cells.

(115d) Kinetic and Environmental Impact Modeling of Reaction Products From Supercritical Water Gasification of Phenol

Chad M. Huelsman and Phillip E. Savage, Chemical Engineering, University of Michigan, Ann Arbor, MI

Biomass gasification is a potential source of sustainable fuel gas that can help mitigate growing global energy needs and concerns over fossil fuel depletion and greenhouse gas emissions. Hydrothermal gasification in supercritical water offers improved thermal efficiencies for wet biomass, which otherwise requires energy intensive dewatering. This technology has been demonstrated on many starting feedstocks, including seaweed, microalgae, wood, straw, switchgrass, corn stover, cellulose, and lignin, but, in each case, phenol and/or other aromatic compounds persisted in the liquid phase as undesired byproducts. Phenolic compounds are notoriously difficult to gasify and represent one of the main obstacles to complete gasification of biomass, but little is known about the chemistry that leads to such liquid phase byproducts. In order to better understand this underlying chemistry, kinetic and temperature studies were conducted using phenol as a model compound. Reactions were carried out in quartz batch reactors to eliminate catalytic effects from metal walls. These experiments led to the production of little gas, but instead a swath of heavier molecular weight compounds formed. A reaction network was constructed based on additional experiments and careful analysis of chemical pathways. Many of the non-gaseous byproducts identified in these reactions are polycyclic aromatic hydrocarbons (PAHs), EPA priority pollutants, and/or have known human health and environmental effects. The UNEP/SETAC environmental toxicity model "USEtox" was employed to characterize the human and ecotoxic impacts due to a hypothetical emission of this byproduct stream into freshwater. Coupling environmental impact modeling with kinetic modeling of the formation chemistry of these byproduct molecules can provide the insight needed to minimize their production, decrease the risk associated with an unintended environmental release, and simultaneously increase gas yields.

(115e) Alternatives for Anhydrous Ethanol Production an Economic and Environmental Approach

Michel Kahwaji Janho[1], Jorge E. Gatica[2], Fernando Daniel Mele[3], María Rosa Hernández[4] and Mauricio Colombo[3], (1)Chemical And Biomedical Engineering, Cleveland State University, Cleveland, OH, (2)Chemical and Biomedical Engineering, Cleveland State University, Cleveland, OH, (3)Ingenieria de Procesos y Gestion Industrial, Universidad Nacional de Tucuman, Tucuman, Argentina, (4)Chemistry and Chemical Engineering, Universidad Nacional de Tucuman, Tucuman, Argentina

The production of sugar is one of the oldest agro-industrial activities. It includes the production of sugar cane, sugar and dairy products: paper, alcohol, energy, etc. In addition to the production of sugar for domestic and international markets, one of the activities that is growing in importance is the production of alcohol from molasses. Energetic integration has also resulted in bringing power generation from bagasse as a secondary activity in modern sugar-cane processing industries.

A new chapter for bio-energy products associated with the manufacture of sugar has gained relevance as blending gasoline with ethanol is becoming part of energetic policies in most industrialized countries. Indeed, energy products from renewable sources are receiving increasing attention, and this has generated a renewed stimulus to examine alternative processes for bio-ethanol production. It has become common for sugar mills to have complementary distilleries with production capacities ranging from 50 to 400 liters per day of non-anhydrous alcohol.

At sugar plants, steam is generated in boilers using as fuel bagasse-cellulosic fiber waste obtained crushing the sugar cane- and it is used to give useful heat and shaft work to the plant, but it can also be used to generate electricity with export opportunities to the electrical network. The great number of process alternatives outlines a serious decision making problem in order to take advantage of the resources.

At steady state conditions the amount of bagasse obtained is more than enough to supply the demands of useful heat and energy. Thus these industries can advantage of the emergence of a new potential market, as they have an increased potential for generating a surplus of electricity for sale to the grid.

In this paper, a classic bio-ethanol distillery is adopted as case-study and simulated using AspenPlus® and UniSim ® commercial simulators. Alternative process flow sheets of varying complexity are analyzed aiming to reduce power consumption and effluent generation.

The process simulation allows a systematic analysis that leads to a good combination of all the values of process variables. In particular, since they demand significant energy influx, an important subsystem for analyzing is the separation of alcohol-water mixtures, There are also different technologies for this dehydration. In this paper we consider azeotropic distillation and hybrid schemes with membrane separation. The development of a simulation module for calculating the membranes performance, typically not included in commercial process simulators,

The consumed electric energy and steam depends on the adopted scheme and they have, depending to their source or generation policy, an associated cost and environmental impact. Among the energy systems that deserve special attention, cogeneration in the sugar industry must be pointed out. This part of the process makes efficient use of a common fuel for the generation of heat and power. As a result, the environmental impact and costs associated with the use of electricity and useful heat change appreciably, such as anhydrous alcohol production schemes can be achieved with a lesser environmental at the expense of decreasing the economic benefit which would result from selling electricity to the grid.

Cogeneration is not new in the sugar-cane processing industry. Indeed, typically co-generation refers to the generation of heat as the main objective and the generation of electrical and mechanical power as secondary objectives. To date, however, no versatile tool that would allow analyzing economical feasible alternatives of anhydrous alcohol production, bearing in mind environmental issues, is available.

In addition, the explicit consideration of environmental issues offers an alternative route to explore trade-offs. Life Cycle Analysis (LCA) is used in this work. Unlike most environmental approaches to design that focus on reducing the generation of residuals, but without taking into account the impact associated to related processes, LCA allows considering environmental impact as an integral part of the problem. The software used to evaluate the environmental impact is SimaPro®.

Alternative schemes that meet the same specifications of purity and recovery in effluent streams, are examined in this paper. Of particular relevance is the assessment of costs and environmental impact of these processes. A baseline case is defined in our case-study, and structural and parametric changes are made. Different scenarios with respect to economic and environmental issues taking into account the specifications of the final product. Near-optimal scenarios are formulated by defining an operating window for some of environmentally friendly technologies available. The impact of different operating policies for the production of anhydrous alcohol on the environment is examined in detail.

The membrane area, the number of plates in the distillation units, the heat exchange area, and the pressure manipulators are considered as structural variables. Among the most relevant operational variables considered one can mention reflux ratio, pressure ratio, as well as compositions and flow rates of streams.

(115f) Comprehensive Utilization of Rice Husks

Haoran Chen[1], Weixing Wang[2], Jarett C. Martin[1], Adam Oliphant[1], Katelyn Deborn[1], Lishan Liang[2], Wenxi Huang[2], Hong Wang[2] and **Luyi Sun**[1], (1)Department of Chemistry and Biochemistry, Texas State University-San Marcos, San Marcos, TX, (2)Department of Chemical Engineering, South China University of Technology, Guangzhou, China

The annual global production of rice husks (RHs) is ca. 170 million tons. The current application of RHs has been very limited, mainly owing to their tough, woody, abrasive nature, low nutritive properties, great bulk, and high ash content. Therefore, RHs are widely considered as a bio-waste. RHs are mainly composed of lignocellulose (ca. 85 wt%) and hydrated silica (ca. 15 wt%). An approach for comprehensive utilization of RHs has been developed to prepare lignocellulose and silica nanoparticles from this biowaste. Majority of the lignocellulose was firstly extracted from RHs by dissolving in ionic liquids (ILs). The dissolved lignocellulose was subsequently separated and can be used as a starting material for fabric products, biofuel, etc. The remaining RH residues after extraction that contain a high concentration of hydrated silica were thermally treated to synthesize silica nanoparticles with a high purity and surface area. The implication of the present findings is discussed.

(115g) Biomasspyrolyisrefinery: Pyrolysis Oil Hydrodeoxygenation

Hannes Pucher[1], Roland Feiner[1], Nikolaus Schwaiger[1], Peter Pucher[2] and Matthaeus Siebenhofer[1], (1)Institute of Chemical Engineering and Environmental Technology, Graz University of Technology, Graz, Austria, (2)R&D, BDI - BioEnergy International AG, Grambach, Austria

BiomassPyrolyisRefinery:

Pyrolysis Oil Hydrodeoxygenation

H. Pucher, R. Feiner, N. Schwaiger, P. Pucher*, M. Siebenhofer

Graz University of Technology, Institute of Chemical Engineering and Environmental Technology, *BDI-BioEnergy International AG

Non-renewable energy sources such as oil and coal will not suffice the increasing demand of energy in the future. It is assumed that renewable resources such as Biomass are able to fill this dawning gap because they have been considered as the most promising energy sources in the next century [1]. In addition the European Commission has released a directive to increase the volume of renewable biofuels within the field of automotive fuels to 10% by 2020 [2].

To fill the upcoming gap and meet the requirements of the directive a biomass liquefaction concept from renewable resources is presented in this work. The BiomassPyrolysisRefinery concept consists of two main process steps. In the first step lignocellulosic biomass, like wood and straw, is converted in pyrolysis oil and pyrolysis char through liquid phase pyrolysis. In the second step these two products are upgraded. Two promising upgrading technologies are investigated in this concept: hydrodeoxygenation [3] [4] of pyrolysis oil and hydrogenation of pyrolysis char. Through hydrodeoxygenation the polar, highly acidic, water and oxygen rich pyrolysis oil is converted in an appropriate, carbon rich, oxygen and water poor product phase. Then the upgraded product is ready for further processing in a conventional refining process.

The main focus of this project is to investigate and optimize the upgrading of pyrolysis oil to maximize the output of the carbon rich fraction and raise stability. For that purpose, the pyrolysis oil is dehydrated prior to the hydrodeoxygenation step. The water content can be reduced by 90%, (Table 1). Then the operation parameters for hydrodeoxygenation, like hydrogen pressure, temperature, reaction time and catalysts, are determined. Investigations have been carried out in lab size and batch mode.

Table 1: Comparison of Pyrolysis Oil, Pyrolysis Oil dehydrated and Crude Oil [5].

		Pyrolysis Oil	Pyrolysis Oil dehydrated	Crude Oil
Water Content	[%]	60	5	0,1
Density	[kg/m^3]	1130	1200	860
Polarity		polar	polar	apolar

Carbon Content	[%]	22,3	51,6	83-86
Hydrogen Content	[%]	9,6	7,0	11-14
Oxygen Content	[%]	67,7	41,0	<1
Nitrogen Content	[%]	0,4	0,4	<1

A Conversion of pyrolysis oil to carbon rich product phase of about 45% has already been obtained. This results indicate that it is possible to control and supress the unwanted reactions like repolymerisation. The composition and characterisation of the products were determined by GC-MS, elemental analysis, NDIR continuous gas analysis and by a solvent-solvent extraction procedure defined by Oasmaa [6].

[1] Wang, G. et al., "The direct liquefaction of sawdust in tetralin.," *Energy Sources, Part A,* pp. 1221-1231, 2007.

[2] Richtlinie_2009/28/EG, "EU Richtlinie zur Förderung der Nutzung von Energie aus erneuerbaren Quellen und zur Änderung und anschließenden Aufhebung der Richtlinien.".

[3] Mercader, F.M. et al., "Hydrodeoxygenation of pyrolysis oil fractions: Process understanding and quality assessment through co-processing in refinery units," *Energy & Environmental Science* , pp. 985-997, 2011.

[4] Wildschut, J. et al., "Insights in the hydrotreatment of fast pyrolysis oil using a ruthenium on carbon catalyst," *Energy Environ. Sci.,* pp. 962-970, 2010.

[5] Mortensen, P.M. et al., "A Review of catalytic upgrading of bio-oil to engine fuels," *Applied Catalysis A: General,* pp. 1-19, 2011.

[6] Anja, O., Fuel Oil Quality Properties of wood-based Pyrolysis Liquids, 2003.

(115h) Cultivation of a New Butanol-Acetone Producing Clostridium Species Strain G33

Teck Khiang Chua, National University of Singapore, Singapore, Singapore and **Jianzhong He**, CEE, National Univeristy of Singapore, Singapore, Singapore

ABSTRACT

The global energy crisis has renewed interest in producing alternative liquid fuels from clean, renewable sources, e.g., generation of biofules via biological processes. In this study, a *Clostridium* sp. strain G33 was obtained with soil inoculum from grassland in Singapore, which is capable of producing dominant butanol from monosugars (e.g., glucose and xylose). The 16S rRNA gene of strain G33 shows 99% identity with *Clostridium* sp. strain 8052, while both strains possess the ability to produce biosolvents such as acetone and butanol. In the presence of 30 g/L glucose, culture G33 could produce 8.60 g/L butanol in batch cultures with a butanol-to-acetone ratio of 2.7:1 and a butanol yield of 29%, the yield of which is the highest comparing with previous reported strain at similar conditions. No ethanol was detected during the fermentation, implying a simplified purification process of the products (butanol and acetone as end-products) in contrast to that of typical *Clostridium strains* (acetone, butanol and ethanol as end-products). Butanol dehydrogenase (*bdh*) was found to be the functional gene responsible for butanol production. Interestingly, strain G33 could utilize xylan to produce hydrogen and fatty acids under mesophilic conditions, while most clostridium could only feed on simple sugars. When supplemented with 10 g/L xylan, culture G33 generated ~1.4 mmol hydrogen and ~10 mM butyric acid. Direct utilization of hemicellulosic feedstock by culture G33 provides advantages over the traditional physicochemical pretreatment and enzymatic hydrolysis. Therefore, culture G33 shows potential to convert the renewable lignocellulosic biomass into biofuels and serves as a candidate for industrial butanol production.

ID: 275766

INVESTIGATION OF BROWN COAL CHAR-CO_2 GASIFICATION IN A LABORATORY SCALE DROP TUBE FURNACE

Md Saiful Alam, Kyushu University, Fukuoka, Japan
Kazi Bayzid Kabir, Monash University, VIC, Australia
Agung Tri Wijayanta, Kyushu University, Fukuoka, Japan
Koichi Nakaso, Kyushu University, Fukuoka, Japan
Jun Fukai, Kyushu University, Fukuoka, Japan
Sankar Bhattacharya, Monash University, VIC, Australia

Abstract

This paper provides detailed investigation of Victorian brown coal char gasification in laboratory scale drop tube furnace (DTF) under CO_2/N_2 mixtures. The DTF is a 2.0 m long quartz reactor consisting of two concentric tubes (diameters 50 mm and 80 mm). Morwell coal char with diameter of <106μm (mean value = 50μm) is used for this investigation. This char sample was prepared in a tubular furnace under pure N_2 at 823K. The char is gasified under CO_2/N_2 atmosphere at temperatures of 1173 and 1273K. The gas compositions from char gasification in the DTF are measured on-line using micro gas chromatography (GC). The results show that char conversion and CO concentration increases with increasing the inlet gas temperature. At 1273K, CO concentration increases due to increased rate of the char-CO_2 reaction. A 3D comprehensive numerical modeling is employed to investigate the char-CO_2 gasification process under various operating conditions. The calculated results are compared with experiments at two different temperatures of 1173 and 1273K using at 20 vol. % CO_2 /N_2 mixtures. The comparisons show a good agreement confirming the validity of the current model. We predict carbon conversion from char-CO_2 gasification for various char particle sizes at elevated gas temperature (>1273K). The temperature distribution in the furnace is found to be uniform under current operating conditions. It is also found that the char conversion significantly depend on the inlet gas temperature and particle size.

Keywords: Char-CO_2 gasification, char conversion, CO_2 reduction

Introduction

Coal is an important global resource due to its relative abundance. Presently about 75% of electric power generation in China, over 50% in the US and nearly 40% in the world, depend on coal [1]. Victoria in Australia has the second largest reserves of brown coal in the world, with economically recoverable reserves projected to last over 500 years at the current rate of consumption [2]. Victorian brown coals have very high moisture content of up to 70% [3]. Therefore, it becomes highly reactive and spontaneously combustible upon drying. At present, the majority of commercial utilization of this resource is for mine-mouth power generation, which suffers from very low efficiency [4]. Alternative modes of utilization are currently being investigated, including gasification and subsequent liquid fuels production. The high reactivity of dried Victorian brown coal makes it eminently suitable for conversion by gasification in which there has now been considerable interest on entrained flow gasification. However, to date, there has been no published work investigating the entrained flow gasification of Victorian brown coals. This work is first-of-its-kind and part of a wider project investigating entrained flow gasification behavior of Victorian brown coal chars.

Char-CO_2 reaction is considered to be a dominant reaction in oxygen-blown entrained flow gasification. E.S. Hecht et al. [5] explained that the presence of the char-CO_2 gasification reaction increases the char conversion rate for combustion at low O_2 concentrations, but decreases char conversion for combustion at high O_2 concentrations. Wijayanta et al., [6] reported that the formation of soot from coal volatiles under various operating conditions. The paper explained that the CO_2 input had a large effect on PAHs/soot reduction at high temperatures.

The study on gasification of Victorian brown coal char is very limited. Considering the environmental need for CO_2 reduction and higher carbon conversion by utilizing CO_2, the present paper reports some of the results on CO_2 gasification of Victorian brown coal char under entrained flow conditions. To scale-up gasification process and improve carbon conversion from the char gasification, it is necessary to conduct the numerical simulation. Therefore, the main objective of this study is to predict carbon conversion under various operating conditions and make preliminary comparison with experimental data.

Experimental

Char preparation

Coal samples were oven dried at 60 C for over 1 hour. The feed was entrained with 0.5 L/min nitrogen as the primary gas and 4.5 L/min nitrogen as the secondary gas. Pyrolysis experiments were conducted at furnace temperatures of 1000 C to prepare char samples. The resultant char samples were stored at less than 4 C to retard surface oxidation prior to further analysis and gasification experiments. The prepared char was then sieved into three size fractions: < 106 μm, 150-355 μm and 355-500 μm. The elemental composition of chars was determined using a Perkin-Elmer CHNS/O analyzer (Model 2400).

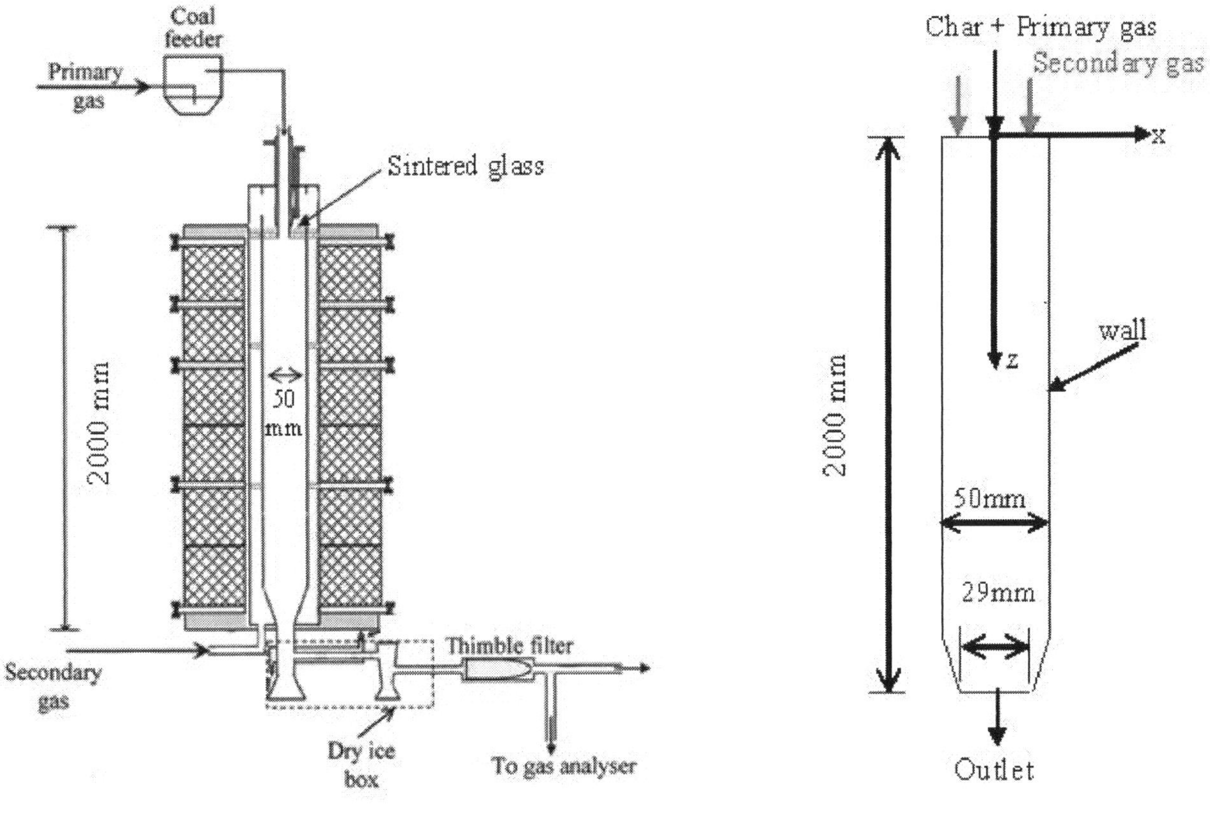

Figure 1(a): Schematic of Drop Tube Furnace (DTF) (b): Computational domain

Char gasification

An electrically heated Drop tube furnace (DTF) coupled with a transparent quartz reactor [7] shown in Figure 1a is employed for char (95.25 wt. % carbon and 4.75 wt. % ash) gasification. The primary gas (10 vol. % of the total inlet gas) at 300K flows through the feeder and entrains the char particles at a feeding rate of ~0.03 g/min before entering the top of the inner chamber (50mm in diameter) of the quartz reactor. The secondary gas (balance of total) is introduced at the bottom of the reactor through the outer annulus (80mm in diameter) of the quartz reactor. This gas is pre-heated to the reactor temperature before entering through a sintered plate and mixing with char and primary gas at the top of the inner tube. Outlet gas from the reactor flows through a char collection system where it is rapidly cooled in a dry-ice trap. Cold gas is then passed through a filter to remove particulates and any potential tars, and sent to a Varian micro-GC (GC-4900) for gas species analysis.

Numerical Study

Mathematical Model

For the fluid phase, the time-averaged steady-state Navier-Stokes equations as well as the mass and energy conservation equations are solved. The numerical domain is shown in Figure 1b. The governing equations for the conservation of mass, momentum, energy and species in 3D Cartesian coordinates are given as:

Mass: $$\nabla.\left(\rho\vec{\upsilon}\right) = S_m \tag{1}$$

Momentum: $$\nabla.\left(\rho\vec{\upsilon}\vec{\upsilon}\right) = -\nabla p + \nabla.\left(\bar{\bar{\tau}}\right) + \rho\vec{g} + \vec{F} \tag{2}$$

Energy: $$\nabla.\left(\vec{\upsilon}(\rho E + p)\right) = -\nabla.\left(\sum_j h_j J_j\right) + S_h \tag{3}$$

Species: $$\nabla.\left(\rho\vec{\upsilon}Y_i\right) = -\nabla.\vec{J_i} + R_i + S_i \tag{4}$$

where S is source term to the continuous phase from the dispersed char particle, $\rho\vec{g}$ is the gravitational body force, $\vec{F}$ is external body forces that arise from interaction with the dispersed phase, h_j is the sensible enthalpy of gas species *j and* J_j is mass flux of species *j*. S_h represents energy source term due to heat of chemical reaction and radiation. R_i is the net rate of production of species i by chemical reaction. ∇ and $\bar{\bar{\tau}}$ are define as:

$$\nabla = \frac{\partial}{\partial x}\vec{i} + \frac{\partial}{\partial x}\vec{j} + \frac{\partial}{\partial x}\vec{k} \tag{5}$$

$$\bar{\bar{\tau}} = \mu\left[\left(\nabla\vec{\upsilon} + \nabla\vec{\upsilon}^T\right) - \frac{2}{3}\nabla.\vec{\upsilon}I\right] \tag{6}$$

Laminar finite rates model is used for the gas phase reactions (Table 1). Laminar finite rate computes only Arrhenius rate. The net rate of production of species i is computed as :

$$R_i = M_{w,i}\sum_{r=1}^{N_R}\hat{R}_{i,r} \tag{7}$$

where $M_{w,i}$ is the molecular weight of species i and $\hat{R}_{i,r}$ is the Arrhenius molar creation/destruction of species i in reaction r. The molar rate of creation/destruction of species i in reaction ($\hat{R}_{i,r}$) is written as follows:

$$\hat{R}_{i,r} = \left(v''_{i,r} - v'_{i,r}\right)\left(k_r\prod_{j=1}^{N}[C_{j,r}]^{(\eta'_{j,r}+\eta''_{j,r})}\right) \tag{8}$$

where $C_{j,r}$ = molar concentration of species j in reaction r [kg mol/m^3

$\eta'_{j,r}$ = rate exponent for reactant species j in reaction r

$\eta''_{j,r}$ = rate exponent for product species j in reaction r

Table 1: Reactions considered in calculation

Gas phase reactions		
$CO + \frac{1}{2}O_2$	$\rightarrow CO_2$	(R1)
$CO + H_2O$	$\rightarrow CO_2 + H_2$	(R2)
$CO_2 + H_2$	$\rightarrow CO + H_2O$	(R3)
$CO + 3H_2$	$\rightarrow CH_4 + H_2O$	(R4)
$CH_4 + H_2O$	$\rightarrow CO + 3H_2$	(R5)
$CH_4 + \frac{1}{2}O_2$	$\rightarrow CO + 2H_2$	(R6)
$H_2 + \frac{1}{2}O_2$	$\rightarrow H_2O$	(R7)
Surface reactions		
$C + \frac{1}{2}O_2$	$\rightarrow CO$	(R8)
$C + CO_2$	$\rightarrow 2\,CO$	(R9)
$C + H_2O$	$\rightarrow CO + H_2$	(R10)

In discrete phase modeling, char particles of known properties are injected into the furnace and tracked in a Lagrangian fashion throughout the computational domain. The dispersed phase is solved by tracking a large number of char particles through the calculated flow field.

The rate of depletion of carbon from coal char due to a surface reaction in Table 1 is calculated by using Smith model which is expressed as [8, 9]:

$$\bar{R} = A_p \eta\, Y_j p_j \frac{kD}{D+k} \tag{9}$$

where $\bar{R}$ represents rate of particle surface species depletion (kg/s).

The diffusion coefficient D is computed as:

$$D = C_{1,r} \frac{\left[\left(T_p + T_\infty\right)/2\right]^{0.75}}{d_p} \tag{10}$$

The kinetic rate of the reaction, k is defined as:

$$k = Ae^{-(E/RT_p)} \tag{11}$$

The char surface reaction consumes the oxidant species in the gas phase; i.e., it supplies a (negative) source term during the computation of the transport equation for this species. The surface reaction consumes or produces energy, in an amount determined by the heat of reaction and the heat balance equation is used as follows:

$$m_p c_p \left(dT_p / dt\right) = hA_p (T_\infty - T_p) - f_h \left(dm_p / dt\right) H_{reac} \tag{12}$$

where H_{reac} is the heat released by the corresponding surface reaction.

The heat transfer coefficient, h, is evaluated using the correlation of Ranz and Marshall [10-12] as follows:

$$hd_p / k_\infty = 2.0 + 0.6\,\mathrm{Re}_d^{1/2}\,\mathrm{Pr}^{1/3} \tag{13}$$

where

k_∞ = thermal conductivity of the continuous phase [W/m-K]

Pr = Prandtl number of the continuous phase ($c_p \mu / k_\infty$)

The Discrete Ordinates (DO) radiation model is used to solve the radiative heat transfer equation. The DO radiation model solves the radiative transfer equation (RTE) for eight numbers of discrete solid angles, each associated with a vector direction fixed in the global Cartesian system (x, y, z). The DO model considers the RTE in the direction as a field equation which is written as:

$$\frac{dI(\vec{r},\vec{s})}{ds} = -(a+\sigma_s)I(\vec{r},\vec{s}) + an^2\frac{T^4}{\pi} + \frac{\sigma_s}{4\pi}\int_0^{4\pi} I(\vec{r},\vec{s}')\Phi(\vec{s}.\vec{s}')d\Omega' \tag{14}$$

where I is the radiation intensity, $(\vec{r},\vec{s})$ are the unit vectors in the directions of propagation, a is the local absorption coefficient, σ_s is the scattering coefficient, Φ represents a phase function used to characterize the nature of the scattering media

Boundary Conditions

Uniform distributions of inlet mass flow rate and constant temperature are given for inlet primary gas and secondary gas boundary surfaces. The wall is assumed as stationary and smooth with no slip condition. A constant wall temperature is assigned for wall boundary surface. A surface injection model is used for char inlet. The boundary condition of the discrete phase at walls is assigned as "reflect", that means the discrete phase elastically rebound off once reaching the wall. At the outlet, the discrete phase simply escapes/exits the computational domain.

Numerical Modeling

The computation is carried out using the finite-volume-based commercial CFD software FLUENT (version 12.1) under steady state condition. The simulation uses the pressure-based solver, which employs an implicit pressure correction scheme and decouples the momentum and energy equations. SIMPLEC algorithm is used to couple the pressure and the velocity. First order upwind scheme is selected for spatial discretization of the convective terms.

Results and Discussion

Experiment and calculation

To investigate the char gasification under CO_2 atmosphere, char particles diameter of 50 μm are used in the experimental work. The char is gasified with 20 vol. %CO_2 /N_2 mixtures at three different secondary gas temperatures of 1073, 1173 and 1273K. It is found that at lower temperature of 1073K, the char-CO_2 gasification doesn't proceed. As a result, carbon conversion is very low at this temperature. However, a significant increase in carbon conversion from 24% to 68% for 50μm particle has been obtained if the secondary gas temperature was ranged between 1173 and 1273K. 3D numerical simulation is conducted to investigate the char-CO_2 gasification process. The inlet conditions for calculation are shown in Table 2. The calculated outlet carbon conversion and gas species concentration are compared with experimental result in Figure 2. During experiment, it is impossible to maintain a perfectly constant char flow rate. A minute quantity of char remaining in the char hopper is difficult to collect resulting in discrepancy between experimental data and predictions.

Table 2: Experimental conditions

Primary gas temperature [K]	300	300
Secondary gas temperature [K]	1173	1273
Diameter of char particle [μm]	50	50
Flow rate of char [kg/s]	3.675×10^{-7}	4.72×10^{-7}
Flow rate of primary gas [kg/s]	2.65×10^{-6}	2.65×10^{-6}
Flow rate of secondary gas [kg/s]	2.39×10^{-5}	2.39×10^{-5}
Concentration of CO_2 [vol%]	22.52	21.2

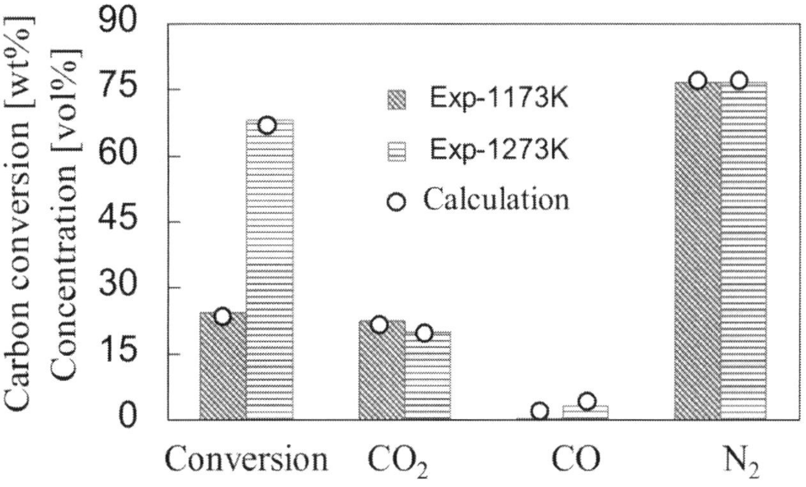

Figure 2: Comparison between experiment and calculation

Contours of gas temperature and species concentration

The contours of gas temperature and gas species concentrations produced from two different temperatures are shown in Figure 3. The temperature contour shows that there is no significant change in the furnace gas temperature. This is because of small furnace diameter and constant temperature of the furnace wall. The char is fed with primary gas at 300K and then mixed with high temperature secondary gas. The conversion of carbon by char-CO_2 reaction (Boudouard reaction) started after the char particle enters into the high temperature gas region. As a result, CO_2 concentration decreases while the CO concentration increases. It is found that the concentration of CO increases at center region because of char injection at center of the furnace where carbon reacts with CO_2.

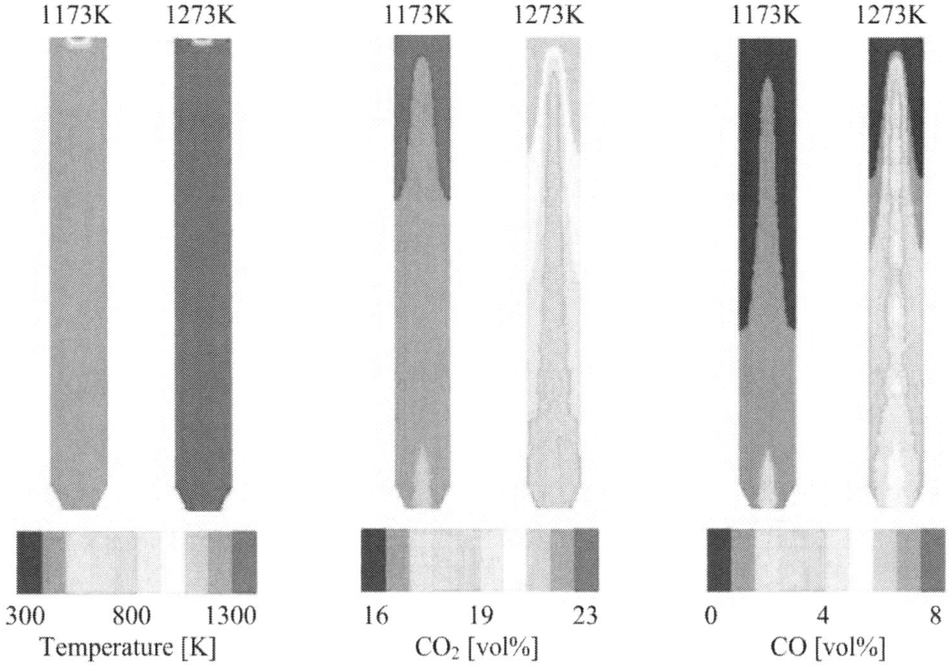

Figure 3: Contours of gas temperature and species concentration for diameter of 50 μm char at secondary gas temperature of 1173 and 1273K

Prediction of carbon conversion

A number of numerical simulations are carried out to predict the carbon conversion of char. The range of particles diameter and secondary gas temperatures considered in this study are 50-425 μm and 1073-1573K, respectively. Figure 4 shows the change of carbon conversion with secondary gas temperature for particle diameters of 50, 250 and 425 μm. At 1073K, the conversion of carbon for 50 μm particle is only 5% whereas the larger particles show no gasification at this temperature. The conversion for 50μm particle increases linearly with increasing the secondary gas temperature. But, the conversion of larger particles significantly changes at higher temperature. However, a complete conversion of carbon can be achieved with increasing the gas temperature. The calculation shows that the conversion of 50, 250 and 425 μm particles is 100% if the secondary gas temperatures > 1373K. The changes in carbon conversion with particle diameter at different temperatures are shown in Figure 5. It is found that at lower temperature (1073K), the conversion of carbon decreases gradually with increasing the particles diameter. But at higher temperature, the carbon conversion changes rapidly.

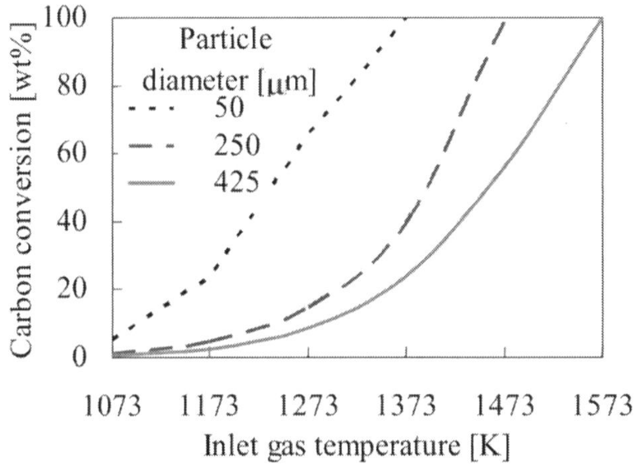

Figure 4: Effect of secondary gas temperature on carbon conversion for various sizes of char particle

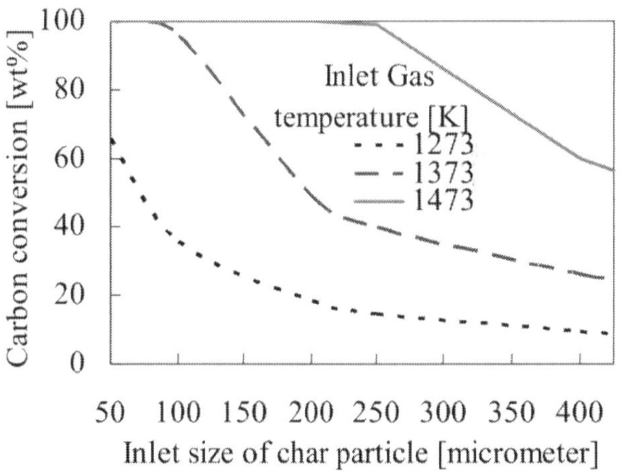

Figure 5: The changes of carbon conversion with inlet diameter of char particle at various secondary gas temperatures

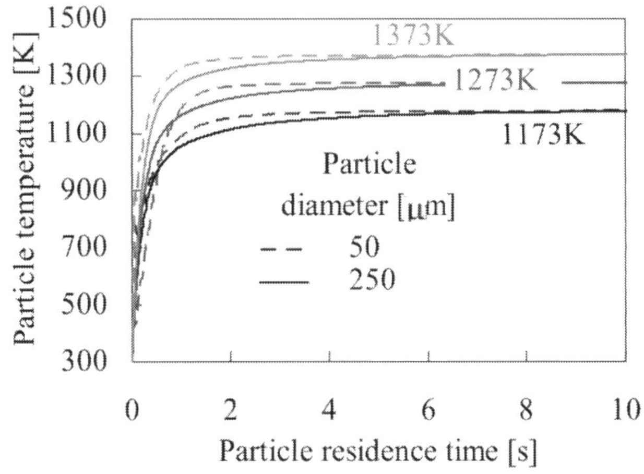

Figure 6: The change of particle temperature with particle residence time

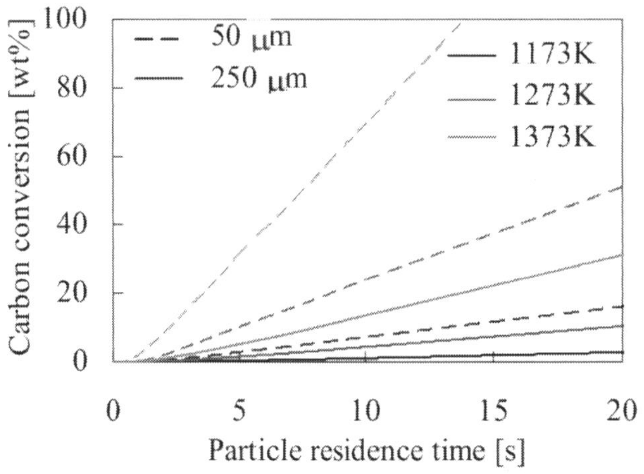

Figure 7: The change of carbon conversion with particle residence time

Change of particle temperature and conversion

The changes of char particle temperature with residence time for various sizes at various secondary gas temperatures are shown in Figure 6. It is found that the particle temperature also increases with increasing the gas temperature. However, at the same temperature, initially the rise in temperature for smaller particle is higher than the larger one. After certain time, the temperatures for both smaller and larger particle become same. The rapid increase in temperature for smaller particles makes the conversion of smaller particle faster than the larger one. As a result, at 1373K 100% carbon conversion of 50μm particle can be achieved after 15s (Figure 7). On the other hand, only 20% conversion is observed for the 250μm particle at that time. The trend is the same for all calculated conditions carried out in this study.

Concluding comments

The char-CO_2 gasification of Morwell coal char is studied at 1173 and 1273K under CO_2/N_2 atmosphere. Char conversion increases with increasing gas temperature and decreasing particle size. The comparison between experiment and calculation shows that the numerical modeling employed in this calculation is able to predict the char-CO_2 gasification accurately. Therefore, the predicted CO_2 gasification results can be used to design a new furnace for the desired gasification. Further work is in progress to refine the model with wider range of data from CO_2 and steam gasification.

Acknowledgement

This research is supported by Japan Society for the Promotion of Science (JSPS) doctoral fellowship for young researchers and GCOE, Novel carbon Resource Sciences, Kyushu University, and the Brown Coal Innovation Australia (BCIA) under the Victoria-Kyushu collaboration.

References

[1] Chui E.H. et al., Simulation of Entrained Flow Coal Gasification / Energy Procedia 1 (2009) 503–509.
[2] ABARE. Australian Energy Resource Assessment. Canberra: Geoscience Australia and ABARE; 2010.
[3] Allardice D.J. et al., Chapter 3 - Water in Brown Coal and Its Removal. Advances in the Science of Victorian Brown Coal, Amsterdam, Elsevier Science, 2004, 85-133.
[4] ABARE. Australian Energy Resource Assessment. Canberra: Geoscience Australia and ABARE; 2010.
[5] Ethan S. et al., Effect of CO_2 gasification reaction on oxy-combustion of pulverized coal char, Proceedings of the Combustion Institute 33 (2011) 1699–1706.
[6] Wijayanta A.T. et al., Numerical investigation on combustion of coal volatiles under various O_2/CO_2 mixtures using a detailed mechanism with soot formation. Fuel 93(2012), 670–676.
[7] Lian Z. et al., Experimental Investigation of the Combustion of Bituminous Coal in Air and O_2/CO_2 Mixtures: 1. Particle Imaging of the Combustion of Coal and Char, Energy Fuels, 24(2010), 4803–4811.
[8] A. Silaen and T. Wang. Int. J. Heat Mass Transfer, 53(2010)2074–2091.
[9] Smith I.W., The combustion rates of coal chars: a review, Proc. Combust. Inst. 19 (1982) 1045-1065.
[10] ANSYS FLUENT 12.0 User's Guide, ANSYS Inc. 2009-01-29
[11] Ranz W.E. and Marshall W.R., Chem. Eng. Prog., 48(3), 141-146, 1952.
[12] Ranz W.E. and Marshall W.R., Chem. Eng. Prog., 48(4), 173-180, 1952.

(128b) Kinetic and Mechanistic Studies of Microbial and Chemical Enhancement of Carbon Mineralization

Huangjing Zhao, Earth and Environmental Engineering and Chemical Engineering, Columbia University, New York, NY, Edris Taher Sr., Earth and Environmental Engineering, Columbia University, New York, NY, Kartik Chandran, Earth and Environmental Engineering, Columbia University and Ah-Hyung Alissa Park, Earth and Environmental Engineering & Chemical Engineering, Columbia University, New York, NY

Carbon mineralization is one of the safest methods of sequestering anthropogenic carbon dioxide. It is based on the exothermic reaction between carbon dioxide and the metal ions present in silicate minerals, such as antigorite, to form geologically and thermodynamically stable mineral carbonates. However, several challenges need to be overcome to successfully deploy this technology. In particular, the acceleration of the extremely slow mineral weathering step along with process optimization is essential to ensure economic feasibility as well as overall sustainability. In this study, the effect of various organic chelating agents on the dissolution rate of antigorite was explored to accelerate the weathering rate. First, the chelating agents were produced via the anaerobic digestion of food waste in a microbial reactor. As these catalysts which are enhancing the carbon storage are derived from a waste stream, their employment in the proposed carbon mineralization technology enhances the overall sustainability of the process. Magnesium extracted from antigorite was then reacted with CO_2 to form precipitated magnesium carbonates to mimic commercially available $CaCO_3$-based filler materials. The effects of various chelating agents, pH, reaction time, and reaction temperature on the mean particle size, particle size distribution, composition, and particle morphological structures of precipitated magnesium carbonates were also investigated.

(128c) CO2 Utilization: Combined Dry and Steam Reforming As an Alternative to Steam Methane Reforming

Preeti Gangadharan, Krishna Chaitanya Kanchi and Helen H. Lou, Dan F. Smith Department of Chemical Engineering, Lamar University, Beaumont, TX

Syngas platform for the production of fuels and chemicals, including methanol, dimethyl ether, ethylene, propylene and Fischer-Tropsch fuels, has been a topic of great interest in industry and academia in recent years. Steam methane reforming (SMR) which is the most widely applied method of producing syngas from natural gas, reacts natural gas with steam. Dry reforming of methane (DRM) uses waste carbon dioxide to produce syngas from natural gas. Dry reforming alone has not yet been implemented commercially; however, a combination of steam methane reforming and dry reforming of methane (SMR+DRM) has been used in certain plants for several years.

This paper deals with the simulation of both the SMR and SMR+DRM processes, and an evaluation based on economic and environmental criteria to determine whether the SMR+DRM process is competitive with the more popular SMR process. From the results, it is observed that the SMR+DRM process has a lower carbon footprint. And, with the development of more efficient DRM catalysts, this process could be economically competitive with steam methane reforming.

(128d) Liquid Fuels From Alternative Carbon Sources Minimizing Carbon Dioxide Emissions

Matthew J. Metzger[1], Benjamin J. Glasser[2], Bilal Patel[3], James A. Fox[4], Baraka Celestin Sempuga[5], Diane Hildebrandt[6] and David Glasser[6], (1)Chemical and Biochemical Engineering, Rutgers University, Piscataway, NJ, (2)Chemical and Biochemical Engineering, Rutgers, the State University of New Jersey, Piscataway, NJ, (3)Centre of Material and Process Synthesis, University of the Witwatersrand, Johannesburg, South Africa, (4)University of the Witwatersrand, Center of Material and Process Synthesis, Johannesburg, South Africa, (5)School of Chemical and Metallurgical Engineering, Centre of Material and Process Synthesis, University of the Witwatersrand, Johannesburg, South Africa, (6)Centre of Material and Process Synthesis (COMPS), University of the Witwatersrand, Johannesburg, South Africa

The energy needs of the world continue to grow, as does the resulting environmental impact. Policy makers continue to call for alternative energies to replace today's petroleum based liquid fuels. However, liquid fuels have significant advantages and it is probably unwise to abandon the existing infrastructure without appropriately exploring alternatives to lessen the environmental burden of producing liquid fuels. Biomass and coal are often proposed as alternatives to petroleum based carbon sources, but we demonstrate that those processes, by themselves, lose a significant amount of their potential product to unwanted carbon dioxide emissions. However, combining biomass and coal with cleaner natural gas yield processes with less environmental impact to produce liquid fuels with small, zero, or even negative, carbon dioxide emissions. Our process synthesis approach is applied to commonly encountered liquid fuel production methods in order to identify promising routes and to establish feasibility limits on those less promising alternatives.

(128e) Co-Pyrolysis/Gasification of Coal and Biomass

Oisik Das and Foster Agblevor, Biological Engineering, Utah State University, Logan, UT

Coal is a major primary energy source for USA, however, coal is also a major source of carbon dioxide and other associated pollutants. On the other hand, biomass is nearly carbon neutral and contain less pollutants than coal. Thus, by combining coal and biomass, potential both carbon dioxide and pollutant emissions may be reduced considerably. We investigated the pyrolysis/gasification of various biomass feedstocks (corn stover, hybrid poplar, and switchgrass) and sub-bituminous coal in both fluidized bed reactor, thermogravimetric analysis (TGA) and pyroprobe gas chromatographic analysis. Various ratios of coal to biomass were investigated. The TGA analysis showed that whereas the corn stover and and switchgrass appeared to influence the maximum weight loss at 400°C,the polar wood had minimal influence on this peak. In the case of the poplar, the coal and the wood appeared to behave independently. Data will also be presented on the co-pyrolysis/gasification in a fluidized bed reactor.

(136a) A Shell-and-Tube Direct Steam Generator for Concentrated Solar Power Generation

Theodore Wiesner and Rong Xu, Chemical Engineering, Texas Tech University, Lubbock, TX

"Making Solar Energy Economical" is one of the Grand Challenges posed by the National Academy of Engineering. However, innovation in concentrating solar power technologies (CSP) lags by considerable margins that of all other renewable energy technologies. The levelized cost of electricity (LCOE) for solar thermal electricity ranges from *ca.* 13¢ to 21¢/kWh. Among energy sources alternative to fossil fuels, it is the most expensive, with the exception of crystalline and thin-film photovoltaic technologies. The solar thermal LCOE is currently considerably above electricity derived from coal combustion (8¢ to 14¢/kWh). For a benchmark 100 MW_e parabolic trough power plant, the solar field capital cost contributes 4.5 to 6¢/kWh of the LCOE. Reductions in the capital costs and improvements in the efficiency of the collector field are necessary to make solar thermal power more competitive with both fossil fuel electricity and with power derived from other alternative energies.

On the other hand, solar-derived electricity is highly desirable, due to its carbon-free nature and the essentially infinite solar resource.

Presently, parabolic trough solar energy generation systems (SEGS) feature directly irradiated tubular heat collection elements (HCE) that transfer enthalpy to a heat transfer medium (HTM). The state-of-the-art design of a parabolic trough receiver-collector is illustrated in the left panel of Figure 1. An SEGS based upon a parabolic trough collector field is depicted in Figure 2. The collector field consists of many parabolic troughs and receivers, and the hot HTM is collected in a manifold. Steam is generated by exchange with the HTM in a power block separate from the collector field.

To reduce the capital cost of the parabolic trough collector field, we propose a new shell-and-tube direct steam generator (STDSG, Figure 1). The shell-and-tube receiver combines multiple pipe bores with a single receiver surface. In the same fashion as current HCEs, sunlight is absorbed by the outer surface of the receiver while molten salt circulates within the shell. However, the shell also contains a tube bundle, within which steam is directly generated. The unit is sized such that enough heat is absorbed and stored in the salt during times of insolation to generate steam off-sun, increasing the dispatchability of the collected power. Figure 1 illustrates the proposed technical advancement of the HCE technology, from a directly irradiated receiver, to a double pipe receiver-generator, and ultimately to a shell and tube receiver-generator. As indicated in Figure 2, we envision replacing the collector field and steam generator with a smaller number of parabolic reflectors and shell-and-tube receivers incorporating direct steam generation.

A shell-and-tube direct steam generator consolidates two major functions in a solar electric generation system: 1) It combines multiple heating tubes into a single irradiated receiver; 2) It eliminates the separate steam generation block. *Such process intensification directly reduces the capital and operating costs of the parabolic trough collector field, the single most expensive component of an SEGS,* as well as the cost of the power block. An STDSG appears to be a completely novel concept. A thorough review of the open literature reveals no prior work on a directly irradiated shell-and-tube heat exchanger. A review of the US patent literature reveals some devices for solar water distillation that are remotely related. However, none of these patents concern power generation, and none feature a shell-and-tube design with combined energy capture, steam generation, and energy collection for times of low insolation.

A preliminary mathematical model demonstrates the potential of the shell-and-tube direct steam generator. Consider a countercurrent double-pipe heat exchanger that is irradiated on its outer surface. The diameters of the outer and inner pipes are D_o and D_i, respectively, and the diameter ratio $r = D_o/D_i$. Steam condensate perfuses the bore while the molten salt heat transfer fluid flows in the annulus. We now define the *solar multiple* as the ratio of energy generated by the solar field to the energy required for peak output by the power block. The value of the solar multiple f_s should be approximately three in order to sustain 24/7 steady state power production without backup from another fuel source.

The model predicts the results in Figure 3. (Note: In the preliminary model, we have simulated heating the condensate only to saturation temperature due to the complexity of modeling vaporization.) In the left panel, one sees that the temperature of both the annular and the bore fluids increase simultaneously. Therefore, one can heat the bore fluid while on-sun, and accumulate energy in the annular HTM for use off-sun. In the right panel, the solar multiple monotonically increases as the diameter ratio r increases. The desired value of the solar multiple, three, occurs for a diameter ratio of approximately two. This suggests that reasonable geometries are attainable with an STDSG.

We are presently extending the preliminary double-pipe model to a shell-and-multi-tube model in *COMSOL®*, a finite-element simulation package. In the latter, we are incorporating tube-side vaporization, unsteady-state behavior, and the more intricate shell-and-tube geometry (Figure 4). Figures

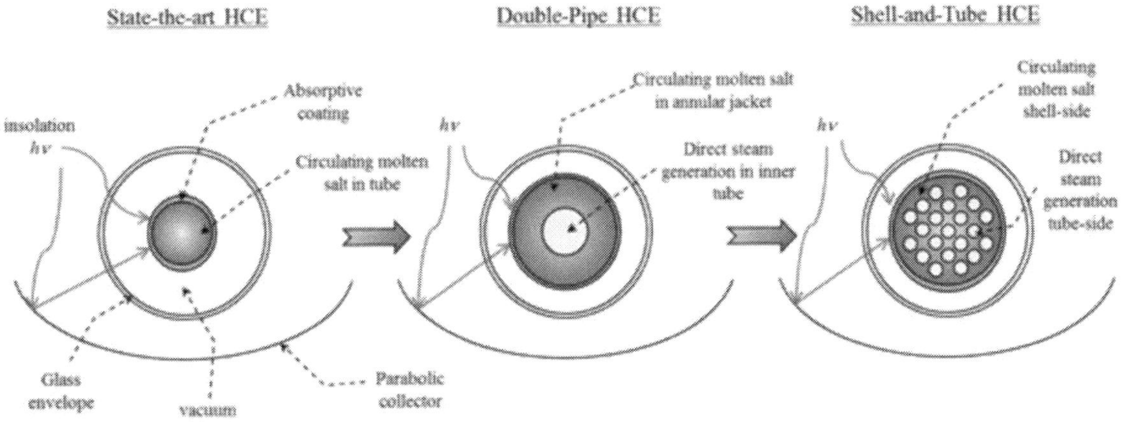

Figure 1. Proposed Advancement in Solar Receiver Technology

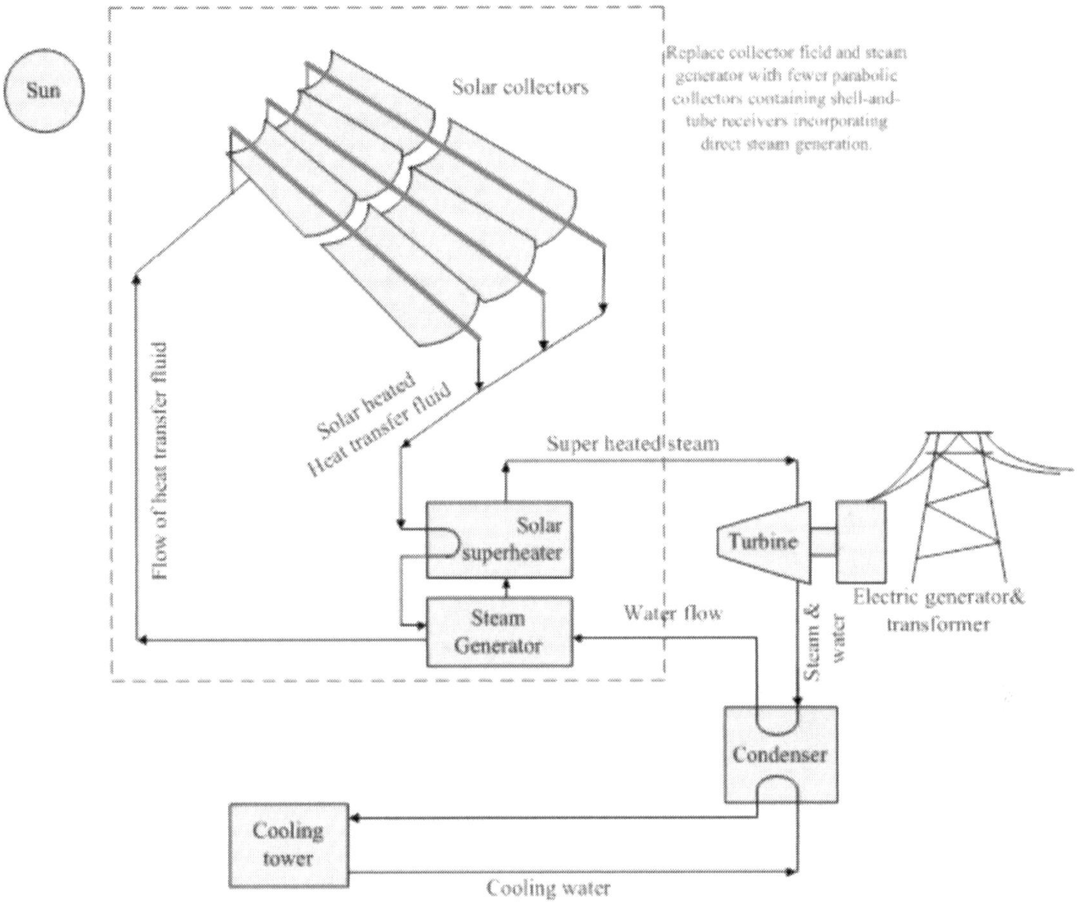

Figure 2. Parabolic Trough Solar Electricity Generating System. Adapted from KJC Operating Company, Boron, CA.

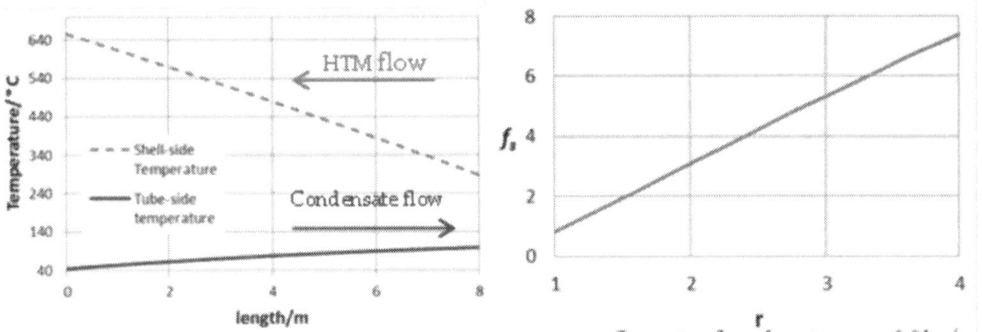

diameter ratio $r = 4$

insolation $q_{solar} = 0.5\,\mathrm{MW/m^2}$

exchanger length $L = 8\,\mathrm{m}$

mass flow rate of condensate $w_i = 0.8\,\mathrm{kg/s}$

mass flow rate of molten salt $w_o = 1.5\,\mathrm{kg/s}$

bore diameter $D_i = 2.54\,\mathrm{cm}$

Figure 3. Left-Simulated shell-and tube-side temperature profiles in a double-pipe solar receiver. Right-solar multiple as a function of the diameter ratio. Values of key simulation variables are below the plots.

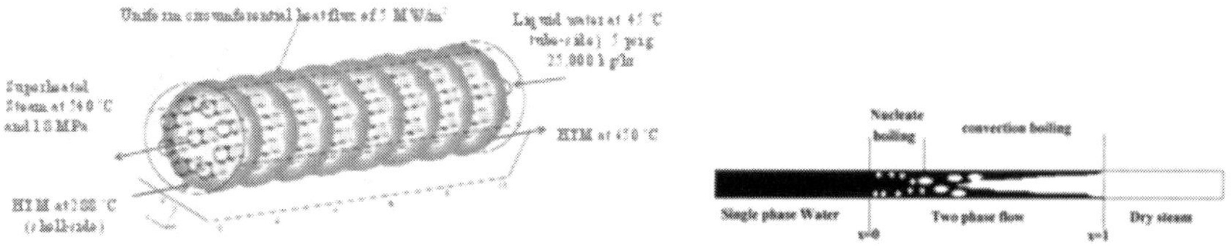

Image from Odeh et al. (1998)

Figure 4. Left-Temperature Profiles in a Solar Shell-and-Tube Solar Receiver. The target conditions for using it as a STDSG are superimposed. Right-Two-phase flow regime in a direct steam generation tube

(136b) Energy Storage Layouts for Concentrated Solar Plants – A Comparison of Process Control

Flavio Manenti, Dipartimento di Chimica, Materiali e Ingegneria Chimica , Politecnico di Milano, Milano, Italy, Zohreh Ravaghi-Ardebili, Dipartimento di Chimica, Materiali e Ingegneria Chimica, Politecnico di Milano, Milano, Italy and Giovanni Manenti, R&D Department, FBM Hudson Italiana SpA, Terno D'Isola (BG), Italy

Energy Storage Layouts for Concentrated Solar Plants – A comparison of Process Control

Flavio Manenti[1,], Zohreh Ravaghi-Ardebili[1], Giovanni Manenti[2]*

[1]Politecnico di Milano, Dipartimento di Chimica, Materiali e Ingegneria Chimica "Giulio Natta", Piazza Leonardo da Vinci 32, 20133 Milano

[2]FBM Hudson Italiana SpA, R&D Dept., Via Valtrighe 5, 24030 Terno d'Isola (BG), Italy

Energy storage is a major process design and control issue in concentrated solar plants. The intrinsic discontinuous nature of solar energy forces to install units able to store energy under favourable conditions, then to release energy during night or unfavourable conditions. Accordingly, smoothed operations and continuous energy production can be guaranteed. Several technologies are today available for storage; tanks of molten salts, steam accumulators, high thermal capacity solids are among the most important.

Regardless the technology adopted for storage, the plant layout plays a key role in the process performance and control philosophy. Control of storage unit and steam generation station under steep load changes, as during sunrise and clouds passage, depends appreciably on layout. Storage devices can be installed either on the main process stream or be part of a secondary circuit linked to the main one; devices can be arranged either in series or in parallel and, again, devices of different type can be grouped. Fig. 1 shows different layouts for the specific case of molten salts.

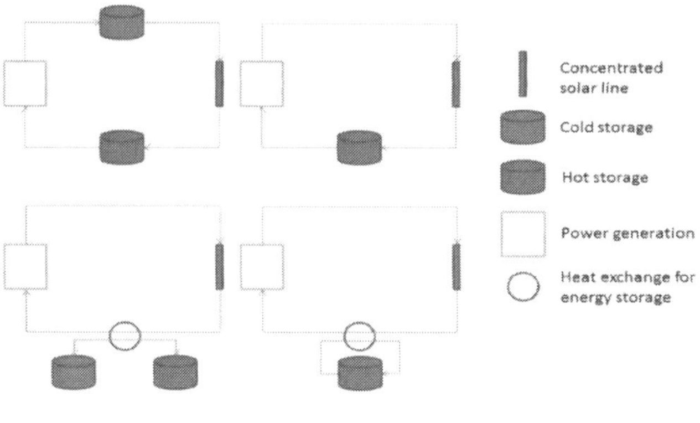

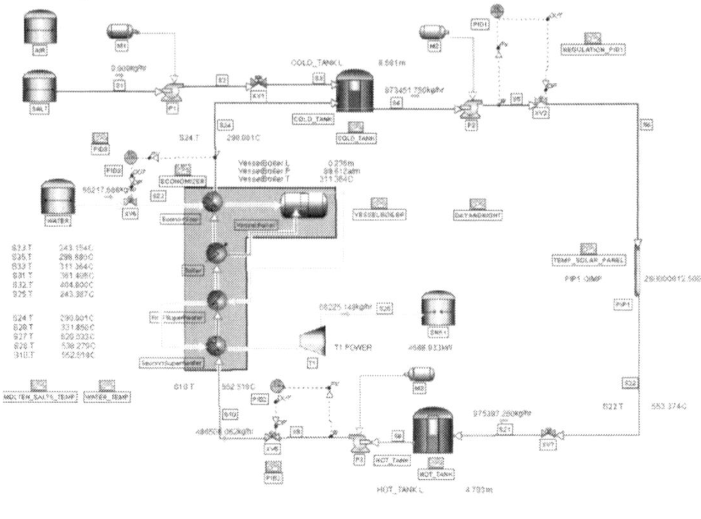

Fig.1

Fig.2

This work deals with the comparison of process control for different storage layouts based on molten salts. Results coming from process simulations of unsteady loads are reported and a general discussion on operability and controllability is done. For this purpose, the different layouts have been modelled by means of a commercial process simulator (Fig. 2), where specific fluid properties and units have been implemented. Same investigating approach can be adopted for storage systems based on different technologies.

(136c) Limiting Efficiencies for Solar Thermal Energy Conversion to Fuels

Dharik S. Mallapragada, Fabio H. Ribeiro, W. Nicholas Delgass and Rakesh Agrawal, School of Chemical Engineering, Purdue University, West Lafayette, IN

The conversion of solar energy into chemical energy in the form of a fuel derived from the raw materials of air, water and soil, is critical for energy storage and transportation in a future solar energy driven economy. A key metric of interest here is the fraction of incident sunlight recovered as fuel, or the process sun-to-fuel (STF) efficiency. The importance of the STF efficiency metric is seen in the context of the large magnitude of global primary energy consumption relative to the land resource available to harness solar energy.

In this presentation, we explore the STF efficiency limits of thermochemical processes producing carbon-based fuels and H_2 from sunlight starting with the raw materials of CO_2 and H_2O. Thermochemical STF processes absorbing the entire solar spectrum as heat at high temperatures (achieved through the use of concentrators) are expected to have higher STF efficiencies compared to photochemical processes relying on a specific wavelength band. Our analysis finds that the theoretical STF efficiency remains nearly unchanged whether using single or multiple blackbody absorbers for solar heat recovery.

For feasible solar concentrations of 2000-8000, the theoretical maximum Sun-to-H_2 (STH_2) efficiency ranges from 72-79%, corresponding to operating temperatures of 1200-1600 K. The use of multiple reaction steps during the water-splitting process is shown to further lower the theoretical efficiency. The corresponding STF efficiency range for methane and methanol is 70-76% and 65-71% respectively.

To approach real processes, we discuss a "practical" STF and STH_2 efficiency limit, which considers the irreversible heat integration between the process streams, heat sources and sinks. A notable observation here is that the practical maximum efficiencies derived using irreversible heat integration are only 5-10% lower in value than the corresponding theoretical efficiencies.

(136d) Concentrated Solar Energy Combined with Thermal Storage for 24/7 Operation of a High-Temperature Thermochemical Process

Anna S. Wallerand[1], **Zoran R. Jovanovic**[1] and Aldo Steinfeld[1,2], (1)Department of Mechanical and Process Engineering, ETH Zürich, Zürich, Switzerland, (2)Solar Technology Laboratory, Paul Scherrer Institute, Villigen, Switzerland

A concept of a high-temperature endothermal chemical reactor is modelled to assess its 24/7 operation utilizing concentrated solar energy as the source of process heat. The concept is based on three main constituents: a concentrating solar tower system with high-temperature receiver, a thermal storage unit, and an endothermal chemical reactor. In this work, the chemical reactor is represented by an allothermal fluidized bed producing syngas via coal gasification with steam.

During daytime, absorbed heat is transferred to air flowing through a structured volumetric receiver exposed to concentrated solar radiation. The hot air is split into two streams. One stream charges a thermal storage unit envisioned as a packed bed of spherical ceramic particles; the other stream is passed directly through a multitude of heat exchanger tubes immersed in the fluidized bed gasifier. At night, cold air is heated by flowing through the packing in the thermal storage prior to being passed through the reactor heat exchanger tubes, thereby assuring 24/7 operation of the coal gasification and uninterrupted production of syngas.

The governing steady-state and transient mass and energy conservation equations are formulated and solved numerically for a 1 MW gasifier. The system is examined in terms of the dimensions of the constituent components (storage/reactor volume, area of heliostat field), the influence of specific design parameters (heat transfer coefficients), its overall response to seasonal and daily variations in solar irradiance, scale-up potential (10, 50 MW), material strains (volumetric expansion due to high temperature differences), and the overall solar to chemical/electrical efficiencies.

(137a) Group Contribution+ (GC+) Based Estimation of Environment-Related Properties for Design of Sustainable Processes: Development of Property Models and Uncertainty Analysis

Amol Hukkerikar, CAPEC, Department of Chemical and Biochemical Engineering, Technical University of Denmark, Kgs. Lyngby, Denmark, Bent Sarup, Vegetable Oil Technology Business Unit, Alfa Laval Copenhagen A/S, Søborg, Denmark, Douglas Young, National Risk Management Research Laboratory, US EPA, Cincinnati, OH, Gürkan Sin, Department of Chemical and Biochemical Engineering, Technical University of Denmark, Kgs. Lyngby, Denmark and Rafiqul Gani, CAPEC, Department of Chemical and Biochemical Engineering, Technical University of Denmark (DTU), Kgs. Lyngby, Denmark

Currently there is a great deal of interest in the development of methods and tools for the process synthesis, design, and analysis of sustainable processes. The design of sustainable processes require that it should satisfy various conditions (or constraints) such as, increased productivity, minimum energy consumption, reduction in raw materials, recovery of products, minimum generation of pollution etc. (Carvalho et al., 2006). This task can be performed by using a chemical process simulator (to perform mass and energy balance for the concerned process) together with the waste reduction (WAR) algorithm (Young and Cabezas, 1999) to obtain a quantitative measure of the potential environmental impacts which, as part of LCA analysis of process synthesis and design, contributes to identify sustainable processing paths and design alternatives. The basis for the quantification of the environmental impacts is a set of environment-related properties (such as, fathead minnow LC50 (96-hr), oral rat LD50, global warming potential etc.) of chemical substances involved in the process. For many chemicals of interest the experimental data of environment-related properties is not available since the measurement of these properties are extremely time consuming and expensive. Also, processes that deal with synthesis of new chemicals require a suitable property prediction method in order to obtain reliable estimates of environment-related properties of new chemicals. In such cases, group contribution (GC) methods are generally suitable to obtain the needed property values since these methods provide the advantage of quick estimates without requiring substantial computational work. In GC methods, the property of a chemical is a function of structurally dependent parameters which are determined as a function of the frequency of the groups representing the chemical and their contributions. Among the GC methods for estimation of properties of chemicals, the Marrero and Gani (MG) method (Marrero and Gani, 2001) is well-known. The MG method allows estimation of properties based exclusively on the molecular structure of the chemical and exhibits a good accuracy and a wide range of applicability covering chemical, biochemical, and environment-related chemicals. For the case where the molecular structure of a given chemical is not completely described by any of the available groups, the atom connectivity index (CI) method has been employed together with MG method to create the missing groups and to predict their contributions (Gani et al., 2005). This combined approach has led to the development of group contribution^{+} (GC^{+}) method of a wider application range than before since the missing groups and their contributions can now be easily predicted through the regressed contributions of connectivity indices. In addition to the accuracy of predicted property values, it is also important to know the uncertainties of the estimated property values that arise due to uncertainties in the regressed parameters of the selected property model.

The main objective of this work is to develop property prediction models based on the GC^{+} approach (combined GC method and CI method) to provide reliable estimates of environment-related properties together with uncertainties (for example, prediction error in terms of 95% confidence interval) of the estimated property values. For this purpose, a systematic methodology for property modeling and uncertainty analysis developed by Hukkerikar et al. (2012) is used. The methodology includes a parameter estimation step to determine parameters (group/atom contributions, adjustable parameters, and a universal parameter) of property models, and an uncertainty analysis step to establish statistical information about the quality of parameter estimation, such as the parameter covariance, the standard errors in predicted properties, and the confidence intervals. For property modeling with a GC method, the MG method has been considered. For property modeling with a CI method, the models proposed by Gani et al. (2005) have been considered. For parameter estimation, large data-sets of experimentally measured property values of wide range of chemicals (hydrocarbons, oxygenated chemicals, nitrogenated chemicals, poly-functional chemicals, etc.) taken from the database of US Environmental Protection Agency (EPA) is used. In total 10 environment-related properties, which include the fathead minnow LC50 (96-hr), *daphnia magna* LC50 (48-hr), oral rat LD50, aqueous solubility, bioconcentration factor, permissible exposure limit (OSHA-TWA), photochemical oxidation potential, global warming potential, ozone depletion potential, and acidification potential have been modeled and analysed. The application of the developed property models for estimation of environment-related properties and uncertainties of the estimated property values is highlighted through several illustrative examples. The developed models for environment-related properties have been implemented in ProPred, a property estimation toolbox of ICAS ® (Integrated Computer Aided System) software developed by CAPEC, DTU. The developed property models provide reliable estimates of environment-related properties needed to perform design and analysis of sustainable processes and allow one to evaluate the effect of uncertainties of estimated property values on the calculated PEI index of the process giving useful insights into quality and reliability of the design of sustainable processes. The use of the developed property models for environment-related properties in the design of sustainable process is highlighted through a case study.

References:

1. A. Carvalho, R. Gani, and H. Matos, 16th European Symposium on Computer Aided Chemical Engineering, (2006) 817-822.

2. D. M. Young, and H. Cabezas, Computers and Chemical Engineering, 23 (1999) 1477–1491.

3. J. Marrero, R. Gani, Fluid Phase Equilibria, 183-184 (2001) 183–208.

4. R. Gani, P.M. Harper, M. Hostrup, Industrial and Engineering Chemistry Research, 44 (2005) 7262–7269.

5. A.S. Hukkerikar. B. Sarup, A. Ten Kate, J. Abildskov, G. Sin, and R. Gani, Fluid Phase Equilibria, 321 (2012) 25-43.

6. Integrated Computer Aided System (ICAS version 14.0), Department of Chemical Engineering, Technical University of Denmark, Lyngby, Denmark, 2011.

(137b) Implementation of the Greenscope Design Tool to Develop Sustainable Processes

Gerardo J. Ruiz-Mercado, Raymond L. Smith and Michael A. Gonzalez, U.S. Environmental Protection Agency, Cincinnati, OH

This contribution describes and demonstrates the implementation of the GREENSCOPE design tool, for the assessment and modeling of process sustainability. GREENSCOPE (Gauging Reaction Effectiveness for the Environmental Sustainability of Chemistries with a multi-Objective Process Evaluator) provides an evaluation and diagnostic of process performance sustainability in the areas of environment, efficiency, energy, and economics (i.e., four Es).

GREENSCOPE is constituted of a set of sustainability indicators categorized according to each of the four E bases. This design/evaluation tool provides individual indicator scores on a practical and dimensionless sustainability measurement scale. The setting of best and worst-case scenarios, as reference states for each indicator allows the user to recognize the sustainability performance status of the process under study in relation to feasible and realistic boundary values. Thus, existing or in-development new processes can be evaluated according to their relative indicator scores, and process modifications can strive towards achievable targets.

In addition, this evaluation/design for sustainability tool clearly identifies and guides the designer on data requirements when calculating the process performance assessment. The user can visualize data-indicator relationships on the way to choosing selected indicators for evaluation.

In this presentation GREENSCOPE is applied to a practical case study for the production of biodiesel, thus aiding the designer in localizing the most relevant process design aspects that can be manipulated for improving the overall process performance sustainability. These performance improvements will be reflected in reducing the quantity of goods and services needed, and preventing and minimizing releases, while increasing the overall economic and social benefits.

Estimation of Air Handling Units Condensate Recovery in the Medical Campus Buildings at University of Alabama at Birmingham

[1]Atul Kajale
Doctoral Student
Department of Civil, Construction, and Environmental Engineering
University of Alabama at Birmingham
1075 13th Street South
Birmingham, Alabama 35294-4440

[2]Robert W. Peters, Ph.D., P.E.
Professor of Environmental Engineering
Department of Civil, Construction, and Environmental Engineering
University of Alabama at Birmingham
1075 13th Street South
Birmingham, Alabama 35294-4440

[2]Matthew Winslett
Engineering Manager, Energy Management Division
Facilities Management Department
University of Alabama at Birmingham
1530 3rd Avenue South
Birmingham, Alabama 35294-4564

Abstract

According to the U.S. Drought Monitor (U.S. Drought Mitigation Center December 13, 2011), the majority of the Southeastern states are classified as being in a state of "moderate", "severe", "extreme", and or "exceptional" drought. Due to the prevailing drought conditions in the southeastern U.S., it is necessary to conserve water resources by using them as efficiently and responsibly as possible. The increasing population, socioeconomic developments, and industrialization will have significant impacts on the way we consume water today. There will be an imbalance in the water supply and demand if water is not used more efficiently. Water and energy are interdependent and hence it is necessary that building operators and industry leaders realize that water conservation is as important as energy conservation. Building operators of many industrial and commercial buildings prefer energy conservation as a priority due to the high savings in electricity bill, water conservation is often considered as an afterthought. However, as far as power generation is concerned, both energy and water are interdependent; water is used for power generation and energy is used in harvesting water. There is a need to grow water conservation awareness, and a need for more research to develop effective water conservation techniques that are unique to the local conditions of the area.

Air Handling Units (AHUs) are most commonly used for space conditioning in commercial, industrial, and institutional buildings and take advantage of large amounts of chilled water internally for cooling their coils as warmer air is forced across them. During the hot summer months, a large quantity of water is condensed when hot air passes through the cooling coils. In the past, condensate formed during this process was dumped into storm drains resulting in discarding large quantities of usable water. The Energy Management Division (EMD) of the Facilities Management Department at University of Alabama (UAB) has taken the initiative to collect this water and reduce the make-up water requirement of their three major centralized cooling plants on campus. EMD is currently conducting a pilot study in five campus buildings to investigate the performance of the condensate recovery system.

Alabama, Georgia, Tennessee and other Southeastern States have experienced exceptional drought during summer of 2007. The intensity of drought in the southeastern states for August 2007 is shown in Figure 1.

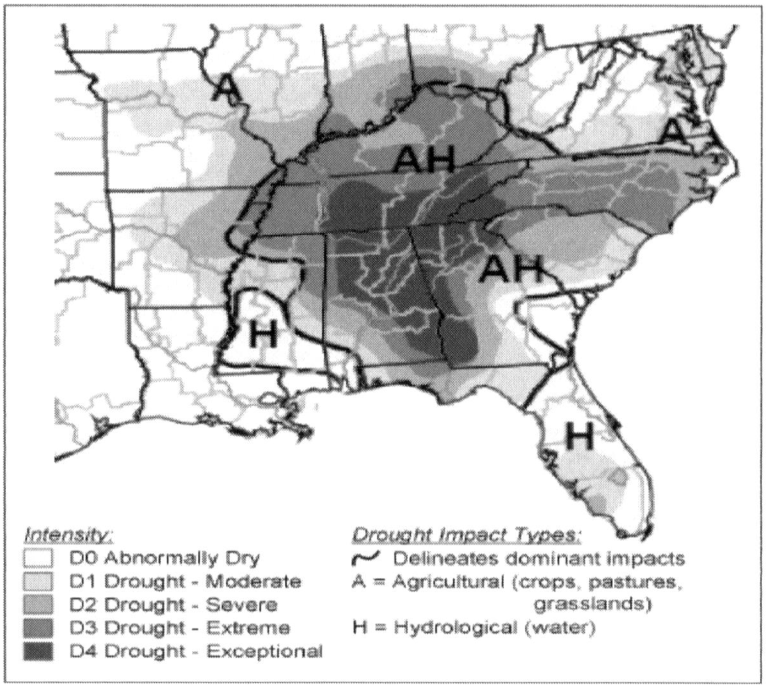

Figure 1. Intensity of Drought in Southeastern U.S. in August 2007.

"Drought is a natural event that, unlike floods or tornadoes, does not occur in a violent burst but gradually happens; furthermore, the duration and extent of drought conditions are unknown because rainfall is unpredictable in amount, duration and location" [Office of Water Resources, 2004].

Previous droughts experienced in the State of Alabama, had social, economic, and environmental effects and these effects were not mitigated in its full capacity due to the slow response, and unpreparedness [Office of Water Resources, 2004]. According to the Alabama Drought Management Plan of 2004, the industries and the commercial water users are advised to lower their consumption, develop and implement water conservation by limiting unnecessary consumption [Office of Water Resources, 2004]. Thus, the facility operators in Alabama need to implement the water conservation techniques in their facilities to lower the water consumption in response to the severe drought conditions in the region.

Figure 2 shows the population and water withdrawal trends in the State of Alabama from the years 1960-2005 [courtesy of U.S. Geological Survey, 2005]. It can be seen that the population is increasing steadily from year 1970 onwards. The highest water withdrawals were observed during the year 1980, afterwards it decreased by 20% in 1985. In 2000 and 2005, the water withdrawals for thermoelectric power generation have again increased over the previous years. This graph indicates that thermoelectric power generation plants are the largest consumers of water in the state of Alabama.

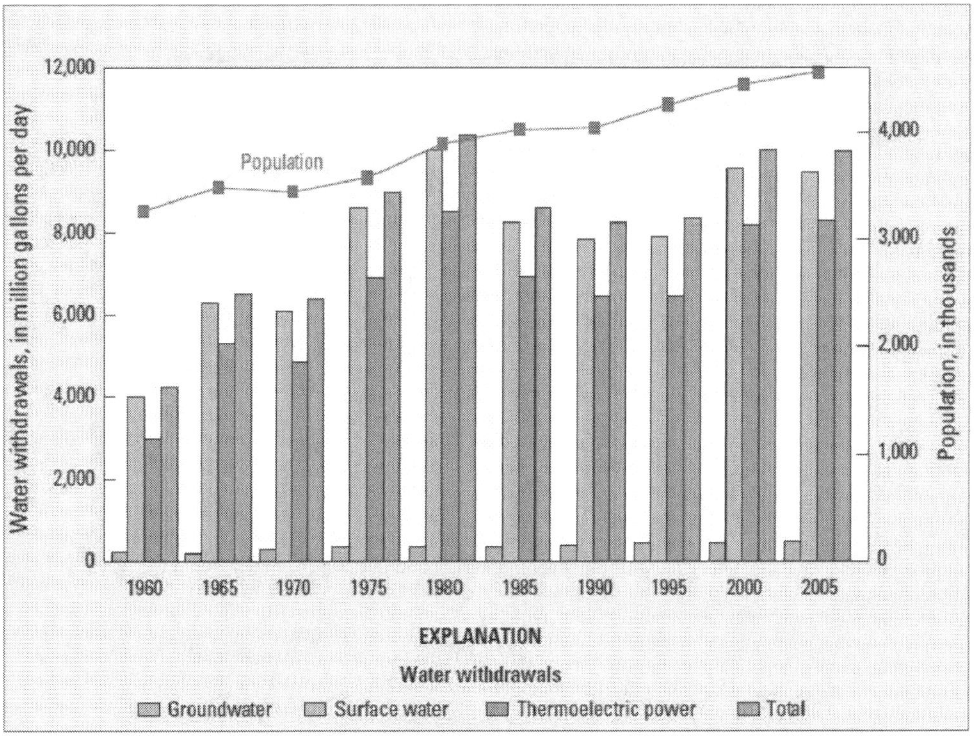

Figure 2. Effect of population on fresh water withdrawals in Alabama.
[U.S. Geological Survey, 2005].

Problem Statement

Due to the prevailing drought conditions and increasing pressure on the potable water supplies in the Southeast region of the United States, it is necessary to research more advanced technologies to enhance water efficiency and sustainability goals. One way of conserving water in large buildings is to install condensate recovery systems and reuse it for applications other than drinking water. Several studies have been conducted to predict the amount of condensate that can be captured from AHUs working at specific conditions (indoor and outdoor climatic conditions) and geographical locations such as in Athens, Georgia [Lawrence *et al.*, 2010] and Doha, Qatar [Bryant and Ahmed, 2008]. These models are validated but do not indicate the exact quantities of condensate. To our knowledge, no research has been conducted for the Birmingham, Alabama location to evaluate the feasibility of a condensate recovery system. It is intended to investigate the performance of condensate collection system at Birmingham, Alabama and how it differs from studies conducted in other cities. The cost of retrofit can be obtained easily from any contractor but due to the limited research in this field, it is difficult to estimate the payback period of the system. As a result, the focus of this research investigation is to develop a model/estimation formula to predict the amount of condensate from the AHUs for Birmingham, Alabama that can be used for buildings on the UAB campus as well as the throughout the State of Alabama and likely southeastern U.S.

Objectives

The primary goal of this research is to develop a model that helps to estimate the amount of condensate from the AHUs on the University of Alabama at Birmingham campus. Secondary goals include the estimation of potential water savings using the Condensate Collection System (CCS), and analysis of the performance of the developed model under different scenarios and for different cities. In cases when the quantity of the condensate water exceeds the make-up water requirement, the excess condensate will become waste if not recovered. Therefore, it is necessary to analyze both the quantity and quality of the condensate water to determine its potential uses in other applications which includes irrigation of campus gardens (lawns and grass), in supplying to campus fountains, and in flushing toilets.

Literature Review

There are several CCS projects installed in the United States but only three research studies involved prediction of quantity of condensate volumes using theoretical formulas which are explained in detail in the next section of this report. Many CCS were installed without realizing water and money savings associated with it. Some of the examples are listed in this paragraph.

Bahrain Airport Services in the Middle East recycles 2.3 million gallons of condensate water every year for toilets flushing, washing, and landscaping [Guz, 2005]. EPA's Environmental Services Branch Laboratory in Houston, Texas, collects 832,000 gallons of water every year, project cost of $6,000, saved $20,000 over past 6 years. National Air and Radiation Environmental Laboratory (NAREL), Montgomery, Alabama installed CCS, help facility to reduce water use by 800,000 gallons annually, 65% of facility's water use. Winship Cancer Institute of Emory University, Atlanta, Georgia – saves 900,000 gallons per year which is $4,860, cost of project $45,000 – payback period 9.25 years. Several CCS projects have been installed in the recent past, only 3 studies performed detailed analysis.

In a related study, the EPA's Science and Ecosystem Division (SESD) installed a condensate recovery system in May 2008 on the roof of their Athens, Georgia facility to reduce the potable water usage [Betty, 2006]. The total cost of this project was $24,000 which saved 540,000 gallons of water for the period of May through December 2008 resulting in a 16% reduction in SESD's water use [Betty, 2006]. The estimated payback period for this project is less than 6 years based on the savings recorded in the first year of operation.

- **Study Conducted by Guz [2005]**

The increasing cost of water and sewer services were the driving forces for the first successful installation of a condensate recovery system in San Antonio, Texas [Guz, 2005]. The study highlights similar projects conducted in the San Antonio area which includes the downtown mall which generates 7604 gallons per month and the central library that produces 43,200 gallons per month. The equation developed by [Guz, 2005] is shown by Equation 1. During the summer, peak condensate production can be in the range of 0.5 to 0.6 gph/1000 ft^2 of cooled area.

$$Condensate\ water\ production\ (gal\ gal/hr/1000\ ft^2)$$

$$= (tons\ of\ capacity) \times (load\ factor) \times ((0.2\ gallons) \qquad Equation\ (1)$$

Guz [2005] found that buildings with area greater than 100,000 ft^2 can generate enough condensate water volumes allowing lower payback periods for the CCS [Guz, 2005].

- **Study conducted by Lawrence *et al.* (2010)**

Lawrence *et al.* [2010] were involved in a study of a CCS at the University of Georgia at Athens (UGA) campus while working with the American Society of Heating, Refrigerating, and Air-Conditioning Engineering (ASHRAE). The formula used by the Lawrence *et al.* [2010] is stated as follows:

$$Q = airflow \times density \times 60\frac{min}{h} \times \Delta\omega \qquad Equation\ (2)$$

where: $\Delta\omega$ is the difference in absolute humidity between incoming outdoor air and supply air leaving the unit. During the cooling season, the total amount of condensate collected was 171,793 gallons but using this model, the amount was predicted to be 134,021 gallons. The amount of condensate predicted using Equation 2 differs from the actual amount of condensate collected by 28%.

- **Study Conducted by Bryant and Ahmed [2008]**

Bryant and Ahmed [2008] investigated the sustainability of a condensate collection system installed in institutional buildings located on the Education City campus of Texas A&M University at Doha, Qatar [Bryant and Ahmed, 2008]. According to Bryant and Ahmed [2008], this formula can be used for commercial buildings in hot and humid areas to estimate the amount of condensate water. Their formula is given below:

$$Q = \frac{1.08\, CFM \times \Delta T}{12000} \qquad Equation\ (3)$$

where: CFM=volumetric air flow in cubic feet per minute, and ΔT= temperature difference across the chilled water cooling in °F.

However, Bryant and Ahmed [2008] did not provide any information about the actual value of condensate collected to the compared values predicted using Equation 3. The average daily condensate production was about 145 gallons per day for a period of 28 days.

All the studies above show that the geographical location of the building is very important to generate more condensate volume. Hot and humid weather is very favorable for the condensate collection. Figure 3 indicates that the geographic location is very important in considering a condensate collection system. According to this map, Birmingham falls on the border of 15+ to 10–15 gallons of water per cubic feet per minute (cfm) of air. Birmingham's hot and humid climate has the potential of capturing approximately 10–15 gallons of condensate water per cubic feet meters of the air [Lawrence *et al.*, 2010].

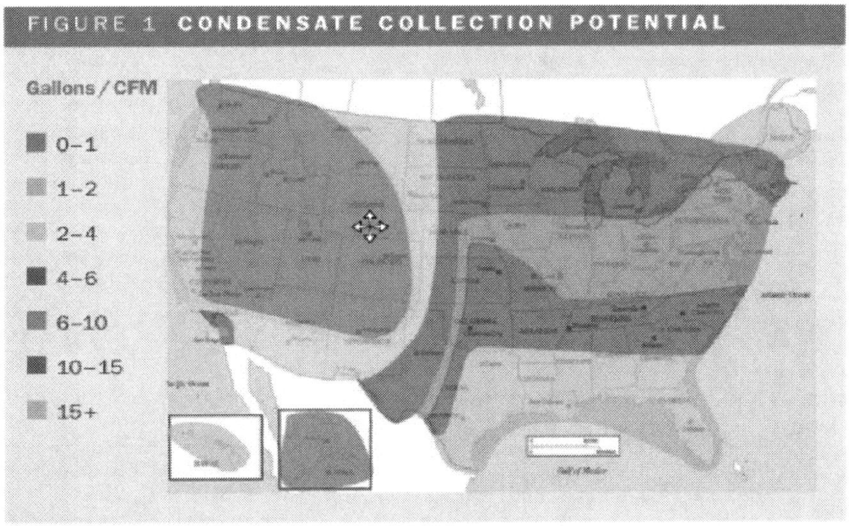

Condensate collection potential varies by regional climates. The map represents the potential amount of condensate that can be collected from each cfm of outdoor air over one year (8,760 hours).

Figure 3. Condensate Collection Potential in United States.
Lawrence and Perry, [2010].

The type of building, its usage and occupancy have a big impact on the amount of condensate collected [Lawrence and Perry, 2010]. Large commercial buildings are primarily used for office, educational, retail, research laboratories and healthcare purposes and require large amounts of outside air which are cooled in the AHUs and distributed throughout the building. Laboratory and clinical buildings require 100% outside air in each air exchange process whereas other buildings generally recirculate some part of the conditioned air along with the outside air [Lawrence *et al.*, 2010]. The dedicated outside air systems utilize 100% outside air which can produce more condensate [Lawrence *et al.*, 2010]. Higher amounts of moisture in the outside air generate higher condensate volumes [Lawrence *et al.*, 2010]. Buildings with high occupancy require multiple air exchanges per hour and hence can produce large condensate volumes.

Design of Condensate Collection System (CCS)

Condensate is formed on the outside surfaces of the pipes and other surfaces in the close proximity of the cooling coils (heat exchanger) in the AHU. Condensate from each AHU is then taken to small tanks from which it is pumped to a 500-gallon storage tanks. Figure 4 shows a schematic of the condensate recovery system at UAB.

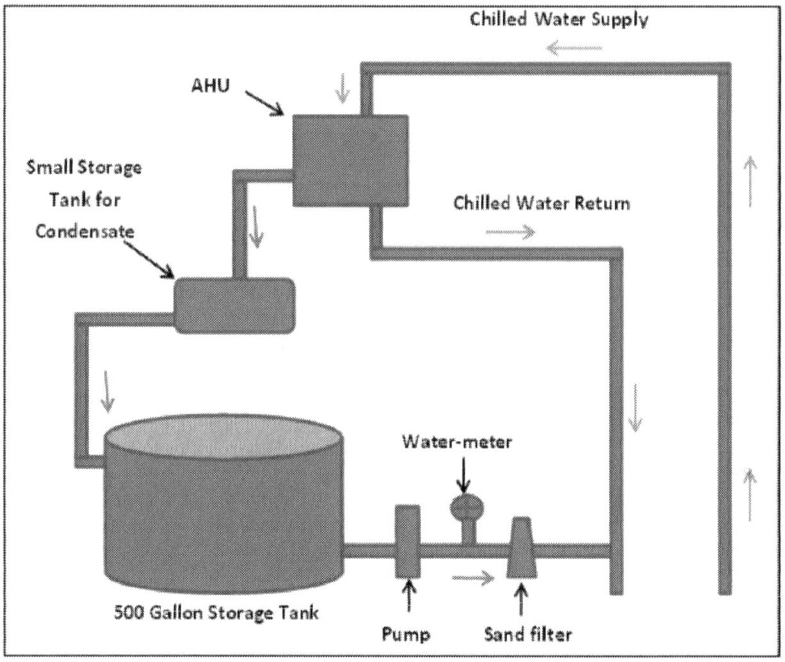

Figure 4. Schematic of the Condensate Recovery System.

Figure 5 shows a picture of a small storage tank located in the Zeigler Research Building. The condensate from the AHU is collected here. The water can be seen flowing into the tank from the inlet on the left side of the tank.

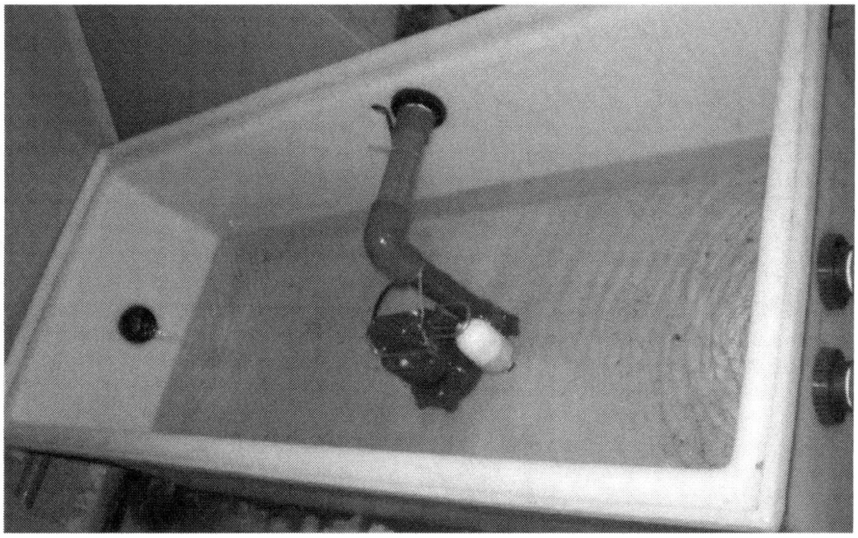

Figure 5. Small Storage Tanks to Collect Condensate Water from AHUs (Zeigler Research Building).

These tanks have sump pumps with level switches in the floating devices (white in color), depicted in the middle of the tank. When the water level in these small tanks reaches a certain level, the float activates the motor and pumps the water to the 500-gallon tank. The level switch is installed to prevent the tanks from overflowing. These devices also help in stopping the pump when the water level in the tanks reaches the bottom. The condensate from the small tanks is directed to the storage tanks. During hot summer days, the condensate recovery is very high. An additional pipe is installed at the top of the tank to remove any overflowing water which is then directed to the sewer drain.

Figure 6 shows the picture of a 500-gallon storage tank installed in the Zeigler Building. Condensate collected in each small tank is pumped into this tank for storage. As the demand of chilled water increases at the chilled water plant, the water from the storage tank is injected in to the "chilled water return" pipe at a higher pressure to prevent backflow.

Figure 6. 500-gallon Storage Tank.

The Energy Management Division of the UAB Facilities Management has successfully installed pilot condensate recovery systems in five buildings on the UAB campus: McCallum Building (MCM), Kaul Building (KB), Zeigler Building (ZB), Tinsley Harrison Tower (THT), and Bevill Building (BB). Table 1 summarizes building information, the condensate collection system, and initial cost of installation. The installation of CCS in the MCM, KB, ZB, and THT was completed well before the cooling season of 2011. Therefore, there was an opportunity to collect data during this period except for the BB where system installation work was finished in August of 2011.

Table 1. Summary of the Pilot Condensate Recovery System at UAB.

Building Name	System	Installation Cost	Remarks
McCallum Building (MCM)	2 AHUs drains into 2 storage tanks	$25,000.00	McCallum and Kaul share one 500-gallon tank; started in May 2011
Kaul Building (KB)	5 AHUs drains into one 500-gallons tank	$16,220.00	
Zeigler Building (ZB)	3 AHUs drains into two storage tanks	$5,500.00	Zeigler and Tinsley Harrison Tower share a one 500-gallon tank; started in May 2011
Tinsley Harrison Tower (THT)	10 AHUs drain into a 500-gallon tank	$2,475	
Bevill Building (BB)	6 AHUs drains into two 500-gallon tanks	$29,965.00	Started in August 2011
Total		$79,160.00	

Beginning June 14th of 2012 through December 31th of 2012, temperature, relative humidity, dew point temperature, dry bulb temperature, and wet bulb temperature will be collected from the online sources such as www.wunderground.com. Condensate data will be obtained every minute using waterlevel datalaggers for the abovementioned period. This timeframe is chosen in order to collect data for both cooling and non-cooling seasons. The data from the data-loggers is stored on a computer and then transferred to Excel spreadsheets for further analysis. The primary variables that will be analyzed in this study are as follows:

- Condensate water collected
- Temperature, dry bulb temperature, and dew point temperature,
- Relative humidity,
- Air supply rate,
- Intake air volume per exchange,
- Area of the building served by the air handler,
- Capacity of the air handler.

Results and Discussion

The preliminary analysis of the condensate data obtained during June 14th, 2012 through August 30th, 2012 is presented in the next paragraphs. Figure 7 shows the monthly volume of condensate water from each Kaul air handler #4, MCM air handler #1, and MCM air handler #2. Over the period of 75 days, air handlers Kaul #4, MCM #1, and MCM #2 produced 117903 gallons, 188612 gallons, and 207673 gallon of condensate water respectively which represents 0.5 million gallons of condensate water.

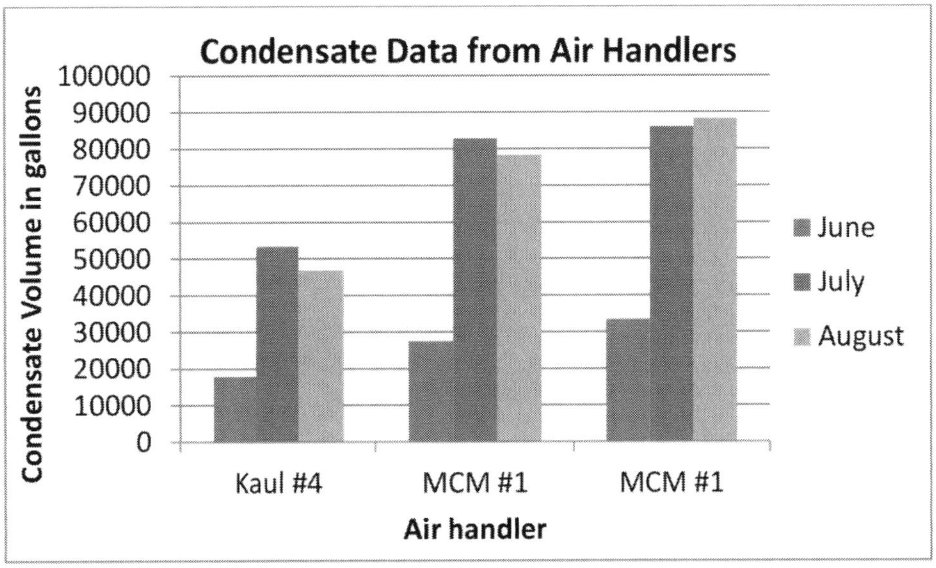

Figure 7. Condensate Water Collection during 2012.

Figure 8 shows the condensate data resulting from the storage tanks.

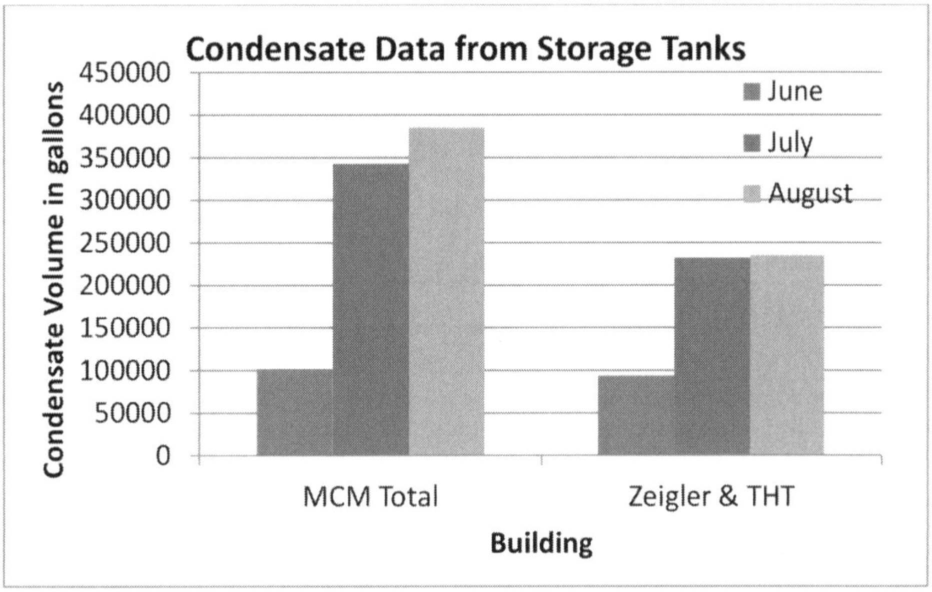

Figure 8. Condensate Data from Storage Tanks during 2012.

The data of condensate water collected in the 500-gallon storage tank is also recorded during this time and is shown in Figure 7. More than 1.5 million gallons of condensate water was collected in 75 days from both the tanks. The descriptive statistics of the condensate data collected from Kaul #4, MCM #1, and MCM #2 is displayed in Table 2 through 4.

Table 2. Condensate Data of Kaul Air Handler #4.

Kaul #4	Volume	Average	Std. Dev.	Maximum
June	17776.03	45.23163	9.512099	90.94465
July	53309.86	71.65304	27.08663	116.7918
August	46817.23	65.75454	27.97125	109.2404
Total	117903.1			

Table 3. Condensate Data of MCM Air Handler #1.

MCM #1	Volume	Average	Std. Dev.	Maximum
June	27540.39	70.04769	36.53066	142.4811
July	82795.32	111.284	43.6994	172.2053
August	78276.2	109.9385	35.2959	167.8352
Total	188611.9			

Table 4. Condensate Data of MCM Air Handler #2.

MCM #1	Volume	Average	Std. Dev.	Maximum
June	33421.75	84.99098	34.28401	160.3069
July	85981.64	115.5667	58.57804	223.176
August	88269.82	123.9745	46.05853	216.9447
Total	207673.2			

When the condensate water reaches the maximum level in the storage tanks, this water is pumped and injected in to the chilled water return pipe. The datalogger has successfully recorded this process and hence the collected data can be said to be fairly accurate. Figure 9 shows the mass balance of the condensate water (for Kaul #4) coming in to the tank and the water pumped out of the tank. The R^2 value is 0.96 which represents minimum error in data collection. Similar trends have been observed for other storage tanks too.

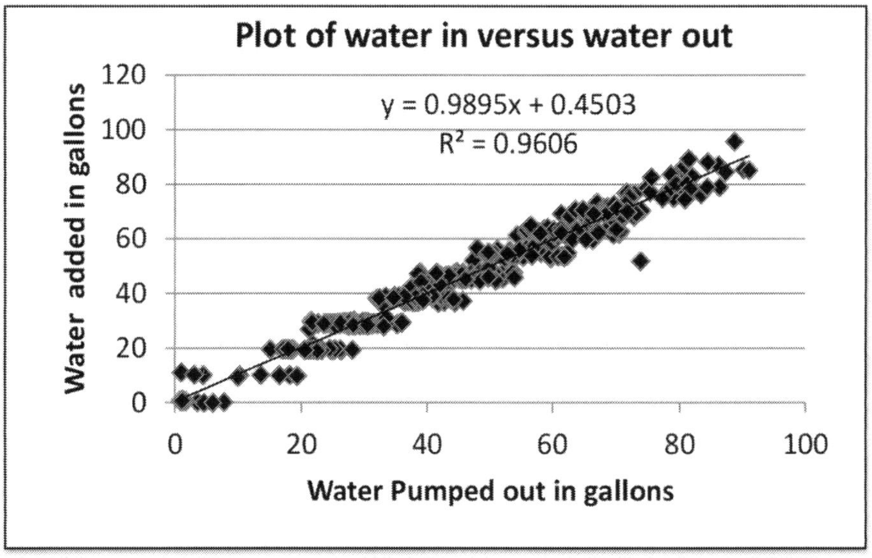

Figure 9. Plot of water in versus water out.

Regression Analysis

Regression analysis will be performed to understand the effect of each variable on the condensate recovery. Multiple regression analysis will also be performed to verify the combined effect of the variables on the condensate recovery. The individual effect of outside temperature, dew point temperature, and humidity on condensate collection for air handler Kaul #4 is analyzed using regression analysis and is shown in Table 5. The R^2 values in the table indicate that the dew point temperature has the strong correlation with the condensate production.

Table 5. Effect of Weather Parameters on Condensate Collection (Kaul #4).

Model	R^2	Std Error	P	Intercept	Coefficient
Temp	0.11	27.65	0.019	28.55	0.47
Dew Point Temp	0.85	10.66	4.5E-298	-271.328	4.9
Humidity	0.22	24.45	0.0196	9.32	76.62

Multiple regression analysis was performed to examine the combined effect of weather parameters on the condensate production. Table 6 shows the results of a multiple regression analysis for air handler Kaul #4.

Table 6. Combined Effect of Weather Parameters on Condensate Production.

Model	R^2	Std Error	P	Intercept	Temp	Dew	Humidity
June	0.82	8.72	1.72E-54	-251.968	1.13	2.88	48.82
July	0.8	6.13	5.16E-39	-182.92	-0.17	4.12	-13.62
Aug	0.86	10.5	6.8E-63	-355.45	1.7	3.49	70.76

Future Work

In order to develop a regression model to predict the amount of condensate water from any air handler, effect of different parameters will be analyzed using linear regression.

- Analysis of Variance (ANOVA): One-way ANOVA test will be performed to examine the effect of each variable on the outcome (condensate recovery Q) where Q is explained as follows:

$$Q = f_1\left(T_{dry}\right), f_2\left(RH\right), f_3\left(P_{baro}\right), f_4\left(CPM_{Main\ Supply\ Rate}\right)$$

Two-way analysis will be performed to study the effect of interactions of two or more variables on the condensate recovery where Q can be explained as follows:

$$Q = f_5\ Combination\ of\ \left(T_{dry}\right), \left(RH\right), \left(P_{baro}\right), \left(CPM_{Main\ Supply\ Rate}\right)$$

Multiple linear regression analysis will be performed to understand the effect of primary variables on the condensate recovery. The equation of the multiple linear regression analysis which represents the two-way ANOVA test mentioned above can be stated as follows:

$$Q = aT + bRH + c\left(P_{baro}\right) + d\left(CPM_{Main\ Supply\ Rate}\right) + error$$

- Sensitivity Analysis and Uncertainty Analysis: Sensitivity analysis will be performed to determine which key parameters influence the performance of the model developed under this study. It is necessary to understand behavior of different parameters on models output. The variation in the results due to error in the measurement of any data will be calculated to examine if the output vary slightly or significantly. The effect of changing one of the parameters (relative humidity, temperature, air supply), say 10% of original value, in predicting condensate will be analyzed under this test.

- Payback Period Calculations: The payback period calculation will be based on comprehensive analysis of savings that are possible by recycling the condensate water. Following savings are possible by recycling the condensate water for different applications: savings in water bill, savings in sewer bills, savings in treatment cost of effluents, and savings in energy that would have been used to chill the water. For calculation of payback period of the CCS, all the savings were calculated separately and used to determine the actual payback period.

- Life Cost Analysis (LCA) and Life Cost Analysis (LCC): The CCS involves storage tanks, new piping, motors, meters, plumbing fixtures, and some modifications to the building structure. It is necessary to maximize the environmental performance of the CCS by determining which phase of the process cause greatest harm to the environment. This can be done by performing Life cost analysis and inventory and impact analysis.

Discussion

It is expected that model developed during this study will provide more accurate predictions of condensate volumes. This model will be useful in estimating how much water can be recovered annually and hence the payback period of the system. The CCSs installed at UAB medical campus buildings have collected more than 1.4 million gallons of water in 75 days which is equivalent to a saving of $5900. Comprehensive analysis of water, energy, and sewer savings will help in more accurate payback period calculations which will be helpful for the building managers and facility operators in installing CCSs.

Acknowledgement

The authors of this paper would like to acknowledge the Facilities Management Department and their Energy Management Division for funding this project and providing the technical information related to the research. We would also like to thank The Sigma Xi-The Scientific Research Society for the Grants-in-Aid of Research for equipment purchase.

References Cited

Betty, K., 2006. *Water Management Plan USEPA Region 4 Science and Ecosystem Support Division.* U.S. Environmental Protection Agency, Athens, GA.

Bryant, J.A., and T. Ahmed, 2008. *Condensate Water Collection for an Institutional Building in Doha, Qatar: An Opportunity for Water Sustinability.* Proceedings of the 16th Symposyum on Improving Building Performance Systems in Hot and Humid Climates, TX.

Guz, K., 2005. *Condensate Water Recovery.* American Society of Heating, Refrigeration and Air-Conditioning Engineers (ASHRAE), Atlanta, GA..

Lawrence, T.M., J. Perry, and P. Dempsey, 2010. "Predicting Condensate Collection from HVAC Air Handling Units. *American Society of Heating, Refrigerating, and Air-Conditining Engineering (ASHRAE).*

Lawrence, T., and J. Perry, 2010. *Capturing Condensate.* Retrieved June 13, 2011, from High Performance Buildings 2010: http://www.nxtbook.com/nxtbooks/ashrae/hpb_2010fall/index.php?startid=56

Office of Water Resources, 2004. *Alabama Drought Management Plan.* Alabama Department of Economic and Community Affairs, Montgomery, AL.

U.S. Geological Survey. (2005). *Alabama Water Use, 2005.* Montgomery, AL.

(137d) Optimization-Based Methodology for the Development of Wastewater Facilities for Energy and Nutrient Recovery

Channarong Puchongkawarin[1], David C Stuckey[1] and **Benoit Chachuat**[2], (1)Department of Chemical Engineering, Imperial College London, London, United Kingdom, (2)Centre for Process Systems Engineering, Department of Chemical Engineering, Imperial College London, London, United Kingdom

A paradigm shift is currently underway from an attitude that considers wastewater streams as a waste to be treated, to a proactive interest in recovering materials and energy from these streams [Guest et al., 2009]. Until recently a majority of the activities related to resource recovery from wastewater have focused on waste sludge streams, which are a by-product of biological treatment. Because these streams have relatively low flows in comparison to the main wastewater streams, and are more concentrated, resources can be recovered from them with minimal changes to the wastewater treatment infrastructure. In contrast, the main focus of this presentation is on direct resource recovery from municipal and industrial wastewater (although the proposed methodology can also be extended to encompass wastewater sludge as well).

A diverse toolkit is available and is increasingly being applied by practitioners to promote sustainability of wastewater facilities. The increasing market value of wastewater components, such as ammonia and phosphorus, are acting as key drivers for resource recovery from wastewater. Organic carbon too can be recovered from wastewater using anaerobic digestion to produce a methane-rich biogas that can be combusted on-site for heat or electricity generation, or cleaned-up and sold. Other examples of newer technologies for carbon recovery include microbial fuel cells as well as the production of bioplastics.

Although there seems to be a general consensus that wastewater (and wastewater sludge) is a potential source of valuable resources, and that the technology needed for such resource recovery is maturing, it is the lack of decision-making tools and design methodologies that is the primary problem to identifying the most sustainable solutions in a given geographic and cultural context [Hamouda et al., 2009]. For the synthesis of sustainable resource recovery facilities, one must account for the trade-offs between capital and operating costs and sales on the one hand, and water quality and other environmental and social considerations on the other hand. But as the array of technical options grows, a simple enumeration of all possible alternatives quickly becomes unmanageable, let alone the fact that each technology has its own parameters to specify or optimize. Technical insights are useful to reduce the combinatorial problem and often allow us to arrive at promising solutions as in [Sutton et al., 2001], but they do not always provide all the information that is required for an optimal (or near optimal) system.

In this presentation, we advocate the use of systems engineering methods and tools to address this problem in a systematic way. A superstructure modelling and optimization approach [Biegler et al., 1997; Kokossis and Yang, 2010] is considered, whereby the objective is to determine an optimal resource recovery facility in terms of (i) its units, (ii) the piping interconnections between the units, and (iii) the flow rates and compositions in the interconnections. The optimal system configuration is one that maximizes a certain sustainability or economical index of the facility, for instance a weighted sum of LCA impacts or the net present value (NPV). Regarding constraints, material balances on flows and concentrations around the sources, the units, and the sinks are to be obeyed in addition to the discharge limits and certain design and structural specifications. This formulation leads to nonconvex MINLP problems for which state-of-the-art global optimization software such as BARON [Tawarmalani and Sahinidis, 2004] can be used.

Key to the success of this methodology is the development/selection of mathematical models for the units that are simple enough, yet provide reliable estimates of their performance and associated costs. The simplest approach considers fixed conversion, removal or split ratios in the treatment and separation units, as is typically assumed in water network synthesis problems [Ahmetovic and Grossmann, 2011; Khor et al., in press]. In this work, we advocate an alternative, potentially more accurate, approach that relies on detailed first principle models, as implemented in wastewater treatment simulators such as GPS-X (www.hydromantis.com/GPS-X.html). Following a surrogate-modeling approach, simple linear or piecewise-linear models can be fitted to the predicted performance from a process simulator at a large number of influent flow rates and compositions. Likewise, reliable capital and operating costs can be obtained from preliminary costing software, such as CapdetWorks (www.hydromantis.com/CapdetWorks.html).

To illustrate this methodology, we showcase the synthesis of a resource recovery facility for the treatment of 100 m3/h of an industrial wine distillery effluent. A small superstructure that consists of 2 biological treatment units (UASB, SMBR), 2 filtration units (sand filter, membrane unit), and 2 nutrient recovery units (struvite crystallizer, zeolite adsorber) is investigated. Besides the treatment/separation units, an auxiliary piece of equipment is the generator producing electricity from biogas. The objective is to maximize the NPV accounting for electricity, struvite and ammonia sales, while satisfying maximum discharge requirements as defined by the EU Directive 91/271/EEC on Urban Wastewater Treatment. Several scenarios are discussed, which demonstrate that significant improvements in the NPV can be achieved through resource recovery.

By and large, the proposed methodology should be seen as a decision support system for isolating, among hundreds or even thousands of alternatives, those promising resource recovery systems whose development is worth pursuing. Based on this preselection, further simulation and optimization studies can then be undertaken to refine the performance and cost prediction by taking into account detailed design and operation considerations, as well as process integration. Such decomposition is indeed warranted as current computational capabilities and available algorithms do not allow for the optimal design and operation to be solved in a single step due to complex unit configuration, multiple scales, time dependence, and uncertainty. Besides providing a rational tool for comparing alternatives, this approach can also be useful for resource allocation, for instance in determining which technologies are critical and deciding where further research and development effort should focus.

References:

- E. Ahmetovic and I.E. Grossmann (2011). Global superstructure optimization for the design of integrated process water networks. AIChE journal, 57(2):434-457.

- L.T. Bielger, I.E. Grossmann, and A.W. Westerberg (1997). Systematic Methods of Chemical Process Design. Prentice Hall.

- J.S. Guest, S.J. Skerlos, J.L. Barnard, M.B. Beck, G.T. Daigger, H. Hilger, S.J. Jackson, K. Karvazy, L. Kelly, L. MacPherson, J.R. Mihelcic, A. Pramanik, L. Raskin, M.C. Van Loosdrecht, D. Yeh, and N.G. Love (2009). A new planning and design paradigm to achieve sustainable resource recovery from wastewater. Environmental Science & Technology, 43(16):6126-6130.

- M.A. Hamouda, W.B. Anderson, and P.M. Huck (2009). Decision support systems in water and wastewater treatment process selection and design: A review. Water Science & Technology, 60(7):1757-1770.

- C.S. Khor, B. Chachuat, and N. Shah (in press). A superstructure optimization approach for water network synthesis with membrane separation-based regenerators. Computers & Chemical Engineering.

- A.C. Kokossis and A. Yang (2010). On the use of systems technologies and a systematic approach for the synthesis and the design of future biorefineries. Computers & Chemical Engineering, 34(9):1397-1405.

- P.M. Sutton, H. Melcer, O.J. Schraa, and A.P. Togna (2011). Treating municipal wastewater with the goal of resource recovery. Water Science & Technology, 63(1):25--61.

- M. Tawarmalani and N.V. Sahinidis (2004). Global optimization of mixed-integer nonlinear programs: A theoretical and practical study. Mathematical Programming, 99(3):563-591.

(137e) Sustainable Condensations On the Early Earth: Prebiotic Peptide Coupling in Formamide

Ryan Clairmont[1], Elizabeth Cope[2], Charles Liotta[2] and Andreas Bommarius[3], (1)School of Chemical & Biomolecular Engineering, Georgia Institute of Technology, Atlanta, GA, (2)School of Chemistry and Biochemistry, Georgia Institute of Technology, Atlanta, GA, (3)Georgia Institute of Technology, Atlanta, GA

The transition from a chemical inventory of diverse monomers to functional biopolymers is an important topic in understanding the emergence of life on Earth. Chemical evolution can also yield valuable clues for Green Chemistry. Alternative solvents (solvents other than water) have been proposed as a means of accessing the condensation reactions necessary for monomer coupling by both concentrating reactants and providing a dehydrated environment. Formamide,prebiotically abundant, has been proposed as an alternative solvent and was used to investigate amino acid condensations using glycine and β-alanine. Three conditions (open, closed, and cycling) were studied to understand the impact of being exposed to different environments on the early Earth. The cycling system was found to best promote oligomer formation, but with low conversions 1-2% dimer. Formylation is a competing side reaction that occurs in formamide, blocking the N-terminus of the amino acid and preventing condensation, especially to diketopiperazines. Estimated thermodynamic and kinetic parameters can be used to model the formation of larger oligomers and show the importance of free amino acid equilibrium in producing longer oligomers.

(137f) Aqueous Surfactant Solutions As Smart Solvents for Catalytic Reactions

Michael Schwarze, Tobias Hamerla, Anke Rost, Katja Seifert and Prof. Dr. Reinhard Schomäcker, Department of Chemistry, Berlin Institute of Technology, Berlin, Germany

With the growing need for sustainability, the aim of chemical processes is not only to provide a product; moreover it should follow the principles of *Green Chemistry* formulated by Paul Anastas and John Warner [1]. The practical use of these principles should help to prevent pollution, save energy, and reduce the potential for chemical accidents. One decisive approach is the replacement of harmful or toxic organic solvents by water. An essential problem of this approach is that many reactants are poorly soluble in water, but solubility can be significantly increased by adding surfactants. In this context, the application of surfactant-based reaction media has increased during the last years and it is shown in the literature that many noble-metal catalyzed homogeneous reactions can be successfully performed in these media, e.g. hydrogenation reactions [2], coupling reactions [3], hydroformylation reactions [4] or methathesis reactions [5]. Moreover, surfactant systems allow for catalyst recovery which is a major problem in homogenous catalysis.

For aqueous-micellar solutions catalyst recovery is possible by micellar enhanced ultrafiltration (MEUF), a well known technique from the field of waste-water purification.

Figure 1: Aqueous-micellar reaction solution (dissolved rhodium catalyst) before (yellow) and after (colorless) ultrafiltration process (membrane MWCO=10 kDa).

In the case of microemulsion systems, a mixture of water/organic solvent/surfactant (here the organic solvent is only partially replaced by water), catalyst recovery can be performed by phase separation. In Figure 2, the different states of a microemulsion system are shown. Under optimized conditions, the catalyst can be recycled after the reaction from a 2- or 3 phase system.

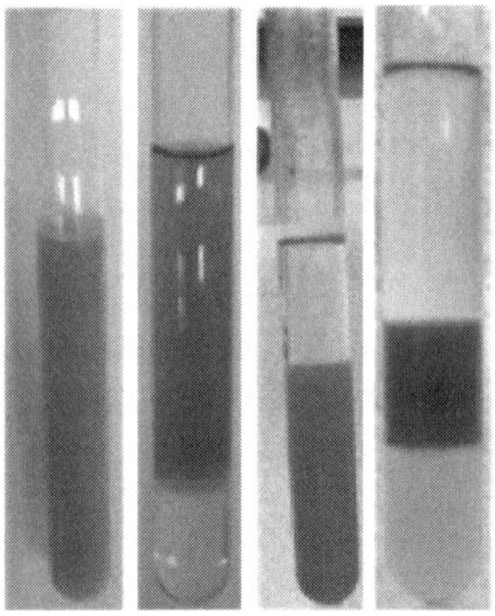

Figure 2: States of a microemulsion system with dissolved rhodium catalyst.

Although there are many examples of reactions in these media, the development of a combined reaction-separation-process is no easy task. In this contribution we will show results for the rhodium catalyzed hydrogenation of itaconates in aqueous-micellar solutions and discuss the main parameters which are important with respect to catalyst recycling and product isolation using MEUF, e.g. the partition coefficient of the reactants and the influence of the surfactant and surfactant concentration.

Table 1: Partition coefficients of itaconates in aqueous-micellar TX-100 solutions [6].

	Itaconic acid	Dimethyl itaconate	Diethyl itaconate	Dibutyl itaconate
log K_{MW}	2.3	2.3	3	5.3

Furthermore, we will show examples for noble-metal catalyzed reactions in microemulsions, e.g. coupling reactions and hydroformylation reactions. In the case of microemulsions, a detailed analysis of the phase behavior is required to find the best conditions for reaction and catalyst recycling by phase separation. If this information is available, several combined reaction and separation cycles can be performed as recently reported by H. Nowothnick et al. [7].

[1] P.T. Anastas, J.C. Warner, Green Chemistry: Theory and Practice., Oxford and New York: Oxford University Press, **1998**; [2] M. Schwarze et al., *RSC Adv.* 1 (**2011**) 474; [3] B.H. Lipshutz et al., *Org. Lett.* 10 (**2008**) 1333; [4] M. Gottardo et al., *Adv. Synth. & Cat.* 352 (**2010**) 2251; [5] B.H. Lipshutz et al., *Org. Lett.* 10 (**2008**) 1325; [6] M. Schwarze et al., *Ind. & Eng. Chem. Res.* 51 (**2012**) 1846; [7] H. Nowothnick et al., *Angew. Chem. Int. Ed.* 50 (**2011**) 1918.

(137g) Modeling a CO_2 Refrigeration Cycle Using an Ionic Liquid Co-Fluid

George Mozurkewich, Luke D. Simoni, William F. Schneider and Mark A. Stadtherr, Department of Chemical and Biomolecular Engineering, University of Notre Dame, Notre Dame, IN

A number of popular vapor-compression refrigerants, which have been used successfully for decades, have come under environmental challenge for their potential to deplete stratospheric ozone and/or to contribute to global climate change. While current refrigerants have little or no ozone depletion potential, many of the materials used still have a quite significant global warming potential (GWP) relative to carbon dioxide (for which GWP = 1 by definition). CO_2 itself is in fact an attractive refrigerant, and has been suggested and used as an environmentally benign replacement for high GWP refrigerants. Vapor-compression cycles based on CO_2, however, have very high operating pressures and limited coefficients of performance. Pressures can be reduced and efficiency increased by circulating CO_2 in tandem with a liquid (a "co-fluid") into which the CO_2 can be absorbed. The resulting co-fluid cycle is analogous to the vapor-compression cycle with condensation replaced by absorption and evaporation replaced by desorption [1]. Ideal co-fluids will have low volatility, good chemical and thermal stability, and a tunable strength of CO_2 absorption. Previous work was based on physical absorption into organic liquids, providing only a limited range of absorption strength.

Recently, aprotic heterocyclic anion (AHA) ionic liquids (ILs) have been proposed for use as co-fluids, due to their extremely low volatility, stability over a wide temperature range, and capacity for CO_2 absorption. Furthermore, since these materials absorb CO_2 both chemically and physically, their affinity for CO_2 can be finely tuned through appropriate chemical functionalizations. To analyze their performance in a refrigeration system, we have developed a co-fluid cycle model that incorporates both physical and chemical absorption of CO_2 by ILs. The model describes the compression, expansion, and heat-exchange processes in terms of basic physical and chemical principles and predicts optimal operating pressures and resulting coefficients of performance. Principal inputs are the enthalpies and entropies of solution as may be derived from experimental absorption isotherms or predicted by *ab initio* quantum chemical calculations. By identifying desirable ranges for the chemical properties of the co-fluid, the model provides theoretical guidance for the design of suitable ILs.

[1] See citations in Mozurkewich et al, *Int. J. Refrig.* **25**, 1123 (2002).

***** Formatting Issue - Best Available Paper/Graphic *****

EXAMINATION OF THE DIFFERENCES IN ENERGY EFFICIENCY OPPORTUNITIES IN LIGHTING RETROFITS OF FOUR DIFFERENT TYPES OF CAMPUS BUILDINGS

[1]Dana Lackey and [2]Robert W. Peters, Ph.D., P.E.
[1]Graduate Student
[2]Professor of Environmental Engineering
Department of Civil, Construction, and Environmental Engineering
University of Alabama at Birmingham,
Birmingham, Alabama 35294-4440

Abstract

From June 2009 through November 2011, a series of walkthrough lighting surveys were conducted on the University of Alabama at Birmingham (UAB) campus. These lighting surveys involved the use of a ballast sensor, which flashed "green" when fluorescent ballasts were energy-conserving electronic, or "red" when they were less-efficient magnetic. Researchers proceeded through the rooms of each building recording data of sensor result, fixture size, and number and type of bulbs (primarily T12 or T8). The buildings could be categorized into four different types based on usage as observed by researchers:

- Academic buildings (5): buildings in which classes are taught and professors have offices and laboratories.
- Community Service buildings (4): examples include health clinics and office buildings with little space devoted to laboratories.
- Medical Research buildings (6): most floors devoted to research laboratories and offices.
- Parking Decks (2): building's only use is parking of vehicles.

Raw data from the ballast sensors determined that 72% of academic buildings, 68% of community service buildings, 81% of research buildings, and 90% of parking decks were using magnetic ballasts. Various assumptions regarding wattage per fixture with respect to number and type of bulbs, as well as costs of equipment, were taken into account. Using these assumptions, it could be determined that while differences were small between wattage reduction potential when expressed as a percentage (47.5-54.5%), there were major differences those actual wattage reductions, as well as in payback periods (0.9-1.7 years) for the different building types. Despite a greater percentage of magnetic fixtures in parking decks than any other building type, the most beneficial type of building for these types of retrofits was determined to be academic buildings, as they had the highest potential wattage reduction (118.9 kW), as well as the lowest payback period (0.8 years), but only once an outlier was removed.

Introduction

Despite some degree of economic recovery over the past four years, the administration at UAB is still seeking to reduce energy costs as much as possible. Prior to this study in Spring 2009, the campus cost over a million dollars a week to operate. Around this time, the Engineering Management Division of the Facilities Management Department (hereafter referred to as "Energy Management") began a project whose aim is to determine how much energy and resulting expenditures can be reduced by upgrading older lighting fixtures to newer models.

In general, two chief upgrade types could be made to the lighting fixtures. Magnetic ballasts could be replaced with electronic ones, and T12 fluorescent light bulbs could be replaced with T8 fluorescent bulbs. Electronic ballasts are more energy-efficient than magnetic ballasts because they more easily convert the 60 Hz power input to 20-40 kHz of high-frequency alternating current [1]. In fact, they do this so efficiently that their magnetic counterparts are no longer being produced as of July 2010 [2]. T8 bulbs are 1" in diameter (eight-eighths of an inch), and are more energy-efficient than their older T12 counterparts, which are 1.5" in diameter (twelve-eighths of an inch) because they have a reduced interior volume V for a common length L:

This interior volume that is 44.4% of their T12 counterparts enables the gases and rare-earth phosphors contained within to emit a comparable amount of light for a lesser power input [3]. In addition to this, T8 fluorescent bulbs also tend to require replacement less often than T12 bulbs and are often more aesthetically pleasing than T12 bulbs.

Introduction and Background

Buildings and Building Types Surveyed

The buildings which were surveyed were chosen by Energy Management between June 2009 and November 2011. They can be sorted out into four basic types of use as described above in the abstract: academic, community service, research, and parking decks, each denoted with a shape-code that is first displayed in each bullet point below. The survey of each building was completed on the month and year listed below in parentheses:

▲ Academic Buildings: buildings in which classes are taught and professors have offices and laboratories.
1. Business-Engineering Complex (BEC; June 2009);
2. School of Dentistry (July 2009);
3. Volker Hall (August 2009);
4. Education Building (August 2011); and
5. Humanities Building (October 2011).

■ Community Service Buildings: examples include health clinics and office buildings with little space devoted to laboratories.
1. General Services Bldg (August 2009);
2. 19th Street Community Health Services Building (CHSB-19; June 2010);
3. 20th Street Community Health Services Building (CHSB-20; July 2010); and
4. Community Care Building (July 2010).

● Medical Research Facilities: most floors devoted to research laboratories and offices.
1. Boshell Diabetes Research Bldg. (August 2009);
2. Bevill Biomedical Research Bldg. (December 2009);
3. McCallum Bldg. (May 2010);
4. Lyons-Harrison Research Bldg. (June 2010);
5. Tinsley-Harrison Tower (June 2010); and
6. Zeigler Research Building (July 2010).

★ Parking Decks: building's only use is parking of vehicles.
1. 16th Street Parking Deck (July 2009); and
2. 19th Street Parking Deck (June 2009).

The locations of the buildings on the UAB campus are given in Figure 1, the map below. The shape-coded bullets for each building type are used in it as well as throughout this paper.

*** Formatting Issue - Best Available Paper/Graphic ***

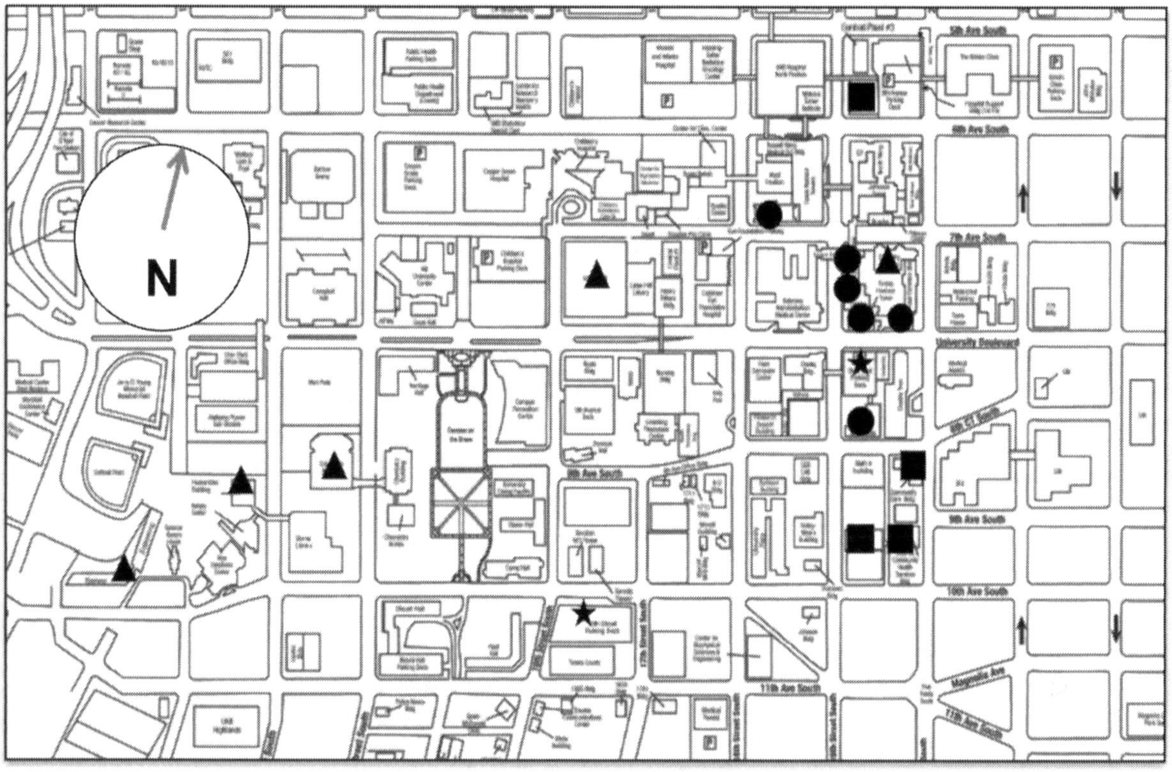

Figure 1. Map of locations of surveyed buildings.

Project Tasks and Goals

The goal of the project was to determine how much could be saved by retrofitting each building with energy-efficient lighting. It was the task of the research team, composed of graduate students in the Civil, Construction, and Environmental Engineering Department at UAB, to proceed through the buildings using a ballast sensor, which flashed "green" when operated under an electronic ballast and "red" when operated under a magnetic one, and record observations regarding the result of the sensor, the fixture size, and number and type of bulbs.

After walkthrough survey data were collected, assumptions were made and in some circumstances, additional data collected from other sources to determine wattage as each building is now (current scenario), then wattage under a condition where all retrofits were made (projected scenario), then the wattage from the projected scenario was subscribed to the current scenario, and multiplied by a cost per kWh to determine savings, then finally the costs of the implements of the retrofit compared with the savings per kWh to determine a payback period. A flowchart of the process is given in Figure 2.

*** Formatting Issue - Best Available Paper/Graphic ***

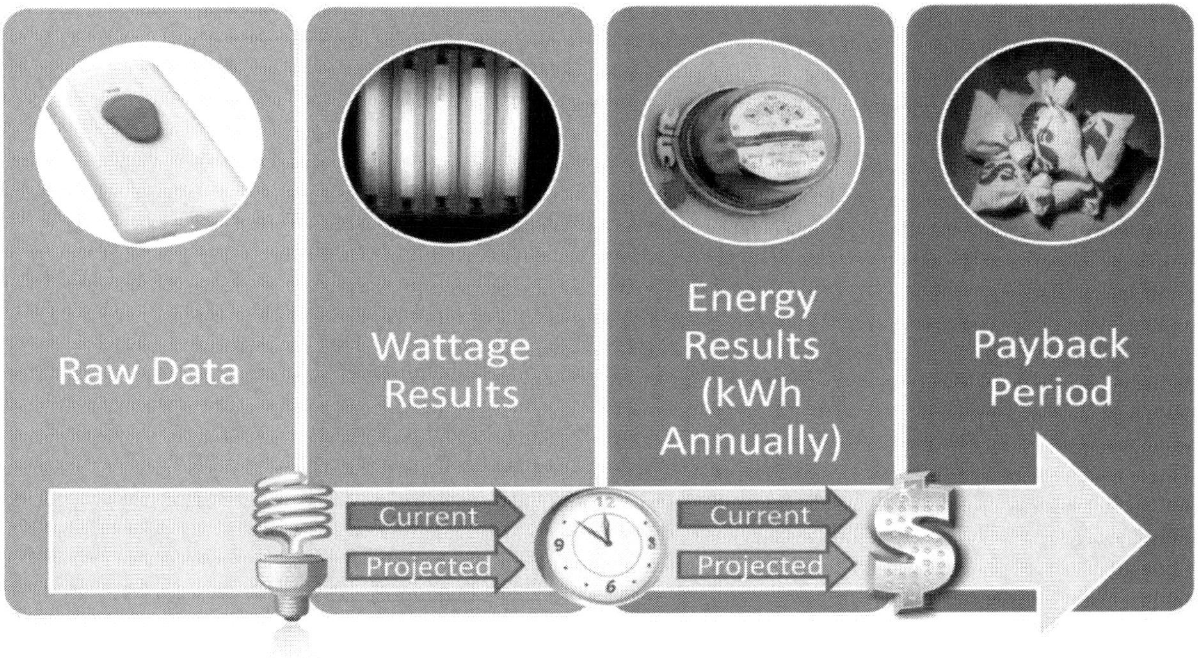

Figure 2. Flowchart of process to determine payback period.

Raw Data Results

Table 1 below shows the four building types' values of results of the raw data from the ballast sensors in each building, listed as a count of fixtures that were detected by the sensors as electronic and magnetic, and the percentage of fixtures that were magnetic.

Table 1. Raw data results of the four building types.

Building type shape-code and name	Electronic	Magnetic	% Magnetic
▲ School of Dentistry	3,765	8,831	70.11%
▲ BEC	123	3,916	96.95%
▲ Volker Hall	3,010	3,389	52.96%
▲ Education Building	395	1,315	76.90%
▲ Humanities Building	522	916	63.70%
■ General Services Bldg.	1,100	2,979	73.03%
■ CHSB-19	313	962	75.45%
■ CHSB-20	433	582	57.34%
■ Community Care Building	340	623	64.67%
● Boshell Building	1,046	4,702	81.80%
● Bevill Building	183	2,457	93.07%
● McCallum Building	161	3,013	94.93%
● Lyons-Harrison Research Building	300	1,269	80.88%
● Tinsley-Harrison Tower	490	1,446	74.69%
● Zeigler Research Building	501	727	59.20%
★ 16th Street Parking Deck	90	509	84.97%
★ 19th Street Parking Deck	61	1,430	95.91%

***** Formatting Issue - Best Available Paper/Graphic *****

Means and standard deviations for each of the four building types are given below in Table 2.

Table 2. Means and standard deviations of percentage magnetic ballasts in each of the four building types.

Building Type	Sample Mean	Standard Deviation
Academic Buildings	72.1%	16.4%
Community Service Buildings	67.6%	8.3%
Medical Research Facilities	80.8%	13.1%
Parking Decks	90.4%	7.7%

The results show that the community service buildings were most up-to-date, while parking decks were the least. One explanation for this is that community service buildings and academic buildings have to be attractive to clients and students, so their building managers are more likely to demand attractive lighting. On the other hand, medical research facilities are rarely visited by non-employees and parking decks merely function as a place for motorists to park their cars, so their building managers are least likely to demand attractive lighting. It should also be noted that the 16th street parking deck did not actually contain any T12's; it was instead fitted largely with metal halide bulbs in magnetic ballasts.

Building Power for Lighting and Determination of Scenarios

Wattage Determination of Accessible Rooms

The observation regarding number and type of bulbs in each fixture contributed greatly to determining the actual wattage of each fixture. In some cases, there was no cover on the fixture and the bulb was bare; then the bulb type could be determined visually. When it was hidden by a diffuser or other cover type, electronic tube-type ballasts, whether straight or U-shaped, were assumed to contain T8 bulbs, while magnetic tube-type ballasts were assumed to contain T12 bulbs. In most cases, the number of bulbs could easily be counted whether or not a cover was used. Prior to the launch of the surveys in June 2009, Energy Management tested the power input of configurations of T12 fluorescent bulbs in magnetic ballasts and T8 bulbs in electronic ballasts of 1-4 bulbs per fixture. The results of their experiment in watts are provided below in Table 3.

Table 3. Wattages of fixtures determined by Facilities Management.

No. Bulbs	Electronic (T8)	Magnetic (T12)
1 Bulb	27	50
2 Bulbs	52	87
3 Bulbs	77	137
4 Bulbs	103	174

In addition to this, compact fluorescent bulbs were assumed to be 15 watts in both magnetic and electronic ballasts, while incandescent bulbs, which function only in magnetic ballasts, were assumed to be an average of 75 watts. If an electronic ballast containing T12 bulbs was encountered, which was relatively rare, it was considered to be the same wattage as one containing T8 bulbs. These numbers were summed up for each fixture in each room, but another assumption had to be made for the total wattage of each building to be determined.

Estimation of Wattage of Similar Rooms and Upscale Factor

Occasionally the research team would come across a room that was inaccessible for whatever reason. The gross square footage (GSF) of each room was often given in floor plans provided by Energy Management. For cases in which a room was inaccessible, if the room's GSF was similar to a nearby room that was accessible, the room was estimated to contain the same lighting configuration as that nearby room. If the room was inaccessible and not similar in GSF to any nearby room, it was determined

*** Formatting Issue - Best Available Paper/Graphic ***

inaccessible. Finally, the total GSF for each floor, which was always given on floor plans, was divided by the sum of the GSF in all rooms that were either accessible or estimated to determine that floor's upscale factor. Multiplying this number by the total wattage of the floor gave the estimated wattage the floor, and then adding up the wattages for each floor of the building gave the wattage for the entire building as it currently stands.

Current and Projected Scenarios

The above steps are used to achieve the "current" scenario, the building's power input as it is today. However, as the aim of the project is to determine how much power can be saved by retrofitting the lighting fixtures, so, a "projected" scenario was also necessary to calculate. The assumptions of this scenario are as follows:

1. All magnetic ballasts would be replaced with electronic ballasts;
2. Fixtures using 1-2 T12 bulbs would be replaced with the same number of T8 bulbs;
3. Fixtures using 3-4 T12 bulbs would be replaced with fixtures using only 2 T8 bulbs;
4. Existing T8 ballasts would be unchanged; and
5. Incandescent bulbs would be converted to compact fluorescent bulbs.

Once these steps were taken, the wattages were recalculated as described in the above two subsections; the results of each scenario and the difference between them are in Table 4.

Table 4. Wattages calculated for each building and the differences between scenarios.

Building type shape-code and name	Current	Projected	Difference
▲ School of Dentistry	542,116	259,427	282,689
▲ BEC	171,015	69,517	101,498
▲ Volker Hall	217,232	139,332	77,900
▲ Education Building	155,813	57,095	98,718
▲ Humanities Building	103,875	69,992	33,883
■ General Services Bldg.	151,912	81,620	70,292
■ CHSB-19	156,007	71,535	84,472
■ CHSB-20	98,063	58,666	39,397
■ Community Care Bldg.	88,608	48,094	40,514
● Boshell Building	256,273	115,662	140,610
● Bevill Building	240,346	106,774	133,572
● McCallum Building	300,346	164,406	135,940
● Lyons-Harrison Building	173,728	83,345	90,383
● Tinsley-Harrison Tower	219,817	118,704	101,113
● Zeigler Research Building	125,420	78,930	46,490
★ 16th Street: MH → HPNa	45,087	17,780	27,307
★ 19th Street: T12 → T8	63,646	38,376	25,270

Means and standard deviations for the reduction in wattage from the current scenario to the projected scenario for each of the four building types are given below in Table 5.

*** Formatting Issue - Best Available Paper/Graphic ***

Table 5. Means and standard deviations of wattage reduction between current and projected scenarios in each of the four building types.

Building Type	Sample Mean	Standard Deviation
Academic Buildings	118,938	95,454
Community Service Buildings	58,669	22,375
Medical Research Facilities	108,018	36,440
Parking Decks	26,288	1,440

Academic buildings appear to have the greatest potential wattage reduction, but also the highest standard deviation, which indicates they may be most unpredictable in determining the potential benefit of a retrofit. The next-highest potential is found in medical research facilities, and the lower standard deviation indicates they are much more reliable in determining the benefit of a retrofit. Also reliable are community service buildings and parking decks, but the low sample mean differences indicate a relatively low benefit of retrofitting them.

Converting Gross Wattage to Annual Energy and Payback Period

Time Assumptions and Annual Energy Reduction (kilowatt-hours, kWh)

Once each scenario's wattage was determined, a number of assumptions could be utilized to determine the annual energy usage for each building. They are as follows:

1. Hours of Operation: 12 hours for each building
   a. Daytime % Lit: 80% - Assumed because roughly one in five rooms goes unused even during hours of full operation [Lackey, 2010][4].
   b. Nighttime % Lit: 30% - This was actually based on the results of an older study [5] by one of the authors [Lackey, 2010][4].
2. Workweek: 5 days of 12-hour operation per week.
   a. 4-day scenario also investigated, but not listed in this report.
3. Number of workweeks per year: 50

In the interest of brevity, these results will not be listed in this conference proceedings paper, as they did little to change the proportionalities between the scenarios, but the means and standard deviations of each building's reduction in annual kilowatt-hours are given below in Table 6.

Table 6. Means and standard deviations of annual kWh reduction between current and projected scenarios in each of the four building types.

Building Type	Sample Mean	Standard Deviation
Academic Buildings	547,522	403,548
Community Service Buildings	268,690	106,855
Medical Research Facilities	514,369	153,967
Parking Decks	210,346	56,161

Cost /Savings Assumptions and Payback Period

Once these values were determined, a few assumptions were used to determine the cost of the retrofit; this is where the data taken during the walkthrough regarding bulb type, including whether the bulb was straight or U-Shaped, comes into play. The assumptions are as follows:

1. Costs of Equipment
   a. \$3 for straight T8;
   b. \$6 for U-shaped T8;
   c. \$25 for electronic ballast;
   d. \$15 per man-hour, at 15 minutes to replace each fixture; and

***** Formatting Issue - Best Available Paper/Graphic *****

e. Based on consensus of Internet searches.

2. Electricity Savings: 10¢ per kWh as determined in the previous subsection.

After the cost for each building in terms of total number of items that needed to be purchased, plus 10%, rounded up to avoid any shortages, was determined, it was then divided by the savings per year (the annual kWh divided by $0.10 per kWh) to determine a payback period. The results of this analysis are given below in Table 7.

Table 7. Payback period for the retrofits for each building in years.

▲ School of Dentistry	0.84
▲ BEC	0.69
▲ Volker Hall	0.84
▲ Education Building	0.78
▲ Humanities Building	4.38
■ General Services Bldg.	1.09
■ CHSB-19	0.94
■ CHSB-20	1.25
■ Community Care Bldg.	1.29
● Boshell Building	0.90
● Bevill Building	1.40
● McCallum Building	2.90
● Lyons-Harrison Building	1.03
● Tinsley-Harrison Tower	2.03
● Zeigler Research Building	2.01
★ 16th Street: MH → HPNa	1.31
★ 19th Street: T12 → T8	1.00

Note that Humanities Building appears to be an outlier among the Academic buildings, as it has a payback period of over four years, while the others are less than a year. One reason for this is that it contains predominantly two-bulb T12 magnetic figures, and the benefits of switching from those to 2-bulb T8 electronic fixtures are not as great as they are for 3-4 bulb T8 electronic fixtures. The means and standard deviations of the payback periods of each building type, including a special entry for academic buildings except for Humanities, are given in Table 8.

Table 8. Means and standard deviations of payback period for each building type.

Building Type	Sample Mean (years)	Standard Deviation (years)
Academic Buildings (AB)	1.51	1.62
AB without Humanities outlier	*0.79*	*0.07*
Community Service Buildings	1.14	0.16
Medical Research Facilities	1.71	0.75
Parking Decks	1.15	0.22

Despite the fact that community service buildings and parking decks have payback periods that are just over a year, the average for Academic Buildings is even lower than that, once the statistical outlier of the Humanities building is removed. In practice this cannot always be done, but there is a good chance that if a building contains large numbers of 2-bulb T12 fixtures, under the rubric explained in this report, the payback period will be relatively high.

*** Formatting Issue - Best Available Paper/Graphic ***

Summary and Conclusions

A summary of the results of the previous sections is provided below in Table 9.

Table 9. Summary of results for each building type, with most desired values bolded.

Building Type	% Magnetic	Wattage Reduction	Payback Period (years)
Academic Buildings	72.1%	**118,938**	1.51 ***(0.79)***
Community Service Buildings	67.6%	58,669	**1.14**
Medical Research Facilities	80.8%	108,018	1.71
Parking Decks	**90.4%**	26,288	1.15

This indicates that a thorough analysis to the payback period should be done. If the result of any preliminary step, either percentage of ballasts which are magnetic, or gross wattage reduction, is taken as the key indicator of which building type would benefit the most from a lighting retrofit, the actual payback period might be seriously underestimated. Only by calculating the payback period does it become apparent that community service buildings will pay for themselves the fastest and most relatively reliably. With academic buildings, one will usually get an even shorter payback period, but one runs the risk of running into that outlier which takes much longer to pay for its retrofits.

Acknowledgement

The authors of this paper would like to acknowledge the Facilities Management Department and their Energy Management Division for funding this project and providing the technical information related to the research.

References Cited

1. http://ateam.lbl.gov/Design-Guide/DGHtm/electronicvs.magneticballasts.htm

2. http://arrowelec.wordpress.com/2010/03/03/effective-july-2010-magnetic-ballasts-will-be-banned-from-production/

3. http://www.mge.com/business/saving/detail/t8.htm

4. Lackey, D., 2010. "Determination of the Effects of the Effects of University Building Usage on Energy Efficiency and Recommendations for Improvement", M.S. Thesis, University of Alabama at Birmingham, Birmingham, AL.

(139a) Engineering Metabolic Modules for Electrofuels Production in Proteobacteria

Curt R. Fischer, JT Sauls, John Sexton, Reshma Shetty and Jason Kelly, Ginkgo Bioworks, Boston, MA

Electrofuels as a concept seeks to produce liquid transportation fuels from carbon dioxide and renewable electricity. Ginkgo Bioworks's approach to electrofuels is to use formic acid salts as a vector of electricial energy for biology-based fuels production. Unlike other vectors such as hydrogen, formate is water-soluble and easy to handle safely, which simplifies process scale up. We have designed and are building a set of metabolic modules aimed at (i) converting the electrons carried by formate into central metabolic intermediates such as NAD(P)H, (ii) converting the carbon carried by formate into central metabolic intermediates such as pyruvate, and (iii) producing liquid, water-immiscible hydrocarbons from pyruvate. Together these modules enable the biosynthesis of liquid fuels from formate salts.

Recently we have also begun to implement these modules in alternate host organisms (alphaproteobacteria instead of *E. coli*). These hosts offer a number of potential advantages over *E. coli* but require the fabrication and characterization of synthetic biology "parts" including promoters, plasmid replication and selection elements, etc. that will also be discussed.

(139b) Metabolic Engineering of Clostridium Ljungdahlii for Production of Fuels and Other Chemicals From Carbon Dioxide Via Microbial Electrosynthesis or with Carbon Monoxide or Hydrogen As the Electron Donor

Toshiyuki Ueki, Ching Leang, Kelly P. Nevin and Derek R. Lovley, Department of Microbiology, University of Massachusetts, Amherst, MA

Clostridium ljungdahlii is an acetogenic microorganism that has the natural ability to use hydrogen as the electron donor to reduce carbon dioxide as its terminal electron acceptor with the production of acetate that is excreted from the cell. Carbon monoxide can also serve as an electron donor for acetate production. Under some conditions ethanol and butanediol are produced in addition to acetate, but usually in lesser quantities. Recent studies in our laboratory have also demonstrated that *C. ljungdahlii* can directly accept electrons from a negatively poised electrode in order to reduce carbon dioxide to acetate. The ability of microorganisms like *C. ljungdahlii* to convert carbon dioxide to organic products with electrons derived from an electrode as the driving force is known as microbial electrosynthesis. Carbon dioxide is reduced to organic products in *C. ljungdahlii* via the Wood-Ljungdahl pathway. Acetyl-CoA is a key intermediate in the Wood-Ljungdahl pathway, which offers the possibility of genetically modifying *C. ljungdahlii* to make a diversity of products more valuable than acetate. Techniques for genetic manipulation of *C. ljungdahlii* were developed which permitted deleting genes from, or inserting genes into, the chromosome. Initial gene manipulation studies focused on the mechanisms for hydrogen uptake and identified a single hydrogenase gene that was required for growth on hydrogen. This hydrogenase-deficient strain continued to reduce carbon dioxide on negatively poised electrodes, demonstrating that hydrogen was not an intermediate in electron transfer from electrodes to *C. ljungdahlii.* Introducing the appropriate genes and/or deleting genes for unwanted pathways made it possible to divert metabolism in *C. ljungdahlii* away from acetate and ethanol production and toward the production of more desirable products. A strain for the production of the transportation fuel butanol was developed, as were strains that can produce other multi-carbon organic commodities from carbon dioxide with an electrode or hydrogen serving as the energy source. Furthermore, these strains could ferment sugars to desired products. These results demonstrate that *C. ljungdahlii* is a genetically tractable chassis organism for the production of fuels and other commodities from a diversity of renewable energy sources.

(139c) Bioelectrosynthetic Reduction of Carbon Dioxide to Biofuel Precursors Using a Mixed Microbial Consortium Isolated From Bovine Rumen

Ryan Hammonds[1], Kristi Whitehead[1], Stephen Creager[2], Jung Oh[3] and J. Michael Henson[1], (1)Biological Sciences, Clemson University, Clemson, SC, (2)Clemson University, Anderson, SC, (3)Chemistry, Clemson University, Clemson, SC

Microbially-catalyzed reduction of carbon dioxide to higher-molecular weight compounds through consumption of electrical current offers prospects for electrochemical production of renewable liquid transportation fuels. Recently, pure cultures of chemoautotrophic microorganisms were reported to produce electrofuel precursors, such as acetate, from carbon dioxide and electrical current. A consortium consisting of several different microbial populations acting as a stable community might offer an energetically favorable approach for production of these electrofuels. In order to evaluate this concept, bovine rumen contents were enriched under strict anaerobic conditions for autotrophic, acetogenic microbes for several generations using hydrogen and carbon dioxide. A stable acetogenic consortium was obtained and transferred to bioelectrosynthesis cells supplied with electric current from a potentiostatic controller. Current consumption and biofuel precursor production were measured. Initial electron recovery as biofuel precursors exceeded 80% efficiency. To further improve rates and efficiency of biofuel precursor production, a combinatorial approach to evaluate growth conditions, active microbial community structure, and reactor geometry is in progress.

(139d) Electroautotrophic Synthesis of Acetate and Methane

Christopher W. Marshall[1], Edward V. LaBelle[1], Erin J. Fichot[2], Daniel E. Ross[2], R. Sean Norman[2] and **Harold D. May**[1], (1)Marine Biomedicine & Environmental Science Center, Medical University of South Carolina, Charleston, SC, (2)Environmental Health Sciences, University of South Carolina, Columbia, SC

Electroautotrophic synthesis of chemicals from CO_2 offers an attractive way of producing fuels that would be sustainable, domestic and environmentally benign, particularly in relation to climate change. Autotrophic rates of acetate and methane production by microbes with hydrogen gas as an electron donor have been demonstrated in the multiple kg m^{-3} h^{-1} range. However, this approach requires a source of hydrogen. Generation of hydrogen by electrolysis, or direct delivery of electrons through an electrode, could serve as the source of hydrogen if sustainable energy, such as solar or wind, is used to drive the production of electrical energy. The overall method becomes even more attractive if the use of rare earth elements is avoided and only microbial catalysts are employed. This approach is especially advantageous if the microbes operate within the electrolysis/electrochemical cell. Using a community of microorganisms selected from a brewery waste we have produced high amounts of methane and acetate from reductive electrical current in electrochemical cells with only microbial catalysts. The data thus far indicate that the generation of both products is dependent upon the biological catalysts associated with the electrodes, and the production rates observed for both products exceed any published rates as of the time of this submission. The electromethane could readily be dropped into existing natural gas infrastructure and used as a transportation fuel. Acetate is a commodity chemical but is also a building block that may be used to produce other fuels. We are attempting to convert that acetate into alcohols (ethanol and butanol) with communities of microbes selected within the electrochemical cells. The rates of production, electrochemical data associated with the catalysis, reactor design, microbial catalyst identification and community characteristics (examined metagenomically) will be discussed.

Integrated Electromicrobial Conversion of CO_2 to Higher Alcohols

Han Li,[1,2] Paul H. Opgenorth,[3] David G. Wernick,[1] Steve Rogers,[4] Tung-Yun Wu,[1] Wendy Higashide,[4] Peter Malati,[5] Yi-Xin Huo,[4] Kwang Myung Cho,[4] James C. Liao[1,2,3,6]*

A man-made photovoltaic device is relatively efficient in converting sunlight to electricity, but the electrical energy generated is difficult to store. The biological photosystems, on the other hand, are limited by the intrinsic design and biomaterials available, for which no near-term improvements are in sight (*1*). One way to circumvent both problems is to link man-made solar cells to biological CO_2 fixation and fuel production. Theoretically, H_2 generated by solar electricity can drive CO_2 fixation in lithoautotrophic microorganisms engineered to synthesize high energy-density liquid fuels. However, the low solubility, low mass transfer rate, and safety issues of H_2 in microbial cultures limit the efficiency and scalability of such processes.

Compared with H_2, formic acid is a favorable energy carrier. Electrochemical production of formic acid from CO_2 and H_2O can achieve relatively high efficiency (*2*, *3*). Formate is highly soluble and is readily converted to CO_2 and NADH in the cells, providing a safe replacement for H_2. However, the high solubility of formate increases the cost of separation. Accumulated formate will decompose at the anode, decreasing the yield of the process (*2*). Therefore, simultaneous electrochemical formate production and biological formate conversion to higher alcohols is desirable (Fig. 1A). Unfortunately, introduction of electricity to microbial cultures may impede cell growth.

As such, an integrated process for reduction of CO_2 to liquid fuel powered by electricity requires (i) metabolic engineering of a lithoautotrophic organism to produce liquid fuels, (ii) electrochemical generation of formate from CO_2 in fermentation medium, and (iii) enabling microbes to withstand electricity. In this work, we chose *Ralstonia eutropha* H16 as the production host and isobutanol and 3-methyl-1-butanol (3MB) as the target fuels, which can be used in the internal combustion engines. We introduced the set of genes previously reported (*4*, *5*) for isobutanol and 3MB production into *R. eutropha* H16 (fig. S1, A to E). We also disrupted polyhydroxybutyrate synthesis (fig. S1, A and B) and used the heterologous isobutanol and 3MB production pathway as the new metabolic sink for carbon and reducing equivalents. In a pH-coupled formic acid feeding fermentor, the engineered strain LH74D produced fuels with the final titer of over 1.4 g/l (~846 mg/l isobutanol and ~570 mg/l 3MB) (fig S2B).

Next, we used an In foil cathode and a Pt anode in the culture medium bubbled with air containing 15% CO_2 to produce formate electrochemically (Fig. 1A). However, growth of *Ralstonia* was inhibited when electric current was introduced. Studies using *Escherichia coli* showed transient inhibition of electric current on cell growth (fig. S3). When the current stopped, cell growth resumed, suggesting that unstable compounds such as reactive oxygen or nitrogen species might be responsible for the growth inhibition.

To test this hypothesis, we constructed three *Ralstonia* strains with different reporter plasmids to detect the presence of reactive oxygen and nitrogen species. The plasmids each contained a *lacZ* gene driven by a promoter of *Ralstonia katG*, *sodC*, or *norA* gene, genes which have been shown to be induced by hydrogen peroxide (H_2O_2), superoxide free radicals (O_2^{-}), and nitric oxide (NO), respectively (*6*–*8*). When the plasmid-bearing strains were exposed to electrolysis, expression of β-galactosidase from *sodC* and *norA* promoters were induced but not from the *katG* promoter (Fig. 1B). These results suggested that O_2^{-} and NO trigger a stress response in *Ralstonia* cells and inhibit growth. To circumvent this toxicity problem, a porous ceramic cup was used to shield the anode (Fig. 1A). This inexpensive shield provides a tortuous diffusion path for chemicals. Therefore, the reactive compounds produced by the anode may be quenched before reaching the cells growing outside the cup. Using this approach, healthy growth of *Ralstonia* strain LH74D and production of over 140 mg/l biofuels were achieved with the electricity and CO_2 as the sole source of energy and carbon, respectively (Fig. 1C).

This integrated process to convert CO_2 to liquid fuels does not depend on biological "light reactions." Electricity generated from photovoltaic cells, wind turbines, or off-peak grid power sources can be used to drive CO_2 fixation and fuel production. Thus, this process provides a way to increase photosynthetic efficiency by coupling man-made photoelectric generation device with biological CO_2 fixation and fuel production capability. The approach demonstrated here can also be applied to produce other compounds, thus opening the possibility of electricity-driven bioconversion of CO_2 to a variety of chemicals. Furthermore, because formate is a by-product of chemical dehydration and conversion of biomass (*9*), transformation of formate into liquid fuel will also play an important role in the biomass refinery process.

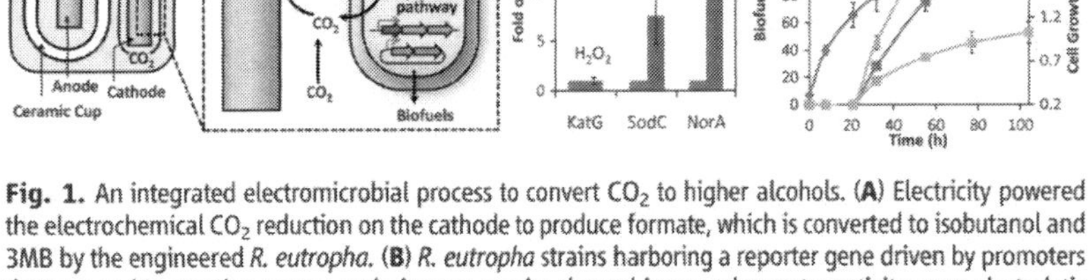

Fig. 1. An integrated electromicrobial process to convert CO_2 to higher alcohols. (**A**) Electricity powered the electrochemical CO_2 reduction on the cathode to produce formate, which is converted to isobutanol and 3MB by the engineered *R. eutropha*. (**B**) *R. eutropha* strains harboring a reporter gene driven by promoters that respond to reactive oxygen and nitrogen species showed increased reporter activity upon electrolytic exposure. (**C**) Engineered strain *R. eutropha* LH74D (supporting online material) showed healthy growth and produced over 140 mg/l biofuels in the integrated electromicrobial reactor with electricity and CO_2 as the sole energy and carbon sources, respectively. Error bars indicate SD (n = 3).

References and Notes

1. R. E. Blankenship *et al.*, *Science* **332**, 805 (2011).
2. K. S. Udupa, G. S. Subramanian, H. V. K. Udupa, *Electrochim. Acta* **16**, 1593 (1971).
3. S. Ikeda, T. Takagi, K. Ito, *Bull. Chem. Soc. Jpn.* **60**, 2517 (1987).
4. S. Atsumi, T. Hanai, J. C. Liao, *Nature* **451**, 86 (2008).
5. S. Atsumi *et al.*, *Appl. Microbiol. Biotechnol.* **85**, 651 (2010).
6. D. Touati, *J. Bacteriol.* **170**, 2511 (1988).
7. C. Michán, M. Manchado, G. Dorado, C. Pueyo, *J. Bacteriol.* **181**, 2759 (1999).
8. A. Pohlmann, R. Cramm, K. Schmelz, B. Friedrich, *Mol. Microbiol.* **38**, 626 (2000).
9. H. Mehdi *et al.*, *Top. Catal.* **48**, 49 (2008).

Acknowledgments: This work is supported by the Electrofuel program of Advanced Research Projects Agency–Energy (ARPA-E).

Supporting Online Material

www.sciencemag.org/cgi/content/full/335/6076/1596/DC1
Materials and Methods
Figs. S1 to S4
Tables S1 and S2
References (*10*–*22*)

8 December 2011; accepted 14 February 2012
10.1126/science.1217643

[1]Department of Chemical and Biomolecular Engineering, University of California, Los Angeles, CA 90095, USA. [2]The Molecular Biology Institute, UCLA, Los Angeles, CA 90095, USA. [3]Department of Chemistry and Biochemistry, UCLA, Los Angeles, CA 90095, USA. [4]Easel Biotechnologies, Los Angeles, CA 90024, USA. [5]Department of Materials Science and Engineering, UCLA, Los Angeles, CA 90095, USA. [6]Institute of Genomics and Proteomics, UCLA, Los Angeles, CA 90095, USA.

*To whom correspondence should be addressed. E-mail: liaoj@seas.ucla.edu

Rhodobacter as a platform for autotrophic fuel production

Nymul Khan[1], Alex Rajangam[1], Eric Nybo[2], Joe Chappell[2], Wayne Curtis[1]

[1] *Department of Chemical Engineering, Pennsylvania State University, University Park, PA*

[2] *Department of Plant and Soil Sciences, University of Kentucky, Lexington, KY*

Overview

Our overall ARPA-E project goal seeks to produce a renewable hydrocarbon feedstock by the genetic engineering of a C_{34} branched hyrdrocarbon of Botrococcus braunii into *Rhodobacter capsulatus*. We have shown the ability of this genetically engineered autotrophic organism to produce C_{30+} isoprenes from H_2, O_2 and CO_2. My specific component of this project has been to examine fundamantal kinetics of autotrophic growth towards the production of bioreactactor systems and multiplexed autotrophic growth capabilities for screening the genetically engineered bacteria.

Introduction

Liquid fuels are an essential part of today's economy. The transportation sector is the largest consumer of liquid fossil fuels (13.6 million bbl/day in 2011, www.eia.gov/forecasts/aeo/MT_liquidfuels.cfm) and the second largest source of CO_2 emission (1.9 billion metric tons in 2008, http://www.eia.gov/oiaf/1605/ggrpt/carbon.html) after power plants. Coal and natural gas account for 67% of US electricity generation and together generate the major portion of the CO_2 (about 2.3 billion tons in 2008, http://www.eia.gov/oiaf/1605/ggrpt/carbon.html). So, besides being an unsustainable source of energy, consumption of fossil fuels has high environmental impact due to the generation of CO_2. Furthermore, it is estimated that the demand for liquid fuels for transportation sector is likely to increase during the next 20-30 years (Figure 1).

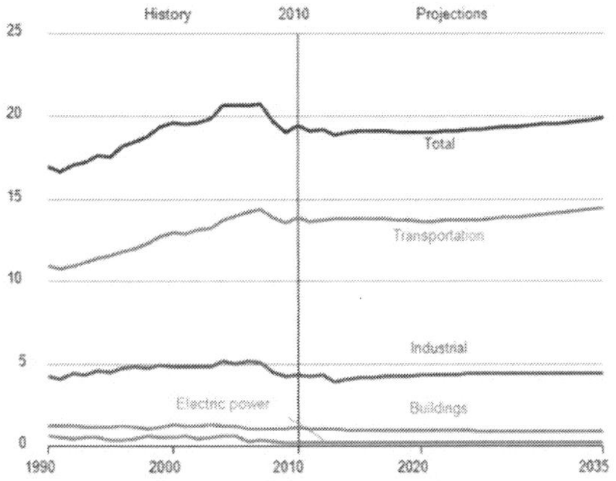

Figure 1: Consumption of petroleum and other liquids by sector, 1990-2005 (million bbl/day). http://www.eia.gov/forecasts/aeo/MT_liquidfuels.cfm#uses

In comparing liquid fuel products to other energy storage forms, it is apparent that a liquid hydrocarbon fuel is a much more efficient energy carrier by weight and volume than thermal, mechanical, battery and hydrogen energy storage. It is probably not possible to replace liquid fuels for many current end uses. Therefore, there is an urgent need to develop a renewable source of liquid fuels. Our proposed process is to take the distributed energy of solar radiation and store it in a liquid fuel molecule that can be easily used in the existing infrastructure of refining, storage and distribution. Figure 2 shows a schematic of the process. The electricity generated by photovoltaics will be used to split water to produce H_2 and O_2. These gases plus the CO_2 from power plants or other point sources will be used as the major substrates to grow an autotrophic bacterium (*Rhodobacter capsulatus*), genetically engineered to produce a liquid fuel, botryococcene.

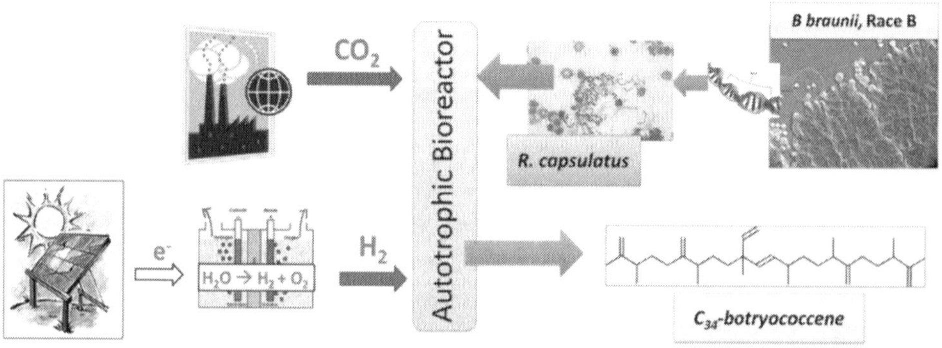

Figure 2: Schematic of the proposed biofuel production process in the context of supporting technologies of photovoltaics (P.V.) and carbon capture.

Botryococcenes are a group of C_{34}-branched hydrocarbons naturally produced by the micro-algae *Botryococcus braunii* Race B. These oxygen-free molecules are extremely energy dense (43.8 KJ/g as compared to ethanol at 29.7 KJ/g). As a hydrophobic long chain hydrocarbon, it easily separates as a distinct layer from water and has characteristics very similar to crude oil (Hillen, Pollard, Wake, & White, 1982). Hillen and co-workers also showed that botryococcene can be hydrocracked with a commercial catalyst to produce 67% gasoline, 15% aviation, 15% diesel and 3% residual oil. The octane rating of the gasoline fraction was much higher than natural crude and very low nitrogen and sulfur levels to affect the catalyst operation. As such, it can be a drop-in feedstock to the current petroleum industry. In fact, *B. braunii* oils appear to have contributed as much as 1.4% to the total hydrocarbons in oil shales (McKirdy & Cox, 1986).

Non-photosynthetic growth of organisms on CO_2 has been observed for a long time (Kelly, 1992) and has been proposed as a platform for the production of single cell protein, including an interface to H_2 production by electrolysis (Aragno & Schlegel, 1992). *Rhodobacter capsulatus* is a purple non-sulfur bacterium which has an extremely versatile metabolic capacity for growth under variety of growth conditions including aerobic heterotrophic, anaerobic photoheterotrophic and dark autotrophic using H_2 as the electron donor, O_2 as the electron source and CO_2 as the carbon source (Madigan & Gest, 1979). It has been chosen as the organism for this work for both its diverse metabolic capacity and our previous experience of working with this organism for membrane protein expression.

An autotrophic process is quite unique in its requirement for multiple gas phase substrates. While traditional aerobic fermentation processes are already limited by gas-liquid mass transfer, it is not difficult to imagine that that will be the limiting factor in this process as well. To make a rough comparison, while it takes ~0.7 L O_2 to grow 1 g of *E. coli*, it would take ~12.5 L of gas mix (in stoichiometric composition) to grow 1 g of *R. capsulatus* (based on yield of *R. capsulatus* on the different gases, Madigan & Gest 1979). This poses some unique restrictions on the bioreactor and genetic engineering of the organism. For example, from a preliminary calculation using the yield of the bacteria on the different gas species, the average cost of electricity and gasoline it was immediately clear that the growth-associated fuel production is not a feasible scenario for this process. Figure 3 shows a comparison of just the electricity cost per gallon of fuel corresponding to the fraction of fuel accumulated per unit biomass. Even at a very high fraction of fuel accumulation the cost is exceedingly high (and this does not include the operating or other downstream processing costs). So, the

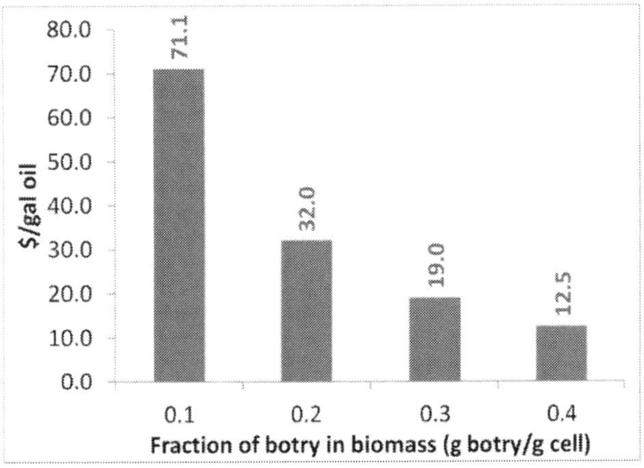

Figure 3: Cost of electricity per gallon of fuel at different fuel to biomass ratio.

fuel production must be growth-dissociated. The parameter that determines the quantity of gas that would be consumed under very low or no-growth condition is the **maintenance coefficient** (and the true yield) of the bacteria. While the maintenance coefficients of bacteria growing heterotrophically are quite abundant in the literature, those for autotrophic bacteria (like *R capsulatus*) are not very common. In this paper, we present our recent work related to the measurement of this parameter for *R capsulatus*.

True yield and maintenance coefficient

Maintenance coefficient of an organism is the amount of energy it needs to stay alive and maintain itself without cell division. On the other hand, true yield of an organism is the amount of substrates it would consume for growth only and not for maintenance. Both of these parameters are very important not only for accurately assessing the baseline economic feasibility of the proposed process but also indicate energetic efficiency of the organism. The yield and maintenance coefficient of *R capsulatus* on H_2, the most expensive substrate, is not available in the literature. In fact, values of these parameters are generally very scarce for any organism under autotrophic conditions.

In principle, maintenance coefficient and true yield can be calculated if the yield on the substrate is known as a function of the growth rate. (Pirt, 1965) derived the following equation as the inter-relationship among specific growth rate (μ) and yield (Y_{XS}) and true yield (Y_{XS}^T) and maintenance coefficient (m_S).

$$\frac{1}{Y_{XS}} = m_S \frac{1}{\mu} + \frac{1}{Y_{XS}^T} \quad (1)$$

Y_{XS} and μ are both experimentally measurable parameters. Therefore plotting $\frac{1}{Y_{XS}}$ against $\frac{1}{\mu}$ a straight line can be obtained whose slope and intercept would be equal to the m_S and $\frac{1}{Y_{XS}^T}$ respectively. This has been traditionally used in the determination of true yield and maintenance coefficient. However, in a batch system the growth rate varies continuously, therefore an accurate measure of instantaneous growth rate and yield are not possible. At the same time, the culture goes through physiological changes as it progresses through the different growth stages and therefore the yields obtained this way may not refer to the same condition. So, true yield and maintenance are measured in continuous systems, in which the growth rate of an organism can be fixed at a particular value by fixing the dilution rate (inverse of residence time in a reactor).

Cell balance on a continuous reactor:

$$V\frac{dX}{dt} = FX_0 - FX \quad (2)$$

where V = volume of reator, $\frac{dX}{dt}$ = growth rate, F = flow rate, X_0 and X = inlet and outlet cell concentration respectively. With 0 inlet cell concentration at steady state,

$$\frac{1}{X}\frac{dX}{dt} = \frac{F}{V} \quad (3)$$

From the definitions of specific growth rate, $\mu = \frac{1}{X}\frac{dX}{dt}$ and dilution rate, $D = \frac{F}{V}$, we have $\mu = D$ in a steady state continuous flow reactor.

Multiple steady states are established by varying the dilution rate and yield values are measured at these steady states. The corresponding yield and dilution rates are plotted on a $1/Y$ vs. $1/D$ graph to get the true yield and maintenance coefficient. However, this method is very time consuming and produces a limited number of data points. These parameters can alternatively be measured in deceleration-stat or acceleration-stat (D-stat or A-stat) (Barbosa et al., 2005) rather than a true chemostat. Instead of going from one dilution rate to the next in a step and waiting for a steady state to be established, the dilution rate is varied smoothly from one to the next. If the rate is slow enough for the cells to adapt physiologically, the culture goes through a succession of pseudo steady-states and the yield and dilution rate in this period is used to calculate the maintenance and true yield. The D-stat method is followed in our determination of the yield and maintenance coefficient of *R capsulatus*.

Materials and methods

The experimental setup and schematic of the autotrophic bioreactor is shown below (Fig. 4). H_2, O_2 and CO_2 gases are mixed via three mass flow controllers (Brooks Instruments, CA) and delivered to the reactor after going through a humidifying column. The outlet gas passes through a drying column, then through a mass flow controller (Omega Engineering) that measures the flow rate and a GC that measures the composition. The composition data is collected once every hour. For safety concerns about using explosive gas mix (H_2+O_2), the reactor used in this work is a 250 ml media bottle. The liquid flow rate into and out of the reactor is controlled via a highly precise peristaltic pump (Watson-Marlow), whose rotation can be controlled to within 0.01 rpm by a 0-5 V analog input. This is very critical in maintaining an accurate and consistent flow rate and smoothly varying it for D-stat. The actual flow rate of the inlet media is calculated by periodically recording the weight of the media bottle which sits on a balance and fitting appropriate regression lines through them dynamics. The culture is recirculated via a second peristaltic pump for online optical density measurement. Samples are taken periodically from the outlet for offline OD and dry weight measurements. The media used for growth is based on a previous work in the literature (Madigan & Gest, 1979) where the autotrophic growth of *R capsulatus* was first demonstrated. A pH probe continuously measures the pH of the culture. As discussed before, the culture

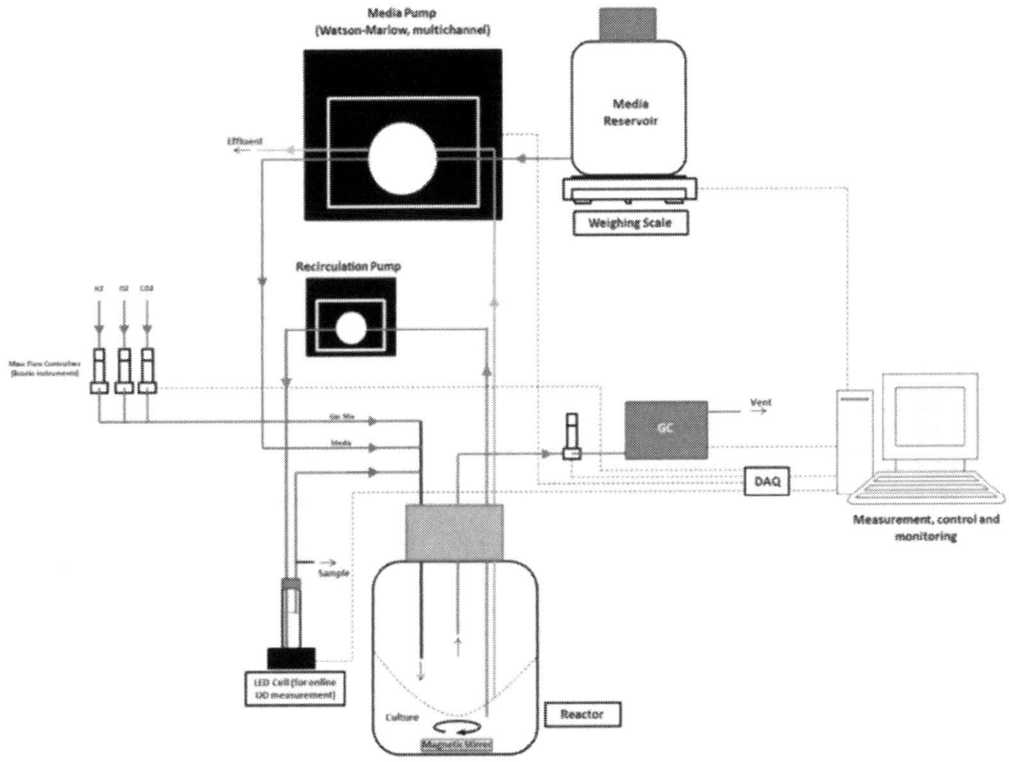

Figure 4: Schematic of the experimental setup.

is allowed to come to a steady state at a particular dilution rate and then the dilution rate is decreased slowly. The deceleration rate is based on previous literature work on *R capsulatus* (Hoekema, Douma, Janssen, Tramper, & Wijffels, 2006) and experience with running the experiment.

Another critical aspect of the experiment is the measurement of gas flow rate and composition, which is much more difficult to do accurately than liquid phase because of strong temperature and pressure dependence. Mass flow meters, although capable of measuring flow rate of a single gas very accurately, are not very accurate for a mixture of gases and at low flow rates. A soap-bubble meter was therefore placed at the end of the line to calibrate the outlet flow based on manual measurements. This was checked against the flow rate based on composition correction of the MFM and they compared very well. Furthermore, a state-of-art GC was used to analyze these gases continuously in a single pass for composition measurement.

Results and discussion

The experimental data is shown from Fig. 5 where the dilution rate was decreased from about 0.08/hr to about 0.06/hr at a rate of $0.094/day^2$. This deceleration rate was based conservatively on a previous work of (Hoekema et al., 2006) where they measured the yield and maintenance of *R capsulatus* in the photoheterotrophic growth mode, producing H_2. Optical density plot (Fig 5A) shows that the expected pattern of variation is followed as the dilution rate is varied. Fig 5C shows the precision of media flow variation. The final calculated results are plotted in Fig 5D. The true yield and maintenance coefficient values are shown below (Table 1).

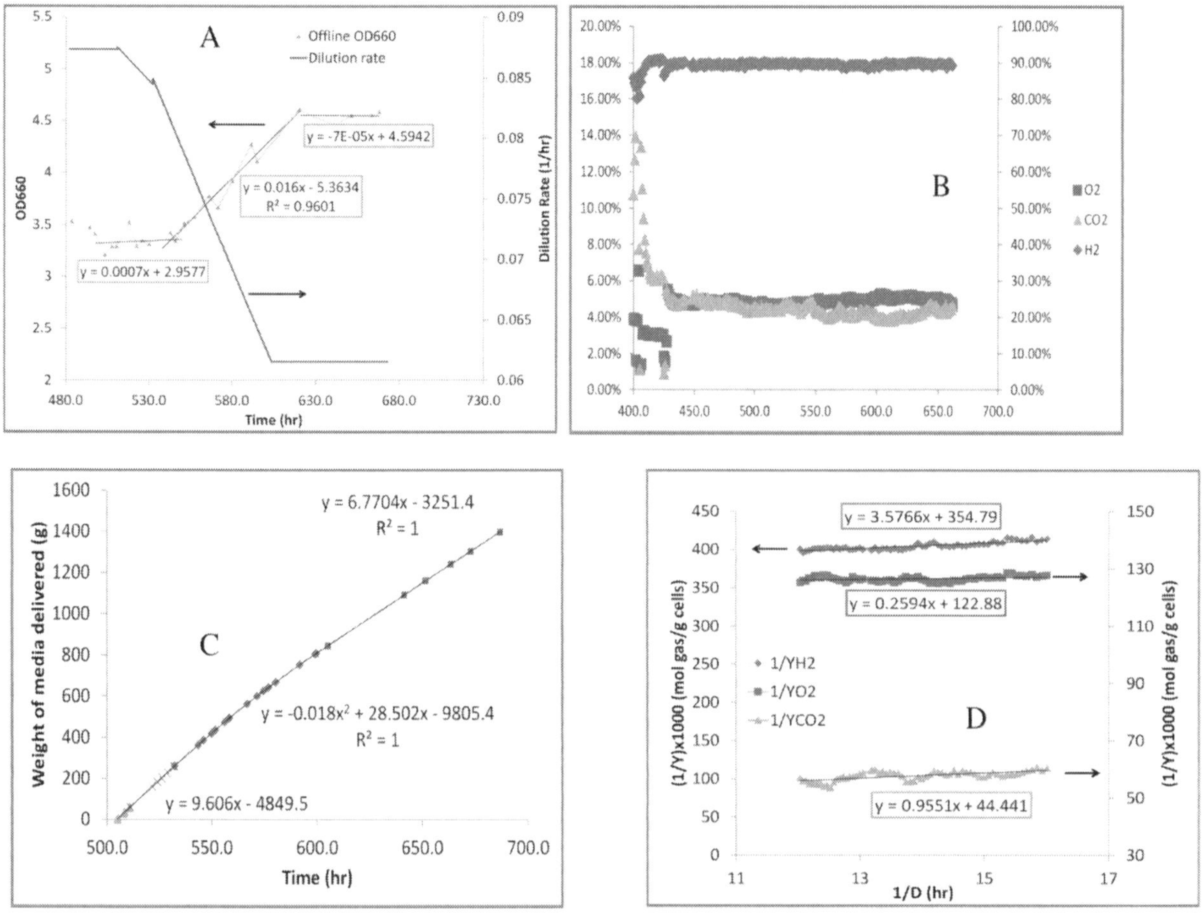

Figure 5: Experimental data for *R. capsulatus* true yield and maintenance coefficient determination. (A) Ramp down of dilution rate (red) and the corresponding OD increase (green), (B) gas phase composition, (C) media bottle weight to measure flow rate, (D) 1/Y vs. 1/D.

These values compare quite well with the values for *R eutropha* reported in the literature. Table 2 lists the true yields and maintenance coefficients of *R eutropha* and a couple of methanogens. Siegel & Ollis (1984) noted that the maintenance coefficient of *R eutropha* is a strong function of dissolved O_2 concentration. As observed by other researchers (Howells, 1982), O_2 can be both inhibitory and limiting for the growth of a bacteria at different concentrations and is strongly strain dependent. Under continuous culture conditions, the steady state in between inhibitory and limiting concentrations is unstable and tends to move either toward washout at higher concentration or towards O_2 limitation (Siegel & Ollis, 1984). We can see from the values that the maintenance coefficient increases significantly under H_2-limited conditions. The explanation offered by the authors is that the NAD-reducing hydrogenase in *R eutropha* is deactivated by O_2 and becomes an additional metabolic burden to replace the deactivated hydrogenases at high O_2 concentration, hence the increase in maintenance cost. High O_2 concentration is also found to be inhibitory to *R capsulatus* (Madigan & Gest, 1979; Siefert & Pfennig, 1979). These facts lead us to believe that this is a general phenomenon in lithotrophic CO_2 fixation and give us an idea of the possible operating regime for such a process. Future studies will examine the effect of these parameters for *R eutrophoa* as well.

Table 1: Experimental values of true yield and maintenance coefficient.

	True yield, Y_{XS} (g cells/mol gas)	Maintenance coefficient, m_S (mol gas/g cells.hr)
H_2	2.82	3.58×10^{-3}
O_2	8.13	2.60×10^{-4}
CO_2	22.52	9.55×10^{-4}

The yield and maintenance coefficient of *M. bryantii* were determined in batch growth experiments. The yield was calculated from q_{max} (maximum specific substrate utilization rate) and μ_{max} (maximum specific growth rate), which were determined under non-limiting H_2 concentration. Also, they treated the maintenance coefficient as "endogenous metabolism" i.e. oxidation of biomass to support cell maintenance, and determined the maintenance rate, b, from a biomass decay experiment where there was no H_2 present. This is clearly different from a true maintenance coefficient where the biomass is not really oxidized to provide for maintenance but the actual substrate is consumed by active cell machinery to maintain itself. Since they have not exactly provided the maintenance coefficient, we have calculated the value from yield (Y) and maintenance rate (b). Again, it is not really compatible since they were calculated under two different physiological conditions. However, it gives us something for comparison. The values for *M arboriphilus* were obtained under H_2-limited continuous culture conditions (Morii, Koga, & Nagai, 1987). The true yield and maintenance values were not given directly in the paper but the dilution rates and the substrate utilization rates were used to calculate the values using method outlined above.

Table 2: True yield and maintenance coefficient of some organisms on H_2.

Organism	True yield (g cells/mol H_2)	Maintenance coefficient (mol H_2/g cells.hr)	Experimental conditions	Reference
Ralstonia eutropha	4.1	0.001	O_2 limited continuous culture	(Siegel & Ollis, 1984)
	4.5	0.025	H_2 limited continuous culture	
Methanobacterium bryantii M.o.H	0.325	0.0115	Non-limited batch culture	(Karadagli & Rittmann, 2005)
Methanobrevibacter arboriphilus A2	0.40	0.0106	H_2 limited continuous culture	(Morii et al., 1987)

The values for these methanogens are included for comparison. It is readily apparent that the yield and maintenance are both much worse for these organisms. It is quite likely because the methanogenic process is anaerobic and the yields are expected to be lower than an aerobic process. It needs to be borne in mind that these organisms are actually producing a product that is much more reduced than biomass and that is also a reason for low yield and high maintenance. The energetic analysis of a fuel producing process therefore must include the metabolic effect of producing the fuel molecule.

Economic analysis

The preliminary economic assessment of the electricity cost per gallon of fuel has been made for the non-growth scenario for a comparison against the growth based scenario. For calculations, literature values for *R eutropha* and the experimental values shown in the previous section for *R capsulatus* were used. Cost of H_2 and CO_2 are taken to be \$3.82/kg and \$87/ton respectively. The H_2 cost is based on the current solar electricity prices but as solar energy becomes more widespread this is expected go down and affect the economics correspondingly.

Baseline scenario. Growth based. Shown in Figure 3.

Proposed scenario. At very low growth rates maintenance energy would be the main form of energy consumption. At this point, we lack the important parameter of fuel production rate by the bacteria to make an accurate estimate. A work-around to provide a quantitative perspective is to use a similar value for another organism from literature. *Chlorella vulgaris* which produces lipid under severely growth-limited conditions was chosen as a reasonable basis. From (Tuerk, 2011) this value is estimated to be 0.001256 g lipid/g cells.hr. This value was used as a baseline fuel production value and varied until 0.1 g borty/g cells.hr to observe the effect on cost of fuel (Fig 6). (Residence time in the reactor was a parameter in this calculation which was set to 10 hr. The cell retention was set to 0.9). The high fuel cost calculated for low production rates demonstrates that even in a non-growth associated production scenario maintenance energy requirements play a significant role in determining the cost of fuel production.

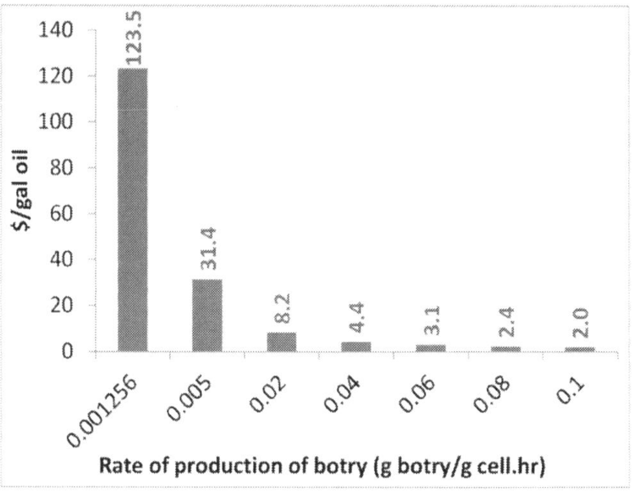

Figure 6: Cost of electricity per gallon of fuel under non-growth associated fuel production.

Future work on this will include studying the effect of various parameters (like O2 tension, nitrogen source etc.) on the maintenance coefficient of *R capsulatus*. Similar studies will be done on *R eutropha* for an unbiased comparison of the two organisms.

References

Aragno, M., & Schlegel, H. G. (1992). The chemolithotrophic prokaryotes. In A. Balows, H. G. Truper, M. Dworkin, W. Harder, & K. Schleifer (Eds.), *The Prokaryotes, 2nd Edition. A Handbook on the Biology of Bacteria: Ecophysiology, Isolation, Identification, Applications.* (pp. 332–343). New York: Springer-Verlag.

Barbosa, M. J., Zijffers, J. W., Nisworo, A., Vaes, W., van Schoonhoven, J., & Wijffels, R. H. (2005). Optimization of biomass, vitamins, and carotenoid yield on light energy in a flat-panel reactor using the A-stat technique. *Biotechnology and bioengineering*, *89*(2), 233–42. doi:10.1002/bit.20346

Hillen, L. W., Pollard, G., Wake, L. V., & White, N. (1982). Hydrocracking of the oils of Botryococcus braunii to transport fuels. *Biotechnology and bioengineering*, *24*(1), 193–205. doi:10.1002/bit.260240116

Hoekema, S., Douma, R. D., Janssen, M., Tramper, J., & Wijffels, R. H. (2006). Controlling light-use by Rhodobacter capsulatus continuous cultures in a flat-panel photobioreactor. *Biotechnology and bioengineering*, *95*(4), 613–626. doi:10.1002/bit

Howells, E. R. (1982). Single-cell protein and related technology. *Chemistry Industry*, (15), 508–511.

Karadagli, F., & Rittmann, B. E. (2005). Kinetic characterization of Methanobacterium bryantii MoH. *Environmental science & technology*, *39*(13), 4900–4905.

Kelly, D. P. (1992). The chemolithotrophic prokaryotes. In A. Balows, H. G. Truper, M. Dworkin, W. Harder, & K. Schleifer (Eds.), *The Prokaryotes, 2nd Edition. A Handbook on the Biology of Bacteria: Ecophysiology, Isolation, Identification, Applications.* (pp. 332–343). New York: Springer-Verlag.

Madigan, M. T., & Gest, H. (1979). Growth of the photosynthetic bacterium Rhodopseudomonas capsulata chemoautotrophically in darkness with H2 as the energy source. *Journal of bacteriology*, *137*(1), 524–30. Retrieved from http://www.pubmedcentral.nih.gov/articlerender.fcgi?artid=218479&tool=pmcentrez&rendertype=abstract

McKirdy, D., & Cox, R. (1986). Botryococcane in a new class of Australian non-marine crude oils. *Nature*, *320*. Retrieved from http://www.nature.com/nature/journal/v320/n6057/abs/320057a0.html

Morii, H., Koga, Y., & Nagai, S. (1987). Energetic analysis of the growth of Methanobrevibacter arboriphilus A2 in hydrogen-limited continuous cultures. *Biotechnology and bioengineering*, *29*(3), 310–5. doi:10.1002/bit.260290304

Pirt, S. J. (1965). The maintenance energy of bacteria in growing cultures, *163*(991), 224–231.

Siefert, E., & Pfennig, N. (1979). Chemoautotrophic growth of Rhodopseudomonas species with hydrogen and chemotrophic utilization of methanol and formate. *Archives of Microbiology*, *122*(2), 177–182. Retrieved from http://www.springerlink.com/index/V323851512727030.pdf

Siegel, R. S., & Ollis, D. F. (1984). Kinetics of growth of the hydrogen-oxidizing bacterium Alcaligenes eutrophus (ATCC 17707) in chemostat culture. *Biotechnology and bioengineering*, *26*(7), 764–70. doi:10.1002/bit.260260721

Tuerk, A. L. (2011). *An assessment of photosynthetic biofuels and electrofuels technologies under rate-limited conditions.* Pennsylvania State University.

(140b) "Green" Solvents: Multi-Criteria Knowledge-Based and Statistical Approaches for Their Evaluation

Stavros Papadokonstantakis, Tianhe Liu, Alireza Banimostafa and Konrad Hungerbühler, Institute for Chemical and Bioengineering, Swiss Federal Institute of Technology, Zurich (ETHZ), Zurich, Switzerland

"Green" solvents: Multi-criteria knowledge-based and statistical approaches for their evaluation

Stavros Papadokonstantakis, Tianhe Liu, Alireza Banimostafa, Konrad Hungerbühler

Institute of Chemical- and Bioengineering, Swiss Federal Institute of Technology (ETH) Zurich, Switzerland

Sustainable process design has been recognized as one of the key research challenges for process systems engineering. In fact, the experience gained so far has pointed out that the applicability of sustainability principles can be more advantageous in earlier phases of process design characterized by more degrees of freedom for decision making. Typical decisions made in these early process design phases involve the selection of chemical synthesis path, chemical auxiliaries, unit operation conditions up to a basic flowsheet design.

Solvents belong to chemical auxiliaries of particular interest for the chemical industry due to their multi-functional use (e.g., to achieve desired reaction rates while protecting from the reaction exothermicity, to ensure solid product quality requirements through effective crystallization, to perform liquid-liquid separations or azeotropic distillations etc.) and their large amounts (e.g., in fine-chemical and pharmaceutical production large amounts are used per mass of final products). Historically, solvents have been designed to maximize technical performance and minimize cost. But nowadays, an effort to find 'green' solvents gains interest, for replacing conventional solvents with environmentally benign substitutes while, of course, maintaining their effectiveness for the intended use. In this context, the concept of ''green" solvents expresses the goal to minimize the environmental impact resulting from their use in chemical production, both from a life cycle impact assessment (LCIA) and a safety, health and environmental (SHE) hazard assessment point of view.

Several shortcut, index-based methodologies have been proposed for evaluating the relevant LCIA and SHE effects, both for independent substances and for chemical production processes. These index-based frameworks encapsulate prior knowledge regarding main SHE hazards, cradle-to-gate resources consumption, and emission related effects. The indices are typically defined on the basis of substance properties, derived from available databases or estimated using available techniques (thermodynamic property estimation methods, QSAR models, etc.), and process features derived either from a preliminary process flowsheet or even on the basis of the chemical synthesis path. On the one hand, the simplicity of the index-based approaches makes them attractive, especially for early stages of process design lacking detailed process information, but on the other hand they face criticism regarding limited coverage of considered effects, subjectivities for the calculation of aggregated end-point indices, and unknown resolution of the scaled final scores according to which the ranking of process alternatives is inferred.

Especially for the problem of subjective aggregation to end-point scores principal component analysis (PCA) has been recently proposed as a potential remedy. In this approach the rows of the PCA input matrix refer to the evaluation objects (e.g., solvents, chemical process steps, chemical process sections or overall chemical processes) and the columns refer to metrics for their assessment (e.g., LCIA and/or SHE hazard assessment indicators). The evaluation objects are ranked according to the obtained principal component (PC) scores, which in the simplest PCA version can be viewed as linear combinations of the assessment metrics. The weights of these linear combinations express the importance of the original assessment metric for a specific PC and are derived in such a way that the first PC covers most of the variability of the PCA input matrix, the second PC most of the remaining variability under the condition that it is orthogonal to the first one, and so on. The concept behind this PCA-based method, e.g., when it is used for the assessment of substances, is that dependencies must exist between the assessment metrics, since the molecular structure of a substance should, in theory, provide (although most of the times via unknown functional relationships) all necessary information for the assessment metrics. The PCA-based method will, therefore, detect these dependencies in a theoretically sound way and this information will be used instead of *a priori* weighting schemes for the diverse metrics. In this way, two objectives are achieved simultaneously, i.e., the inter-relationships between the diverse metrics reveal the "true" dimensionality of the multi-criteria assessment and the level of multi-criteria similarity among individual solvents can be derived. Moreover, the same PCA-based concept can be used in an expectation-maximization iterative procedure for filling data gaps, typically occurring in the assessment metrics.

The proposed framework is demonstrated (see Figure 1) on the basis of 138 organic solvents used in industrial applications. The performance of the PCA-based method for filling data gaps is compared with other approaches based on priority settings between substance properties and reveals in which extent this should be preferred (i.e., relative amount and patterns of data gaps). Moreover, two case studies are presented in which the results of the PCA-based framework are compared to the results of other studies (from the Safety and Environmental Technology group in ETHZ and GlaxoSmithKline, respectively) using different index-based methods with *a priori* postulated weighting schemes. The results indicate that the presented framework is useful for covering most of the information contained in the other index-based methods for the evaluation of "green" solvents, while being inherently flexible with respect to the number and type of assessment metrics, as well as to the existence of data gaps. Finally, the present study discusses the advantages and limitations of the PCA-based approach in terms of simplicity, interpretability and statistical robustness.

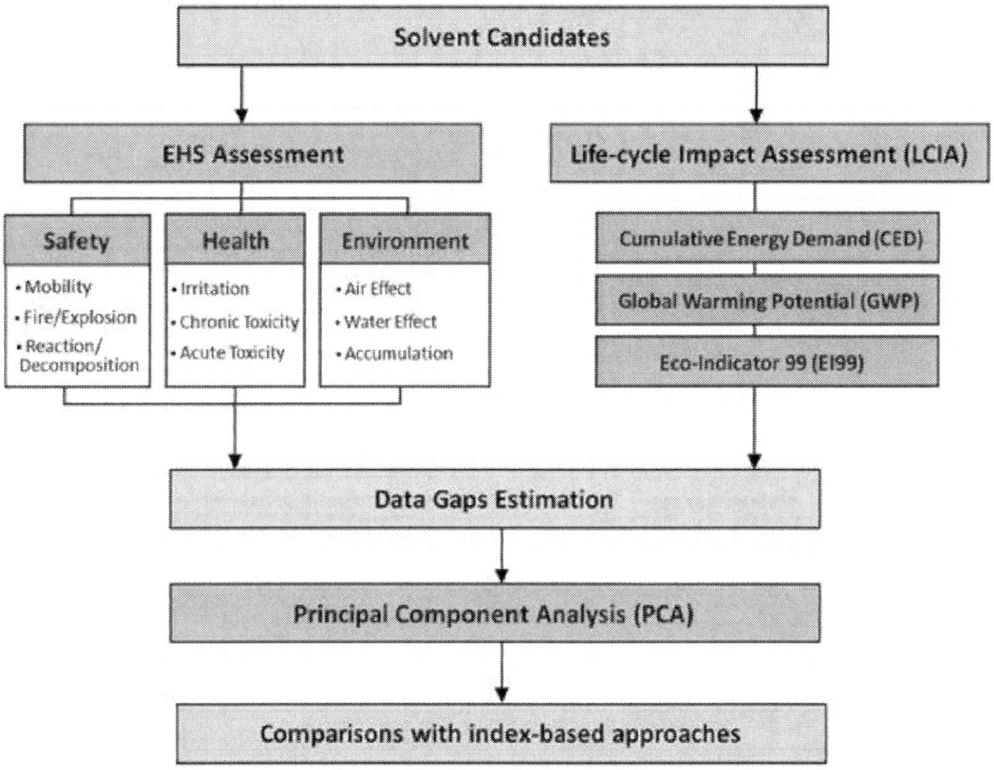

Figure 1: Integrating a PCA-based approach for the evaluation of "green solvents" and comparison with existing index-based approaches.

(140c) Resilience in Complex Economic Networks: Insights From Graph Theory and Input-Output Models

Shauhrat Chopra, Civil and Environmental Engineering, University of Pittsburgh, Pittsburgh, PA and Vikas Khanna, Department of Civil and Environmental Engineering, University of Pittsburgh, Pittsburgh, PA

Comprehensive analysis of complex systems has exposed the demand for bolstering resilience in such systems with respect to sustainable development. Economies are self-organizing, adaptive, complex systems where the constituent components or sectors are connected with each other in myriad ways. Such networks can be modeled as a directed-weighted graph where the nodes represent diverse sectors and the edges represent the monetary transactions between sectors. For example, Input-Output tables that describe the monetary exchanges between different pairs of industries in an economy are available for different countries. Determining the resilience of complex networks, like the economy, enables us to assess the capacity of the network to retain its functionality and structure while adapting to stress, a property vital for a sustainable network.

Employing a systems approach from a resilience perspective is useful for understanding the trade-offs between efficiency and the ability to cope with variations in the environment because of reorganization. Solely, observing simple cause-effect type of computation among the actors may neglect information regarding indirect effects and system-wide properties arising due to interaction. These interactions and their cascading effects can be better assessed through a systems approach. In addition, research on network robustness has elucidated that complex networks could be resilient to random perturbations but vulnerable to explicit attacks on critical nodes. Identifying vulnerable sectors of the economy, and subsequent, reinforcement of the network topology through policy measures can increase resilience of complex systems like the economy.

The goal of this study is to compare the network structure of the United States (U.S.) and the Chinese economies to identify vulnerabilities in the system and their response to disruptions. We also discern the evolution of resilience from the period of late 1980's to the mid 2000's for both economies by utilizing network theory. The Chinese and the U.S. economies represent distinctive categories, with China emerging as one of the biggest manufacturing economies, and the U.S. economy evolving into a service economy. The proposed research integrates concepts of graph theory and Economic Input-Output (EIO) model to elucidate network properties concerning resilience of economic systems. Structural and functional analysis based on graph theory sheds light on vulnerabilities in the network using various centrality and vitality measures. Moreover, using IO analysis and network efficiency performance, we evaluate the direct and indirect impact of disruptions on the economic system. We apply network analysis tools on IO tables for the Chinese and the US economy from 1980's to 2000's with an aspiration to observe the evolution of the basic backbone of these economies. It is noted that on introduction of various stresses, resilient systems exhibit low modularity and percolation. Moreover, on increasing the flexibility in the system, both service and manufacturing economies exhibit higher resilience to disruptions. The implications of the results for developing resilient and sustainable economic systems will be presented and discussed.

(140d) Role of Controllability for LONG TERM Sustainability

Pahola T. Benavides, VRI- Custom, Chicago, IL, Urmila Diwekar, Vishwamitra Research Institute, Center for Uncertain Systems: Tools for Optimization and Management, Clarendon Hills, IL; Center for Uncertain Systems: Tools for Optimization and Management (CUSTOM), Vishwamitra Research Institute, Clarendon Hills, IL; Bioengineering, University of Illinois at Chicago, Chicago, IL and Heriberto Cabezas, Office of Research and Development, U.S. Environmental Protection Agency, Cincinnati, OH

Successful implementation of sustainability ideas in ecosystem management requires a basic understanding of the often nonlinear and nonintuitive relationships among different dimensions of sustainability, particularly the system-wide implications of human actions. This basic understanding further includes a sense of the time scale of possible future events and the limits of what is and is not likely to be possible. With this understanding, systematic approaches based on control theory can then be used to develop policy guidelines for the system. Therefore, controllability of the system is very important to determining long term sustainability of system. In a recent article in nature by Liu et al. [1] presents a new analytical tool to study controllability of complex system. We apply these tools to a complex system consisting of an integrated ecological-economic social model, which comprises various ecological (natural) and domesticated compartments representing species along with a macroeconomic price setting model [2,3]. This model has been used to study sustainability of the planet under different scenarios like population explosion and per capita consumption. We argue that controllability of this system is linked with the long term sustainability and present our arguments in the light of various scenarios studied earlier.

Bibliography

1. Liu Y., J. Slotine and B. Albert-La´szlo, "Controllability of complex networks", 12 may 2011, nature, 473, 167.
2. Shastri Y. , U. Diwekar, H. Cabezas, and J. Williamson " Is Sustainability Achievable? Exploring the Limits of Sustainability with Model Systems", (2008), Environmental Science and Technology, 42, 6710.
3. Shastri Y. ,U. Diwekar, and H. Cabezas, "Optimal Control Theory for Sustainable Environmental Management", (2008),Environmental Science and Technology, 42, 5322.

(140e) Life-Cycle-Based Sustainability Assessment of Nanocoating Material

Rohan Uttarwar and Yinlun Huang, Department of Chemical Engineering and Materials Science, Wayne State University, Detroit, MI

Over the last decade, nanopaint has become a dominant coating material with a wide range of applications in the automotive, aerospace, ship-making, construction, architectural and pharmaceutical industries. Despite the obvious benefits of nanocoatings, there exist unintended health and environmental risks associated with the use of nanoproducts which are not yet fully understood. In order to develop nanocoating systems with improved product performance and reduced impact on environment and health care, a complete assessment over each stage of the life-cycle of nanocoating material is needed. Development of sustainability assessment techniques for the comprehensive evaluation of performance of nanocoating products over their life-cycle becomes urgent.

The life-cycle of nanocoating materials consists of the stages ranging from raw material selection and processing, to nanopaint/nanocoating manufacturing, product use and disposal. In this work, a model-based sustainability assessment methodology is introduced that can be used to assess the economic, environmental, and social aspects in every life-cycle stage. To perform a comprehensive assessment, different sets of sustainability metrics have been identified for use in different life-cycle stages. These metrics are analyzed to ensure the consistency of the assessment. Due to the lack of sufficient data from open resources, a number of computational models are developed that can generate data that is sound from engineering perspective. The methodology has been used to study the sustainability performance of nanopaint and its application to automotive coatings. A comprehensive case study will highlight critical issues concerning the material's development and nanoparticles emission to the environment and health impact, economic incentive and social satisfaction.

(140f) The Effect of CO2 Adsorbent to the H2 Production in Biomass Gassifier

KuZilati KuShaari, Chemical Engineering Department, Universiti Teknologi PETRONAS, Seri Iskandar, Perak, Malaysia

Biomass gasification has become one of the most popular methods to produce biohydrogen. Hydrogen yield has been proven to further increase by adding CO_2 adsorbent into the biomass gasifier. The proposed CO_2 adsorbent for biomass gasification system is Calcium Oxide (CaO) as it is converted to solid Calcium Carbonate ($CaCO_3$) as it reacts with CO_2 in the gasifier. Due to its high density, the solid $CaCO_3$ produced can be easily removed from the system and it will accumulate at the bottom of the gasifier. This paper highlights the study on the effect of CO_2 adsorbent to the H_2 production from the gasifier using CFD approach. The results show that the addition of CO_2 adsorbent helps to increase the H_2 production. The simulation was done for two different cases, which are gasification process with in-situ CO_2 adsorption and without adsorbent. The reaction temperature for both cases ranging between 500-900 ^{0}C. From the results obtained, it is observed that the addition of CO_2 adsorbent does improve the H_2 production by 12%.

(167a) Design and Operation of Bioethanol Production Process for Small-Scale Biorefinery

Kazuhiro Mochidzuki[1], Shin-ichi Kobayashi[1], Akiyoshi Sakoda[2], Tran Phuoc Nhat Uyen[3], Vu Le Van Khanh[3], Nguyen Dinh Quan[3], Le Thi Kim Phung[4] and Phan Dinh Tuan[5], (1)IIS, Univ. of Tokyo, Tokyo, Japan, (2)Institute of Industrial Science, University of Tokyo, Tokyo, Japan, (3)HCMUT, HCM City, Vietnam, (4)CHMUT, HCM City, Vietnam, (5)Hochiminh City Univ. of Tech., Hochiminh, Vietnam

To establish a model of sustainable integration of local agriculture and biomass industries in rural areas in South-East Asia, e.g., Vietnam, we are developing the biomass utilization technologies which meet with the satiation of actual communities. This presentation deals with a pilot-scale work of lingocellulosic bioethanol production, focusing on small-scale local biorefinery systems. We built a pilot plant which consists of bioethanol production process (with an 800 L fermentation tank) and biomass boiler. In this pilot plant, we developed the design of small-scale bioethanol production processes which utilized locally available biomass feedstock, including pretreatment system, small fermentor operation and use of biomass boiler, through the operation of the pilot process. For example, rice straw was converted into ethanol and rice husk was used for biomass boiler. Also, some byproducts from local industries were tested as nitrogen source for yeast. As wall as the technological development, the material and energy balances in the total system were analyzed to evaluate its features and benefit as a core facility in a small-scale biorefinery, focusing on the community-level availability biomass feedstock and demands of bioenergy/biofules.

(167b) Effects of Pellet Feed and Air Supply Rates On the Combustion Behavior of Spruce Pellets in a Grid-Type Stove

Bawadi Abdullah[1], Yoshimitsu Uemura[2], Noridah Osman[1], Takami Kai[3], Hirokazu Takanashi[4] and Yukio Teraoka[5], (1)Centre for Biofuel and Biochemical Research, Universiti Teknologi PETRONAS, Tronoh, Malaysia, (2)Centre for Biofuel and Biochemical Research, Universiti Teknologi PETRONAS, Tronoh, Malaysia, (3)Chemical Engineering, Kagoshima University, Kagoshima, Japan, (4)Chemistry and Biotechnology, Kagoshima University, Kagoshima, Japan, (5)Environmental Sciences and Technology, Kagoshima University, Kagoshima, Japan

The growing attentions to the utilization of biomass as a heat and power productions via conventional burning has initiated this study of using spruce materials as a feedstock. Unlike other types of biomass i.e. olive cake, apricot and peach stones, spruce is abundantly available and thus, suitable to use as a source of feedstock. In this study, the pellet of spruce was combusted in a grid type combustion chamber (5 kW stove). The stove consists of electric heater igniter, a screw feeder (2-18 g/min) and an air blower (0.6-1.0 NTP-m^3/min). To observe the combustion behavior, the concentration of O_2, CO, NO and NO_2 were measured continuously using a combustion gas composition sensor (Hodaka HT-2000). For each run, approximately 20-25 min of combustion under constant pellet feed and air supply was conducted. It is interesting to observe that combustion showed two phases cyclically with one of the cycles is flaming while other char is glowing. It is expected that the behavior of gas composition in the form of cyclic period and amplitude due to the combustion condition. The details would be discussed in the full paper.

(167c) Is Elevated Pressure Required to Achieve a High Fixed-Carbon Yield of Charcoal From Biomass?

Liang Wang[1], Oyvind Skreiberg[1], Morten Gronli[2] and **Michael J. Antal**[3], (1)SINTEF Energy Research, Trondheim, Norway, (2)Energy and Process Engineeringt, Norwegian University of Science and Technology, Trondheim, Norway, (3)Hawaii Natural Energy Institute, University of Hawaii, Honolulu, HI

Elevated pressure secures the highest fixed-carbon yields of charcoal from corncob. Operating at a pressure of 0.8 MPa a Flash-Carbonization reactor realizes fixed-carbon yields that range from 70 to 85% of the theoretical thermochemical equilibrium value from Waimanalo corncob. The fixed-carbon yield is reduced to a range from 68 to 75% of the theoretical value when whole Waimanalo corncobs are carbonized under nitrogen at atmospheric pressure in an electrically heated muffle furnace. The lowest fixed-carbon yields are obtained by the standard proximate analysis procedure for biomass feedstocks: this yield falls in a range from 49 to 54% of the theoretical value.

A round-robin study of corncob charcoal and fixed-carbon yields involving three different thermogravimetric analyzers (TGAs) revealed the impact of vapor-phase reactions on the formation of charcoal. Deep crucibles that limit the egress of volatiles from the pyrolyzing solid greatly enhance charcoal and fixed-carbon yields. Likewise, capped crucibles with pinholes increase the charcoal and fixed-carbon yields compared with values obtained from open crucibles. Large corncob particles offer much higher yields than small particles. These findings show that secondary reactions involving vapor-phase species (or nascent vapor-phase species) are at least as influential as primary reactions in the formation of charcoal.

Our results offer considerable guidance to industry for its development of efficient biomass carbonization technologies. Size reduction handling of biomass (e.g. tub grinders and chippers), which can be a necessity in the field, significantly reduces the fixed-carbon yield of charcoal. Fluidized bed and transport reactors, which require small particles and minimize the interaction of pyrolytic volatiles with solid charcoal, cannot realize high yields of charcoal from biomass. When a high yield of corncob charcoal is desired, whole corncobs should be carbonized at elevated pressure. Under these circumstances, carbonization is both efficient and quick.

(167d) Understanding the Reaction Network and Kinetics of the Hydrothermal Liquefaction of Microalgae

Peter J. Valdez, Chemical Engineering, University of Michigan, Ann Arbor, MI and Phillip E. Savage, Chemical Engineering, University of Michigan, Ann Arbor, Ann Arbor, MI

Hydrothermal liquefaction can convert wet algal biomass, at elevated temperatures and pressures, into an energy-dense biocrude oil without the need for dewatering the algae or extracting the lipids. Liquefaction occurs as the biomacromolecules react in and with water at elevated temperatures (e.g., 350 °C) and pressures to form smaller molecules that are closer to being in the range required for liquid fuels. Recent studies of microalgal liquefaction, examining some process variables, provide some clues about the reaction pathways, but not enough information to develop an entire reaction network. As the field progresses, quantitative reaction models based on the governing reaction network will be needed for process design and optimization. Developing such a model requires data from a thorough systematic study of the influence of all the relevant process variables on liquefaction of a single alga.

We present results from the liquefaction of *Nannochloropsis* sp. and the product intermediates to determine reaction pathways. From the experimental results, we determined the reaction network and constructed a model to predict the yields of the product fractions. We used first-order kinetics to describe the formation and consumption of the product fractions from liquefaction. The abundance of different compounds and the complexity of the numerous reactions necessitated the use of a global kinetic analysis to make the problem tractable. We applied this model to the liquefaction of other alga species. We will also show the fate of phosphorus in the product fractions.

Inhibition of Catalyst Poisoning in Supercritical Water Gasification of Sewage Sludge: Technical Study on Reactor Design

Osamu Sawai, Teppei Nunoura and Kazuo Yamamoto
Environmental Science Center, The University of Tokyo, Bunkyo-ku, Tokyo, Japan

Abstract

A lab-scale batch reactor with two compartments, so called two-bed reactor, was introduced to the supercritical water gasification of sewage sludge. With Ni catalyst presence, the reactor was capable of suppressing deposition of byproducts and residuals on catalyst surface attaining higher gasification efficiencies. Moreover, acceleration of hydrogen gas production was achieved when K_2CO_3 was loaded with the presence of Ni catalyst. Further studies have revealed that K_2CO_3 loading location influenced the gasification behavior dramatically. Besides Ni catalyst, RuO_2 and Ru catalyst were also employed using the reactor. In prospect of future application, discussion on heavy metal concentration in the effluent is made as well.

Introduction

Supercritical water gasification is a thermal method which enables us to transform biomass to hydrogen rich gas with reduced tar and char formation [1,2]. Easy gas separation and fast reaction rate (compared to anaerobic digestion) leading to compact equipment as well as elimination of post-drying step unlike conventional thermo-chemical treatment processes [3] are also likely to be featured as significant advantages of this technology. Although the process seems to be promising, supercritical water gasification for real biomass is still under development. The main challenges are problems concerning with salts originating from biomass. Alkaline salts tend to increase hydrogen yield, but at the same time it is associated with the risk of plugging [4] and catalyst poisoning [5]. The issue still remains as a challenge being the major obstacle to technical application.

This paper focuses on the geometry of the gasification reactor that will lead to the inhibition of catalyst poisoning in supercritical water. Experimental investigations were conducted under lab-scale batch reactors in order to exclude the plugging issues, thus focusing on the catalyst poisoning inhibition. Corresponding to the new reactor design, combination of metal supported catalyst and alkaline additives was introduced.

Methods

Materials

In this work, sewage sludge was used as real biomass feedstock. Used sewage sludge classified as excessive sludge, was supplied from the conventional activated sludge processes operated by a waste water treatment plant in Japan. Sewage sludge was dried completely in an oven kept at 105°C for at least 6 hours, and subsequently sieved into fine particles (<180 μm). All the experiments and results shown in this work were conducted using this post-treated sewage sludge (cited as "initial sewage sludge" further on). Precise profile of the initial sewage sludge has been described in our previous work [6].

Nickel (65 wt.%) on silica/alumina (35 wt.%) catalyst (cited as "Ni/SiO_2-Al_2O_3" catalyst further on) and ruthenium (5 wt.%) on alumina (95 wt.%) catalyst (cited as "Ru/Al_2O_3 catalyst" further on) used were purchased from Sigma-Aldrich, USA. Ruthenium oxide and potassium carbonate used were purchased from Wako Pure Chemical Industries Ltd., Japan. Deionized water was prepared in our laboratory by Milli-Q A10; Millipore and all materials were used without further treatment.

Experimental Set Up

All the experiments were performed by two-bed reactor, a lab-scale batch reactor manufactured from stainless steel (SUS316) tube and Swagelok® SUS316 fittings/valves. As shown in figure 1, the reactor consists of two compartments, A and B, with volume ratio of approximately 4.2:1. This reactor configuration aims to reduce poisoning caused by the solid residue. The compartment is expected to physically separate or limit direct contact of biomass/water feedstock and catalyst. SUS316 frit filter with pore size of 2 μm, purchased from Valco Instruments Co. Inc., USA is used to separate the two compartments. An identical frit filter is inserted on the other side of compartment B to limit the access of solids to equipped pressure transducer and the needle valve.

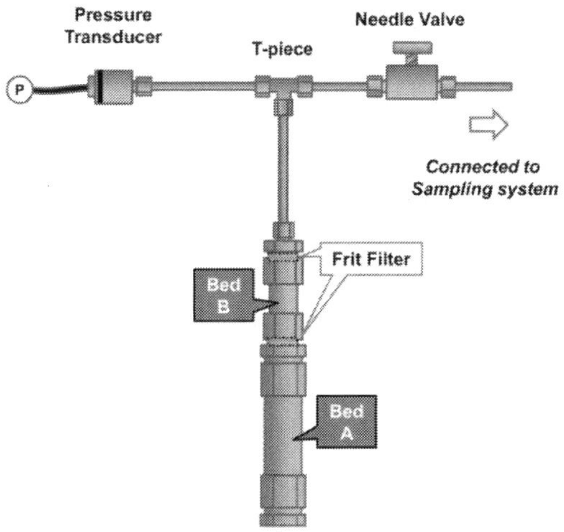

Figure 1. Schematic diagram of two-bed reactor

Char and tar which are likely to pass the filter causing catalyst poisoning, can be resolved by using an additive inhibiting their formation. In our experiments, K_2CO_3 which is reported to function as a char and tar formation inhibitor [7-9], was added to either of the compartment. Deionized water, initial sewage sludge (5 wt.% of deionized water), catalyst and K_2CO_3 additives (10 wt.% of sewage sludge respectively) were loaded in either of the bed. The gasification was conducted under the condition of 400°C and 27 MPa in standard (cited as "standard SCWG condition" further on).

The gasification reaction was initiated by immersing the two-bed reactor into temperature-controlled molten salt bath (mass ratio of salt was adjusted to K_2CO_3:$NaNO_2$:$NaNO_3$ = 6:5:1). After lapse of predefined time, the reactor was taken out of the bath and subsequently quenched by a water bath to stop the reaction. Reaction time, defined as time from which the reactor was immersed into molten salt bath until it was taken out, was set to 30 minutes. During the reaction, pressure was monitored by the transducer to make sure that the reaction media was kept at supercritical conditions.

Reaction products were collected after the reactor was cooled to ambient condition. Gas products were first transferred to the initially evacuated sampling loop with known volume. Amount of gas product was determined from pressure monitored by pressure transducers connected to the sampling loop. Produced gas was sampled from one of the sampling loop ports using a gas-tight syringe for gas analysis injection. Liquid and solid residue collected as mixtures were subsequently separated by centrifugation run at 2500 rpm for 5 minutes. Liquid was filtered by 0.45 μm pore size syringe filter (Millex LH, Millipore) and diluted by deionized water prior to the analysis. Separated solid residues were dried in an oven kept at 105°C for at least 6 hours.

Analytical Equipments

Gas analysis was carried out with gas chromatograph (GC) GC-2014; SHMADZU equipped with Shincarbon ST 50/80 column and thermal conductivity detector (TCD) to separate hydrogen, methane, carbon monoxide, and carbon dioxide. Total organic carbon (TOC) content in the liquid phase was determined by total organic carbon analyzer TOC-V; SHMADZU. Concentrations of heavy metals were analyzed by inductively coupled plasma mass spectrometry (ICP-MS) ICP-8500; SHIMADZU. Microscopic observation of solid residues was carried out by scanning electron microscope (SEM) JSM-5600; JEOL ltd. equipped with energy dispersive X-ray spectroscopy (EDX) Oxford Link ISIS; Oxford Instruments. Prior to SEM analysis, Au was spattered on the sample as a pretreatment.

Results and Discussion

Validity of Two-bed Reactor

Loading location of Ni catalyst was varied in order to demonstrate the validity of the reactor. Figure 2 shows the effect of Ni catalyst loading location to gas yield and gas efficiency. Citation such as "(SS+Ni Cat.)|None" indicates that sewage sludge and Ni catalyst is loaded in compartment A, and that compartment B is vacant. In this paper, hydrogen gas efficiency (HGE) and carbon gas efficiency (CGE) are defined as shown below.

HGE = (mol of H atom in gas products) / (mol of H atom in sewage sludge loaded) (1)
CGE = (mol of C atom in gas products) / (mol of C atom in sewage sludge loaded) (2)

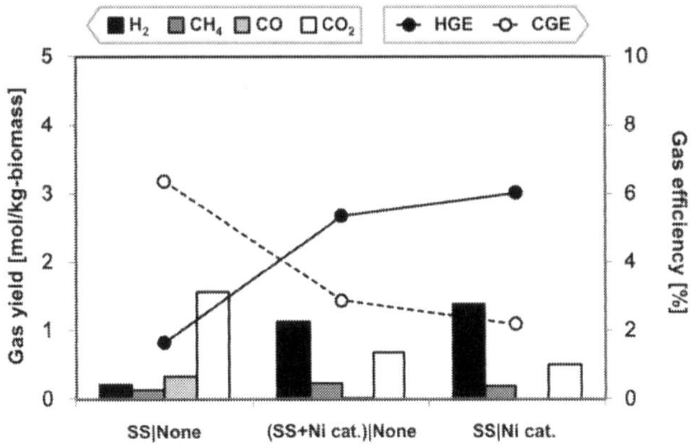

Figure 2. Effect of two-bed reactor to gas yield and efficiency

Addition of Ni catalyst increased hydrogen gas yield and decreased carbon monoxide and carbon dioxide gas yield significantly as shown in figure 2. Hydrogen gas yield increased and carbon dioxide gas yield decreased furthermore when the sewage sludge and Ni catalyst was loaded in different compartments. On the other hand, methane gas yield was not affected by the loading location of Ni catalyst. Gasification efficiency curves behaved corresponding to the component gas behavior. From this trend, it is clear that the reactor geometry adopted in this work did not limit catalyst activation, but contributed to attain higher hydrogen gas efficiency.

The two-bed reactor not only exhibited higher hydrogen gas efficiency but also inhibited both inorganic salt and byproduct deposition. SEM image of Ni catalyst after run "SS|Ni cat." is shown in figure 3. As shown in the SEM images, salts and byproducts deposited on neither of the support surfaces. Although SiO_2 support (figure 3 (a)) avoided the deposition of byproducts, its porous structure was destroyed during the supercritical gasification.

Figure 3. SEM image of Ni catalyst after reaction using two-bed reactor; (a) SiO_2 (b) Al_2O_3 support.

Effect of K_2CO_3 Addition on Gas Yield and Metal Elution

To attain higher gasification efficiency, addition of K_2CO_3, well known for accelerating water-gas shift reaction [10-13] and suppressing formation of char [7,8] and tar [9] was taken into consideration. Figure 4 shows the effect of K_2CO_3 addition or/and Ni catalyst to gas yield and gas efficiency. From the figure, addition of K_2CO_3 shows a significant increase of hydrogen gas yield in the case of Ni catalyst absence. Methane and carbon dioxide gas yield increased slightly whereas carbon monoxide gas yield decreased. HGE and CGE curve behaved accordingly to the corresponding gas component yield.

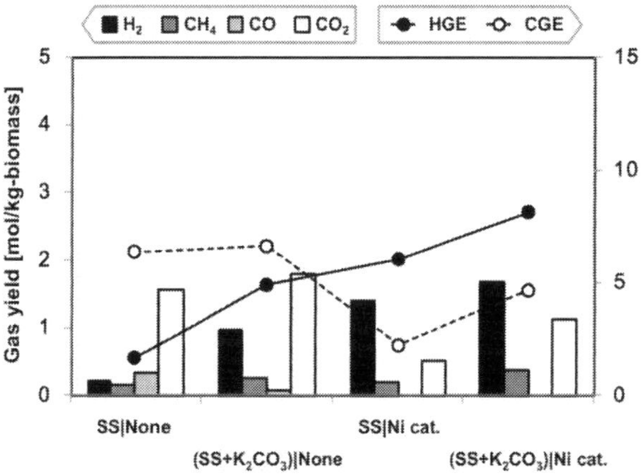

Figure 4. Effect of K_2CO_3 addition to gas yield and efficiency

Effect of K_2CO_3 addition under the presence of Ni catalyst is also shown in the figure 4. Combination of Ni catalyst and K_2CO_3 [(SS+K_2CO_3)|Ni Cat.] presented a positive effect on the hydrogen gas yield exhibiting higher yield than when K_2CO_3 or Ni catalyst was loaded alone [(SS+K_2CO_3)|None, SS|Ni Cat.]. Similar behavior was observed for methane exhibiting highest gas yield when Ni catalyst was used with K_2CO_3. HGE increased rapidly corresponding to the behavior of hydrogen and methane gas yield. The figure also indicates that Ni catalyst suppresses the formation of carbon monoxide production. K_2CO_3 addition presented a positive effect on carbon dioxide production under Ni catalyst presence, resulting in a higher CGE. Compared to the works of D'Jesus *et al.* or Schmieder *et al.* in which they reported that potassium addition did not influence the gasification yield in case of real biomass [12,14], sewage sludge presented both positive and negative effect on the yield of component gas.

In parallel, effect of K_2CO_3 addition to metal elution was investigated. From figure 5 comparing the effect of K_2CO_3 addition to metal elution, Mo exhibited higher degree of elution when K_2CO_3 was present. Elution of Ni and Mo was induced when the gasification is conducted under the presence of Ni catalyst with K_2CO_3. Zn was eluted when the gasification is conducted under catalytic conditions. On the contrary, species such as Fe and Se tended to increase when Ni catalyst was used regardless of K_2CO_3 addition.

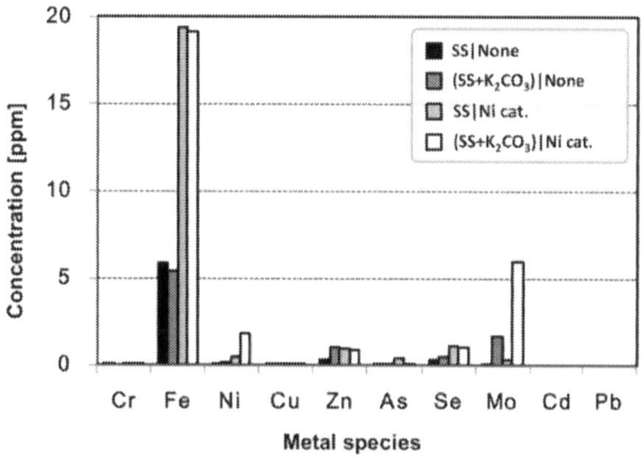

Figure 5. Effect of K_2CO_3 addition to metal concentration in liquid phase

Result shown here indicates that the enhancement of gas yield by Ni catalyst and/or K_2CO_3 addition is likely to be associated with elution of heavy metal species originating from reactor material and metals deposited on the catalyst.

Effect of K_2CO_3 Loading Location

Effect of K_2CO_3 loading location on gas yield and efficiency was explored. K_2CO_3 was loaded with either sewage sludge or Ni catalyst. As shown in figure 6 comparing the effect of K_2CO_3 loading location to gas yield and efficiency, the highest gas yield was achieved when Ni catalyst was separated from K_2CO_3 and sewage sludge loaded compartment, i.e., [(SS+K_2CO_3)|Ni Cat.] condition. On the other hand, gasification was suppressed when sewage sludge was separated with Ni catalyst and K_2CO_3 loaded compartment. Hydrogen, methane and carbon dioxide gas yield was lower than the case in which sewage sludge, Ni catalyst, and K_2CO_3 was loaded together in the same compartment. Furthermore, carbon monoxide yield showed a slight increase which implies that the catalytic activity was not efficiently exerted at the condition where Ni catalyst and K_2CO_3 were loaded in the same compartment.

In addition, elution of metal species was investigated. Effect of K_2CO_3 loading location to metal elution is shown in figure 7. As shown in the figure, metal elution can be inhibited by loading Ni catalyst and sewage sludge in separate compartments. Especially Ni concentration decreased drastically when the Ni catalyst and sewage sludge were loaded separately regardless of K_2CO_3 loading location. Elution of Mo was inhibited drastically when Ni catalyst was loaded separately from both K_2CO_3 and sewage sludge. Zn and Se showed similar trend to Mo but with moderate degree. Interaction between Ni catalyst and K_2CO_3 can assumedly provoke metal elution via behaviors such as drastic alteration of pH or dissolution of produced gas. Therefore, limitation of interaction by separating Ni catalyst and K_2CO_3 has possibly relieved the metal elution. The result obtained here implies that the two-bed reactor can also contribute to the inhibition of metal elution as well as gasification enhancement and inhibition of catalyst poisoning.

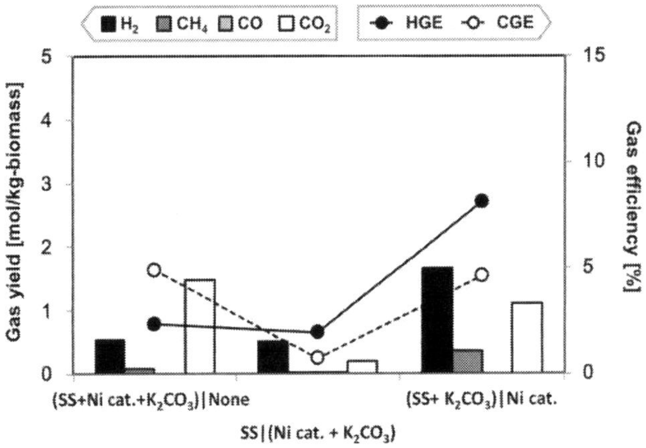

Figure 6. Effect of K_2CO_3 loading location to gas yield and efficiency

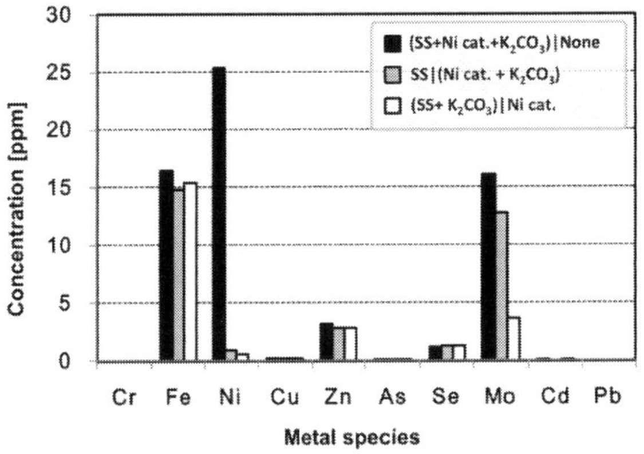

Figure 7. Effect of K_2CO_3 loading location to metal concentration in liquid phase

Variation of catalyst types

Metal supported catalysts reported to perform high catalytic activity were employed. Figure 8 compares the gas yield and efficiency of sewage sludge SCWG using RuO_2, Ru and Ni catalyst. Nickel apparently presented the highest catalytic activity in this work unlike in previous reports [15-17], since the catalyst amount (including support weight) added was fixed in this work. RuO_2, also reported to enhance biomass gasification under SCW condition, presented lower hydrogen gas yield than Ru catalyst. However, HGE was higher due to the production of methane. Comparative high methane yield is also reported by Yamamura *et al.* in which they applied RuO_2 to sewage sludge SCWG [17]. They have concluded that RuO_2 not only enhanced the gasification reaction but at the same time accelerated the methanation reaction. Meanwhile, the addition of RuO_2 in our work also enhanced the production of carbon dioxide and carbon monoxide gas yield, which as a result increased the CGE significantly.

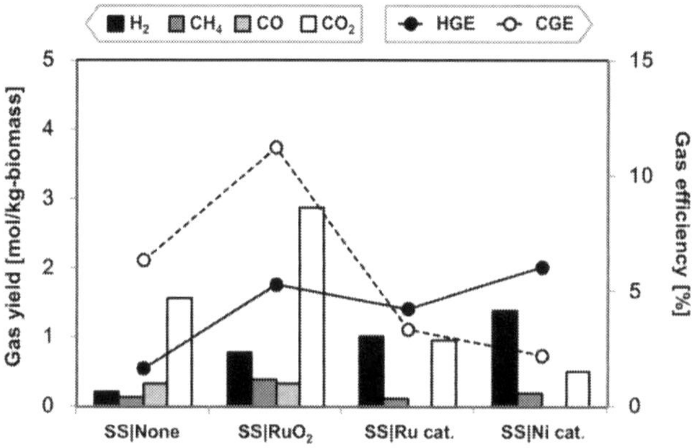

Figure 8. Comparison of gas yield using various catalysts

Conclusions

In this work, a lab-scale batch reactor with two compartments was designed to inhibit catalyst poisoning. Under the condition of 400°C, 27 MPa and reaction time of 30 minutes with Ni catalyst presence, the reactor was capable to suppress the deposition of byproducts and residuals with higher gasification efficiencies. Furthermore, acceleration of hydrogen gas production was achieved when K_2CO_3 was added to sewage sludge compartment, loaded separately from Ni catalyst. Further studies have revealed that the loading location of K_2CO_3 influenced the gasification behavior dramatically.

RuO_2 and Ru catalysts reported to promote SCWG of various biomass were also employed. Due to low metal loading of Ru, Ni catalyst tended to exhibit slightly higher hydrogen yield than Ru catalyst. RuO_2 exhibited lower catalyst activity for hydrogen production but promoted the production of carbon dioxide compared to Ni catalyst and Ru catalyst. Metal elution was also investigated throughout this work. The results have revealed that enhancement of gas yield by catalyst and/or K_2CO_3 addition is likely to be associated with elution of heavy metal species (trade-off relationship between gasification yield/efficiency and metal elution), originating from reactor material, sewage sludge and supported catalyst particles. From the view point of effluent management, elution of metal should also be taken into account along with the catalytic activity of gasification in case of selecting a catalyst in a real process.

Acknowledgments

The authors acknowledge the support of New Energy and Industrial Technology Development Organization (NEDO) and Japan Society of the Promotion of Science (JSPS) through the Funding Program for World-Leading Innovative R&D on Science and Technology (FIRST). The authors wish to gratefully appreciate the Materials Design and Characterization Laboratory, Institute for Solid State Physics (ISSP), University of Tokyo for providing SEM-EDX used in this work.

References

1. L.J. Guo, Y.J. Lu, X.M. Zhang, C.M. Ji, Y. Guan, A.X. Pei, Hydrogen production by biomass gasification in supercritical water: A systematic experimental and analytical study, Catalysis Today 129 (2007) 275–286.
2. A. Demirbas, Hydrogen-rich gas from fruit shells via supercritical water extraction, Int. J. Hydrogen Energy 29 (2004) 1237–1243.

3. D.T. Furness, L.A. Hoggett, S.J. Judd, Thermochemical treatment of sewage sludge, Water Environ. J. 14 (2000) 57–65.
4. M.J. Antal, S.G. Allen, D. Schulman, X.D. Xu, R.J. Divilio, Biomass gasification in supercritical water, Ind. Eng. Chem. Res. 39 (2000) 4040–4053.
5. M.H. Waldner, F. Krumeich, F. Vogel, Synthetic natural gas by hydrothermal gasification of biomass: Selection procedure towards a stable catalyst and its sodium sulfate tolerance, J. Supercritical Fluids 43 (2007) 91–105.
6. O. Sawai, Y. Tanaka, T. Nishihara, T. Nunoura, K. Yamamoto, Application of Supercritical Water Gasification to Municipal Wastewater Treatment Plant: Experimental Studies and Operational Challenges, 10th International Symposium on Supercritical Fluids, San Francisco, California, USA (2012) L-139.
7. A. Sinag, A. Kruse, V. Schwarzkopf, Key compounds of the hydropyrolysis of glucose in supercritical water in the presence of K_2CO_3, Ind. Eng. Chem. Res. 42 (2003) 3516-3521.
8. T. Minowa, F. Zhen, T. Ogi, Cellulose decomposition in hot-compressed water with alkali or nickel catalyst, J. Supercritical Fluids 13 (1998) 253–259.
9. A. Kruse, T. Henningsen, A. Sinag, J. Pfeiffer, Biomass gasification in supercritical water: influence of the dry matter content and the formation of phenols, Ind. Eng. Chem. Res. 43 (2004) 502-508.
10. J. Yanik, S. Ebale, A. Kruse, M. Saglam, M. Yüksel, Biomass gasification in supercritical water: II. Effect of catalyst, Int. J. Hydrogen Energy 33 (2008) 4520–4526.
11. J.A. Onwudili, P.T. Williams, Role of sodium hydroxide in the production of hydrogen gas from the hydrothermal gasification of biomass, Int. J. Hydrogen Energy 34 (2009) 5645-5656.
12. H. Schmieder, J. Abeln, N. Boukis, E. Dinjus, A. Kruse, M. Kluth, G. Petrich, E. Sadri, M. Schacht, Hydrothermal gasification of biomass and organic wastes, J. Supercritical Fluids 17 (2000) 145–153.
13. X.H. Hao, L.J. Guo, X. Mao, X.M. Zhang, X.J. Chen, Hydrogen production from glucose used as model compound of biomass gasified in supercritical water, Int. J. Hydrogen Energy 28 (2003) 55–64.
14. P. D'Jesus, N. Boukis, B. Kraushaar-Czarnetzki, E. Dinjus, Gasification of corn and clover grass in supercritical water, Fuel 85 (2006) 1032-1038.
15. P. Azadi, K.M. Syed, R. Farnood, Catalytic gasification of biomass model compound in near-critical water, Applied Catalysis A: General 358 (2009) 65–72.
16. A. Yamaguchi, N. Hiyoshi, O. Sato, K. Bando, M. Osada, M. Shirai, Hydrogen production from woody biomass over supported metal catalysts in supercritical water, Catalysis Today 146 (2009) 192–195.
17. T. Yamamura, T. Mori, K.C. Park, Y. Fujii, H. Tomiyasu, Ruthenium (IV) dioxide-catalyzed reductive gasification of intractable biomass including cellulose, heterocyclic compounds, and sludge in supercritical water, J. Supercritical Fluids 51 (2009) 43-49.

(167f) Effect of Acid and Alkali On Supercritical Water Gasification of Glucose

Yukihiko Matsumura, Hokuto Takai and Chutinan Promdej, Hiroshima University, Higashi-Hiroshima, Japan

Supercritical water gasification, which is the gasification of biomass and biomass component compounds in hot compressed water above critical temperature (647 K) and critical pressure (22.1 MPa) of water, is expected to be an effective gasification technology for wet biomass. However, its reaction kinetics is still under investigation. Especially, char and tarry material formation is a big problem, and its reaction kinetics elucidation is needed. During the study of glucose decomposition in supercritical water, the authors' group found that some of the reactions behave as Arrhenius reaction rate, and others do not, and dsitinguished ionic and radical reactions. It was then found that the char formation is ionic reaction. Char formation was expected to be enhanced by the organic acids produced by glucose decomposition. If it is the case, addition of acid or alkali should affect the char formation. In this study, the authors present the results of HCl and NaOH addition on the supercritical water gasification of glucose, clarifying the effect of acid and alkali on the reaction kinetics.

(174a) Reaction Route Selection Based On Sustainability Criteria for Glutaraldehyde Production

Preeti Gangadharan and Helen H. Lou, Dan F. Smith Department of Chemical Engineering, Lamar University, Beaumont, TX

Challenges in the specialty chemicals industry are enormous. Companies need to deliver the most profitable product mix to keep up with global competition, while also keeping in mind long term environmental goals. In most cases, problems such as waste disposal or treatment may be avoided, or their intensity reduced, by considering their effects prior to plant design. Hence, sustainability criteria should form an integral part of process design, starting from the early design stage. In this stage, the sustainability performance of each potential reaction pathway is evaluated in terms of possible profit, driving force of the pathway (Gibbs energy), inherent safety index, environmental analysis, and atom economy. For specialty chemicals, particular emphasis has to be given on safety and environmental criteria.

A structured methodology incorporating all the major criteria is applied to reaction pathway selection for producing Glutaraldehyde, a specialty chemical primarily used as a bactericide, tanning agent, and fixative. Three alternative routes are considered. The fast and comprehensive assessment is used to quantify the desirability of each route, so that inferior routes can be screened out systematically, while also identifying areas for improvement, prior to initiating detailed design. In the detailed design stage, the sustainability performance of the selected process is improved through sequential sustainability assessment covering economic, safety, environmental, and efficiency criteria.

(174b) Testing Room Temperature Ionic Liquid Solutions

Paul Brezovec, Concurrent Technologies Corp., Johnstown, PA

To avoid legacy problems associated with cadmium's carcinogenicity and high toxicity, some new Air Force weapon system steel component designs by original equipment manufacturers (OEMs) specify an aluminum plating instead, to satisfy weapon system corrosion protection requirements. When these weapon systems come to the Air Logistics Centers (ALCs) for maintenance, repair and overhaul (MRO), repair-scale processes need to be in place to refurbish the coating that is on the exterior and recesses of the components. The new technologies used by the OEMs to plate these components with aluminum require extensive infrastructure and capital expense.

Technologies that electroplate aluminum from non-aqueous chemistries such as room temperature ionic liquid solutions are being marketed to replace cadmium electroplating for MRO needs, with fewer infrastructure and capital expenses than the OEM's outlays. In general, room temperature ionic liquids are salts that remain liquid near room temperatures and have a very low vapor pressure unlike organic solvents. Room temperature ionic liquids can dissolve metal salts from which the metals such as aluminum may be electrodeposited, and have attracted significant laboratory research efforts, because they can be energy efficient and some are considered "green."

The applicability of this technology to DoD plating shops, Air Force ALC plating shops in particular, is of interest. Despite their promise, there are caveats associated with room temperature ionic liquids, including concerns with the toxicity of a variety of them, and processing difficulties. In particular, processing must consider the effects of additives on the physical properties of the ionic liquid-based electrolyte. A review of room temperature ionic liquid additives and their effects on the electrodeposition process will be presented, and screening testing results to date for potential applications will be described.

(174c) Looped Carbon Capturing and Environmental Remediation: Case Study of Magnetic Polypropylene Nanocomposites

Jiahua Zhu[1], Hongbo Gu[1], Sowjanya B. Rapole[1], Minjiao Chen[1], Suying Wei[2] and Zhanhu Guo[1], (1)Integrated Composites Laboratory (ICL), Dan F Smith Department of Chemical Engineering, Lamar University, Beaumont, TX, (2)Chemistry and Biochemistry, Lamar University, Beaumont, TX

A waste-free process to recycle $Fe@Fe_2O_3$/polypropylene (PP) polymer nanocomposites (PNCs) is introduced to synthesize magnetic carbon nanocomposites (MCNCs) and simultaneously produce useful chemical species which can be utilized as feedstock in petrochemical industry. The magnetic nanoparticles (NPs) are found to have an effective catalytic activity on the pyrolysis of PP. The PNCs (with a NP loading of 20.0 wt%) undergo a complete degradatisxcsxaxazxasxzasason with 2-hour pyrolysis at 500 ^oC in the H_2/Ar atmosphere and the degradation components exhibit a distribution of different carbon number species, while only 40% of pure PP is decomposed after applying the same pyrolytic conditions. The coked solid waste in the conventional process has been utilized as carbon source to form a protective carbon shell surrounding the magnetic NPs. The magnetic carbon nanocomposites (MCNCs) pyrolyzed from PNCs containing 20.0 wt% NPs demonstrate an extremely fast Cr(VI) removal from the wastewater with an almost complete removal of Cr(VI) within 10 min. The pH effect on the Cr(VI) removal efficiency is investigated with a preferable value of 1-3. The adsorbent exhibits much higher adsorption capacity in the acidic solutions than that in the alkali solutions. The large saturation magnetization (32.5 emu g^{-1}) of these novel magnetic carbon nanocomposites allows fast recycling of both the adsorbents and the adsorbed Cr(VI) from liquid suspension in a more energetically and economically sustainable way by simply applying a permanent magnet. The significantly reduced treatment time required to remove the Cr(VI) makes these MCNCs promising for efficient removal of the heavy metals from the wastewater. Kinetic investigation reveals a pseudo-second-order adsorption of Cr(VI) on these novel magnetic carbon nanocomposite adsorbents.

(174d) Transformation of Carbon Dioxide Into Boron-Graphene Oxide Nanocomposites Using a Reducing Agent

Jae W. Lee[1], Junshe Zhang[1], Yu Zhao[2], Daniel L. Akins[2], Ruth Stark[3], Xudong Guan[4] and Ah-young Byun[5], (1)Department of Chemical Engineering, The City College of New York, New York, NY, (2)Department of Chemistry, The City College of New York, New York, NY, (3)Chemistry, CCNY, New York, NY, (4)Chemistry Department, CCNY, New York, NY, (5)Chemical Biomolecular Eng, KAIST, Daejeon, South Korea

Carbon dioxide (CO_2), a readily available, low-cost and sustainable carbon source, could be converted to liquid fuels, useful chemicals, and carbon materials. Among the several alternatives of reducing CO_2, attractive options include transforming CO_2 to diamonds and nanotubes, either through direct CO_2 splitting or via reaction with metals subject to pressures higher than 70 MPa. Nonetheless, the possibility of CO_2 reduction to other carbon nanomaterials like graphene oxide without using metal-containing compounds and without employing extreme conditions has not been demonstrated. Graphene oxide is not only a suitable and scalable precursor of graphene, but also a flexible platform for biological, chemical, electronic, medical, and optical applications. This talk reports the conversion of CO_2 to graphene oxide nanocomposites using ammonia borane (NH_3BH_3), which is a well-known candidate material for hydrogen storage. The conversion involves two consecutive steps: carbon fixation and then graphenization. The carbon fixation is the reaction of CO_2 with NH_3BH_3 under mild temperature and pressure conditions ($T < 100\ ^oC$ and $P < 3$ MPa) to produce a solid compound that contains methoxy (OCH_3), formate (COOH) and aliphatic groups, whereas the graphenization is the pyrolysis of this solid compound at high temperatures ($T > 600\ ^oC$) and atmospheric pressure of nitrogen to generate graphene oxide-boron oxide nanocomposites. The generation of the nanocomposites is confirmed by micro-Raman spectroscopy, solid state ^{13}C and ^{11}B magic angle spinning-nuclear magnetic resonance (MAS-NMR), transmission electron microscopy (TEM), and atomic force microscopy (AFM).

(174e) Environmental Life Cycle Assessment of New Energetic Materials

Amirhossein Mehrkesh and Arunprakash T. Karunanithi, Department of Civil Engineering, University of Colorado Denver, Denver, CO

There is growing interest in the development of new energetic ionic liquids for use as aerospace propellants and explosives. Due to their negligible vapor pressure ionic liquids are usually considered as "green" alternatives to volatile organic compounds. In addition, ionic liquids and salts have tunable physical and chemical properties. Energetic materials release large amount of energy when they decompose. In the case of explosives all energy is released rapidly while in the case of fuels the energy is released in a controlled manner. These materials derive their energy content from oxidation of the carbon backbone or from their high positive heats of formation. The general requirements for energetic materials are high energy and density, thermal stability, low sensitivity to impact and low toxicity. Traditional energetic materials that are commonly used in explosive formulations are HMX (1,3,5,7-tetranitro-1,3,5,7-tetraazacyclooctane), RDX (1,3,5-trinitro-1,3,5-triazacyclohexane) and TNT (2,4,6, trinitrotoluene). Hydrazine derivatives are widely used as energetic fuel in rocket propulsion systems. Most recently several energetic ionic liquids have been proposed as high-energy materials, propellants and explosives. A large number of ionic salts and ionic liquids that are based on a triazole derivative have been proposed as energetic materials. 1,2,4 triazole and 1,2,3 triazole have heats of formation values of 109 KJ/mol and 272 KJ/mol respectively.

In this work we use environmental life cycle assessment method to study the environmental impact associated with the production of ionic liquids and ionic salts to compare with the environmental impact of common energetic material such as TNT. This will provide insights into whether the newer energetic materials that are currently being designed are indeed environmentally friendly as being assumed.

(174f) Economic Comparison of Processes for the Recovery of Lactic Acid by Reactive Distillation with Different Alcohols

Chien-Yuan Su, Cheng-Ching Yu and Jeffrey D. Ward, Department of Chemical Engineering, National Taiwan University, Taipei, Taiwan

Poly lactic acid (PLA) is a biodegradable, environmentally-friendly polymer that can be made from renewable resources and has many potential applications. Production of lactic acid (LAC) from renewable resources is usually accomplished by fermentation which has the additional advantage that an enantiomerically pure product is produced. Therefore, separation and purification of lactic acid from fermentation broth is important. However, it is also challenging because LAC is relatively nonvolatile and tends to oligomerize at high temperature. The most attractive method for purification of LAC is a two-step reaction process in which the acid is first reacted with an alcohol to form a more volatile ester which can be more easily purified by distillation. Then pure LAC is recovered by hydrolysis of the ester (Kumar et al., 2006; Kumar and Mahajani, 2007).

In this work we determine the optimal designs of esterification/hydrolysis lactic acid recovery processes using different alcohols and compare the cost of each process. We consider the presence of a heavy organic impurity and LAC oligomerization reactions in order to make the design problem more realistic. Processes are simulated using the commercial simulation package ASPEN PLUS. The processes are optimized to minimize the Total Annual Cost (TAC), with an overall lactic acid with lactic acid esterification and hydrolysis conversions of 99 mol%.

Different alcohols lead to different process configurations due to different component characteristics (especially boiling point ranking). The methanol system includes one pretreatment column, two reactive distillation columns and one separation column. The advantage of the methanol system is that methyl-lactate is the lightest compound among all lactic esters, so it is easier to separate methyl-lactate from other species. The butanol system includes one pretreatment column, two reactive distillation columns, two decanters and two separation column. The advantage of the butanol system is that there is a two-liquid phase zone, which facilitates separation of the species. It also may reduce the energy consumption of the separation system if the system is well designed. Designs using ethanol and isopropanol are also briefly discussed.

The butanol system has the lowest TAC among all alcohols due to low separation cost in the esterification and separation column because of the two-liquid-phase region. Furthermore, the methanol system has lower TAC than the ethanol and isopropanol systems due to the absence of an azeotrope between the alcohol and water.

References

Kumar R., Nanavati H., Noronha S. B. and Mahajani S. M. A continuous process for recovery of lactic acid by reactive distillation, *Journal of Chemical Technology and Biotechnology*, **81**, 1767-1777, 2006.

Kumar R. and Mahajani S. M. Esterification of Lactic Acid with n-Butanol by Reactive Distillation, *Ind. Eng. Chem. Res.*, **46**, 6873-6882, 2007.

(174g) Synthesis of Organosilicon Complexes From Rice Husk Derived Silica Nanoparticles

Weixing Wang[1], Jarett C. Martin[2], Rongcai Huang[2], Wensi Huang[1], Anhua Liu[3], Aijie Han[4] and **Luyi Sun**[2], (1)Department of Chemical Engineering, South China University of Technology, Guangzhou, China, (2)Department of Chemistry and Biochemistry, Texas State University-San Marcos, San Marcos, TX, (3)Materials Science and Engineering, South China University of Technology, Guangzhou, China, (4)Department of Chemistry, University of Texas-Pan American, Edinburg, TX

Organosilicon compounds are a group of important materials with widespread applications. However, their synthesis has been relying mainly on the carbothermal reduction of silica to silicon and subsequent reaction between silicon and alkyl or aryl chloride. This multi-step approach involves high temperature, high pressure, and strong acidity, which is energy-intensive, expensive, and eco-hazardous. An alternative approach was explored to synthesize organosilicon complexes via a green chemical process using silica nanoparticles derived from rice husk (RH) biomass as the starting material. By controlling the pre-treatment of RHs and pyrolysis conditions, silica samples with various surface areas and levels of crystallinity were synthesized. Such silica can be converted to silicon complexes via a low temperature approach. The synthesized silicon complexes were characterized by ^{1}H, ^{13}C, ^{29}Si nuclear magnetic resonance and elemental analysis. Overall, the biogenic silica nanoparticles with high surface area and low crystallinity exhibit high reactivity. Thanks to their high reactivity and low cost, such biogenic silica nanoparticles from RHs can serve as an ideal starting material for producing organosilicon complexes.

(174h) A Novel Synthesis of Odd-Numbered Fatty Acids From Corresponding Natural Fatty Acids and the Potential Application in Soap-Based Detergents

Xiaoxiao Huang[1,2] and Yun Fang[1], (1)School of Chemical and material engineering, Jiangnan University, Wuxi, China, (2)The Key Laboratory of Food Colloids and Biotechnology, Ministry of Education, Jiangnan University, Wuxi, China

The odd-numbered fatty acids (OFAs), almost negligible in nature, are potential precursors, such as lubricant agents, medicines for their special properties. OFAs were occasionally prepared in small amounts by oxidation of aldehyde, olefin, or by fermentation route. Those traditional routes have many drawbacks, such as lower conversion, more difficult separation, higher energy consumption, environmental pollution, which hinders the process in large scale. A novel synthesis route of OFAs is developed in this paper from natural fatty acids of one more carbon atom through chlorination, hydrolysis and followed by decarboxylation. The structure of OFAs was confirmed by gas chromatography-mass spectrum (GC-MS) and fourier transform infrared spectrum (FTIR). The corresponding sodium soaps were prepared from both the newly launched odd-numbered fatty acids and the parent even-numbered fatty acids. The experimental results showed that the Krafft points of the odd-numbered soaps were lower or even far lower than that of the parent fatty acids, which means the former has better solubility in cold water. This discovery shed light on formulations of soap-based detergents which were bothered longtime by poor low-temperature solubility of sodium soap derived from natural fatty acids.

(191a) Systems Biology Analysis of the Effect of Acetic Acid On S. Cerevisiae 424A(LNH-ST) and Resistant Strains During the Co-Fermentation of Glucose and Xylose

Nathan S. Mosier, Agricultural and Biological Engineering & Laboratory of Renewable Resources Engineering (LORRE), Purdue University, West Lafayette, IN, Miroslav Sedlak, Agricultural & Biological Engineering & Laboratory of Renewable Resources Engineering (LORRE), Purdue University, West Lafayette, IN and Nancy W. Y. Ho, ChemE/LORRE, Purdue University, West Lafayette, IN

Efficient conversion of hemicellulose-derived sugars to ethanol at high yields and titers are goals toward commercializing cellulosic ethanol production. *S. cerevisiae* 424A (LNH-ST) developed at Purdue University can efficiently ferment glucose and xylose. However, inhibitors present in cellulosic feedstocks (acetic acid) and the desired fermentation product (ethanol) reduce yeast growth rate and fermentation rates, especially during xylose fermentation.

Through selective culturing, we have developed new, evolved strains with improved xylose fermentation compared to the original strain. Under control fermentations (no acetic acid), the new strain consumes xylose (g/g cells/hr) 2.2 times faster than the original strain. When challenged with acetic acid (10 g/L), the rate of xylose consumption drops almost 90% in the control strain compared to a drop of only 47% in the evolved strain. We present the results from a system biology approach to analyzing differences between our original strain and newly developed strains. We report differences between the strains and the effect of acetic acid on both. Data from the concentrations of intracellular metabolites as well as gene expression for multiple points during batch fermentations are analyzed to give insights into changes in the evolved strain.

(191b) Enhanced Ethanol Tolerance in Yeast by Micronutrients

Felix Lam, Chemical engineering, Massachusetts Institute of Technology, Cambridge, MA, Gerald Fink, Whitehead Institute, Cambridge, MA and Greg Stephanopoulos, Chemical Engineering, Massachusetts Institute of Technology, Cambridge, MA

Ethanol will continue to be the dominant mass-market biofuel in the near to intermediate term. The yeast *Saccharomyces cerevisiae* is the catalyst of choice in the conversion of plant sugars to ethanol. One of the primary bottlenecks to higher ethanol production is the toxicity of ethanol itself as well as compounds present in the fermentation medium. Cell death occurs at ethanol concentrations greater than 15-20% (v/v), and hydrolysates of cellulosic biomass (widely considered to be the next generation feedstock for renewable fuels) contain a variety of organic molecules inhibitory to growth. Conditions that increase ethanol tolerance could improve both product yield and the viability of yeast for reuse in subsequent fermentations. However, the biological basis for ethanol tolerance remains elusive: whole genome studies have shown that tolerance is influenced by many genes, and thus far, none have elicited significant improvement when manipulated alone. An alternative approach is to alter extracellular conditions such that cells become less sensitized to high ethanol concentrations. To this end, we have screened and identified several micronutrients which, when added to a fermentation culture, can increase ethanol production in *S. cerevisiae* by ~50%. These modifications to the medium do not require any genetic manipulation of strains and, in fact, lead to increased ethanol output in both laboratory and industrial strains. Furthermore, we have found that the boost in ethanol production is attributable directly to an enhancement of population viability (i.e., tolerance), and occurs in both glucose- and xylose-fermenting strains alike. Our findings simultaneously hint at a novel, univariate determinant of ethanol tolerance and are a step forward in unlocking the economic potential of cellulosic biomass.

(191c) The Impacts of Glucose Repression on Invertase Expression in Various Yeast Strains During the Ethanol Fermentation of Sucrose-Containing Substrates

Lei Zhang, Institute of New Energy Technology, Tsinghua University, Beijing, China and Shi-zhong Li, Institute of Nuclear and New Energy Technology, Tsinghua University, Beijing, China

Recently, the sweet sorghum ethanol using advanced solid-state fermentation (SSF) developed by Tsinghua University in China has been applied in the largest solid-state fermenter (550 m^3) of the world, which makes sugar-based fuel ethanol production more cost-competitive, first demonstrating the SSF for fuel ethanol production is feasible in industrial scale. Although the relatively high ethanol productivity have been achieved, seeking the yeast strains with high fermentation rate is still an important means to further strengthen the cost competitiveness of the process, due to the inherently insufficient mass and heat transfer of SSF. Due to the high contents of sucrose in sweet sorghum and sugarcane, the hydrolysis rate of sucrose by yeast into fermentable sugars, including glucose and fructose, significantly determines the ethanol fermentation rate of sucrose-containing substrates. Invertase is the key enzyme for yeast involved in the cleavage of sucrose, which is subjected to the glucose repression that poses significantly adverse effects on the yeasts' fermentation of mix-sugar substrates by extension of fermentation time. The dynamic changes of glucose levels during the utilization of sucrose-containing substrates make the regulation of invertase expression in yeast more complicated. Thereby, in this study we investigated the impacts of invertase expression on the final fermentation performances of the various *Saccharomyces cerevisiae* strains through monitoring the dynamic variation of specific invertase acivities of yeasts during the fermentation of sucrose (120 g l^{-1} of sucrose) or mixed sugars (70 g l^{-1} of sucrose, 30 g l^{-1} of glucose, and 23 g l^{-1} of fructose). During the cell growth in YPD media with glucose as sole carbon source, the both commercial strains CA and AF showed much higher induction levels of invertase than both CL and AL did when the glucose levels in the media decreased to the threshold value of triggering the invertase induction. In both sucrose and mix sugars media, surprisingly, the both originally invertase-repressed CA and AF showed an obviously increasing trends of invertase activities after 7 h of fermentations even under the glucose repression conditions at the time. Moreover, the both CA and AF showed highest ethanol productivities in sucrose and mix sugars media. Meanwhile, AL showed a stable and relatively high background invertase activity during the whole process of ethanol fermentation even under the impact of glucose repression, which led the high sucrose hydrolysis rate of AL at the early stage of fermentation. On the other hand, ethanol inhibits the induction of invertase to a great extent, and the invertase induction of both CA and AF showed much stronger tolerance to ethanol. All these results suggested that the glucose repression might be not the only mechanism for the regulation of the invertase involved in the process of ethanol fermentation in *S. cerevisiae* but a net effect of multi-factors, such as glucose repression, stress response. Meanwhile, these results can provide us some useful indicators for isolating or developing the yeast strains with high ethanol productivities with sucrose-containing feedstocks as sustrates.

(191d) Effects of Dissolved Inorganic Carbon and Mixing On Growth and Lipid Formation of Chlorella Vulgaris

Jinsoo Kim, School of Energy, Environmental, Biological and Medical Engineering, University of Cincinnati, Cincinnati, OH, Zhouyang Liu, School of Energy, Environmental, Biological, and Medical Engineering , University of Cincinnati, Cincinnati, OH, Joo-Youp Lee, School of Energy, Environmental, Biological, and Medical Engineering, University of Cincinnati, Cincinnati, OH and Ting Lu, Wastewater Treatment Division, Metropolitan Sewer District of Greater Cincinnati, Cincinnati, OH

Microalgae have recently received great interests as a potential alternative biofuel source because they can generate a large amount of triacylglycerols (e.g. 20-50 % dry cell weight) for biodiesel production. *Chlorella vulgaris* has been extensively studied due to its relatively rapid growing rate and reasonable quantity of triacylglycerols (e.g. 14-22 % dry cell weight). Dissolved inorganic carbon including carbon dioxide (CO_2(aq)) and bicarbonate ion (HCO_3^-) is an important carbon source for the autotrophic growth of microalgae. However, the dissolved inorganic carbon concentration in culture mediums is not usually enough for the growth of microalgae in typical open ponds when CO_2(g) is added to the medium due to a narrow pH window suitable for the growth. In this study, sodium bicarbonate ($NaHCO_3$) was used to increase the inorganic carbon concentration. Then, different mixing speeds were employed to investigate the effect of dissolved inorganic carbon concentrations on the algal cell surface. During the entire growth phase, the lipid content in algal cell was also monitored using fluorescence spectroscopy *in vivo* to trace neutral lipid formation.

(191e) High Level Malic Acid Production From Polymalic Acid Fermentation by Aureobasidium Pullulans and Acid Hydrolysis

Xiang Zou Sr., College of Pharmaceutical Sciences, Southwest University, Chongqing, China, **Yipin Zhou**, William G. Lowrie Department of Chemical and Biomolecular Engineering, The Ohio State University, columbus, OH and S.T. Yang, Chemical Engineering, The Ohio State University, Columbus, OH

High level malic acid production from polymalic acid fermentation by *Aureobasidium pullulans* and acid hydrolysis

Xiang Zou[a], Yipin Zhou[b], Shang tian Yang[b*]

(a. *College of Pharmaceutical Sciences, Southwest University, Chongqing 400715, P.R China;*

(b. *William G. Lowrie Department of Chemical and Biomolecular Engineering, The Ohio State University, Columbus, OH 43210, USA*)

Abstract

Malic acid, a C4-dicarboxylic acid, is widely used in the food industry and a potential C4 platform chemical in biorefinery. Malic acid is commercially produced either through hydration of fumaric or maleic acids under high temperature and pressure yielding a racemic mixture or via the enzymatic transformation of fumarate producing L-malic acid by immobilized bacterial cells. However, these processes require expensive raw materials such as petroleum-derivative maleic anhydride or costly enzymes, which have limited their economical competitiveness. Compared with these methods, microbial fermentation using relatively inexpensive substrates is a promising way for economical production of malic acid. However, the low product yield from conventional malic acid fermentation limits further industrialization efforts due to product inhibition. In this work, we developed a novel process for malic acid production from fermentation produced polymalic acid (PMA) followed with acid hydrolysis. First, a high-yield strain *A. pullulans* ZX10 was isolated, which produced PMA at a high titer equivalent to 87.6 g/l malic acid and high productivity of 0.61 g/l h in free-cell fed-batch fermentation. Repeated fed-batch fermentations with cells immobilized in a fibrous-bed bioreactor achieved the highest product titer of 144.2 g/l and productivity of 0.74 g/l h. The fermentation produced PMA was separated by ethanol precipitation and then purified by adsorption with IRA-900 resin, achieving a ~100% purity. Pure malic acid was then produced from PMA by acid hydrolysis. This process provides an efficient and economical way for PMA and malic acid production, and is promising for industrial application.

(191f) Economical Production of Propionic Acid from Renewable Feedstocks by Propionibacteria in a Fibrous-bed Bioreactor

Zhongqiang Wang, Chemical & Biomolecular Engineering, The Ohio State University, Columbus, OH and Shang-Tian Yang, Department of Chemical and Biomolecular Engineering, Ohio State University, Columbus, OH

Propionic acid is an important chemical with various industrial applications. *Propionibacterium freudenreichii* subsp. *shermanii*, typically used as ripening culture for Swiss-type cheese, has been studied for its ability to produce propionic acid and vitamin B_{12}. A *P. shermanii* strain with high propionic acid yield was used for propionic acid production by cells immobilized in a fibrous-bed bioreactor (FBB). Batch and fed-batch fermentations with glycerol and glucose as carbon sources were investigated first. The results showed that the strain was able to convert both glucose and glycerol simultaneously to propionic acid at a high rate and with a high propionic acid yield. Then, the fermentation was studied using cassava bagasse hydrolysate and crude glycerol from biodiesel wastes to demonstrate the feasibility of using low-cost feedstocks to economically produce propionic acid.

(196a) Catalytic Filtration of Pyrolysis Vapor Via Moving Bed Granule Filter

Lysle Whitmer, Center for Sustainable Environmental Technologies, Iowa State University, Boone, IA and Robert C. Brown, Mechanical Engineering, Iowa State University, Ames, IA

Catalytic Filtration of Pyrolysis Vapor via Moving Bed Granule Filter

Lysle Whitmer and Robert C. Brown

Center for Sustainable Environmental Technologies

Department of Mechanical Engineering

Iowa State University

Ames Iowa – 50010 U.S.A.

Abstract

The objective of this project is to simultaneously reduce the amount of particulate matter inherent in bio-oil vapors through hot gas filtration while catalytically deoxygenating the vapors to form stable bio-oil intermediates. Current filtration techniques employed in the production of bio-oil only filter particulate down to the 5-10 micron level, leaving the very fine particles in the vapor stream. Particulate matter is thought to catalyze the polymerization of bio-oil; a process that increases the viscosity and water content of the oil. Additionally, the oxygen content of bio-oil contributes to a number of secondary reactions that also result in the polymerization of bio-oil.

This work focuses on the utilization of a moving bed granule filter to remove fine particulate from a pyrolysis vapor stream using catalytically active filtration media at process temperatures reaching 500°C. The moving bed design features adjustable catalyst residence time allowing for continuous vapor processing under stable operating conditions. This research evaluates the effectiveness of the filter at reducing particulate content in the pyrolysis vapor stream and its ability to deoxygenate the resulting bio-oil.

(196b) A Response Surface Methodology Study On Catalytic Pyrolysis of Biomass with FCC Catalyst

Ofei Mante[1], Foster Agblevor[1] and Ron McClung[2], (1)Biological Engineering, Utah State University, Logan, UT, (2)BASF Catalysts LLC, Iselin, NJ

A response surface methodology was used to investigate the influence of temperature, weight hourly space velocity (WHSV) and vapor residence time on the catalytic pyrolysis of hybrid poplar wood using Y-zeolite based FCC catalyst in a 50 mm bubbling fluidized bed reactor. The levels of the independent variables were: temperature (400 - 600 °C), WHSV (1 - 3 h^{-1}), and vapor residence time (3 - 6 s). The Box-Behnken experimental design was used and a total number of 15 experimental runs including 3 center runs were generated. The analysis of variance (ANOVA) at 95 % confidence interval was performed with Minitab 16 software package and reduced models were generated for the prediction of the responses. It was found that higher levels of WHSV (> 1 h^{-1}) with shorter vapor residence time and moderate temperatures (500 ≤ Temperature < 600) favor higher yield of organic liquid fraction. Also, the employment of higher temperature (> 500 °C) and WHSV (> 2 h^{-1}) reduces the formation of coke/char. The catalytic pyrolysis with the Y-zeolite based FCC catalysts at high temperatures (> 500 °C) and long residence time enhanced the cracking of the primary vapors and increased the formation of CO, H_2 and CH_4. The operation at higher temperatures (> 500 °C) and lower WHSV (≤ 2 h^{-1}) decreased the oxygen content and increased the viscosity of the bio-oil. The reaction temperature was the most influential significant independent variable on char/coke yield, concentration of non-condensable gases, carbon content, oxygen content, pH and viscosity of the bio-oils. The WHSV was found to be the most important significant independent variable that affected the yield of organic liquid and the formation of water. The models for the prediction of the responses with the exception of viscosity were adequate and statistically significant.

(196c) Properties of Olivine Catalysts for Indirect Gasification of Biomass

Remco J. Lancee, Hans O.A. Fredriksson, Peter C. Thüne, Hubert J. Veringa and J.W. (Hans) Niemantsverdriet, Eindhoven University of Technology, Eindhoven, Netherlands

To meet the increasing demand for renewable, CO_2-neutral energy, biomass (in particular waste products) can be used for the production of fuels such as hydrogen, synthetic natural gas or Fisher-Tropsch diesel [1]. Conversion of biomass to such fuels can be performed using an indirect gasification process, where gasification and partial combustion (necessary to sustain the endothermic gasification process) take place in separate chambers. Heat generated by combustion of chars and tars derived from the biomass is transferred to the gasification chamber by a bed material, conventionally sand, that is circulated between the two chambers. By exchanging the inert sand with a catalytically active bed material it is possible to improve the efficiency of the gasification process and to reduce the content of polluting tars in the product gas. A widely used and studied catalytic bed material is the naturally occurring mineral olivine ($(Mg,Fe)_2SiO_4$). Although the exact catalytic function of the olivine is not yet fully understood, it has been observed that pretreatments can improve the performance of this catalyst[2] and that the phase composition of the mineral as well as its activity change over time, under typical process conditions[3].
We have used a flow reactor to mimic the conditions that olivine catalysts are exposed to during gasification of biomass. Oxidation (20% O_2 in Ar) and reduction (CO, H_2 and H_2O mixtures) treatments at 750°C were applied. The resulting changes in the olivine catalyst's bulk composition were characterized using X-ray Absorption Spectroscopy (XAS) and X-Ray Diffraction (XRD). Surface chemistry and morphology changes were characterized using X-ray Photoelectron Spectroscopy (XPS) and electron microscopy. The main changes in the catalyst after exposure to high temperatures, necessary for biomass gasification, concern the iron present in the olivine. After exposure to the oxidizing environment, the olivine catalysts exhibit a significant increase of surface Fe and a simultaneous decrease of Si. Furthermore, Fe_2O_3 and Fe_3O_4 phases are observed to form from Fe segregated out of the bulk mineral. On the catalyst surface, formation of crystallites is observed and the amount of Fe, the most dynamic element, at the surface is more than doubled. The catalyst entering the gasification zone is thus best described as a largely Fe-depleted olivine with the inclusion of FeOx phases at the surface.
After exposure to reducing conditions (H_2+CO+H_2O) the FeOx phases are reduced to FeO and Fe^0, both in the bulk and at the surface. The surface content of Fe is reduced to roughly half the amount present after oxidation and the surface crystallites decrease in size. In CO containing gases significant carbon deposition is observed, some of this in the form of nanofibers.
These results suggest that Fe segregates reversibly to and from the surface under oxidizing and reducing conditions and on time scales which are relevant in commercial plants.
In addition, characterization of the olivine samples using XPS depth profiling and Thermo Gravimetric Analysis (TGA) provided further insight in the dynamic behavior of iron in olivine and its chemical looping abilities under relevant process conditions.

[1] A. V. Bridgwater, A. J. Toft and J. G. Brammer, Renewable & Sustainable Energy Reviews 6, 181-248 (2002)
[2] L. Devi, K. J. Ptasinski and F. Janssen, Industrial & Engineering Chemistry Research 44, 9096-9104 (2005)
[3] J. N. Kuhn, Z. K. Zhao, L. G. Felix, R. B. Slimane, C. W. Choi and U. S. Ozkan, Applied Catalysis B-Environmental 81, 14-26 (2008)

THERMOCHEMICAL BIOMASS PROCESSING FOR H_2 AND BIODIESEL PRODUCTION

Richa Tungal, Rajesh V. Shende
South Dakota School of Mines & Technology, Rapid City, SD, USA

Abstract

This paper reports subcritical aqueous phase reforming of pine wood in presence of homogeneous $Ni(NO_3)_2$ for the production of H_2 and oxygenated hydrocarbons. Slurry of 10-20 wt% pine wood was prepared in distilled water containing 7 wt% $Ni(NO_3)_2$ and loaded in high-temperature high-pressure PARR reactor and heated to 250°C. The reaction progress was monitored for 60 min at 250°C and while the reaction was in progress, gas samples were withdrawn at regular time intervals and analyzed using Chrompak capillary column on the GC equipped with thermal conductivity detector. The analysis of gas samples indicated the presence of H_2, CO_2, CO, and CH_4. Specifically, after 60 min of reaction 13.24 vol% of H_2 was observed. The liquefied biomass (biocrude) obtained was analyzed using HPLC and found to contain sugars (glucose and xylose), HMF, furfurals and oxygenated hydrocarbons such as aldehydes, ketones and organic acids. Among organic acids produced, lactic acid was produced in higher amounts as compared to propionic, formic and acetic acids. The bio-oil was recovered and further reacted with methanol in presence mesoporous NiO material prepared by sol-gel.

Introduction

H_2 and biocrude can be produced from thermochemical treatment of biomass using several technologies such as gasification [1], pyrolysis [2], and aqueous phase reforming using supercritical [3] and subcritical [4] water. Among all the above mentioned processes for converting biomass to valuable fuels, use of subcritical or supercritical water for the aqueous phase reforming of biomass is very advantageous as it eliminates the need for the drying of the biomass. Supercritical water operates at high temperatures of > 374°C and high pressures of > 22.1 MPa. Therefore this process is less attractive for the conversion of biomass into fuels on industrial scale. Contrasting to this, subcritical water processing of biomass is getting more attention as it operates at low to moderate temperatures of 125-350°C and pressures of 0.5-21 MPa [5]. Subcritical water is a non-toxic, inexpensive and environmentally friendly solvent and has been used as the reaction medium for the liquefaction of biomass. Subcritical water has higher ionization constant ($Kw = 10^{-11}$) near the critical point and behaves as acid/base catalyst. Ionization constant of water increases with temperature and hence helps in the solubilization of lignocellulosic biomass components [6].

Recently, Cheng et al., [7] investigated rapid conversion of switchgrass in subcritical water under pyrolysis conditions at 250- 350°C and 20 MPa and reported significant H_2 (0.17 mol %) in the gaseous products. In another study, paper waste was processed under subcritical water conditions in presence of charcoal supported ruthenium catalyst at 250-300°C. This study reported production of CH_4, H_2 and other gases. Many other investigations reported production of different fuels under various hydrothermal conditions. Minowa et al., [8] investigated gasification of cellulose at 350°C in presence of a Ni catalyst and observed product gas containing H_2. Antal et al., [9] used glucose as potential feedstock for H_2 production via steam reforming in presence of different metal oxide supported metal catalyst (Ni/SiO_2) and reported H_2 rich syn gas. Hydrothermal liquefaction of empty palm fruit bunch was performed in alkaline hot compressed water and found to contain oxygenated hydrocarbons such as benzoic acid, methyl ester and phenols in the biocrude [10]. Pourali et al., [11] reported production of organic acids from rice bran at 360°C.

Most of these studies were focused on H_2 generation from model compounds, however, very little effort is noticed in the literature on the production of high energy biocrude from real biomass at low temperatures. In this paper subcritical aqueous phase reforming of pine wood was performed using homogeneous Ni-catalysts at 250°C. The reaction was monitored for 60 min and while the reaction was in progress, the gas and liquid samples were withdrawn and analyzed using GC and HPLC, respectively. The results obtained on H_2 volume generation and biocrude analysis are presented and discussed.

Experimental Methods and Analysis

Materials

$Ni(NO_3)_2.6H_2O$ was purchased from Alfa Aesar and and used without purification. N_2 from a cylinder with the minimum stated purity of 99% was purchased from Linweld Inc., SD. Pine wood was made into slurry in distilled water and used for subcritical hydrothermal processing in N_2 environment.

Experimental Set-up and Procedure

High-temperature high-pressure PARR SS316 reactor of 300 ml capacity equipped with heating and cooling assembly, gas inlet and outlet valves, sampling valves, pressure gauge, safety rupture disc, thermocouple, magnetic stirrer and temperature indicator controller was used to perform subcritical hydrothermal processing of pine wood. This reactor is rated for maximum working temperature and pressure of 350°C and 3000 psi, respectively. An additional SS gas sampling vessel was mounted on the reactor. The entire reactor set-up is shown in Figure 1.

Accurately weighed 15.0 g of pine wood was made into slurry in 150 ml distilled water containing 7wt% $Ni(NO_3)_2$ (taken on a biomass weight basis) and loaded in the high-temperature high-pressure PARR reactor. The contents in the reactor were flushed continuously with N_2 for 10-15 min to replace air and pressurized to 40 psi. The reactor was heated to 250°C under continuous magnetic stirring of biomass slurry at 1300 rpm. It took approx. 45 min to attain the set point of 250°C where the total reactor pressure was about 700 psi. Once the reactor temperature was attained, a gas sample was withdrawn. This time was taken as 'zero' time for the reaction. The reaction was continued for 60 minutes and during the course of the reaction, gas samples were withdrawn at regular time intervals and analyzed using GC.

Product Analysis

Gas samples were withdrawn with a gas tight syringe during the subcritical aqueous phase reforming of pine wood and analyzed using Agilent 6890A GC equipped with Chrompak capillary column and thermal conductivity detector. A sample injection volume of 100 μl was used for the analysis. The GC analysis was performed under the isothermal conditions where column and detector temperatures were maintained at 105°C and 200°C, respectively.

At the end of the reaction, the reactor contents were cooled down to room temperature and centrifuged to collect biocrude and biochar. The filtrate was analyzed with HPLC to analyze sugar and oxygenated hydrocarbons.

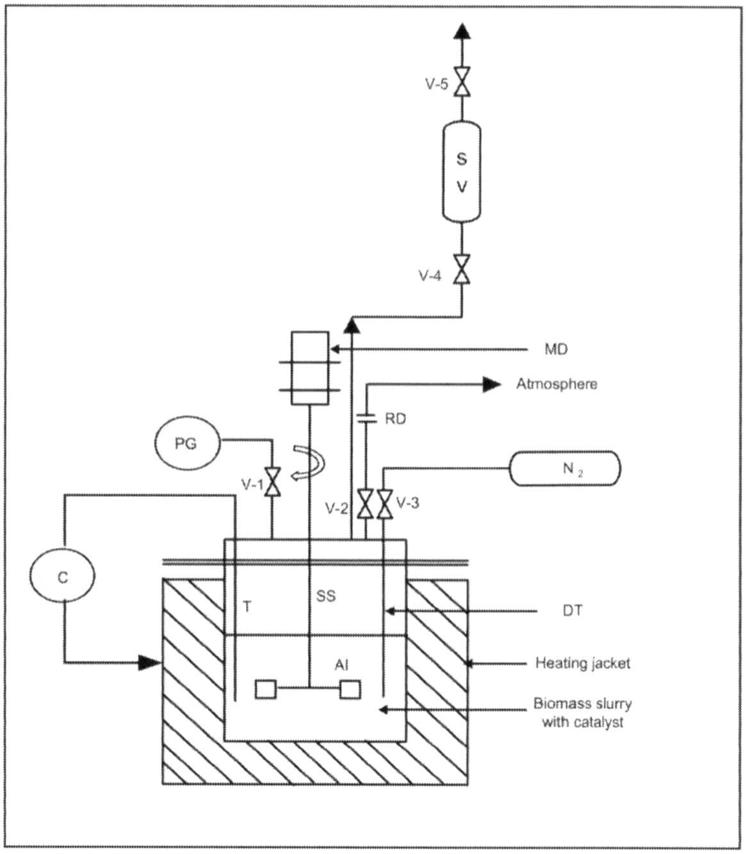

Fig. 1: Reactor set-up for subcritical hydropyrolysis of a biomass for H_2 generation. (RD-rupture disc, T-thermocouple, AI-adjustable impeller, SS-stirring shaft, PG-pressure gauge, MD-magnetic drive, V-valve, DT-dip tube, C-controller, SV-gas sampling vessel).

Results and Discussion

The product gas obtained at 250°C indicated primarily H_2, CO, CH_4 and CO_2 gases. In addition to these gases, H_2O vapor should also be present in the gas samples, however, as the gas sampling vessel was held at ambient conditions, the water vapors might have been already condensed. Moreover the Chrompak capillary column that was used for the analysis of product gases was not capable of detecting water vapors. Therefore, the concentrations were determined without accounting for H_2O vapor in terms of vol% of total headspace volume, which was 150 ml. It was observed that H_2 production from pine wood was negligible at 225°C whereas as temperature increased to 250°C, the H_2 production was observed at 'zero' reaction time. The H_2 vol% obtained is shown in Figure 2, which indicated 13.24 vol% after 60 min reaction time. The product gas contained significant amount of H_2 and CO_2 gases, lower amount of CO and negligible amount of CH_4. Along with H_2 in product gas, CO_2 was also increased with the increase in reaction time up to 60 min during aqueous phase reforming of pine wood in presence of 7 wt % Ni^{2+} catalyst at 250°C. This trend of H_2 production from pine wood suggested that H_2 levels will further increase linearly if the reaction would have been continued beyond 60 min.

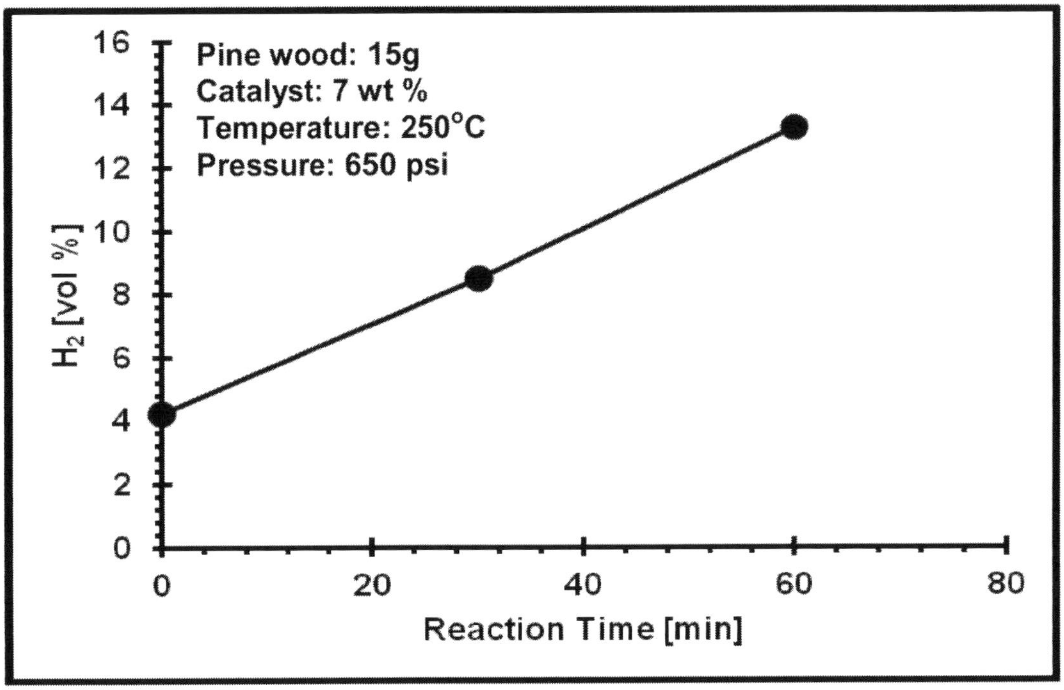

Fig. 2: Production of H_2 during subcritical aqueous phase reforming of pine wood at 250°C in presence of 7 wt % $Ni(NO_3)_2$.

After the completion of reaction, the reactor was cooled down to room temperature. The liquid slurry left after the reaction was centrifuged and the supernatant was separated from biochar. The char formation, however, was not studied in detail as it was not the goal of this investigation. The supernatant is called as biocrude (Figure 3); these biocrude samples were collected at an interval of 30 min. It can be seen from Figure 3 that at 250°C and 'zero' reaction time, pine wood was solubilized in presence of 7 wt % $Ni(NO_3)_2$ catalyst during aqueous phase reforming of pine wood. This shows that $Ni(NO_3)_2$ was not only capable of producing H_2 but was also effective in breaking down the complex wood structure. The biocrude samples collected at different time intervals were filtered and the supernatant was extracted in different solvents such as ethyl acetate, THF and ethyl ether. These extracts were evaporated on rota evaporator and the residue obtained was analyzed using FT-IR. The FT-IR results showed the presence of oxygenated hydrocarbons such as alcohols, aldehydes, ketones, phenols, esters and carboxylic acids in all the extracts.

Fig. 3: Biocrude obtained during subcritical aqueous phase reforming of pine wood at 250°C in presence of 7 wt % $Ni(NO_3)_2$ at different time intervals.

The biocrude obtained during aqueous phase reforming of pine wood at 250°C in presence of $Ni(NO_3)_2$ was analyzed using HPLC and the results obtained are shown in Table 1. The HPLC analysis of the biocrude showed the production of sugars such as glucose and xylose, HMF, furfural and oxygenated compounds such as lactic acid, formic acid and propionic acids. Among various organic acids produced, the production of lactic acid was higher as compared to formic, acetic and propionic acids in the biocrude. This shows that $Ni(NO_3)_2$ catalyst was effective for the production of H_2 as well as valuable chemicals such as oxygenated hydrocarbons in the biocrude obtained during catalytic liquefaction of pine wood.

Table 1. HPLC analysis of the biocrude obtained after aqueous phase reforming of pine wood at 250°C in presence of 7 wt % $Ni(NO_3)_2$

Compounds	Concentration (mM)
Glucose	0.146
Xylose	0.130
Lactic acid	84.626
Formic acid	6.446
Propionic acid	54.452

These organic acids are usually obtained from the catalytic hydrolysis of carbohydrates or sugars under subcritical water reactions. Glucose and fructose have been known to produce organic acids such as acetic, formic at 300-400°C [12]. Bicker et al., [13] reported 42 to 86 % conversion of aqueous sugar solution into lactic acid. The reaction for the production of lactic acid proceeds through a series of reactions where sugars (glucose and fructose) first get converted into glyceraldehydes, later to pyruvaldehyde and finally to lactic acid. Lactic acid, due the presence of bifunctionality, can be converted into other compounds such as propionic acid and H_2 through a series of reactions such as acetaldehyde can be produced via decarboxylation/decarbonylation, propionic acid via reduction, 2,3-pentanedione via condensation etc. [12]. This infers that the production of H_2 might have been routed from the degradation reactions of sugars as well as from organic acids present in the biocrude obtained during the aqueous phase reforming of pine wood under subcritical water in presence of 7 wt% $Ni(NO_3)_2$ catalyst. Based on these findings, the reaction pathways can be proposed as shown by scheme 1.

Biocrude obtained from the catalytic aquoeus phase reforming of biomass under subcritical water conditions can be converted to liquid fuels such as biodiesel. Many researchers have performed the preliminary studies on the hyrothermal treatment of biomass in hot compressed water and indicated that it is more beneficial for the production of liquid fuels as compared to conventional pyrolysis. The liquid obtained after aqueous phase reforming of biomass consists of a complex mixture of more than 300 oxygenated compounds, carbohydrates (sugars) and lignin fragments. During aqueous phase reforming, a large portion of the oxygen from biomass is removed in the form of CO_2 and the biocrude left after the reaction contains only 10-15 % of oxygen as compared to woody biomass where it contains almost 40 % of the oxygen content. However, due to the high oxygen content of biocrude, it is required to be upgraded further and the deoxygenation of biocrude can be performed over zeolites or mesoporous metal oxide catalyts to produce liquid fuels. In one of the studies by Brahmkhatri et al., [14] free fatty acids were used for the production of biodiesel in presence of heterogeneous MCM-41 catalyst. The biocrude obtained during aqueous phase liquefaction of pine wood in presence of homogeneous $Ni(NO_3)_2$ that contained fatty acids can be converted into biodiesel range hydrocarbons using heterogeneous catalyst such as NiO.

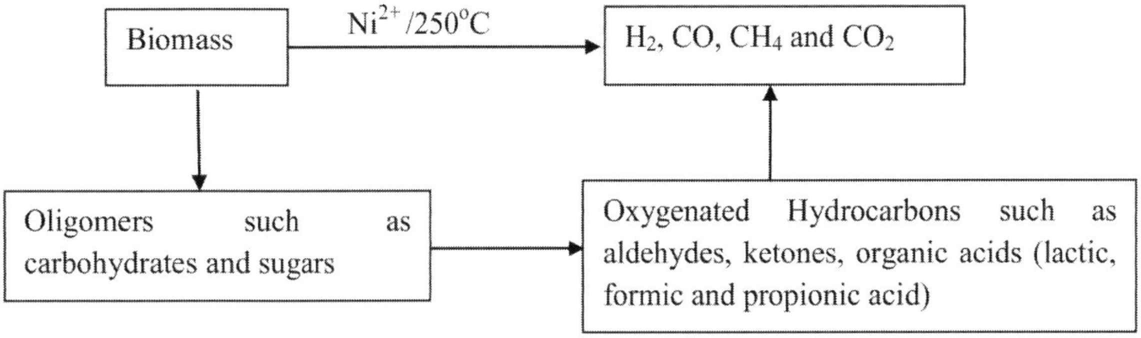

Scheme 1. Possible reaction pathways during catalytic aqueous phase reforming of pine wood at 250°C.

NiO synthesis involved dispersion of metal salt ($NiCl_2$) in ethanol containing brij-76 surfactant. To this visually clear solution, water and propylene oxide were added to achieve the gel formation. The gel was aged, dried, and calcined at 600°C. The calcined powder was characterized using SEM and TEM and found to contain mesopores in the range of about 6-50 nm. This mesoporous NiO catalyst will be used for the conversion of acids present in the biocrude obtained from the subcritical aqueous phase reforming of biomass for the production of liquid fuels such as biodiesel range hydrocarbons. The detailed study of the production of biodiesel range hydrocarbons from biocrude in presence of heterogeneous mesoporous NiO catalyst is in progress. The SEM of mesoporous NiO is shown in Figure 4.

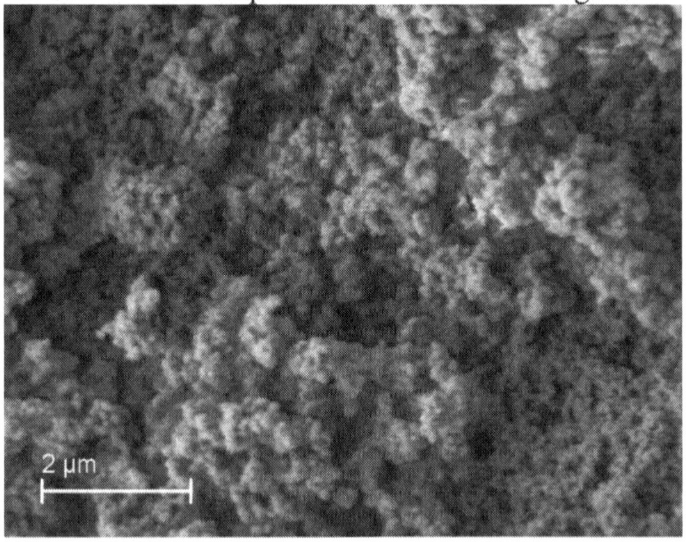

Fig. 4: SEM of NiO made with sol-gel using brij-76 surfactant and calcined at 600°C.

Conclusion

The subcritical aqueous phase reforming of pine wood was successfully performed in high-temperature and high-pressure PARR reactor at 250°C in the presence of homogeneous Ni-catalyst $Ni(NO_3)_2$. Their product gas profiles indicate H_2 rich gas containing CO_2 and CO and CH_4 gases. The concentration of CH_4 was found to be negligible. The maximum H_2 of 13.24 vol% was produced from pine wood using $Ni(NO_3)_2$ catalyst after 60 min of reaction at 250°C. The liquid phase obtained after the completion of reaction was centrifuged and the supernatant was analyzed using HPLC and FT-IR. The liquid phase was found to contain sugars, HMF, furfural and oxygenated hydrocarbons such as lactic, formic and propionic acid where lactic acid was found to be in higher amounts as compared to other acids. It is believed that the production of H_2 has been routed from the production of sugars as well as organic acids during aqueous phase reforming of pine wood under subcritical water reaction at 250°C in presence of $Ni(NO_3)_2$ catalyst. It can be concluded that $Ni(NO_3)_2$ is effective not only in the decomposition of complex lignocellulosic biomass but also the generation of H_2 as well as organic acids in the biocrude which could be converted into liquid fuels such as biodiesel range of hydrocarbons using heterogeneous catalyst such as NiO.

Acknowledgement

The authors wish to acknowledge the financial supports by the Chemical and Biological Engineering Ph.D. support grant and Tyndall AFRL.

References

1. Minowa, T., Inoue, S., (1999), "Hydrogen production from biomass by catalytic gasification in hot compressed water," *Renewable Energy* 16, pp. 114-117.
2. Dalai, A.K., Bakhshi, N.N, Valliyappan, T., (2008), "Pyrolysis of glycerol for the production of hydrogen or syn gas," *Biresource Technology*, 99, pp. 4476-4483.
3. Davda, R.R., Dumesic, J.A., (2002), "Hydrogen from catalytic reforming of biomass-derived hydrocarbons in liquid water," *Nature*, 481, pp. 964-967.
4. Onwudili, J.A., Williams, T.P, (2007), "Hydrothermal catalytic gasification of municipal solid waste, '' *Energy & Fuels,* 21, pp. 3676-3683.
5. Khajavi, S.H., Kimura, Y., Oomaori, T., Matsuno, R., Adachi, S., (2005), ''Degradation kinetics of monosaccharides in subcritical water,'' *Journal of Food Engineering,* 68, pp. 309-313.
6. Kumar, S., Gupta R.B., (2009), "Biocrude Production from Switchgrass Using Subcritical Water," *Energy Fuels*, 23, pp. 5151–5159.
7. Cheng, L., Ye, P. X., He, R. and Liu, S. (2009), "Investigation of rapid conversion of switchgrass in subcritical water," *Fuel Processing Technology,* 90, pp. 301-311.
8. Minowa, T. and Inoue , S. (1999), "Hydrogen production from biomass by catalytic gasification in hot compressed water," *Renewable Energy*, 16, pp. 1114 1117.
9. Antal, M.J, Jr., Manarungson, S, Mok, WS-L. (1994) "Hydrogen production by steam reforming glucose in supercritical water," *Adv. Thermochemical Biomass Conversion.,* pp. 367-77.
10. Akhtar, J., Kuang, S.K., Amin, A.S., (2010) *Renewable Energy*, 35, pp. 1220-1227.
11. Pourali, O., Asghari, F.S., Yoshida, H., (2009), "Sub-critical water treatment of rice bran to produce valuable materials," *Food Chemistry*, 115, pp. 1-7.
12. Serrano-Ruiz, J.C., West, R.M., Dumesic, J.A., (2010), "Catalytic conversion of renewable biomass resources to fuels and chemicals," Annu. Rev. Chem Biomol. Eng., 1, pp. 79-100.
13. Bicker, M., Endres, S., Ott. L., Vogel, H., (2005), "Catalytic conversion of carbohydrates in subcritical water: A new chemical process for lactic acid production," *Journal of Molecular Catalysis A: Chemical*, 239, pp. 151-157.
14. Brahmkhatri, V., Patel, A., (2011), "Biodiesel Production by Esterification of Free Fatty Acids over 12-Tungstophosphoric Acid Anchored to MCM-41," *Ind. Eng. Chem. Res.,* 50, pp. 6620–6628.

(196e) In-Situ Treatment of Pinewood Pyrolysis Vapors Using Zeolites

Vamshi Krishna Guda[1], Hossein Toghiani[2], Rebecca K. Toghiani[2] and Charles U. Pittman Jr.[3], (1)Mississippi State University, Mississippi State, MS, (2)Dave C. Swalm School of Chemical Engineering, Mississippi State University, Mississippi State, MS, (3)Department of Chemistry, Mississippi State University, Mississippi State, MS

Pine wood pyrolysis vapors were treated in-situ using zeolite catalysts. An auger fed reactor was used for the pinewood pyrolysis, while a packed bed reactor mounted on the top of auger reactor housed the catalyst was used to treat the pyrolytic vapors. The pyrolytic vapors produced at 450 ^oC were passed through zeolite catalysts maintained at 450 ^oC at a weight hourly space velocity (WHSV) of 12 h^{-1}. Seven zeolites, including three HZSM-5 catalysts with differing SiO_2/Al_2O_3ratio (23, 55, and 280), as well as mordenite, ferrierite, HY, and beta, were used as catalysts. The effect of catalyst type on deoxygenation of pyrolysis vapors was studied. Chemical characterization of the upgraded bio-oils was performed using GC-MS. Water content, density, viscosity, acid value, pH, and elemental composition of each upgraded bio-oil was also measured. Catalyst characterization was performed using XRD, TEM, and TGA.

Catalytic treatment of pyrolysis vapors decreased the liquid yield at the expense of gaseous product. While all the zeolites deoxygenated the pyrolysis vapors, HZSM-5 catalysts were most effective. The HZSM-5 upgraded bio-oils were less viscous, had low acid number, high pH, and an oxygen content half that of untreated pyrolysis oil. HZSM-5 catalysts also produced the highest amount of phenolics and aromatic hydrocarbons. Acidity, pore size, and the shape selective catalysis of HZSM-5 catalysts were responsible for their high activity. TG analysis revealed that the pore size of the catalysts highly influenced the coking reactions. Catalysts were successfully regenerated.

(196f) Mixed Rare Earth Oxides for Desulfurization and Tar Removal From Gasifier Effluents

Rui Li[1], Michael J. Janik[2], Matthew D. Krcha[2] and Kerry M. Dooley[1], (1)Chemical Engineering, Louisiana State University, Baton Rouge, LA, (2)Chemical Engineering, Pennsylvania State University, University Park, PA

The removal of H_2S and tars from syngas generated in biomass or coal/biomass gasification plays an important role in syngas cleanup. Rare earth oxides (REOs, e.g., Ce/LaO_x) mixed with transition metals (e.g., Mn, Fe, Ni) are promising materials for both sulfur removal and for tar reforming in the presence of some sulfur. We have synthesized such materials by a variety of methods, including templated sol-gel, hydrothermal, incipient wetness coimpregnation, and evaporation-induced self assembly (EISA). These materials have been investigated for either H_2S adsorption or tar reforming at high temperatures (~630-750°C), using model syngas mixtures, a goal being to improve heat integration throughout the syngas process. Successive desulfurization-regeneration cycles using $Mn/Ce/LaO_x$ demonstrated good sulfur capacities can be obtained even in the presence of water and CO_2, using just air for the regenerations. The limits of Mn incorporation into the REOs without bulk phase separation have been studied, and the sulfur adsorption has been modeled using DFT. Naphthalene was used as a model compound for tar reforming/cracking, and the effects of H_2S, CO, water and CO_2 on its reactions have been examined. While H_2S, water and CO_2 are inhibitors, CO is not. Transition metal-doped REOs are more effective than the pure REOs, suggesting an important role of the transition metal in increasing the number of oxygen vacancies. We have also examined propane reforming in model syngas, because its surface behavior is easier to model by DFT. Fresh, used and regenerated sorbents and catalysts have been characterized by BET, TPR, XRD, XANES and EXAFS.

(203a) Thermal Reduction of Magnetically Stabilized Ferrites Under Reduced Pressure

Nicholas AuYeung, Ayyoub Mehdizadeh, Fadi Alnaimat, Fotouh Al-Raqom, Benjamin Greek, Kyle Allen, Amey Barde, Rishi Mishra, Nikhil Seghal, Jinchao Lu, Abhishek Singh and James Klausner, Mechanical and Aerospace Engineering, University of Florida, Gainesville, FL

Oxidation of a metal (or metal oxide) by H_2O and CO_2 followed by a regenerative high temperature thermal reduction of the resulting metal oxide is a promising method for using concentrated solar energy to produce syngas, a precursor for liquid fuels. For such a process to become viable, the metal oxide material that is formed must have the ability to maintain surface area and retard sintering over thousands of redox cycles. Previous experiments have already shown that a magnetically stabilized iron-silica porous structure can significantly enhance hydrogen production via oxidation by H_2O and chemical reduction by CO at 800°C. [1]

Here, the feasibility of thermally reducing the mixed-metal ferrite structure at temperatures up to 1400°C under vacuum is investigated. Reducing the reactor pressure effectively decreases the partial pressure of O_2, resulting in a higher extent of reduction than is possible under atmospheric conditions. Since the need for a diluting inert gas to increase product yield is eliminated, a higher theoretical higher heating value efficiency is possible.

In addition to laboratory experiments, a reduced pressure solar thermochemical reactor capable of utilizing the novel mixed-metal ferrite structure has been designed. Results of testing performed using the University of Florida 42 kW solar simulator will be presented.

[1]Mehdizadeh AM, et al., Enhancement of thermochemical hydrogen production using an iron-silica magnetically stabilized porous structure, International Journal of Hydrogen Energy (2012), doi:10.1016/j.ijhydene.2012.02.189

(203b) The Solar Thermochemical Reduction of CO_2 to CO Via the Heterogeneous Oxidation of Zn(g)

Luke J. Venstrom and Jane H. Davidson, Department of Mechanical Engineering, University of Minnesota, Minneapolis, MN

The heterogeneous oxidation of Zn(g) is a promising means of sustainably reducing CO_2 to CO in the two-step Zn/ZnO solar thermochemical cycle. Prior efforts on CO_2 reduction in the cycle have focused on the oxidation of solid or liquid Zn, and elucidate the need for faster kinetics. The rate of the oxidation of solid and liquid Zn is limited by solid-state diffusion through the ZnO layer that forms on the Zn surface. To eliminate the solid-state diffusion barrier, we propose reducing CO_2 to CO via the heterogeneous oxidation of Zn(g). The Zn(g) oxidation pathway promises complete and raid conversion of Zn, which ultimately increase cycle thermal efficiency. Cycle efficiencies of 30% are possible even when additional energy must be supplied to vaporize zinc, and the cycle efficiency could be higher if heat recuperation was implemented.

Experiments were conducted in a tubular flow reactor maintained at atmospheric pressure to elucidate the kinetics of Zn(g) oxidtion. The rate of the heterogeneous oxidation of Zn(g) was measured gravimetrically. Temperatures were varied between 800 and 1150 K, and Zn(g) concentrations up to 57 mol% were achieved, extending the range previously considered in the literature. A range of reacting gas compositions were considered with CO_2-to-Zn molar ratios ranging between 1 and 60. CO production was monitored with a Raman laser gas analyzer. A numerical model decouples the surface kinetics and the transport of reacting species in the gas phase from the global reaction rate.

Figure 1 shows the measured and predicted rates of the heterogeneous oxidation of Zn(g) along the axis of the tube flow reactor for temperatures between 800 and 1150 K. The oxidation of Zn(g) by CO_2 is rapid, on the order of 10^{-8}–10^{-5} mol cm^{-2} s^{-1}, representing at least an order of magnitude increase over the rates typically observed in the interface-controlled regime of Zn(s) particle oxidation. The best fit between the reaction rate data and the numerical model is obtained using second-order kinetics, with Zn(g) and CO_2 both behaving as first-order reactants. The activation energy for the Arrhenius reaction rate constant is 44±3 kJ/mol and the preexponential factor is $(92\pm6)\times10^{-3}$ mol cm^{-2} s^{-1} atm^{-2}. As a result of the rapid rate of oxidation of Zn(g), less than one second is required to convert more than 85% of the Zn to ZnO.

The present work confirms that substantial improvements in the thermal efficiency of the Zn/ZnO thermochemical cycle for reducing CO_2 to CO are possible if the critical Zn oxidation step were to utilize the heterogeneous oxidation of Zn(g). Additionally, the kinetic expression obtained for the heterogeneous oxidation of Zn(g) is suitable for reactor design and scale-up since it reflects the intrinsic surface kinetics, and is not influenced by mass transfer.

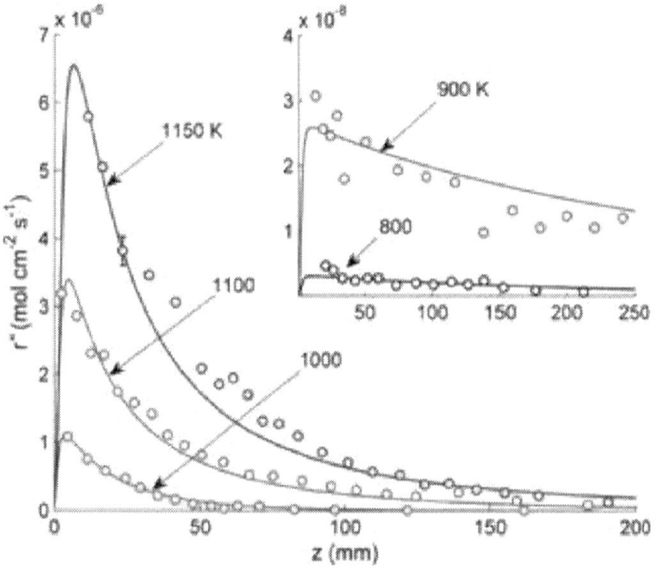

Figure 1 The rate of the heterogeneous oxidation of Zn(g) as a function of axial position in the tube flow reactor. The points are the experimental data and the lines are the optimized model.

(203c) Two Step Solar Thermochemical CO2 Splitting with CoFe2O4 On Porous Al2O3 Nanostructure

Darwin Arifin[1], Victoria J. Aston[2], Xinhua Liang[2], Anthony H. McDaniel[3] and Alan W. Weimer[2], (1)Chemical and Biological Engineering, University of Colorado Boulder, Boulder, CO, (2)Chemical and Biological Engineering, University of Colorado at Boulder, Boulder, CO, (3)Sandia National Laboratories, Livermore, CA

Projected growth in global population and continued industrialization of developing countries will increase total world energy consumption by 50% in the next three decades. This increased demand for energy will be largely met by increased fossil fuel consumption, thereby increasing anthropogenic carbon in the atmosphere and further fuelling geopolitical conflicts over control of dwindling energy resources. Recycling CO_2 by splitting it in a solar-based thermochemical process is an attractive solution to both of these 21st century problems.

In this work, we examine a novel chemistry for a two-step, non-volatile metal oxide CO_2 splitting cycle that utilizes the change in iron oxidation states ($Fe^{2+/3+}$) between $CoFe_2O_4$ and $FeAl_2O_4$ spinel compounds within a nano-engineered material. This chemistry is dramatically different than current metal oxide cycles that exploit oxygen non-stoichiometry in ceria or solid solution behavior in ferrites. The engineered material was prepared using atomic layer deposition and maintained structural integrity over 10 heating cycles under conditions that mimic a concentrated solar power application, namely an oxidation temperature of 1000°C, reduction at 1460°C, and a heating rate of 16°C/s from low to high temperature. Oxygen uptake and release behavior was similar to that of ceria. Raman spectroscopy was used to verify cycle chemistry.

(203d) Vision-Based Solar Irradiation Prediction for the Control of Solar-Thermal Reactors

Elizabeth Saade Saade[1], David E. Clough[1] and Alan W. Weimer[2], (1)Department of Chemical and Biological Engineering, University of Colorado, Boulder, CO, (2)Department of Chemical and Biological Engineering, University of Colorado at Boulder, Boulder, CO

One of the biggest challenges that solar thermochemical processes face is dealing with the intermittent nature of solar irradiation. Changes in solar irradiation directly affect the performance of solar-thermal reactors causing unnecessary shutdowns and start-ups, complications in the purification processes downstream of the reactor, and damage to the reactor materials due to thermal shock. Being able to maintain a continuous high performance operation, even in the presence of passing clouds, is one of the main concerns of the feasibility of solar-thermal processes.

In order to address these issues, a control system that manipulates the flow rates into the reactor is required. The proposed control system is a model predictive controller, which makes use of a model of the system in order to determine the manipulation required to minimize the error between the process and a reference trajectory. In addition to a model of the system and current measurements of the controlled variables, the model predictive control system requires knowledge of the disturbances. In this case, the available solar irradiation is treated as a measured disturbance and it is estimated using images of the sky. In order to do this, a digital charge coupled device (CCD) camera (YES Total Sky Imager TSI-880) captures images of the sky and analyzes them to determine the presence of clouds and their characteristics (density of white, speed at which they are moving, time before they cover the sun, etc.). This information is combined with other atmospheric measurements, such as wind speed and relative humidity, and used to estimate the solar irradiation at future steps. The predictions of the disturbances are used, in combination with a dynamic model of the process, to determine the required manipulation at every time step. In this paper, the algorithm used to determine the presence of clouds is described, as well as the method used to estimate the solar irradiation at future time steps.

(203e) Transient Heat and Mass Transfer in a Reacting Cerium Dioxide Medium Exposed to Direct High-Flux Irradiation

Daniel Keene, Jane H. Davidson and Wojciech Lipinski, Department of Mechanical Engineering, University of Minnesota, Minneapolis, MN

A numerical heat and mass transfer model is presented for a porous medium of cerium dioxide exposed to direct high-flux irradiation and undergoing thermal reduction. The model couples conduction, convection and radiation heat transfer modes to the chemical kinetics of the non-stoichiometric reduction of cerium dioxide while allowing for local thermal non-equilibrium between the solid and gas phases. The Rosseland diffusion approximation is applied to solve for internal radiative transport in the medium, with the spectral radiative characteristics obtained for selected morphological configurations of the medium with high porosities and polydispersed cerium dioxide grains. The finite volume method with an implicit time integration scheme are applied to solve the mass and energy equations in space and time to obtain transient temperature distribution and the extent of the chemical reaction. The results show the impact of grain/pore size distribution on heat and mass transfer rates, and consequently on the temperature distribution and the extent of the chemical reaction.

(219a) "Science Looks to Unlock Potential In the Undomesticated Tree Through Precise Genetic Breeding"

Lawrence Sullivan, Bioenergy, ArborGen, Inc., Ridgeville, SC

Scientific advances are unlocking ever more significant uses for the components of trees. In addition to the wood, pulp for paper and wood for fire, increasing demands are being made of trees for fine grade cellulosic sugars that can be converted to biofuels, polyethylene and other compounds for industrial use. Yet the tree itself remains an undomesticated plant. Far from its agricultural counterparts which have been bred for cold tolerance, drought resistance, improved yields and improved food quality, the tree has only been bred for a few generations and predominately to improve form for processing lumber.

ArborGen uses advance breeding techniques to shorten the time required to reach domesticated species for industrial purposes. Varietal propagation allows us to replicate one species to improve consistency. Mass Controlled Pollination allows us to improve a line by controlling one parent in a breeding pair. However, demands for wood cellulose are growing at a rate of acceleration that can never be matched at the conventional breeding level for a species that takes more than 20 years to mature. ArborGen scientists use precision breeding techniques to look at specific ways of improving and domesticating trees for industrial uses.

One excellent example is a lignin-modified *Eucalyptus* species that has proven to release more than twice the usual amount of sugar, making it a promising option as a biomass feedstock for liquid fuel. The tree is being studied by National Renewable Energy Lab (NREL). Using plant biotechnology the modifications were made at two points in the lignin biosynthetic pathway, with the largest increase in sugar release coming from cinnamate-4-hydroxylase (C4H) down-regulation. Although some "low recalcitrance" plant lines suffer from reduced growth, many of the C4H down-regulated lines from the *E. grandis x E. urophylla* cross grow well. C4H lines have an estimated biomass productivity of ten dry-tons per acre per year, with the potential to produce about 1,000 gallons of liquid biofuels per acre.

Scientists from NREL have characterized the C4H lines as containing half the lignin of the unmodified lines. Using a high throughput sugar release assay developed at NREL (Selig et al, 2011 *Biotech. Letters* 1-7), the modified lines were found to release up to 99 percent of their sugars, up from 40 percent in the non-modified plants. This result translates to an improvement of 150 percent, a dramatic demonstration of the impact of lowering recalcitrance.

Further work by NREL and ArborGen on these and similar lines will aim to understand exactly how the recalcitrance is lowered and how this knowledge can be used to develop healthy low recalcitrance lines in an array of species.

(219b) Process Synthesis and Global Optimization of Biomass to Liquid Transportation Fuels (BTL) Systems

Richard C. Baliban, Josephine A. Elia and Christodoulos A. Floudas, Chemical and Biological Engineering, Princeton University, Princeton, NJ

The transportation sector relies heavily on petroleum as a primary energy source, and currently faces challenges over high crude oil prices, volatility of the global oil market, and high levels of greenhouse gas (GHG) emissions [1]. As a result, a substantial effort has been taken in the United States to focus on greater energy independence through the utilization of more domestic energy sources. A recent review highlights the single and hybrid feedstock design alternatives that can produce gasoline, diesel, and kerosene from coal, biomass, or natural gas [2]. Of particular interest are liquid fuels that can be derived from biomass, since the feedstock has the potential to be both a renewable energy source and a large contributor to a reduction in the well-to-wheel GHG emissions of the transportation sector. Though corn-based ethanol and soybean-based diesel comprise a majority of the biofuels manufactured today, their use for fuel production has led to concerns regarding the impact on the price and availability of these feedstocks as sources of food [3]. Lignocellulosic plant sources including dedicated energy crops, agricultural crop residues, and forest residues are capable of being sustainable harvested without affecting the food chain and therefore expected to be a more considerable source of biofuels in the future. The development of processes that can economically utilize these feedstocks to produce synthetic liquid fuels will be of significant importance.

We present an optimization based framework to perform a comprehensive technoeconomic, environmental, and societal assessment of an alternative energy refinery that will convert biomass to the liquid transportation fuels (BTL) gasoline, diesel, and kerosene. Rigorous process design, global optimization, and process synthesis strategies will be utilized [4-11] to determine the optimal plant topology of a BTL refinery that minimizes the overall cost of liquid fuels production. The cost of utility generation and wastewater treatment is treated using a simultaneous heat, power, and water integration within the process synthesis model. Multiple case studies are presented to investigate the effect of biomass type, biomass moisture content, plant capacity, and product distribution on the optimal process topology, total plant cost, and break-even oil price. Major topological decisions from the optimization model and the result on plant capital cost, total operational cost, and the process utility network will be discussed.

[1] National Academy of Sciences, National Academy of Engineering, and National Research Council. Liquid Transportation Fuels from Coal and Biomass: Technological Status, Costs, and Environmental Issues. Washington, D. C., EPA, 2009.

[2] C.A. Floudas, J.A. Elia, R.C Baliban. Hybrid and Single Feedstock Energy Processes for Liquid Transportation Fuels: A Critical Review. *Comp. Chem. Eng.*, 2012: 41(11), 24-51.

[3] L.R. Lynd, E. Larson, N. Greene, M. Laser, J. Sheehan, B.E. Dale, S. McLangulin, M. Wang. The role of biomass in America's energy future: framing the analysis. *Biofuels, Bioprod. Biorefin.*, 2009:3(2), 113.

[4] R.C. Baliban, J.A. Elia, C.A. Floudas. Toward Novel Hybrid Biomass, Coal, and Natural Gas Processes for Satisfying Current Transportation Fuel Demands, 1: Process Alternatives, Gasification Modeling, Process Simulation, and Economic Analysis. *Ind. Eng. Chem. Res.,* 2010:49(16), 7343-7370.

[5] J.A. Elia, R.C. Baliban, C.A. Floudas. Toward Novel Hybrid Biomass, Coal, and Natural Gas Processes for Satisfying Current Transportation Fuel Demands, 2: Simultaneous Heat and Power Integration. *Ind. Eng. Chem. Res.,* 2010:49(16), 7371-7388.

[6] R.C. Baliban, J.A. Elia, C.A. Floudas. Optimization Framework for the Simultaneous Process Synthesis, Heat and Power Integration of a Thermochemical Hybrid Biomass, Coal, and Natural Gas Facility. *Comp. Chem. Eng.*, 2011:35(9), 1647-1690.

[7] J. A. Elia, R.C. Baliban, X. Xiao, C.A. Floudas. Optimal Energy Supply Network Determination and Life Cycle Analysis for Hybrid Coal, Biomass, and Natural Gas to Liquid (CBGTL) Plants Using Carbon-based Hydrogen Production. *Comp. Chem. Eng.*, 2011:35(8), 1399-1430.

[8] R.C. Baliban, J.A. Elia, C.A. Floudas. Simultaneous Process Synthesis, Heat, Power, and Water Integration of Thermochemical Hybrid Biomass, Coal, and Natural Gas Facilities. *Comp. Chem. Eng.*, 2012:37(10), 297.

[9] R.C. Baliban, J.A. Elia, R. Misener, C.A. Floudas. Global Optimization of a Thermochemical Based Hybrial Coal, Biomass, and Natural Gas to Liquids Facility. *Comp. Chem. Eng.*, 2012, http://dx.doi.org/10.1016/j.compchemeng.2012.03.008.

[10] R.C. Baliban, J.A. Elia, V.W. Weekman, C.A. Floudas. Process synthesis of hybrid coal, biomass, and natural gas to liquids via Fischer-Tropsch synthesis, ZSM-5 catalytic conversion, methanol synthesis, methanol-to-gasoline, and methanol-to-olefins/distillate technologies. *Comp. Chem. Eng.*, submitted.

[11] J.A. Eila, R.C. Baliban, C.A. Floudas. Nationwide Energy Supply Chain Analysis for Hybrid Feedstock Processes with Significant CO_2 Emissions Reduction. *AIChE J.*, submitted.

CONDUCTIVE LAYERS ON SURFACE MODIFIED NATURAL FIBRE BASED SUBSTRATES FOR PRINTED FUNCTIONALITY

Dimitar Valtakari[1], Roger Bollström[1], Mikko Tuominen[2], Hannu Teisala[2], Mikko Aromaa[3], Martti Toivakka[1], Jurkka Kuusipalo[2], Jyrki M. Mäkelä[3], Jun Uozumi[4], and Jarkko J. Saarinen[1,4,]*

[1] Laboratory of Paper Coating and Converting, Center for Functional Materials, Abo Akademi University, Turku, FINLAND
[2] Paper Converting and Packaging Technology, Tampere University of Technology, Tampere, FINLAND
[3] Aerosol Physics Laboratory, Tampere University of Technology, Tampere, FINLAND
[4] Faculty of Engineering, Hokkai-Gakuen University, Sapporo, JAPAN
[*] Corresponding author: jarkko.j.saarinen@abo.fi

Abstract

Formation of conductive surfaces by flexographical printing has been studied using an IGT test printer with PEDOT-PSS and Ag conductive inks on coated papers. Printability of multilayer coated paper and TiO_2 nanoparticle coating generated by the liquid flame spray process are compared to commercial plastic film used in printed electronics applications. The wettability of TiO_2 nanoparticle coating can be altered between superhydrophobic and superhydrophilic states by ultraviolet light. It is observed that superhydrophobicity induced by TiO_2 nanoparticles results in poorer ink setting and hence lower conductivities with water-based PEDOT:PSS ink. Therefore, we observe conductivity only after several successive prints. On contrary, we observe several orders of magnitude better conductivities when using a silver ink in flexography. It is believed that sustainable natural fibre based substrates will find more applications in printed electronics application in the future.

Introduction

Printed electronics and intelligence have been under a growing interest since the start of the 21st century with the market value forecasts up to 300 billion USD by 2025 [1]. Conventionally such applications have been produced on plastic films. However, paper based electronics [2] has been studied recently, and it has been shown that simple, all-printed transistors can be made on multilayer coated paper [3]. Another example is a thin, lightweight, and foldable thermochromic display based on a regular copy paper in which thermochromic ink was utilized with patterned electrodes on the other side of the paper providing extremely low cost displays with price less than 0.10 USD/m^2 [4]. Such displays may find applications, for example, in data reporting on portable point-of-care devices. Natural fibre based substrates are rapidly finding new applications outside of the conventional graphical arts industry.

Paper has many advantages over the plastic films: it is made of renewable materials and it is cheap with tailorable surface properties. Paper can be 1000 times cheaper than glass substrates and 100 times cheaper than plastic films [2]. Cellulose, the major component of plant biomass, is the most abundant biopolymer on Earth with annual production up to 1.8×10^{12} tonnes [5]. However, compared to plastic films, paper is porous, uneven, and rough network of fibres. Typically paper surface is smoothened by dispersion coating consisting mineral pigments and organic binders. Recently, nanoscale coating techniques including layer-by-layer and liquid flame spray coating have been studied that allow surface functionalization with significantly reduced coating amounts. Nanoscale control of surface properties is crucial for achieving good performance in electronics applications.

This study concentrates on printing large area conductive surfaces on various coated paper grades. Such conductive surfaces are needed, for example, in electrochromic displays and in solar cells. Photovoltaic (PV) devices on paper substrates were demonstrated by two groups in 2011: Karen Gleason's group [6] utilized an oxidative chemical vapour deposition (oCVD) for fabrication of PV cells on rough tissue paper. The key for the success of such an approach was perfect contour coating by oCVD preventing formation of electrical shorts. However, the process is a slow batch process and not optimal as paper is produced in practice with line speeds typically above 1000 m/min.

An alternative approach was presented by Arvid Hübler's group [7] based on all-printed process flow utilizing a transfer of bottom electrode from a plastic film. With such an approach they were able to overcome the roughness issues of paper as the plastic film defines the roughness. However, their approach, although based on all-printing processes, did not allow a continuous process flow but rather had up to 12 hour gaps between different prints. The observed power conversion efficiency was 1.31 %. For cost-effective manufacturing of organic electronic devices, it would be beneficial to use an all-in-line printing process with possibilities for significant unit cost reduction of the devices [8].

In current PV applications indium tin oxide (ITO) is typically used as a transparent conductive electrode. Unfortunately, there are two major drawbacks with ITO: first, it is expensive and scarce material. Secondly, the brittle nature of the ITO films makes them not suitable for flexible substrates. These limitations prevent the use of ITO on low-cost, large-scale PV applications. Conductive polymers such as poly(3,4-ethylenedioxythiophene)/poly(styrenesulfonate) (PEDOT/PSS) may be suitable candidates as they can be both transparent and conductive. However, there is a strong correlation between conductivity and opacity *i.e.* the better the conductivity, the more opaque is the film. This has been a bottleneck but recent developments have eased the problem of achieving good transparency with high conductivity [9].

PEDOT:PSS is the most common and affordable conductive polymer currently available on the market. It comes in 1:2.5 (pH 1.5 – 2.5), 1:6 (pH 4.0 – 7.0) or 1:20 (pH 7.0) ratio by weight with a solids content of 0.6 – 3.4 % dispersed in water. Viscosity can be varied from 4 to 350 mPas depending on requirements of the application. PEDOT:PSS typically serves as a transparent conductive electrode or as a buffer layer in between the active layer and metallic electrode. Conductivity of PEDOT:PSS can significantly be increased by addition of secondary dopants such as dimethyl sulfoxide, ethylene glycol, or glycerol. Conductive silver inks are available from very fluid inkjet inks to very viscous pastes for screen printing. Inkjet inks are typically a colloidal silver nanoparticle suspension. The highly customizable, fine resolution printed patterns, grids, and electrodes require thermal sintering for conductivity. Thicker inks and pastes, on the other hand, consist of micrometer sized silver flakes. These are usually used for thick film electronics applications, where print layer reproducibility and bulk silver conductivity are of utmost importance.

Our study concentrates on using flexographical printing with conductive inks on various coated paper grades to form large area conductive surfaces. We use two different natural fibre based substrates: a multilayer pigment coated grade and a nanoparticle coated paperboard in comparison to traditional plastic film. The paperboard surface is functionalized by TiO_2 nanoparticles using a liquid flame spray (LFS) coating process. In LFS process liquid precursor of titanium (IV) isopropoxide

(TTIP) is fed into a high temperature and velocity flame in which the metal salt evaporates and nucleates to form nanoparticles of the metal oxide. These nanoparticles can be collected on a paperboard surface in on-line process flow and they cover the whole surface passing under the flame. For details of the LFS process, see Ref. [10].

It is known [11] that ultraviolet (UV) illumination can be used to change the wettability of TiO_2 surfaces, and similar photocatalytic wettability conversion has been observed on TiO_2 nanoparticle coated paperboard surfaces from superhydrophobic (water contact angle (CA) over 150°) to superhydrophilic (water CA less than 20°). Typical coating amounts on surface by the LFS are between 10 and 50 mg/m^2 [10] being several orders of magnitude smaller than in conventional dispersion coating concepts.

Here we report the effects of coating structure and surface wettability on formation of conductive surfaces. Tunable wettability of TiO_2 nanoparticle coated substrates has an effect on printability with water-based inks such as PEDOT:PSS as ink setting is poorer on superhydrophobic surface that is also observed from measured optical densities of the printed layers. We also report significantly higher conductivities with Ag ink on multilayer coated paper than with PEDOT:PSS conductive polymer.

Materials and Methods

We use three different printing substrates in our study: a commercial poly(ethylene terephthalate) (PET) film (Mylar®A) is used as the reference plastic film, a multilayer coated paper, and a double pigment coated paperboard (200 g/m^2, Stora Enso, Skoghall, Sweden).

The multilayer coated paper contains a commercial finepaper (Lumipress 115, StoraEnso, FI) that is coated by a 10 g/m^2 barrier layer consisting of platy kaolin (Barrisurf HX, Imerys Minerals Ltd, UK) blended with 50 pph ethylylene acrylic latex (Aquaseal 2077, Paramelt B.V., NL). To adjust printability, a 5 g/m^2 topcoating layer consisting of 70 pph fine platy kaolin (Barrisurf FX, Imerys Minerals Ltd. UK) and 30 pph fine blocky kaolin (Alphatex, Imerys Minerals Ltd., UK) blended with 6 pph SB latex (Basonal 2020.5, Basf, DE) was coated on top of the barrier layer. The paper was further calendered three times through a softnip at a line load of 120 kN/m and a temperature of 70 C°.

The paperboard sample was coated by TiO_2 nanoparticles using a coating and lamination pilot line at the Tampere University of Technology (Tampere, Finland) with a 30 m/min web speed. TTIP was dissolved in isopropanol with a metal ion concentration of 11.9 mg/ml with a 29.5 ml/min feeding rate into the nozzle whose distance from the moving web was set to 15 cm. This results in a deposition mass of 12.7 mg/m^2 with approximately 415 TiO_2 nanoparticles/μm^2 [12]. The average diameter of the nanoparticles is between 40 and 80 nm.

Flexographical prints were carried out using a laboratory scale IGT GST 2 printability tester. The anilox cylinder had a cell angle of 45° with 40 lines/cm and volume of 20 ml/m^2. PEDOT/PSS (Clevios P HC V4) conductive polymer dispersed in water with viscosity of 100 - 350 mPas was tested at a print speed of 2 m/s. Solvent based silver ink (Creative Materials 125-06 Extremely Conductive Ink) with viscosity of 2 000 – 3 000 mPas was tested at a print speed of 0.5 m/s. The contact angles of the inks were analyzed using a contact angle goniometer (KSV CAM 200) and the UV-exposure of the TiO_2 coated surfaces was carried out using a Bluepoint 4 Ecocure source with an UVA filter. The print densities were measured using a Techkon optical densitometer. Finally, a digital multimeter (Keithley 2100) was used to characterize the conductivities between two hand-painted contacts of silver conductive paint (Electrolube).

Results and Discussion

Contact angles

The contact angles of water, PEDOT:PSS (solid content 1.0 – 1.4 %), and silver printing inks for flexography are summarized in Table 1. Water contact angles are taken after 1 s for each sample whereas longer times for drop settling were required for more viscous PEDOT:PSS droplets (1 s for multilayer, 10 s for LFS paperboard, and 60 s for other samples). With TiO_2 nanoparticle coated samples we observe switching of superhydrophobic surface into superhydrophilic by UV light irradiation. Similar behavior is observed with water based PEDOT:PSS ink but no effect on silver ink is found. The LFS paperboard sample denotes the double coated paperboard without the TiO_2 nanoparticles.

Table 1. Measured contact angles of water, PEDOT:PSS ink, and silver ink.

	Plastic	Multilayer	LFS TiO_2	LFS TiO_2+UV	LFS Paperboard
H_2O	75	50	153	7	74
PEDOT:PSS	74	62	160	26	69
Ag	27	54	52	50	54

PEDOT:PSS prints

Optical print densities of PEDOT:PSS prints were measured before drying and are given in Table 2. The values for print layers are calculated by subtracting the measured background value for the unprinted substrate by $D_{print} = D_{measured} - D_{unprinted}$. As expected, the print densities increase after each successive print. Furthermore, the smallest values are observed for superhydrophobic surface indicating a poor transfer of ink to the substrate.

Table 2. Measured print densities of PEDOT-PSS before curing at the oven.

	Plastic	Multilayer	LFS TiO_2	LFS TiO_2+UV	LFS Paperboard
1st Print layer	0.02	0.03	0.01	0.07	0.01
2nd Print layer	0.06	0.06	0.04	0.09	0.03
3rd Print layer	0.08	0.09	0.06	0.13	0.09
4th Print layer	0.09				

Table 3 summarizes the measured conductance values with PEDOT:PSS flexographical prints after drying in an oven (1h, 120 °C). It was observed that plastic films require more successive print layers (here 4 layers) to obtain conductance due to poor wetting of smooth surface. The best performance was observed with a multilayer coated paper with controlled ink setting on the porous top layer. With TiO_2 nanoparticle coated samples we observe that surface wettability plays an important role when printing with water based flexographical inks. The conductivity values are lower when printing on superhydrophobic surface that is expected due to poorer wetting of the surface and formation of an uneven print layer. Hence, we can conclude that the superhydrophobicity survives even in the dynamic printing process and is lost only after several successive print layers.

Table 3. Observed sheet resistance of different substrates with PEDOT-PSS.

	Plastic	Multilayer	LFS TiO_2	LFS TiO_2 + UV	LFS Paperboard
1st Print layer	> 1000 MΩ/□	640 kΩ/□	> 1000 MΩ/□	1200 kΩ/□	3570 kΩ/□
2nd Print layer	> 1000 MΩ/□	200 kΩ/□	5250 kΩ/□	290 kΩ/□	270 kΩ/□
3rd Print layer	> 1000 MΩ/□	40 kΩ/□	330 kΩ/□	100 kΩ/□	80 kΩ/□
4th Print layer	620 kΩ/□				

It is worth emphasizing that the substrates here are not optimized for printing conductive layers, which explains the observed low conductivity (= high sheet resistance) values. For example, the LFS paperboard sample is a conventional paperboard used for packaging applications. With multilayer coating one can control ink setting resulting in better conductivities. Moreover, successive prints yield a more even and homogeneous print layer with higher conductivities [13]. The observed values are in agreement with the results shown by Barr *et al.* [6] with oCVD coated thin PEDOT layers.

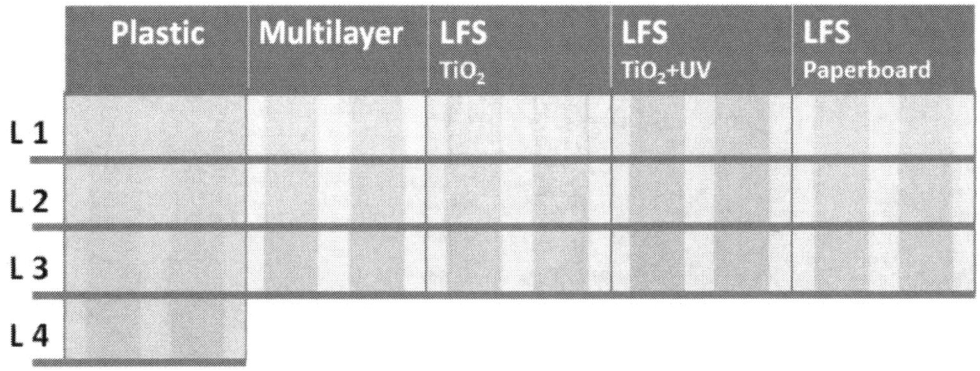

Figure 1. Flexographical PEDOT:PSS prints on different substrates (width of the printed area 7 mm).

Figure 1 displays the scanned images of the flexographical PEDOT:PSS prints. Poor ink setting both on plastic film as well as on superhydrophobic LFS TiO_2 substrate is observed that correlates well with the observed low conductivities. In multilayer coated structure the barrier layer below the thin and porous top coating layer prevents the ink penetration deep into the structure resulting in the highest conductance values.

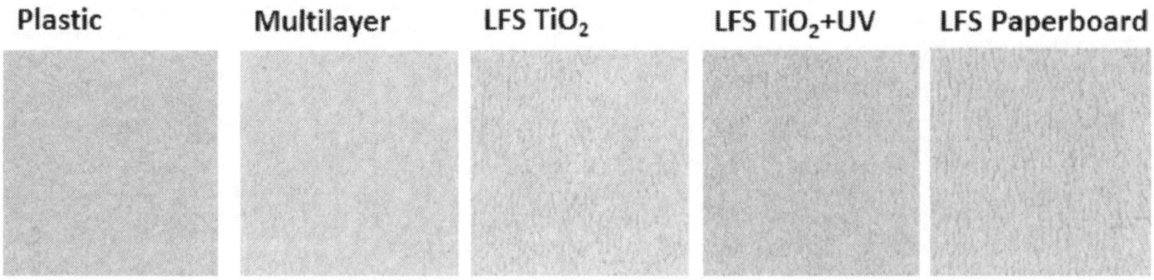

Figure 2. A larger magnification of the flexographical print layer (area 4×4 mm^2, 3 print layers).

Figure 2 displays a magnification of the printed layer structures. We observe a formation of branched clusters of polymer (dark areas), which are especially visible in the LFS paperboard reference sample. Similar behavior has been reported with spin-coated PEDOT:PSS thin films [14] that results in anisotropy in conductivity values in machine and cross-machine direction. This phenomenon was studied with contact angle goniometer as shown in Fig. 3. It is observed that with larger PEDOT:PSS droplets a neck is formed on the superhydrophobic TiO_2 LFS surface as the polymer concentrates on the droplet/substrate boundary. Initially the droplets behave like water droplets but as the polymer settles into the surface, the contact angle grows again with water minimizing the surface energy. The effect is rather slow for the large droplets. However, in flexography the size of the droplets is much smaller and hence, similar behavior may be observed in the print morphology.

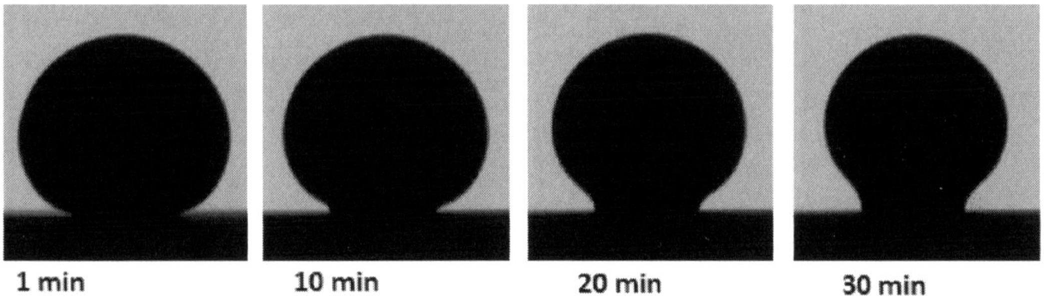

Figure 3. Time series of the PEDOT:PSS droplet on LFS surface.

Ag prints

We have recently demonstrated UV sensor functionality on multilayer coated paper by using flexograhically printed silver electrodes with conductive polymers [15]. Here we concentrate on identifying the properties of the flexographical process to improve conductivity of the prints. All silver samples were printed with 0.5 m/s printing speed. The pressure between the anilox cylinder and the printing plate as well as the pressure between the printing plate and the substrate was set to 100 N. Figure 4 shows microscopic images of three successive prints on multilayer coated paper with their corresponding resistance values. All samples were sintered in an oven at 120°C for 60 min. We observe non-conducting layer with a single print. This is due to two factors: first, the amount of ink transferred from the anilox is too small in order to fill the whole surface. Secondly, the number of cells/cm in our anilox cylinder may be too small compared to the cell volume *i.e.* individual cells are visible in the final print. Hence, no conductivity is observed as there is no connecting path between the individual silver flakes below the threshold value. After the second and third successive prints we observe good conductivity values as flakes become attached to each other.

Figure 4. Microscopic images of the flexographical Ag prints with 9 ml/m^2 and 100 lines/cm (magnification 100x).

The observed conductivity values are shown in Table 4. With each different gravure cell we observe non-conducting layer with a single print – this is expected as the amount of silver transferred to the substrate is too small to reach conductivity threshold. Similar behavior was observed in our earlier study [15] that related the surface coverage values to the observed conductivities.

Table 4. Observed sheet resistance of multilayer substrate with Ag ink using different cell sizes and line counts in the gravure cylinder.

	7 ml/m^2 160 lines/cm	9 ml/m^2 100 lines/cm	11 ml/m^2 80 lines/cm	16 ml/m^2 60 lines/cm	20 ml/m^2 40 lines/cm
1st layer	> 100 MΩ	> 100 MΩ	> 100 MΩ	> 100 MΩ	> 100 MΩ
2nd layer	130 – 156 Ω	35 Ω	1,1 – 1,3 MΩ	14 Ω	126 - 132 Ω
3rd layer	9,3 Ω	6,7 Ω	13 – 15 Ω	9 – 40 Ω	7,5 Ω

As a conclusion we have observed significantly higher conductivities with silver ink than with PEDOT:PSS. The reason for observing conductivity only after two successive prints is most likely due to the metallic anilox roller of our IGT GST2 printing unit, and the need for several successive prints for conductive electrodes is rather ineffective for practical applications. We believe that conductive layers with single print can be obtained by optimizing the printing conditions.

Conclusions

As a summary we have investigated possibilities to form large area conductive surfaces using a flexographical printing process with a laboratory scale test printer. Surface properties play a crucial role in the ink setting on the surface and in observed conductivity values. We have also shown that superhydrophobic surfaces result in poorer conductivity values with water based inks; to observe similar conductivity values as with hydrophilic surfaces several more prints are required. Furthermore, we have explored the optimal performance area for IGT printer with conductive inks: for water based

PEDOT:PSS with viscosity of 100 – 350 mPas faster printing speeds (>1.0 m/s) are required for better reproduction whereas for silver ink lower printing speeds (<1.0 m/s) are optimal. The lower speeds are needed for silver ink due to Ag flakes.

Comparing the observed conductivities between paper substrates to commercial plastic film for PEDOT:PSS we can conclude that the porous paper surface provides a more uniform ink setting, and hence, improved conductivity values. Silver ink provides several orders of magnitude better conductivities but they cannot be used as the semi-transparent top electrode. Further work is needed to optimize the coating structure and ink formulation to achieve high conductivities.

Acknowledgments

This work has been funded by the Academy of Finland (grant no 250 122 & 256 263). JJS wishes to thank the Japan Society for the Promotion of Science (JSPS) for an invitation to perform research at the Hokkai-Gakuen University in Sapporo, Japan in 2011 – 2012 under the "FY2011 JSPS Invitation Fellowship Program for Research in Japan (Long-term)" (grant no L–11537). We wish to acknowledge Imerys Minerals Ltd., UK, Paramelt B.V., NL, StoraEnso, FI, and Basf, DE for providing materials used in the multilayer coated paper.

References

1. Das, R. and Harrop, P. (2008), *Printed & organic electronics forecasts, players & opportunities 2008-2028*, IDTechEx report 7.
2. Tobjörk, D. and Österbacka, R. (2011), Adv. Mater. 23, 1935.
3. Bollström, R., Määttänen, A., Tobjörk, D., Ihalainen, P., Kaihovirta, N., Österbacka, R., Peltonen, J. and Toivakka, M. (2009), Org. Electron. 10, 1020.
4. Siegel, A.C., Phillips, S.T., Wiley, B.J. and Whitesides, G.M. (2009), Lab Chip 9, 2775.
5. Eveleigh, D. E. (1987), Phil. Trans. R. Soc. Lond. A 321, 435.
6. Barr, M. C., Rowehl, J. A., Lunt, R. R., Xu, J., Wang, A., Boyce, C. M., Im, S. G., Bulovic, V. and Gleason, K. K. (2011), Adv. Mater. 23, 3500.
7. Hübler, A., Trnovec, B., Zillger, T., Ali, M., Wetzold, N., Mingebach, M., Wagenpfahl, A., Deibel, C. and Dyakonov, V. (2011), Adv. Energy Mat. 1, 1018.
8. Berggren, M., Nilsson, D. and Robinson, N.D. (2007)., Nature Mater. 6, 3
9. Vosgueritchian, M., Lipomi, D. J. and Bao, Z. (2011), Adv. Funct. Mat. 22, 421.
10. Mäkelä, J. M., Aromaa, M., Teisala, H., Tuominen, M., Stepien, M., Saarinen, J. J., Toivakka, M. and Kuusipalo, J. (2011), Aerosol Sci. Technol. 45, 817.
11. Caputo, G., Nobile, C., Kipp, T., Blasi, L., Grillo, V., Carlino, E., Manna, L., Cingolani, R., Cozzoli, P. D. and Athanassiou, A. J. (2008), Phys. Chem. C 112, 701.
12. Stepien, M., Saarinen, J. J., Teisala, H., Tuominen, M., Aromaa, M., Kuusipalo, J., Mäkelä, J. M. and Toivakka, M. (2011), Appl. Surf. Sci. 257, 1911.
13. Denneulin, A., Blayo, A., Bras, J. and Neuman, C. (2008), Prog, Org. Coatings 63, 87.
14. Nardes, A. M. (2007), *On the conductivity of PEDOT:PSS thin films*, PhD Thesis, Technische Universiteit Eindhoven, The Netherlands.
15. Saarinen, J. J., Ihalainen, P., Määttänen, A., Bollström, R. and Peltonen, J. (2011), Nord. Pulp Paper Res. J. 26, 133.

(219d) Soybean Oil Based Fibers Formed without Solvent or Heat

Dustin W. Janes, Kadhiravan Shanmuganathan, Daniel Y. Chou and Christopher J. Ellison, Department of Chemical Engineering, The University of Texas at Austin, Austin, TX

The vast majority of commercially produced synthetic polymers used for fiber applications are made entirely from non-renewable, petroleum-based feedstocks. Because annual worldwide production of nonwoven fibers is in the billions of kilograms, incorporating even small amounts of biorenewable materials in these products could significantly impact the allocation of non-renewable resources. Additionally, large amounts of solvent and/or thermal energy are usually needed to process polymers into fibers. A safer and more efficient route to fiber production may be to photopolymerize functional monomer liquids during fiber formation processes. In this presentation we show that thiol/-ene chemistry can be harnessed to enable production of fibers containing over 50 wt. % acrylated epoxidized soybean oil without using solvent or heat. In this demonstration, the fibers were made by simultaneous electrospinning and photocuring of a liquid monomer mixture which we believe can be translated to other fiber manufacturing processes such as melt blowing. Scanning electron micrographs illustrate the fiber quality and an average diameter of about 30 μm. Photochemical conversion kinetics of functional groups during light exposure were measured by real-time Fourier transform infrared spectroscopy, providing insight into the advantages of using high-functionality monomers and thiol/-ene chemistry in this application.

VALIDATION OF CALENDERING MODELS FOR COATED PAPERS MADE FROM BLEND OF AGRI-RESIDUE AND SOFTWOOD PULP WITH SOFTNIP CALENDERING TECHNIQUES

A. K. Ray, *Department of Paper Technology, Indian Institute of Technology, Roorkee, Saharanpur, India*

Sanjay Tyagi, *Paper Testing and Energy management Division, Central Pulp and Paper Research Institute, Saharanpur, India*

ABSTRACT

Calendering is the final operation in the manufacturing of papers with the purpose for quality products by densification, and to increase smoothness (or to decrease the surface roughness) and gloss. In order to do so it decreases the thickness and bulk of paper. It is especially important in case of coated papers or paper boards where increase of gloss is essential. Paper is therefore, pigment coated to obtain a surface with improved optical and print performance additionally. However this is done under many modified operations such as supercalendering or softnip calendaring process. Generally, during calendering operation paper is subjected to compressive pressure pulse under the nip. After the nip, there is time dependent partial recovery of the initial thickness. The behaviour of the paper under calender nip is therefore viscoelastic in nature.

In the present paper modelling of stress – strain behaviour of coated paper under soft nip calender and its viscoelastic properties are attempted. The base paper used in this study was collected from the paper mill in reel form and having the furnish composition of 75% agricultural residue (bagasse/wheat straw) and 25% of soft wood chemical pulp. The paper coating trials were conducted on DT pilot coater at the speed of 20- 30m/min. During the run only one top side of paper was coated with the coat weight ranging 10 - 12 g/m^2. The coating formulation consisted of 100 parts pigment (fine ground calcium carbonate and fine clay in ratio of 80:20), 12 parts SB latex, 1 part CMC, 0.12 part OBA and 0.5 part dispersant). Calendering was carried out using soft nip rolls under varying calendering conditions of temperature from 60°C to 85¢ªC and nip load of 84 kN/m to 120 kN/m. The moisture content of paper varies from 4.1% to 5.7 %. For drying of paper two IRT- Monocassettes of infrared units were used to give a temperature of 60 – 70°C. The temperature of air drying unit was kept at 50 °C .The data obtained after softcalendering were subjected to various investigation.

Available models for viscoelastic properties were critically examined and systematic estimation of the interrelated parameters of the viscoelastic models were made. The models were simulated with the data obtained from laboratory soft nip calender for coated paper. The flow chart is drawn to estimate the various parameters to the model simulation. The parametric effects were interpreted through various graphs and figures.

Key Words: Calendering, Softnip, Web temperature, Elastic modulus, Viscoelastic, Bulk

Introduction

Paper is pigment coated to obtain a surface with improved optical and print performance. Calendering is a very important technique for obtaining the above properties. It enhances the surface smoothness as well as gloss of coated paper or paper board. The apparent calendering effect is the surface flatness, the permanent compression and densification of the paper web. The parameters which affect the quality of paper developed through the calendering process are web speed/ speed of roll, pressure or total load, dwell time, calender roll surface temperature, number of nips, nip width, roll diameter, initial web temperature, web moisture, calender roll surface condition and the roll material properties.

Paper can be defined as highly porous random structure of flexible fibers, filler and other particles and can be considered to be elastico- visco- plastic material. As the paper web passes in to the converging nip, it experiences rapid compressive strain until the maximal compression is reached approximately at the nip centre. The main mechanism of paper under calender loading is the compressive deformation. Rodal (1989) pointed out in his model the viscoelastic behaviour of paper during calendering. The polymers in base paper and in the coating layers are either amorphous like lignin, hemicellulose and synthetic binder or partly crystalline like cellulose and starch. Deformation in the structure of polymers are time dependent and partly permanent (plastic) and partly recoverable (viscoelastic). The viscoelastic behaviour of polymer depends on molecular weight and polymer structure as well as external stress, temperature and moisture content. The response of viscoelastic material to the applied stress can be seen a combination of fast, viscous deformation and slower, elastoplastic deformation.

Menard (1999) described the viscoelastic properties of material by storage modulus (E'), loss modulus (E'') and by tanδ, which is the ratio between loss modulus and storage modulus. Browne et. al. (1996, 1994) has studied the importance of viscoelsatic behaviour of paper in hardnip calendering. The viscoelastic behaviour of the paper can be seen as an exponential paper thickness recovery after the nip. At higher linear loads the total recovery is constant and the portion of permanent deformation increases. This means that the whole paper structure deformation is a combination of the immediate or delayed elastic recovery and the permanent defrormation. Browne has shown that at low nip pressure almost all of an- in-nip deformation is recoverable, and the permanent deformation remains low. With increasing nip pressure, the recoverable deformation decreases both as an absolute value as well as compared to permanent deformation. This means that paper shows pure elastic behaviour at low nip pressure, but the importance of plastic deformation increases with increasing nip pressure. In the present study an attempt has been made to validate the viscoelastic models by Browne et. al.

for coated paper, calendered soft nip under laboratory conditions. The results were simulated using the data gathered for coated paper

Models for Viscoelastic behaviour of paper

The common way illustrating the viscoelastic models also termed as rheological models, is to use simple mechanical models involving spring and dashpot elements (Popil, 1989). The spring element is used to describe the idealized elastic behavior and the dashpot for viscous effects. If the model involves only linear behavior, the same is considered as linear viscoelastic model with constant elasticity and viscous parameters. The most commonly used models are

1) Linear Maxwell model
2) Linear Kelvin model (Voight model)
3) Burgers four element model

Linear Maxwell Model: Connecting a spring and a dashpot serially Maxwell model is obtained which characterizes the viscoelastic deformation of fluids which represents the linear Maxwell model where E is the elastic coefficient of spring and η is the coefficient of viscosity.

Linear Kelvin model (Voight model): In Kelvin model spring and a dashpot have been connected in parallel. This model describes the viscoelastic deformation of solids (delayed elasticity).

Burgers model: The four element Burgers model consists of Maxwell component and a Kelvin component. This model accounts for many substantial effects observed in paper like instantaneous deformation and recovery, permanent deformation and delayed recovery.

Systematic Estimation of the interrelated parameters of the viscoelastic model:

The details of derivation of general viscoelastic model are given in reference (Tyagi, 2009). The major equations used for the development of algorithm are given in Table 1.

Experimental Investigation

Base Paper: The base paper used in this study was collected from the paper mill in reel form and having the furnish composition of 75% agricultural residue (bagasse/wheat straw) and 25% of soft wood chemical pulp.

Coating formulation and coating trials:

The paper coating trials were conducted on DT pilot coater at the speed of 30m/min. The distance between the applicator roll and blade was kept at 250mm. The metering blade measured 84 x 315 mm was used in coating trials. The blade angle was kept at

31°. The blade load was adjusted by pneumatic cylinders at the pressure of 1.5 bar. During the run only one top side of paper was coated by the coat weight of 10 - 12 g/m^2. The coating formulation consisted of 100 parts pigment (fine ground calcium carbonate and fine clay in ratio of 80:20), 12 parts SB latex, 1 part CMC, 0.12 part OBA and 0.5 part dispersant). For drying of paper two IRT- Monocassettes of infrared units were used to give a temperature of 60 – 70°C. The temperature of air drying unit was kept at 50 °C

Calendering Conditions

After coating the sheets were calendered in laboratory calender. For paper coated with different coating formulations calendering was carried using soft nip rolls under varying calendering conditions of temperature from 60ºC to 85°C and nip load of 84 kN/m to 120 kN/m. The speed of calendar kept at 22 m/min. The moisture content of paper varies from 4.1% to 5.7 %.

Evaluation of coated papers

Calendered coated/uncoated sheets were conditioned at 27± 1% relative humidity of 65 ± 2% prior to testing. All testing was done according to relevant ISO standard test methods.

Simulation of models for stress-strain behaviour of coated paper under soft nip calendering

The calendering parameters such as roll diameter, roll temperature, nip load, speed, nip intensity and consequent paper bulk, average nip pressure, nip residence time for softnip calenders are given in Table 2.

Using the calendering conditions given in Table 2, experiments were conducted on paper coated in laboratory using pilot blade coater. The regression equations were developed on the model based on Gerspach Equation (1993) and are given below

$0.400 = a_0 + 1.92a_L - 0.68a_\sigma + 0.64at + 60a_T + 5.7a_M$
$0.600 = a_0 + 1.95a_L - 0.66a_\sigma + 0.66a_t + 65a_T + 5.4a_M$
$0.280 = a_0 + 2a_L - 0.64a_\sigma + 0.68a_t + 70a_T + 5.1a_M$
$0.400 = a_0 + 2.04a_L - 0.621a_\sigma + 0.70a_t + 75a_T + 4.7a_M$
$0.333 = a_0 + 2.06a_L - 0.613a_\sigma + 0.70a_t + 80a_T + 4.2a_M$
$0.130 = a_0 + 2.08a_L - 0.602a_\sigma + 0.72a_t + 85a_T + 4.0a_M$

The values of unknown parameters of the calendering equations were calculated through least square techniques so that the sum of the squares of the error is minimized. The values are given in below

$A = -0.102$, $a_0 = 0.2797$, $a_L = 0.2008$, $a_\sigma = 0.2321$, $a_t = 0.0654$, $a_T = -1.4350$

$a_M = -2.4957$

Base on equations 1, 6, 7 and 8, mentioned in Table 1, the relationship was found between the linear load, permanent strain and recoverable strain for coated paper. The in- nip length and stress at nip was also calculated. The values of viscoelastic modulus for coated paper were calculated using the equations 10, 11 and 12. Based on equations and experimental results an algorithm describing the procedure for estimating parameters of viscoelastic models was developed for simulation purpose given in Figure 1. This algorithm is of extreme importance in the development and design of new calenders and in proving their validity for any type of paper.
Based on above algorithm a computer programme was developed in C++ to simulate the above models of stress- strain behaviour of paper coated with the one formulation and calendered with soft nip.

Results and Discussion

The relationship between linear load and permanent strain, in- nip strain and recoverable strain for coated paper calendered under soft nip condition were depicted in Figs.2 – 3. The computer generated regression equations are also given in graphs along with the respective regression coefficients.

Simulation carried out on experimental data generated in laboratory soft nip calendar indicates that model parameters are not merely the material properties but are the function of process parameters. Figure 2 indicate that as linear load varies from 84kN/m to 120kN/m, stress induced in coated paper also increases resulting in a linear increase in recoverable strain. There is little deviation found in the values of permanent and in-nip strain. The model predicted is in excellent agreement with experimental data as R^2 values lies in the range of 0.98 to 0.99
Figure 3 to 6 indicated the relationship of Maxwell elastic and viscoelastic modules and Kelvin elastic and viscoelastic modular relationship with calender loads. Clearly, as the load varies both Maxwell and Kelvin elastic and viscoelastic modulus varies linearly. Again, it is well in accordance to the work done by Corotgino et. al. (8) in the case of hard nip calendering of non coated papers.

Conclusions

Available models for viscoelastic properties of coated paper were critically examined. Systematic estimation of the interrelated parameters of viscoelastic models such as in-nip strain and permanent strain and their relationship are made. Calendering equation for soft nip calendars due to Gerspach was compared for prediction of viscoelastic properties of coated paper. Calculation of in-nip and out-nip width of paper, the values of Kelvin pair G_k and η_k were done. In case of validation of viscoelastic models, results concluded that as the linear load varies the in-nip strain and permanent strain induced in coated paper also increases. Compared to these two the values of recoverable strain found more. Both in-nip and permanent strain exhibits linear relationship with load. The relationship between Maxwell and Kelvin elastic and viscoelastic modules also varies linearly with applied load.

Nomenclature

L	Line load (kN/m)
B_i, B_n, B_f	Initial, in-nip and final bulk (cm^3/g)
L_i, L_n, L_f	Initial, in- nip and final thickness (cm)
ϵ_n, ϵ_p, ϵ_r	In- nip, permanent and recoverable strain
a	In going nip length (cm)
b	Out going nip length (cm)
σ	Stress (kN/m^2)
A_n, a_{0n}, a_{Ln}, a_{sn}, a_{Rn}	Calendering coefficients for hard nip calenders (gm/cm^3)
A_0, a_l, a_σ , a_t, a_T, a_M	Calendering coefficients for soft nip calenders
μ	Nip intensity
S, V	Web speed ((m/min)
T_k	Time constant
R	Roll radius (m)
T	Temperature of web (°C)
M	Web moisture (%)
G_m	Maxwell elastic modulus (MPa)
η_m	Maxwell Viscosity (k Pa s)
G_k	Kelvin elastic modulus (M Pa)
ϵ_k	Kelvin Viscosity (K Pa s)

References

1. Rodal, J., Soft nip calendaring of paper and paper board, TAPPI J., 72 (1989) 5, 77 – 186.
2. Menard, K., Dynamic mechanical analysis – A practical Introduction, CRC Press LLC, 1999.
3. Browne, T. C., Crotogino, R. H., Douglas, W. J. M., Viscoelastic modeling of paper in a calender nip, JPPS Vol. 22, No. 5, May 1996, J170 – J173.
4. Browne, T. C., Crotogino, R. H., Douglas, W. J. M., Measurement of paper strain in the nip of an experimental calender, JPPS vol. 20, No. 9, 1994 J266 – J271.
5. Popil, R. E., The calendaring creep equation – a physical model, Fundamentals of Papermaking Transaction of the ninth fundamentals research Symposium, held at Cambridge and published by Mechanical engineering Publication Ltd., 1989.
6. Sanjay Tyagi, Response of coating and calendaring on quality of papers made from indigenous raw materials, Ph. D. Thesis, submitted to Dept. of paper Technology, IIT Roorkee (Dec 2009).
7. Gerspach, A., Wochenbl Papier 121: 76 (1993). (cf. Mikko Jokio, Papermaking Part 3, Finishing, Book 10, Papermaking Science and Technology, Finish Paper Engineers' Association and TAPPI, 1999)
8. Corotogino, R. H., "Machine calendaring – recent advances in theory and practices", Trans. Tech. Section, Canada P. P. A., 7 Dec, TR 75 (1982).

Table 1: Major equations used for predicting viscoelastic model of coated paper

Eq. No.	Parameter	Equation
1	In- nip strain	$\epsilon_n = \frac{L_i - L_n}{L_i} = \frac{B_i - B_n}{B_i}$
2	In-nip Strain predicted by Calendering Equation	$\epsilon_n = A_n + \mu_n B_i$ Where $\mu_n = a_{0n} + a_{Ln} \log_{10} L + a_{Sn} \log_{10} S + a_{Rn} \log_{10} R$ (for hardnip calenders)
3	Gerspach Equation (for soft nip calendars)	$\mu = a_o + al \log L + a_\sigma \log\sigma + a_t \log t + a_T T + a_M M$
4	Permanent Strain	$\epsilon_p = \frac{L_i - L_f}{L_i} = \frac{B_i - B_f}{B_i}$
5	Permanent Strain Predicted by Calendering Equation	$\epsilon_p = A_p + \mu_p B_i$ Where $\mu_p = a_{0p} + a_{Lp} \log_{10} L + a_{Sp} \log_{10} S + a_{Rp} \log_{10} R$ and B_i is in range $-\left(\frac{A_p}{\mu}\right) \le B_i \le \frac{(1 - A_p)}{2\mu}$
6	Strain predicted by Master Creep Equation	$\epsilon = A_p (1 + \tanh \mu)$
7	Relationship between in nip and permanent strain	$\epsilon_n = C_0 + C_1 \log_{10} \epsilon p$
8	In- going nip	$a^2 = 2RLi\, \varepsilon_n$
9	Recoverable strain	$\varepsilon_r = \varepsilon_n + \varepsilon_p$
10	Kelvin pair G_k	$G_k = \frac{L}{V\tau_k(\varepsilon_R - \varepsilon_{el})}$
11	Strain	$\eta_m = L/(V.\, \varepsilon_p)$
12	Kelvin pair G_k and η_k	$\eta_k = \frac{L}{[V(\varepsilon_R - \varepsilon_{el})]}$
13	Van Haag Model	$\sigma_z = (E_0\, \varepsilon_z)/(1 - \varepsilon_z/\varepsilon_{max})$

Table 2: Calendering conditions for soft nip calendering of coated paper

Roll Temp ,˚C	60	65	70	75	80	85
Nip dwell time t (ms)	4.4	4.54	4.79	5.02	5.13	5.24
Nip load (kN/m)	84	90	100	110	115	120
Nip pressure (kN/mm^2) (σ)	0.208	0.216	0.228	0.239	0.244	0.250
Nip width (mm)	1.61	1.66	1.76	1.840	1.88	1.92
Bulk (cm^3/g)	1.12	1.07	1.00	0.95	0.92	0.88
Nip intensity (μ)	0.448756	0.458577	0.473215	0.486276	0.492239	0.498399

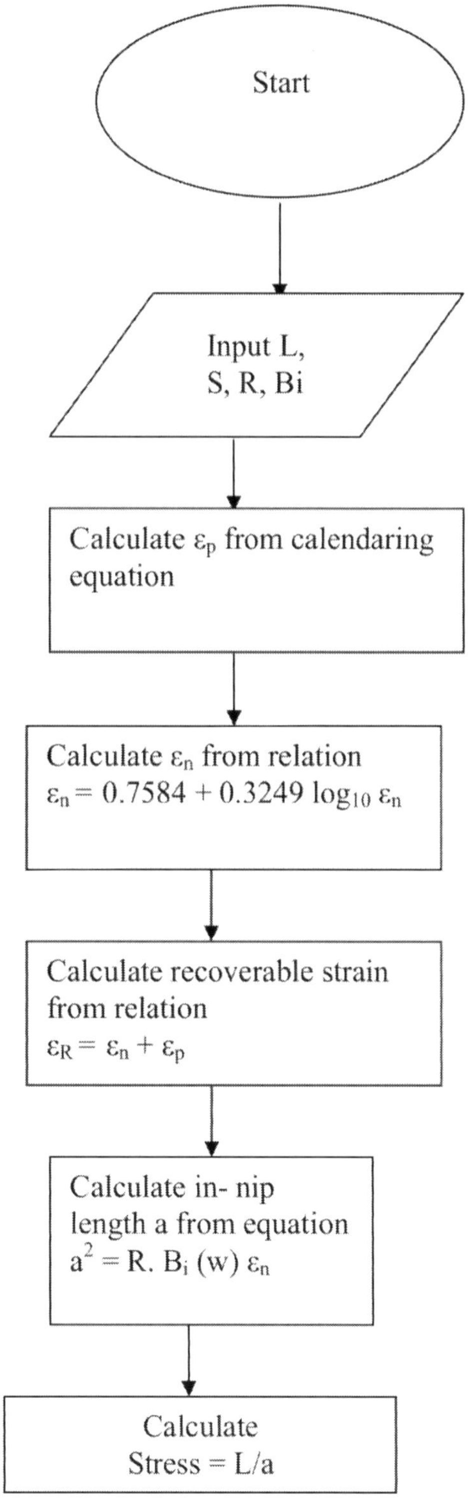
Start
Input L,
S, R, Bi
Calculate ε_p from calendaring equation
Calculate ε_n from relation
$\varepsilon_n = 0.7584 + 0.3249 \log_{10} \varepsilon_n$
Calculate recoverable strain from relation
$\varepsilon_R = \varepsilon_n + \varepsilon_p$
Calculate in- nip length a from equation
$a^2 = R.\ B_i\ (w)\ \varepsilon_n$
Calculate
Stress = L/a

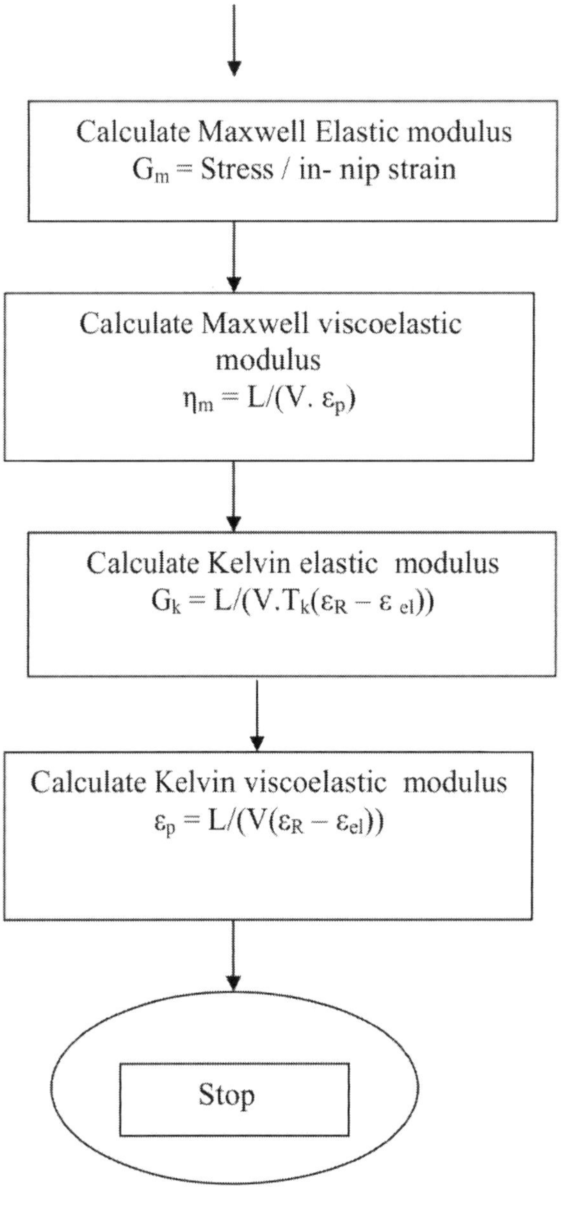

Figure 1: Procedure for estimating parameters of viscoelastic model

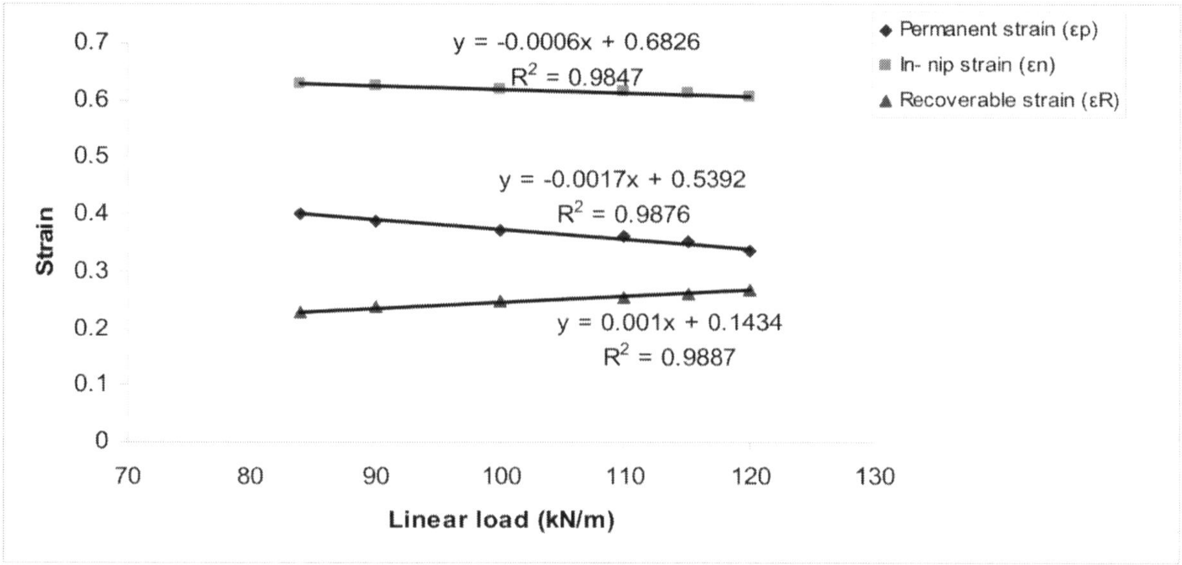

Fig 2: Permanent strain, in- nip strain and recoverable strain of coated paper with varying load (softnip calender)

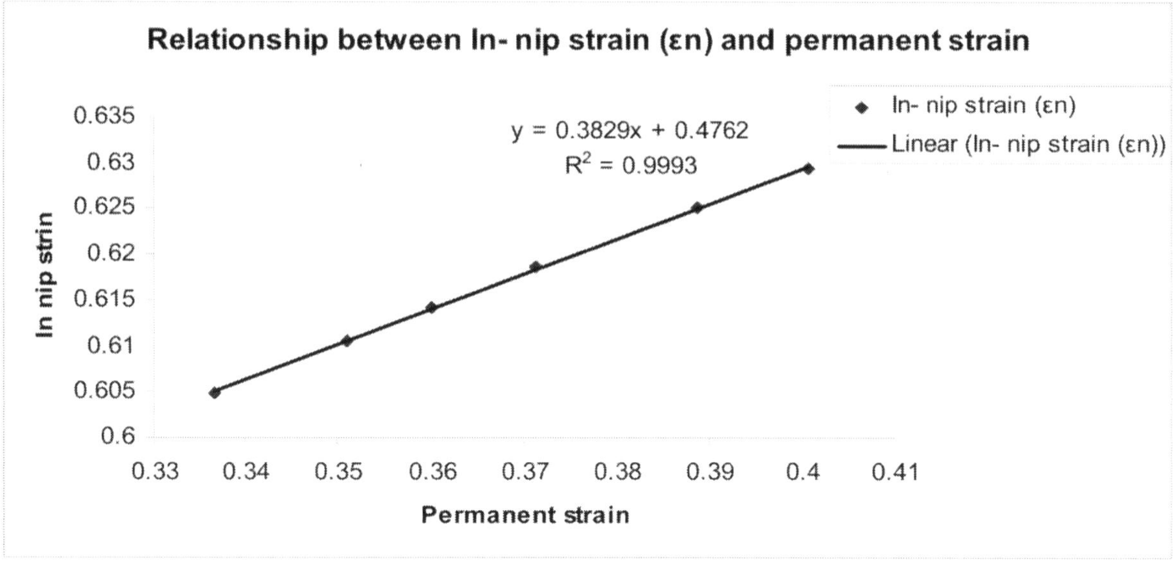

Fig. 3: Relationship between Permanent strains and In- nip strain for coated paper (soft nip calender)

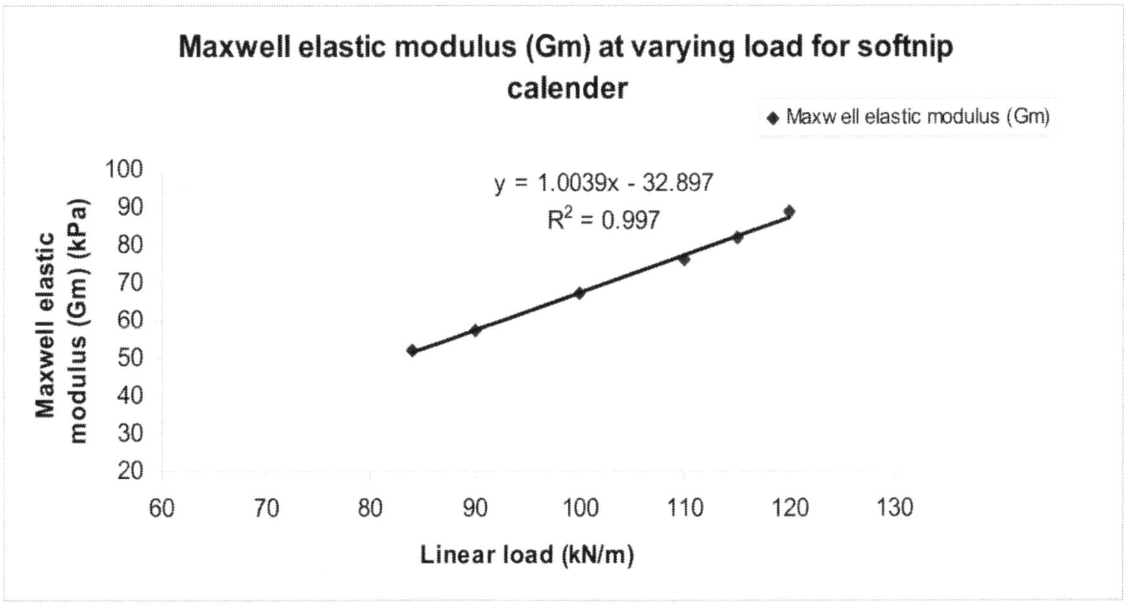

Fig. 4: Maxwell elastic modulus (G_m) of coated paper at varying load (softnip calender)

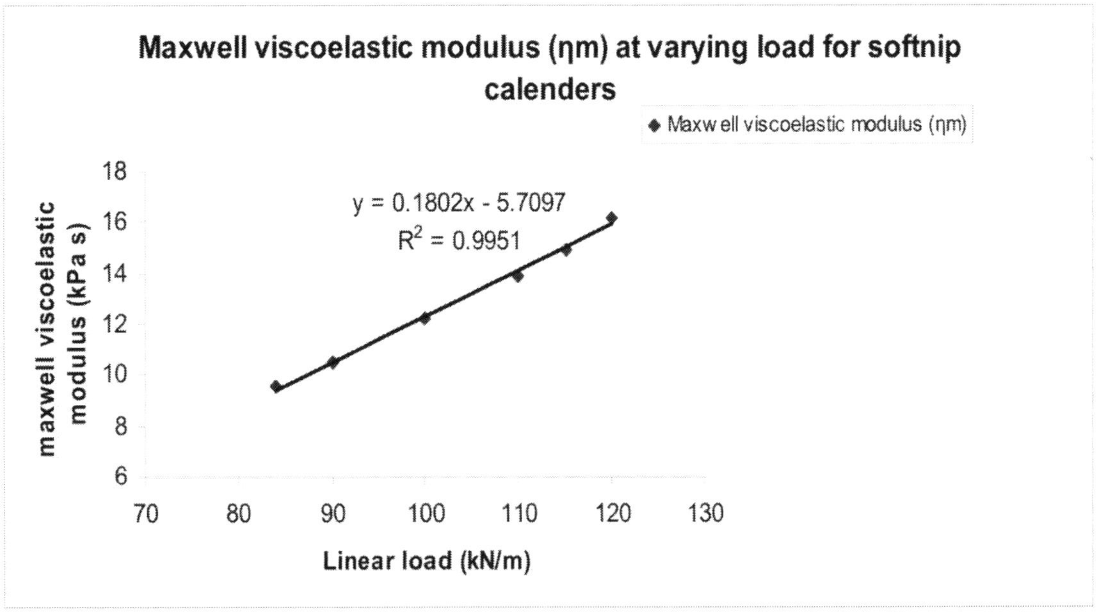

Fig 5: Maxwell viscoelastic modulus (η_m) of coated paper at varying load (softnip calender)

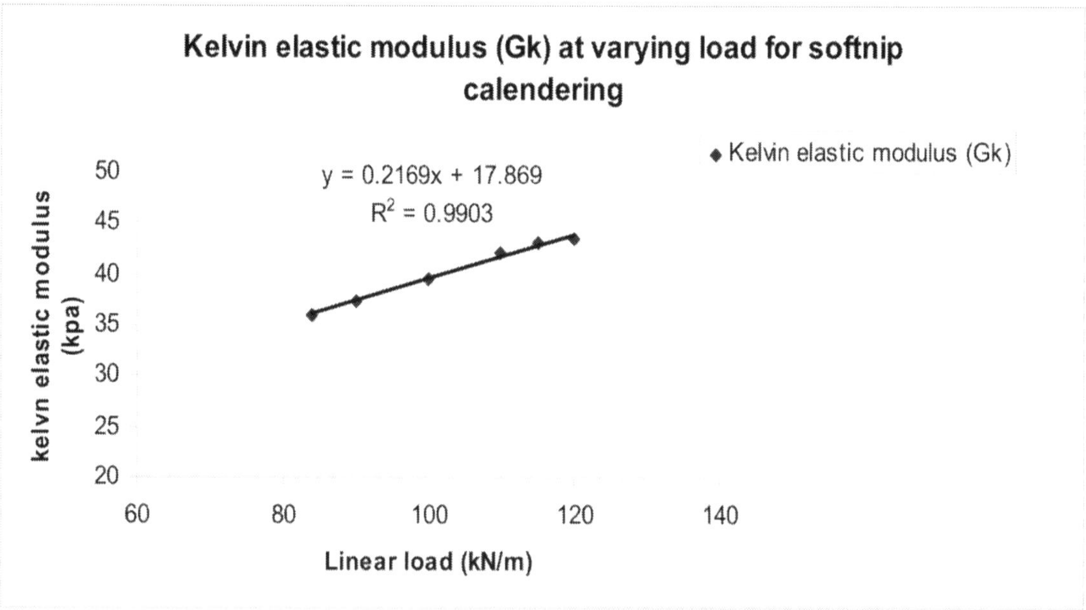

Fig 6: Kelvin elastic modulus (G_k) of coated paper at varying load (softnip calender)

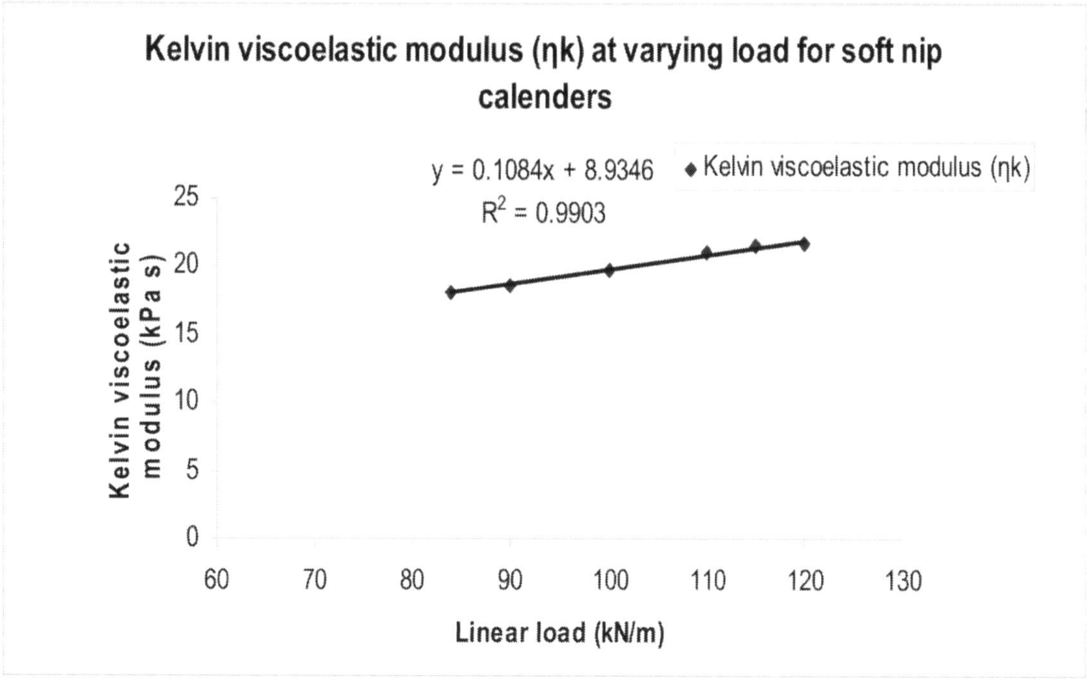

Fig 7: Kelvin viscoelastic modulus (ε_k) of coated paper at varying load (softnip calender)

Characterization of sludge from combined deinking technology

Shilpa Kulkarni[1a], Vivek Kumar[1] and M. C. Bansal[1],
[1]Department of Paper Technology, Indian Institute of Technology- Roorkee, India
Corresponding author: [a]shilpa.ips@gmail.com

1. INTRODUCTION

Recycled fiber has become an important source for the production of the paper. In the early 70's decades, only 7% of waste paper was used as a raw material for the production of the paper which is now increased up to the 47% and acting as a major raw material for paper industry. It is possible due to improved recycling system of wastepaper. Presently almost all grades of paper are being recycled. The share of mixed office waste (MOW) in the recovered wastepaper has increased rapidly during last decade. Non-impact printed paper represents the major share in MOW.

The deinking process is a process which is mainly utilized for removal of paper ink .The sludge generated during this process is called deinking sludge. Its disposal is one of the most important tasks for environmental protection. In recent years, deinking mills have striven to put their sludge by products to positive use [1-3]. The detail characterization of deinking sludge provides important information such as reutilization and proper treatment of sludge before disposal. However, the characteristics of sludge produced by the mill depend on the type of the wastepaper and deinking process. The sludge largely comprise of ink waste, generally with 60% moisture content and varying amounts of Si, Al, Cu, Ti, Fe and Ca as well as organic matters [4-6] and these by-product largely sent to landfill.

The most common options for the processing of paper industry sludge range from their exploitation for agricultural purposes, compositing, or use as a primary material in the manufacture of ceramics and cement [7-9], to energy recovery in biomass boilers or fluidized bed systems.

Each deinking operation produces sludge with somewhat unique characteristics, depending on the type of waste paper processed and deinking process. Waste paper with high amounts of filler, such as clay-coated magazine paper, produces much lower yields of usable fiber than waste paper having most of cellulose fiber, such as computer printout or old newspaper. Lightly printed paper produces less sludge than the heavily printed paper. Similarly alkaline papers contain high filler-to-fiber ratios and, therefore, produce more sludge. In present scenario different deinking process are used. Enzymatic deinking is advantageous for industrial usage because it is efficient, quick and has a low environment impact. Besides, UV-irradiation and Ultrasound technologies have begun to be investigated by some researcher, which is still in its starting stage [10-13]

Characterization of denking sludge generated from chemical deinking process is well reported in literature [12]. Various techniques can be applied to characterize the sludge [6, 11]. The present work focuses on the characterization of deinking sludge generated from photocopier waste paper by utilising various chemical and combined deinking treatment process namely: i) Chemical deinking, ii) Enzymatic deinking, iii) Combined deinking (Chemical + Enzyme + UV-irradiation and Chemical + Enzyme + Ultrasound) technology, iv) Chemical + UV-irradiation, v) Chemical + Ultrasound, vi) Enzyme + UV-irradiation and vii) Enzyme + Ultrasound treatment.Various researchers are investigated on deinking sludge but no result was found on the characterization of combined deinking sludge.

2 Material and Methods

2.1 Raw material

Photocopier grade paper (basis weight 70 gsm) was used to produce waste paper after printing with Xerox Black toner, using Xerox Workcenter pro420 photocopier machine. All sheets were uniformly printed.

2.2 Experimental plan :

Flow chart of experimental plan was show in Figure 1(a) chemical deinking and combined deinking and Figure 1(b) Enzymatic deinking and combined deinking

2.3 Deinking process

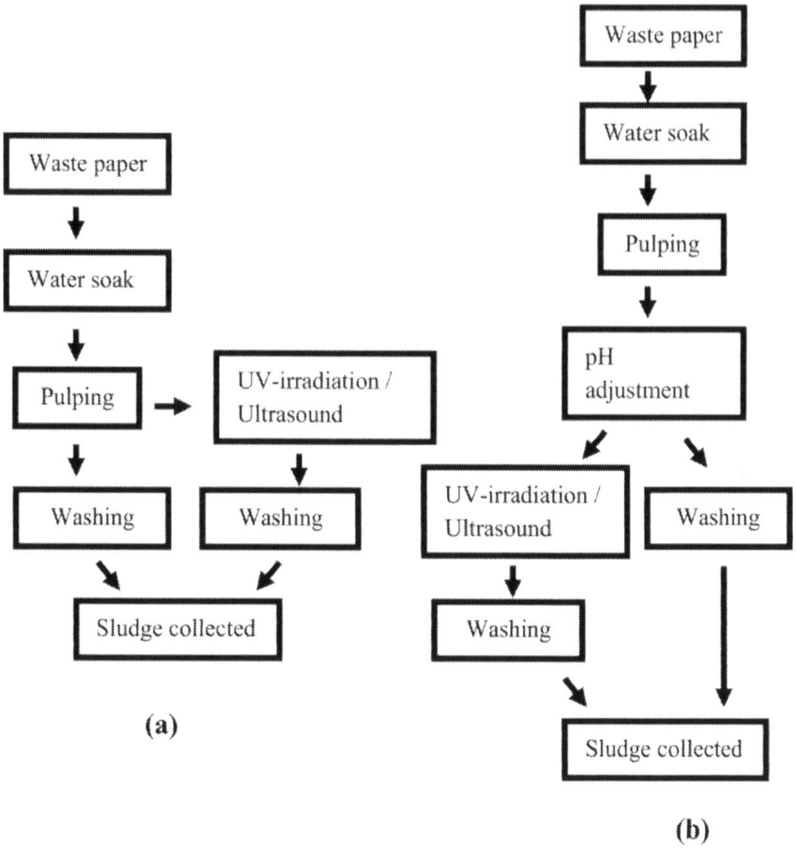

Figure: 1 (a) chemical deinking and combined deinking process **(b)** Enzymatic deinking and combined deinking process

2.3.1 Pulping

Hydrapulper is used for slushing of the xerographic paper used during the process under controlling temperature and at fixed rotor speed 180 rpm. The detail conditions for chemical dosing and enzymatic pulping are shown in **Table 1.**

Table 1: Pulping conditions and chemical dosing

Specification	Chemical deinking	Enzymatic deinking
Hydrapulper consistency	8%	8%
Pulping time of hydrapulper	25-30 min	25-30 min
Temperature	60°C	55°C
Chemicals	Sodium hydroxide- 2.0% Sodium silicate -2.5% Hydrogen peroxide- 0.2% DPTA -0.5%	-
Surfactant	Oleic acid -0.8%	Oleic acid-0.8%
Enzyme dosing	-	Cellulase-10 IU/g O.D paper
pH	-	4-5
Enzyme incubation	-	45 min
UV-irradiation time	25 min	25 min
Ultrasonication time	25 min 15 min	25 min 15 min
Washing time	5 min	5 min

2.3.2 Washing

Washing was carried out by using 60 mesh wire sieves for 5min (~20L water) under running tap water.

2.3.3 Enzyme

Commercial cellulase enzyme (Hi-media) was used for deinking purpose. The enzyme was stored at 4°C prior to use. This enzyme acts mainly on the surface of the fiber in the acidic to neutral pH range (pH 4.0 to 5.0) and between temperatures of 40 and 55°C. The cellulase activity was measured using the carboxymethyl cellulase (CMCase) method [14]. Its highest activity was found to be 1000 IU/g O.D paper at 55°C and pH 4.8. .

The enzyme dose were optimized from the range between (0-15) IU/g O.D paper from this optimization 10 IU/g O.D paper was best. The enzyme was dissolve in citrate buffer having pH (4.8).

2.3.4 UV –irradiation treatment

UV–irradiation treatment was carried out in a plastic container for 25 min.. The plastic container containing the pulp mixture was placed in a timber-framed UV (ultra violet) reactor (77cm ×36cm × 71cm) equipped with (λ =365nm), 72W (Philips) on the top side, located at a distance of 15cm from the sample. The reaction mixture was subjected to irradiation under UV light with continuous stirring. A fan fitted on the side wall was used to lower the heat generated by UV lamps. All the experiments were carried out in completely mixed and batch mode.

2.3.5 Ultrasound treatment

Ultrasound treatment was carried out in an Ultrasonicator (Model no. UTS-120, Unitech Ultrasonics, Mumbai, India) for 15 min and 25 min. The glass container containing the pulp mixture was placed inside chamber. The size of chamber is (150×150×150) in mm, chamber capacity is (3L), and Power of chamber is (120W)

Ultrasound which are converted at high efficiency into mechanical vibrations. This causes violent distension and contraction of cleaning fluids giving rise to millions of microscopic vacuum bubbles which forms and collapse at the same rate and produce a by-product known as "Cavitation " It produces intense scrubbing action throughout the liquid thereby rapidly and efficiently removing contaminants even from the intricate parts of the components.

2.4 Paper Sheet and Sludge characterization

Hand sheets were prepared by using Tappi test method T 205 sp-95. The tensile strength, tear index, and burst strength of paper sheet were determined by the Tappi test method T 494 om -96, T 414 om -98 and T 403 om-97 respectively. Sludge obtained by the washing of the waste paper was characterized by the Thermo gravimetric analyses (TGA) and CHNS analysis. TGA of deinking sludge were carried using EXSTAR TG/DTA 6300. The condition of thermal analysis was from the temperature ambient to 1000°C with the heating rate of 20°C/min under the flow of nitrogen at 200 ml/ min.

3. Result and Discussion

After pulping in presence of various doses of enzyme, we did the flotation of pulp and further determined the brightness of paper sheet which was prepared from enzyme deinked pulp. Table 2 represented the brightness before and after flotation in various doses enzyme deinked pulp. From Table 2 we observed that utilization of 10 IU/ g of enzyme does give the higher brightness. In other word we can say that 10IU/ g of enzyme does removed more ink in comparison to other doses of enzyme.

Table 2: Optimization of enzyme dosing in deinking process

Optimization of enzyme dosing in deinking process		
Optimization of enzyme dosing (IU/g O.D paper)	Brightness Before flotation	Brightness after flotation
0	69.70	76.10
5	73.20	77.40
10	76.90	81.70
15	77.00	81.70

After optimization of enzyme dose we again did the pulping in presence of 10 IU/ g of enzyme dose pulp (contain higher brightness and further treated the pulp under either only by UV irradiation or ultrasonication or by all three combined treatment i.e. chemical, enzyme and UV irradiation/ ultrasonication) . Three other deinking treatments were done separately which was (1) Chemical, (2) Chemical and UV irradiation and (3) Chemical and ultrasonication. The brightness of all the deinking treatments was represented in the Table 3. We observed from Table 3 that (i) combined deinking treatments chemical + enzyme + ultrasound (15 & 25 min.), chemical + US (25min.) and chemical + US (15min.) are better in comparison to other treatments due to higher brightness. (ii) According to brightness results, Ultrasonication was found to be more effective than the only enzyme treatment and enzyme + chemical treatment.

In the Table 4, it shows that tensile index is increases ~26.4 % when combined deinking technologies were used as compared to individual chemical, enzyme and UV-irradiation treatment. The Tear index was increased about to 8.5% in both cases i.e. enzyme with UV-Irradiation treatment for 25 min and enzyme with Ultrasound treatment for 25min. It was more effective then the chemical treatment.

Table 3: ISO Brightness of chemical, enzyme and combined deinking treatments

Deinking Treatments										
Particulars	**Chemical treatment**				**Enzyme treatment**				**Chemical + Enzyme + UV**	**Chemical + Enzyme + Ultrasound**
	Chemical	**Chemical + UV**	**Chemical + US**		**Enzyme**	**Enzyme + UV**	**Enzyme + Ultrasound**			
		25 min	25 min	15 min		25 min	25 min	15 min	25 min	25 min
ISO Brightness (%)	74.3	74.3	75.1	74.3	73.4	74.8	74.9	75.4	68.8	75.7

Table 4: Physical properties of chemical, enzyme and combined deinking treatment

Deinking Treatments										
Particulars	**Chemical treatment**				**Enzyme treatment**				**Chemical + Enzyme + UV**	**Chemical + Enzyme + Ultrasound**
	Chemical	**Chemical + UV**	**Chemical + US**		**Enzyme**	**Enzyme + UV**	**Enzyme + Ultrasound**			
		25 min	25 min	15 min		25 min	25 min	15 min	25 min	25 min
Tensile index (mN/g)	10.8	13.1	16.0	15.0	11.1	13.3	7.7	10.9	10.6	13.4
Tear index ($mN.m^2/g$)	4.6	6.3	4.4	4.7	5.4	6.9	5.3	3.7	4.3	4.7
Burst index ($kPa.m^2/g$)	0.4	0.4	0.5	0.4	0.3	0.4	0.3	0.3	0.3	0.4

3.1 Thermal characterization

Figure 2 is the graph of chemical + enzyme + US (25min) treated sludge sample. The first weight loss in all treatment is very small weight loss which must be due to the presence of the moisture content in the sludge. The combustion of cellulose to form volatile combustion products found in second weight loss in between 100°C-370°C while residual material can be obtained at temperature 900°C. This should be percent of ash. The added fillers % in sludge can be observed by % weight loss between the temperature range 370-750 ºC. Table 5 represents the % of various contents obtained by TGA technique in combined deinking process. From the Table 5 we have found that fiber content in sludge sample was found higher (~55 %) when it was treated by chemical + US , Enzyme + US and only enzyme treatments methods.

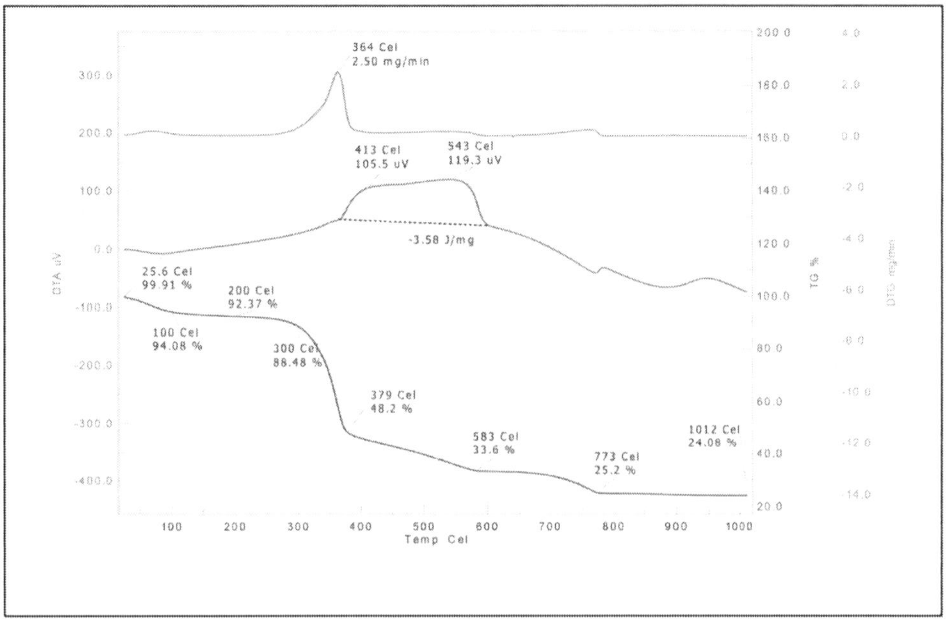

Figure 2: Thermal analysis of chemical + enzyme + US (25min) treated sludge sample

Table 5: TGA Analysis of chemical, enzyme and combined deinking treatment sludge

Percentage of various content on the basis of TGA analysis										
Particulars	**Chemical treatment**				**Enzyme treatment**				**Chemical + Enzyme + UV**	**Chemical + Enzyme + Ultrasound**
	Chemical	**Chemical + UV**	**Chemical + US**		**Enzyme**	**Enzyme + UV**	**Enzyme + Ultrasound**			
		25 min	25 min	15 min		25 min	25 min	15 min	25 min	25 min
Moisture % up to 100°C	4.12	49.10	4.14	6.10	4.09	-	6.80	6.29	6.15	5.90
Volatile combustion products% 100-370 °C	38.68	27.10	55.96	58.10	53.31	-	55.4	55.01	42.15	45.88
Inorganic % 370-750 °C	24.80	12.90	20.20	19.90	18.40	-	20.50	19.60	23.90	23.00
Residual Material at 1000 °C	31.36	10.31	18.67	15.21	23.21	-	16.34	18.01	27.19	24.08

3.2 CHNS Analysis

The CHNS Analyzer find utility in determining the percentages of Carbon, Hydrogen, Nitrogen, Sulphur and Oxygen of organic compounds, based on the principle of "Dumas method" which involves the complete and instantaneous oxidation of the sample by "flash combustion". Results are represented in Table 6. The results shows the higher % of carbon contents in sludge sample obtained by the chemical + US, Enzyme + US, Enzyme + UV and only enzyme treatment. It also confirmed that above said treatment contain higher fiber content in sludge sample.

Table 6: CHNS Analysis of deinking sludge

	Deinking Treatments									
Particulars	Chemical treatment				Enzyme treatment				Chemical + Enzyme + UV	Chemical + Enzyme + Ultrasound
	Chemical	Chemical + UV	Chemical + US		Enzyme	Enzyme + UV	Enzyme + Ultrasound			
		25 min	25 min	15 min		25 min	25 min	15 min	25 min	25 min
C	25.6	-	32.3	34.2	43.2	32.8	32.9	33.5	28.3	28.5
H	2.3	-	3.9	4.7	3.2	3.8	5.2	4.8	2.7	3.5
N	0.2	-	0.2	0.2	0.2	0.2	0.1	0.1	0.1	0.1

4. Conclusion

In denking process, the utilization of 10 IU/ g of enzyme doses removed more ink in comparison to other doses of enzyme. (1) Combined deinking treatments i.e. chemical + enzyme + ultrasound, chemical + US (25min.) and chemical + US (15min.) are better in comparison to other treatments due to higher brightness. Introduction of Ultrasonication treatment was found to be more effective in deinking process. Both tensile strength and Tear index is increases when combined deinking technologies were used as compared to individual chemical, enzyme and UV-irradiation treatment. Fiber content in sludge sample was found higher (~55 %) when it was treated by chemical + US, Enzyme + US and only enzyme treatments methods. In above result, it shows that enzyme treatment is effective and it improves the strength properties with combined technology

5. Acknowledgement

The authors wish to convey their sincere thanks to Dr R.M. Mathur, Director, CPPRI, and Saharanpur for their help and support.

6. References

[1] Kim S, Kim H-J, Park JC. Application of recycled paper sludge and biomass materials in manufacture of green composite pallet. Resources, Conservation and Recycling 2009;53:674-9.

[2] Lou R, Wu S, Lv G, Yang Q. Energy and resource utilization of deinking sludge pyrolysis. Applied Energy;2010; 90:46-50.

[3] Yan S, Sagoe-Crentsil K, Shapiro G. Reuse of de-inking sludge from wastepaper recycling in cement mortar products. Journal of Environmental Management; 2011; 92:2085-90.

[4] Pera J, Amrouz A. Development of Highly Reactive Metakaolin from Paper Sludge. Advanced Cement Based Materials 1998;7:49-56.

[5] Behin J, Vahed S. Effect of alkyl chain in alcohol deinking of recycled fibers by flotation process. Colloids and Surfaces A: Physicochemical and Engineering Aspects 2007;297:131-41.

[6] Méndez A, Fidalgo JM, Guerrero F, Gascó G. Characterization and pyrolysis behaviour of different paper mill waste materials. Journal of Analytical and Applied Pyrolysis 2009;86:66-73.

[7] Ahmadi B, Al-Khaja W. Utilization of paper waste sludge in the building construction industry. Resources, Conservation and Recycling 2001;32:105-13.

[8] Souza PSL, Dal Molin DCC. Viability of using calcined clays, from industrial by-products, as pozzolans of high reactivity. Cement and Concrete Research 2005;35:1993-8.

[9] Conesa JA, Gálvez A, Fullana A. Decomposition of paper wastes in presence of ceramics and cement raw material. Chemosphere 2008;72:306-11.

[10] Gaquere-Parker AC, Ahmed A, Isola T, Marong B, Shacklady C, Tchoua P. Temperature effect on an ultrasound-assisted paper de-inking process. Ultrasonics Sonochemistry 2009;16:698-703.
[11] Gea T, Artola A, SÃ¡nchez A. Composting of de-inking sludge from the recycled paper manufacturing industry. Bioresource Technology 2005;96:1161-7.
[12] Beauchamp CJ, Charest M-Hln, Gosselin A. Examination of environmental quality of raw and composting de-inking paper sludge. Chemosphere 2002;46:887-95.
[13] Zhenying S, Shijin D, Xuejun C, Yan G, Junfeng L, Hongyan W, et al. Combined de-inking technology applied on laser printed paper. Chemical Engineering and Processing: Process Intensification 2009;48:587-91.
[14] Ghosh TK. Measurement of Cellulase Activities. Pure & appli chem 1987;59:257-68.

(245a) Closed-Loop Decision Making Under Uncertainty for Sustainability Enhancement of Industrial Systems

Zheng Liu and Yinlun Huang, Department of Chemical Engineering and Materials Science, Wayne State University, Detroit, MI

Sustainability performance of manufacturing systems could be significantly improved using carefully selected design, manufacturing, and management technologies. However, due to inherent uncertainty associated with accessible data and information relevant to technology candidates, how to identify most desirable technologies is always a challenge. This becomes more challenging for assuring future sustainability performance of the systems as future market conditions, environmental regulations, etc., could change along the time.

In this paper, we introduce a novel decision-making methodology for sustainability performance improvement under severe of uncertainty, where data uncertainty and scarcity leads to no data distribution information derivable. The introduced methodology is developed by resorting to sustainability principles, control science, and process systems engineering fundamentals. In development, we consider industrial sustainability problems as a general system control problem. In this context, a basic closed-loop feedback control scheme is used to characterize an industrial system under investigation, evaluate and analyze its sustainability performance, and identify most suitable technologies based on the preset sustainability goals, for the improvment of the system's sustainability under budget and other constraints. This sustainability decision making and system performance improvement process will continue as a response to the change of external situations that may be related to market, regulation, policy, etc. Since a large amount of information and data is uncertain, imprecise, and incomplete, an interval-parameter-based approach will be used in system characterization, assessment, and decision making. The effectiveness of the derived decisions and system stability will also be evaluated before and after implementation. The proposed decision making methodology has been successfully used to investigate a number of industrial problems. In this presentation, a comprehensive study on sustainable biodiesel manufacturing in a large geographical region will be illustrated.

(245b) Second Law of Thermodynamics and the Notion of Externalities of Macro-Economics: Two Pillars of Assessing the Sustainability of a Manufacturing System

L. T. Fan, Chemical Engineering, Kansas State University, Manhattan, KS and Tengyan Zhang, Western Research Institute, Laramie, WY

Sustainability engineering, an emerging discipline of engineering, can be regarded as an engineering discipline to craft frameworks, or establish paradigms, for attaining sustainability in any social or human endeavor including sustainable manufacturing. No discipline or branch of engineering can be enduring without theoretical principles underlying it or without a firm theoretical foundation. In this regard, two cornerstones of sustainability assessment of any system or process should be the second law of thermodynamics and the notion of externalities of macro-economics.

The inevitable entropy increase attributable to a process of concern manifests itself in exergy dissipation by the process, which can be evaluated by resorting to the available energy balance; this balance results from integrating the first and second laws of thermodynamics. In reality, however, the available energy balance also implicitly embodies the mass conservation law: It entails the detailed accounting of every material species involved in the process in light of the species' matter-energy.

Economic externalities attributable to a system or process of concern manifest themselves in the process' external costs. These external costs inherent to the system can be evaluated through cost accounting or estimation in conjunction with the process' internal costs, such as the labor and equipment costs.

Various methodologies are available for assessing the sustainability of manufacturing or process systems, which take into account exergy dissipation as well as cost, or can possibly take both into account. One of such methodologies is sustainability potential; this hierarchical approach is proposed to estimate the sustainability-potential of a system based on any of the alternative manufacturing or synthetic routes. This novel notion of sustainability-potential is a generalization of the notion of profit-potential, the estimation of which renders it possible to screen out unsustainable processes derived from the alternative synthetic routes at the earliest stage. The efficacy of sustainability potential is illustrated with the manufacture of vinyl chloride.

(245c) A Shortcut Graphical and Spreadsheet Technique for Identifying Mass Exchange Networks

Russell F. Dunn, Chemical and Biomolecular Engineering , Vanderbilt University, Nashville, TN and Kenneth A. Debelak, Chemical and Biomolecular Engineering, Vanderbilt University, Nashville, TN

Cost-effective waste minimization within chemical processes continues to be a critical component for future sustainability of large chemical and polymer manufacturing companies. Several methods have been developed for the synthesis of mass exchange networks (MENs) for waste minimization. In fact, several of the methods previously developed are based on a two stage targeting approach. The first stage uses the rich (waste) and lean (mass separating agents) composite curves to identify the maximum mass integration possible for an undesirable species to be transferred from a group of waste streams to a group of internal process streams that are used as mass separating agents. Furthermore, the second stage identifies the network of mass exchange technologies that meets the target identified in stage one. Several strategies have been employed for identifying the MEN in stage two including, but not limited to, mathematical optimization and state-space structural approaches.

This paper will present a new graphical/spreadsheet technique and design strategy for identifying the MEN in stage two. This technique allows the simultaneous evaluation of multiple diverse mass separation technologies. Advantages of this strategy will be highlighted and several case studies will be discussed that illustrate the effectiveness of this approach. Finally, experiences using this strategy in teaching undergraduate chemical engineering design courses will also be discussed.

(245d) Synthesis and Performance of Sodium 9,10-Dihydroxy Stearic Acid

Xiao Zhao, Shool of Chemical and material engineering, Jiangnan University, Wuxi, China and Yun Fang, School of Chemical and material engineering, Jiangnan University, Wuxi, China

9, 10-dihydroxy stearic acid (DHSA) and its derivatives are of high value due to their wide range of industrial application, including cosmetic, food, lubricant, plastic additives, personal care, pharmaceutial products and etc. It used to be synthesized by introducing hydroxy groups into carbon-carbon double bond of oleic acid with various catalysts. Almost all those process could not be industrialized for the more excessive co-oxidant and the boring separation process. Here we describe a more convenient method to obtain high purity DHSA by purified oleic acid and hydrogen peroxide with the catalyst of formic acid. The conversion of the oleic acid was over 97%. And the crude DHSA product was obtained by decomposition a small amount of excessive hydrogen peroxide over MnO_2, followed by evaporating water and a residue catalyst. The chemical structure of the recrystallised DHSA was characterized by IR, NMR and MS. Sodium 9,10-dihydroxy stearic acid (DHSA-Na) was prepared by neutralization of DHSA with NaOH. Surface tension, wetting property, lime soap dispersing power (LSDP), Krafft point, detergency and other performances of DHSA-Na were well determined. The results indicated a variety of applications in many fields for the excellent surface and interface chemical performance.

(245e) A Framework for Chemical Risk Management in Product Manufacturing and Its Application to Industrial Cleaning

Masahiko Hirao, Emi Kikuchi and Yasunori Kikuchi, Department of Chemical System Engineering, The University of Tokyo, Tokyo, Japan

For sustainable manufacturing, voluntary efforts are needed in industry to manage risks on human health and the environment associated with the use of chemicals. In this study, a practical framework for chemical risk management in product manufacturing based on simultaneous evaluation of functional, economic, and risk aspects is proposed. The framework is embodied through the case study of industrial cleaning processes, where open-top processes with chemical substances such as chlorinated organic solvents, inorganic compounds, and surfactants, are used.

Process and operation design aimed at risk reductions needs two major activities: evaluation of current and alternative processes, and decision-making on the selection of alternatives. Knowledge and tools, such as technical information, risk evaluation method, and process model, are necessary in order to evaluate risks of processes, as well as their functional and economic aspects. Technical information is systematically organized for cleaning processes, where processes have been designed on the basis of heuristics conventionally. Product defects associated with the careless process modifications aimed at risk reductions can be avoided by evaluating function of alternative processes simultaneously on the basis of the technical information. Risk analysis methods using LCA and risk assessment are applied to the evaluation in order to analyse risk trade-off relationships among alternative processes. A model of open-top cleaning processes is developed to estimate process data needed for evaluating alternative processes.

It is necessary to manage conflicts among evaluated functional, economic, and risk aspects in the decision-making phase. Engineers need to select appropriate alternatives for risk reductions, so that the improved process can achieve not only risk reduction targets but also functional requirements and economic viability. The proposed framework including concrete procedure of such multi-objective decision making is developed through the study on actual cases of process improvement in industrial cleaning. It is revealed that the flexibility assessment of constraints on product quality can be effective in managing conflicts between functional and risk aspects. Furthermore, the engagement of up- and downstream manufacturers in product supply chain can contribute to expanding the range of options of risk reduction measures.

(245f) Scheduling for Multiple Chemical Plant Turnarounds to Minimize Regional Air Quality Impacts

Tianxing Cai and Qiang Xu, Dan F. Smith Department of Chemical Engineering, Lamar University, Beaumont, TX

During the chemical manufacturing process, multiple chemicals are released into the atmosphere including benzene(C_6H_6),carbon monoxide(CO), nitrogen oxides(NO_x), sulfur dioxide(SO_2), and volatile organic compounds(VOC_s).Depending on meteorological conditions, emission quantity, and the fate and transport of chemical species, the environmental impact can affect air quality over a great area. This will cause the regional increased exposure to primary pollutants for neighboring populations with the summation of emission pollutant due to the temporal co-occurrence of start-ups and shutdowns among multiple chemical plants. Furthermore, these emission chemical will react with each other to form other secondary pollutants. The mechanism should follow the principle of reaction kinetics. Therefore, from the emission control point of view, the best practices for improving regional air quality should integrate both efforts on plant-wide emission minimization at every industrial point source, relationship and regional-wide multi-plant emission variability control.

In this paper, a scheduling methodology for coordinating multiple chemical plant start-ups is developed to minimize regional air quality impacts by the combination of reaction kinetics, atmosphere dispersion and scheduling will give the practical and feasible scheme to reduce primary and secondary pollutants in the region where multiple plants are co-located. The whole methodology will include two stages: proactive risk assessment and scheduling optimization. During the risk assessment, the focus is put on the changes in the weather conditions (stability class and wind condition), the pollutant and regional emissions from other sources, especially in the extreme weather or other emergent situations. The simulation result will indicate whether it needs to optimize the temporal sequence of multiple plants' activity. If necessary, the optimization modeling result will give an optimized temporal distribution of start-ups and shut-downs to reduce the chance of emission pairing and the concentration of primary and secondary pollutant. The development provide quantitative decision support for multiple stake holders, including government environment agencies, chemical industry, and local communities.

(254a) Integrated Processing of Microalgae for Valuable Food Ingredients, Chemicals and Fuel Production

Kaige Wang[1], Yi Liang[2], Zhiyou Wen[2] and Robert Brown[1], (1)Center for Sustainable Environmental Technologies, Iowa State University, Ames, IA, (2)Food Science and Human Nutrition, Iowa State University, Ames, IA

Lipid-based biofuel production by microalgae has captured interest of researchers due to microalgae's high biomass yield and ability to synthesize and accumulate large quantities of lipid. However, high cost of algal cultivation and processing has limited the commercialization of algal fuel production. Developing a microalgae biorefinery strategy, i.e., using every components of algal biomass for fuels and chemicals production is a feasible approach for commercially utilization of the algal biomass.

This study reports a hybrid processing for treating the microalgae *Schizochytrium limacinum*. Crude glycerol, a major byproduct in biodiesel production plants, was used as substrate for producing high content of algal oil containing DHA (22:6, n-3), a high value omega-3 polyunsaturated fatty acid with beneficial effects on human health. The residues after oil-extraction from *S. limacinum* were then catalytically pyrolyzed on a micro-furnace pyrolyzer with zeolite catalyst being used. Aromatics and ammonium was generated from this catalytic pyrolysis process. Among the aromatics, benzene, toluene and xylene, which are valuable basic petrochemicals,were the dominant ones. Most of the nitrogen in the residues was released in ammonium form, which may be in turn used as nutrient for algal culture. In summary, all the components in the biomass of the *S. limacinum* were utilized for producing chemicals, fuels and value-added products which can be future enhanced the economic for algal industry.

(254b) Pyrolytic Fractionation of Oleaginous Feedstocks

Balakrishna Maddi, Sridhar Viamajala and Sasidhar Varanasi, Chemical & Environmental Engineering, University of Toledo, Toledo, OH

Traditional pyrolysis of oleaginous biomass results in bio-oils that contain a mixture of compounds derived from protein, starch and lipids – constitutive biopolymers of these feedstocks. As a result, such biooils have to undergo energy-intensive denitrogenation and deoxygenation to render them suitable as fuels. We have recently developed a "*pyrolytic fractionation*" technique based on the observation that the protein, starch and lipid portions undergo thermal degradation over distinct non-overlapping temperature intervals. Proteins are most thermally labile followed by starch and lipids. Hence, conducting the pyrolysis in three temperature stages provides an opportunity to recover homogenous and biopolymer-specific biooils. Bio-oils from protein and starch can find other applications, in addition being lower grade fuels. For instance, N-rich biooil from protein can serve as a source of high value N-compounds and levoglucosans from starch pyrolysis could serve as precursors for value-added polymers. Similarly, bio-oil from lipid fraction, predominantly containing free fatty acids, can be converted to fuels (such as biodiesel or hydrocarbons) or other higher value fatty acid derivatives. Finally, the bio-char can be a soil amendment.

In this presentation, results showing component-specific biooil compositions and yields from pyrolytic fractionation of algal biomass and oil seeds will be presented. We will also present a comparison of energy needs for pyrolytic fractionation relative to other themochemical methods for conversion of oleaginous feedstocks.

(254c) Hydrothermal Carbonization and Supercritical Ethanol in Situ Transesterification for the Production of Algal Biodiesel

Robert Levine, Chemical Engineering, University of Michigan, Ann Arbor, MI and Phillip E. Savage, Chemical Engineering, University of Michigan, Ann Arbor, Ann Arbor, MI

The production of transportation fuels from microalgae is complicated by the fact that algae are typically grown suspended in a dilute aqueous media and have a high demand for nitrogen and phosphorous. In this work, we explore a new and environmentally preferable strategy for algal biofuel production. There are three major components of the bio-refinery model we have developed: 1) cultivation of oil-rich algae using recycled water and nutrients; 2) treatment of the harvested algae through hydrothermal carbonization to produce a biochar; and 3) conversion of the carbonized algal biomass into transportation fuels using in situ transesterification with supercritical ethanol.

Lipid-rich *Chlorella* was grown in bioreactors until it contained approximately 50% oil, then carbonized for 15 to 90 min at 180 to 235 °C. Biochar obtained by filtration retained >90% of lipids of the original biomass and was converted into biodiesel by reacting the solids with ethanol at 270-325 °C for 15-180 min. Fatty acid ethyl yields in excess of 90% were obtained while employing 10-100 times less ethanol than previously reported for comparable transesterification reactions. The nutrients liberated during hydrothermal carbonization were used to produce additional algal biomass, thereby recycling N and P. Overall, our work suggests that it is possible to produce net-energy positive algal biofuels with significantly less environmental impact than the petroleum fuels they replace.

(254d) Integrating Photo-Bioreactor and Fermentor to Produce Biofuels and Bioelectricity

Alim Dewan, Marci Kerls and M. Nazmul Karim, Chemical Engineering, Texas Tech University, Lubbock, TX

We have investigated simultaneous production of biofuels and bioelectricity from an integrated bioreactor that combines photo-synthesis and fermentation. In the integrated bioreactor, a fermentor and a photo-bioreactor were combined in a way that the carbon dioxide produced in the fermentation can be used to grow algae in the photo-bioreactor. Also, the reactors were attached to a proton exchange membrane in between so that they form a biological fuel cell. To synchronize the carbon dioxide production in the fermentation and the carbon dioxide utilization in photo-synthesis, the fermentor was run continuously but the photo-bioreactor was run in batch mode. An autotrophic microalgae, *Chlorella vulgaris,* and a bear yeast, *Saccharomyces bayanus,* were grown in the photo-bioreactor and in the fermentor, respectively. The ethanol production by the yeast, oil production by the algae, and electricity generation were monitored over time. The electricity generation was characterized by potentiodynamic polarization. In a two-liter size integrated bioreactor, we observed an ethanol productivity of 0.49 ± 0.37 g l^{-1} hr^{-1}, algae oil productivity 2.5 mg l^{-1} hr^{-1}, power density of 70.61 ± 12.41 mW/m^2 of the electrode, and current density of 155.02 ± 31.23 mA/m^2. Preliminary scale-up studies show a great potential for this integrated system.

(254e) Determination of Lipid Hydrolysis Kinetics in Soybean Oil and Algal Systems

Joshua C. Wissinger[1], Amber L. Bosley[1] and Constance Schall[2], (1)Chemical and Environmental Engineering, University of Toledo, Toledo, OH, (2)Chemical & Environmental Engineering, University of Toledo, Toledo, OH

Determination of lipid hydrolysis kinetics in soybean oil and algal systems

Joshua C. Wissinger, Amber L. Bosley, Constance A. Schall

The third generation of biofuels, algae, is of growing interest as a renewable feedstock due to its high yield of energy per unit area, use of carbon dioxide and other available resources for growth, and minimal competition with the food industry. Biodiesel and jet fuel can be produced from lipid-rich algae. However, extracting lipids from algae is complicated due to the high water content of harvested and mechanically dewatered algae (60 to 90% water). Thus, there is a need for processes that convert wet algal biomass into fuels or fuel precursors directly in the aqueous phase. Such a process alternative is to use the water in the harvested and partially dewatered algae for hydrolysis of lipids to free fatty acids (FFA), forming aqueous and oil rich phases for ease of extraction and separation of the lipid component. Lipid processing techniques include refining FFA by catalytic cracking and hydrotreating for jet fuel production, or the UOP/Eni Ecofining™ process to obtain clean green diesel. In our first step, subcritical water is reacted with lipids producing FFA by lipid hydrolysis eliminating the need for expensive and water sensitive catalysts. In the initial development of an empirical kinetic model, soybean oil, a system much less complex than alga, was used in hydrolysis studies at subcritical water (SCW) conditions. This model was applied to algal lipid hydrolysis. The robustness of our model has been tested by verifying both the autocatalytic behavior of lipid hydrolysis reactions, as well as hydrolysis reactions over a range of temperatures.

(254f) Microwave-Assisted Subcritical Water Extraction of Lipids From Wet Algae

Harvind Kumar Reddy[1], Yingqiang Sun[1], Yin Li[2], Tapaswy Muppaneni[1], Sundaravadivelnathan Ponnusamy[1], Shuguang Deng[1], Tanner Schaub[3], Barry Dungan[4], Francisco Holguin[4], Peter Lammers[5], Wayne Voorhies[6] and Peter Cooke[7], (1)Chemical Engineering, New Mexico State University, Las Cruces, NM, (2)Chemical and Biological Engineering, Zhejiang University, Hangzhou, China, (3)College of Agricultural, Consumer and Environmental Sciences, New Mexico State University, Las Cruces, NM, (4)College of Agricultural, Consumer and Environmental Sciences, New Mexico State university, Las Cruces, NM, (5)Energy Research Laboratory, New Mexico State University, Las Cruces, NM, (6)Molecular Biology, New Mexico State university, Las Cruces, NM, (7)Core University Research Resource Laboratory, New Mexico State University, Las Cruces, NM

Microwave-assisted extraction of algal oil at subcritical water conditions was demonstrated with wet algae samples containing up to 90% of water. The extraction efficiency of this method was compared to those of the conventional solvent extraction methods, Folch and Bligh & Dyer. The extraction parameters including biomass loading, temperature and extraction time are optimized after executing a set of designed experiments. Algal oil was produced after purification of bio crude with activated charcoal. The residual algal biomass was also analyzed for its fatty acid and energy content to validate potential byproduct development as part of algal biorefinery. Energy calculations were performed for microwave assisted subcritical water extraction and compared with those of the conventional methods. Analytical instruments including GC-MS, FT-ICR were used to analyze the biomass and biocrude. Thermogravimetric analysis (TGA), TEM and SEM equipped with EDS were used to characterize the biomass in different stages of the extraction process.

(260a) Characterization of Carbohydrate Accessibility and Enzyme Adsorption Capacity for Diverse Cell Wall Phenotypes Subjected to Alkaline Hydrogen Peroxide Pretreatment

Muyang Li[1], Sivakumar Pattathil[2], Michael G. Hahn[2] and David Hodge[1,3], (1)Biosystems and Agricultural Engineering, Michigan State University, East Lansing, MI, (2)Complex Carbohydrate Research Center, University of Georgia, Athens, GA, (3)Chemical Engineering and Material Science, Michigan State University, East Lansing, MI

This study characterizes the changes to polysaccharides in terms of their abundance, accessibility, and affinity for binding by a range of cellulolytic enzymes and CBMs in the cell walls of grasses and dicots that were alkaline hydrogen peroxide (AHP) pretreated at varied severities or hydrothermally pretreated followed by AHP post-treatment. These bioenergy feedstocks represent relatively diverse cell wall phenotypes, including the monocots switchgrass (*Panicum virgatum* cv. Cave-In-Rock), corn stovers (a commercial hybrid and inbred brown midrib lines bm1 and bm3), and *Miscanthus* spp. and herbaceous and woody dicots including goldenrod (*Solidago* spp.) and hybrid poplar (*Populus* spp.). It was shown that increasing H_2O_2 loading results in increasing digestibility only for monocot grasses, while the digestibility of dicots is saturated at low H_2O_2 loadings. Polysaccharide accessibility in the residual cell wall is characterized through "glycome profile" which involves screening successive cell wall extracts with an array of 156 monoclonal antibodies with affinities for the entire spectrum of non-cellulosic cell wall polysaccharides revealing changes in their accessibility and extractability as a function of pretreatment. Enzyme binding is assessed by characterizing binding isotherms for the proteins contained commercial cellulase enzyme cocktails, purified cellulase components, and a library of GFP-linked CBMs.

(260b) Lignin Structural Changes Associated with Oxidative Pretreatment Catalyzed by Cu-Diimine Complexes

Zhenglun Li, Department of Chemical Engineering and Material Science and DOE Great Lakes Bioenergy Research Center, Michigan State University, East Lansing, MI and David Hodge, Chemical Engineering and Material Science, Michigan State University, East Lansing, MI

A wide range of alkaline-oxidative treatments of lignocellulose are known from the literature and practice of chemical pulping and bleaching and from more recent cellulosic biofuels applications. Alkaline hydrogen peroxide (AHP) has been developed as a pretreatment or post-treatment for increasing biomass digestibility by decreasing the recalcitrance of high-lignin biomass. Traditional AHP pretreatment uses the high pH of 11.5 for deprotonation of hydrogen peroxide which leads to the formation of oxidative radicals (·OH and $\cdot O_2^-$), whereas the low selectivity of the radical reactions and the non-productive decomposition of hydrogen peroxide to oxygen at alkaline pH limit the economic competitiveness of AHP pretreatment. Significantly, oxidative delignification catalyzed by coordinated transition metal ligands within laccases and peroxidases is the route taken for lignin degradation in nature. Catalyzed oxidative pretreatments offer the promise of mild temperatures, high reaction specificities, and high oxidizing efficiency that can simultaneously generate a material that is susceptible to enzymatic carbohydrate depolymerization, be performed at decreased pH with correspondingly less inorganics load, and generate modified low MW aromatics that may be potentially upgraded to petroleum-displacing fuels and chemicals. In this work, the effect of Cu-diimine complexes during catalyzed oxidative pretreatment on model lignins is characterized. Model lignins include a Kraft lignin and dioxane milled-wood and enzymatic lignins from select hardwoods and grasses, while characterization is performed using ^{31}P NMR, ^{1}H-^{13}C HSQC NMR and SEC which show distinct structural changes during pretreatment including, among others, clear evidence of lignin modification and depolymerization and cellulose depolymerization. Additionally, significant improved cellulolytic enzyme digestibility is found for select catalyzed pretreatments in woody dicots.

(260c) Mechanism of Exo-Exo Synergism Between CBH1 and CBH2 During the Hydrolysis of Cellulose

Makoto Ikeo, Shohei Okino, Yoshiki Ueno and Daisuke Taneda, Technology Development Department, JGC Corporation, Higashiibaraki-gun, Ibaraki Prefecture, Japan

In past studies, JGC corporation has reported that the efficiency of enzymatic hydrolysis of cellulose using cellulase with low enzyme loadings can be significantly improved by placing the reaction under static condition (2010 AIChE Annual Meeting, 617c Breaking the Barrier to Obtain High Yields On Enzymatic Hydrolysis Using Low Enzyme Loadings). The hydrolysis of 100ml 10% filter paper slurry at the enzyme loading of 6 mg protein/g-dry substrate was conducted using Cellulase SS (Nagase Chemtex). The agitated condition obtained a hydrolyzate with a final glucose concentration of 60 g/L. In contrast, a hydrolyzate with a final glucose concentration of 90 g/L was obtained under the static condition.

Further investigations found that this improvement occurs by the preservation of CBH2 and the exo-exo synergism between it and CBH1 under the static condition. Cellulase SS solution was agitated without any substrates under 50 °C, and SDS-PAGE analysis was conducted to measure the concentration of individual components. Of all the cellulase components, only the band of CBH2 disappeared within 6 hours of agitation, indicating its deactivation. Further examination using fractionated enzyme solutions, one of which lacked CBH1 and the other lacking CBH2, revealed that both CBH1 and CBH2 are required for the efficient hydrolysis of cellulose.

In this study, the mechanism of the exo-exo synergism between CBH1 and CBH2 was investigated by comparing the characteristics of enzyme adsorption and desorption to the cellulose under agitated and static conditions. The hydrolysis of 10% filter paper slurry with the enzyme loading of 12 mg-protein/ g-dry-substrate under both static and agitated conditions were conducted, and the changes in concentration of each cellulase component in the hydrolyzate was monitored using SDS-PAGE and a densitograph. The glucose concentrations of the hydrolyzates obtained were 88 g/L and 60 g/L under the static and agitated conditions respectively. Under both conditions, the CBH1 and CBH2 concentrations in the hydrolyzate dropped sharply, showing concentration decreases of 100% and 80% respectively, at the initial stage of the reaction, indicating their adsorption to the substrate. Under the static condition in which CBH2 is active, the CBH1 and CBH2 concentrations gradually rose as the reaction progressed, indicating desorption from the substrate. The final concentrations of CBH1 and CBH2 after the reaction were 80% and 90% of the original concentration respectively. Under the agitated condition in which CBH2 is deactivated, no increase in the concentration of CBH1 or CBH2 was observed after desorption. CBH2 is likely to be deactivated by shear stress from agitation. However, CBH1 is stable even under the agitated condition, indicating that, under agitated condition, CBH1 remains adsorbed to the substrate and the hydrolysis remains incomplete.

From this result, the exo-exo synergism was hypothesized as enhancing the hydrolysis by preventing the unproductive binding of CBH1 during the reaction. For further investigation, the adsorption and desorption characteristics of enzymes during the sequential addition of CBH1 and CBH2 fractions was observed. The hydrolysis of 10 wt% filter paper using CBH1 fraction yielded hydrolyzate with glucose concentration of 60 g/L at the end of the reaction, and no desorption of CBH1 from substrate was observed. The addition of CBH2 fraction at the later stage of this reaction reaccelerated the reaction, increasing the final glucose concentration to 90 g/L. Also, CBH1 desorption was observed after the addition of the CBH2 fraction. The results obtained suggested that CBH1 desorbs from the substrate and completes the hydrolysis under the presence of active CBH2, but remains absorbed on the substrate in an inactive form when CBH2 is absent.

(260d) Cellulose Thin Film Hydrolysis by Cellulase Enzymes in the Presence of Inhibitors – a QCM Study

Utshab Chakravorty[1], Stephen E. Rankin[1], Sue E. Nokes[2] and Barbara L. Knutson[1], (1)Chemical and Materials Engineering, University of Kentucky, Lexington, KY, (2)Department of Biosystems and Agricultural Engineering, University of Kentucky, Lexington, KY

Cellulose thin film hydrolysis by cellulase enzymes in the presence of inhibitors – a QCM study

Utshab Chakravorty, Barbara Knutson, Stephen E. Rankin, Sue Nokes

Abstract

Conversion of biomass to chemicals and biofuels using the products of cellulose hydrolysis is an important area of research for the production of the next generation of biofuels. However, the saccharification and biochemical conversion of cellulose from lignocellulosic biomass is potentially limited by the release of inhibitors during the processing steps which are used to make the cellulose accessible. Understanding the role and sensitivity of these inhibitors in altering the activity of cellulase enzymes is necessary to the design of pretreatment and separation approaches that maximize cellulose conversion. Bulk studies of cellulose degradation do not provide sufficiently detailed time-dependent and concentration-dependent information to delineate the effect of inhibitors on the adsorption of cellulases to cellulose surface and the subsequent hydrolytic activity of the cellulases. This study uses the frequency response of model cellulose thin films cast onto Quartz Crystal Microbalance (QCM) sensors to model the binding and hydrolysis of cellulose by cellulases (from *Trichoderma reesei*) in the presence of furfural and acetic acid. The sensitivity of the frequency response during cellulose hydrolysis to the specific inhibitor and inhibitor concentration is demonstrated. The concentration-dependent inhibition frequency response data are modeled using a reaction network that accounts for enzyme adsorption, hydrolysis, and several modes of enzyme inhibition and deactivation. The model results can be used to interpret the mode of inhibition. The rate constants of the binding and hydrolysis reactions determined by regression to sets of QCM data with varying inhibitor concentration indicate a strong binding affinity of the enzyme to the substrate and consistent rate coefficient for hydrolysis. The QCM data are used to understand the possible roles of various modes of active site inhibition and inhibition of binding to cellulose on changes in overall conversion in the presence of the inhibitors.

(260e) Investigation of Hemicellulase Inhibition in the Production of Bioethanol

Erik Kuhn[1], Xiaowen Chen[2], Clare J. Dibble[2], Joseph Shekiro[3], Nicholas J. Nagle[1] and Richard T. Elander[4], (1)Bioprocessing R&D, National Renewable Energy Laboratory, Golden, CO, (2)National Bioenergy Center, National Renewable Energy Laboratory, Golden, CO, (3)National Renewable Energy Laboratory, Golden, CO, (4)National Bioenergy Center, National Renewable Energy Lab, Golden, CO

Many of the lignocellulosic bioconversion routes utilize xylose and/or glucose produced by operations such as pretreatment and enzymatic hydrolysis. The insoluble solids (IS) from pretreatment enter enzymatic hydrolysis and exist either in the presence of the solubilized pretreatment fraction (i.e. whole slurry) or absent of the solubilized fraction (i.e. washed). The difference between washed IS and whole slurry enzymatic hydrolysis is the background concentration of sugars, organic acids, and other compounds prior to enzymatic hydrolysis. Whole slurry enzymatic hydrolysis is preferred as it decreases water and utility use, but higher glucose and xylose yields have been observed utilizing washed solids. Using commercially available hemicellulase and cellulase preparations we have observed little to no xylose production from high solids, whole slurry enzymatic hydrolysis on select dilute acid pretreatment slurries while greater than 80% xylan to xylose yields have been observed from high solids, washed IS enzymatic hydrolysis. Previous work has shown end product inhibition of Aspergillus niger beta-D-xylosidase by D-xylose, and inhibition from phenolic acid compounds on the endoxylanase, XynA, from Clostridium cellulovorans. Organisms that utilize C-5 sugars in fermentation decrease the xylose concentration of the fermentation broth, which may allow for further xylose production during fermentation. In the present work a range of potential xylanase inhibitors observed in dilute-acid pretreatment slurries and fermentation broths are dosed into washed IS and beechwood xylan at varied concentrations prior to enzymatic hydrolysis, and the effect on xylose production is presented.

(260f) Increased Enzyme Binding to Substrates Does Not Always Increase Catalytic Activity

Shishir Chundawat[1], Dahai Gao[1], Anurag Sethi[2], Stephanie Crews[1], Leonardo da costa Sousa[1], Nirmal Uppugundla[1], Venkatesh Balan[1], Sandrasegaram Gnanakaran[2] and Bruce E. Dale[1], (1)Chemical Engineering and Material Science, Great Lakes Bioenergy Center, Michigan State University, Lansing, MI, (2)Theoretical Division, Los Alamos National Laboratory, Los Alamos, NM

Substrate binding affinity is typically thought to be one of the primary rate-limiting steps preceding enzyme catalytic action during most homogeneous reactions. However, interfacial-based enzyme catalysis on insoluble crystalline substrates, like cellulose, has an additional step comprising of individual biopolymer chain decrystallization from the substrate interface. This additional decrystallization step has ramifications on the role of enzyme-substrate binding and its relationship to overall catalytic efficiency. We find that altering the crystalline structure of cellulose from its native allomorphic state called cellulose I to cellulose III results in reduced overall cellulase binding but enhanced hydrolytic activity on the latter substrate. We have constructed a comprehensive kinetic model for processive enzymes acting on insoluble substrates to explain this anomalous finding. Our study indicates that this phenomenon can be explained by the enhanced decrystallization kinetics of individual cellulose chains from the surface of cellulose III that despite reduced overall binding affinity results in enhanced catalytic activity. These findings have implications on our understanding of interfacial enzyme catalysis for biopolymers that incorporate analogous decrystallization events.

(277a) Sustainability in Global Green-Economy Development - Challenges for the Energy and Process Industries

Hebab A. Quazi, Strategic Sustainable Development Institute, AngelNet Energy Water & Environmental Systems, Austin, TX and Ana Guevara, AVENTI Associates, Miami Beach, FL

International Development Organizations such as The World Bank have created international "Clean Technology Fund" that are helping developing nations make greater use of Clean Energy Resources. World Bank's mission is to eradicate global proverty and promote sustainable development. Global climate change poses new development challenges, but opens new economic opportunities. Smart mobilization of international resources including Climate Investment are key to addressing mitigation and adaptation to the Climate Change. World Bank's development and climate change "Strategic Framework" includes support for accelerated development and deployment of new technologies amd steping-up the policy research, knowledge, and capacity building. The Clean Technology Fund assists accelerating transformation to low-carbon economies. Special attention is focused on emissions reduction based clean energy technologies. The paper will addres the challenges faced by and the contributions made so far by new energy and industrial technologies in supporting sustainable global green-economy-development. The technologies includes low-carbon fuels, improved energy conversion, solar, wind, and bio-fuels.

(277b) Sustainable Chemical Processing Combining Concentrated Sunlight and Natural Gas

Alan W. Weimer, Chemical and Biological Engineering, University of Colorado at Boulder, Boulder, CO

"The sun has a unique role in sustainable energy production, in that it is the undisputed champion of energy; the resource base presented by terrestrial insolation far exceeds that of all other renewable energy sources combined"[Lewis and Nocera, 2006]. Solarthermal chemical processing is ideally suited for driving high temperature chemical reactions where products don't have to be stored as intermediates for downstream continuous processing. Major technical challenges revolve around the availability of thermal shock and oxidation resistant high temperature materials for fabricating solar receivers/reactors and the design of scalable reactors having high efficiency. Heat recuperation/integration is a key issue for achieving high thermal efficiency. The efficiencies for such processes are typically higher than those of competing solar technologies because the energy from the sun is used to directly drive chemical reactions instead of first being converted to another energy form, e.g. electricity. According to the U.S. Geological Survey, 1 barrel of oil equivalent is roughly equivalent to 6,000 standard cubic feet (SCF) of natural gas (NG) in terms of energy. Therefore, with oil near $100/barrel and NG near $3/1000 SCF ($3/28.3 SCM), $100 of oil has equivalent energy content as ~$18 worth of NG. This ~ 5X price differential provides for a solarthermal pathway in which low cost NG can be used as a feedstock or, possibly, as a supplement to solar energy used to heat the chemical process 24/7. Major economic challenges include the cost of installed heliostats and competition from low cost fossil fuels such as NG. Three rather near-term opportunities for commercial application include the decarbonization of NG to produce carbon black and hydrogen, the steam-methane reforming of NG for generating peak power, and the gasification of biomass with additive NG.

(316a) Effects of Carbon Sources, Oxygenation and Ethanol On the Production of Inulinase by K. marxianusYX01

Wenjie Yuan, School of life science and biotechnology, Dalian University of Technology, Dalian, China and Jiaoqi Gao, School of Life Science and Biotechnology, Dalian University of Technology, Dalian, China

Inulinase is one of the most important factors of the consolidated bioprocessing (CBP), which combines enzyme production, inulin saccharification, and ethanol fermentation into a single process. In this article, the inulinase production and cell growth on the *K. marxinaus* YX01 under different conditions were studied. Carbon source was shown to be less significant on the production of inulinase although the activity of inulinase was a little higher using inulin as carbon source than using glucose or fructose. The concentration of the carbon sources had a great repression effect on the activity of inulinase, when the concentration is up to 60g/L, the activity of the inulinase was only 50% of that of 20g/L. The enzyme activity was strongly influenced by aeration rate. It has been shown that the activity of inulinase and cell growth under anaerobic conditions both kept at a low level but aeration at 1.0 vvm (air volume/broth volume·minute) led to higher activity. It is interestingly that the inulinase activity per unit biomass did not show significant differences under different aeration rates. Ethanol had great repression on the cell growth. The cell stopped to grow when the addition ethanol was over 9% (v/v), but ethanol did not restrain the activity of secreted inulinase and the enzyme was stable under 15% ethanol addition.

(316b) Effects of Solvent and Enzyme Source for Transesterification of Waste Oils

Michael Gagnon[1], Jonathan Fogel[2], Subhash Minocha[2] and **Palligarnai Vasudevan**[1], (1)Chemical Engineering, University of New Hampshire, Durham, NH, (2)Biological Sciences, University of New Hampshire, Durham, NH

Enzymatic transesterification with lipase as the catalyst eliminates soap formation. Unlike alkali-based reactions, the products can easily be collected and separated. Moreover, enzymes require much less alcohol to perform the reaction, and can be reused despite some loss in activity at the end of each cycle. Several strains of lipases have been found to have transesterification activity, namely *Candida antarctica*, *Pseudomonas cepacia*, and *Thermomyces lanuginosus*. The two primary obstacles in enzyme-catalyzed reactions are i), the immiscibility of the two substrates, namely hydrophilic methanol and hydrophobic triglyceride, resulting in the formation of an interface leading to mass transfer resistance; and ii), the strong polarity of methanol, which tends to strip the active water from the enzyme's active site leading to enzyme deactivation. The addition of an organic solvent as the medium to the reaction system simultaneously overcomes the two limitations by enhancing the solubility of oil and methanol in the solvent, and by limiting the concentration of methanol surrounding the enzyme.

We are genetically engineering plants to constitutively express a lipase for biodiesel production from spent oils. We have cloned the gene of a lipase with known transesterification activity from *Thermomyces lanuginosus*, a thermophylic fungus. Cloning involved isolation of total RNA, reverse transcription of the mRNA into cDNA and PCR amplification of the lipase gene using specific primers. The gene was first inserted into a cloning vector (pCR8/GW/TOPO) and sequenced to confirm its identity. The gene has been inserted into a plant destination vector (pGWB408 and pMDC83) via LR clonase reaction. *Nicotiana tabacum* (tobacco) leaf was transformed with the lipase gene using *Agrobacterium tumefaciens* (strain GV3101) and transferred onto selection media plates. We have also grown *Arabidopsis thaliana* from seeds that were transformed using the floral dip transformation method. The recombinant enzyme was collected from the genetically engineered plants, purified, and tested for both hydrolytic and synthetic activity. The activity results will be compared with enzyme catalysts from a commercial *Thermomyces lanuginosus*, and the effect of solvent will also be presented.

(316c) Production of Renewable Fuel From Enhanced Activated Sludge Through a Fluidized-Bed Catalytic Cracking (FCC) Process

Emmanuel Revellame, William Holmes, Rafael Hernandez and W. Todd French, Dave C. Swalm School of Chemical Engineering, Mississippi State University, Mississippi State, MS

Annual production of activated sludges in the United States ranges from 7 – 8 million tons. Currently, disposal technologies of these sludges include incineration and landfilling, both of which are heavily regulated and could create adverse environmental impacts. A fraction of the activated sludges can be used as fertilizer once it complies with regulations for Class A biosolids. Activated sludges contain microbial consortia known to produce lipidic materials and many other organic compounds, which could be used as fermentation seed for enhancement of its lipid content. Several studies showed that reasonable yields (2 – 6% weight) of biodiesel could be produced from activated sludge. At these yields however, activated sludge biodiesel is less economical than petroleum diesel. By increasing the fuel yield, which could be accomplished by enhancement through fermentation, in combination with an in situ catalytic conversion to fuel, the economics of this feedstock could be improved. This study describes the production of renewable fuel from enhanced activated sludge using Fluidized Catalytic Cracking (FCC). This process could convert most of the compounds present in sludge into fuel, and thus, increase fuel yields and reduce cost.

(316d) Transformative Catalytic Technology for Advanced Biofuel Production

Manhoe Kim, NextCat Inc, Detroit, MI, Like Yang, NextCat INC, Detroit, MI, Steven O. Salley, Chemical Engineering and Materials Science, Wayne State University, Detroit, MI and K. Y. Simon Ng, Department of Chemical Engineering and Materials Science, Wayne State University, Detroit, MI

Biodiesel producers are facing challenges of high cost feedstocks and lack of enabling technology to process low grade (low cost) lipids having high free fatty acid (FFA) and water levels. In order for biodiesel producers to became commercially viable, they must be able to use less expensive lipid feedstocks, a wide range of crude bio-oil feedstocks, with varying FFA content, including palm oil (FFA ~12%), residual corn oil (FFA ~15%) from the distiller dried grains (DDGs), animal fats (FFA 2~35%), yellow grease (FFA 5 - 15%), and brown grease (50~100%). In order to avoid the problems related to homogeneous catalytic system, a transformative heterogeneous catalytic conversion technology using the heterogeneous metaloxides catalyst supported by ZrO_2 has been studied in continuous flow reactor systems. The catalyst processing factors (aging, drying, and extruding, and drying) and the reactor scale-up factors (reactor diameter and length ratio, residence time, superficial flow velocity) which determine catalytic activity and durability has been studied. Downward feed flow has a plug flow feature and gives higher FAME yield than upward flow mode. Superficial velocity of the feed gives effects on both the degree of mixing of the reactants and dissolution of the active sites on the catalyst surface due to shear force induced by the viscous flow. In order to decrease shear force, the void space of the catalyst bed should be increased and the reactor diameter and length ratio should be optimized. This technology can potentially reduce the feedstock, operating and processing costs and reenergize the latent biodiesel production facilities to enable to be used for biodiesel production with various non-edible waste lipids.

(316e) Transesterification of Triglycerides for Biodiesel Over Selected Zeolite Catalysts

Yu-Yuan Wang, Hsin-Yu Chou and Bing-Hung Chen, Department of Chemical Engineering, National Cheng Kung University, Tainan, Taiwan

Transesterification of triglycerides in excess methanol to biodiesel using selected zeolite catalysts has been carried out. Various zeolites, such as beta-zeolite, zeolite Y and Mordenite, doped with alkali metals either by ion-exchange or impregnation, are employed to catalyze the transesterification at a relatively lower temperature around 60°C. A faster transesterification rate and higher conversion efficiency would be achieved with appropriate pretreatment on catalysts. Under a reaction temperature at 60°C, conversion efficiency of triglycerides to biodiesel as high as 97.3% was obtained. Cyclic durability of zeolite catalyst has been examined by running the same zeolite catalysts for 6 cycles of transesterification of triglyceride in methanol. It is found that catalyzed conversion efficiency of triglycerides did not differ significantly during the first 3 cycles. Additionally, regeneration on spent zeolite catalysts has been attempted as well. The zeolite catalyst can be recycled at least for four times without significant deterioration in terms of conversion efficiency of triglycerides.

(316f) Hydrotreating of Soybean Oil Over NiMo Carbide Catalyst On Five Different Supports

Huali Wang[1], Shuli Yan[2], **Elvan Sari**[1], Steven O. Salley[1] and Simon K. Y. Ng[1], (1)Chemical Engineering and Materials Science, Wayne State University, Detroit, MI, (2)NextCAT inc., Detroit, MI

Abstract

As an alternative to conventional sulfide catalysts, NiMo carbide catalysts were prepared by precipitation method and studied for the hydrotreating of soybean oil to produce hydrocarbons in the gasoline to diesel range. The effect of the catalyst supports on activity and selectivity was investigated by using three different types of materials: mesoporous material (Al-SBA-15), alumina (γ-Al_2O_3) and zeolites (ZSM-5, Zeolite β and USY). The structural characterization of the catalysts was performed using XRD, BET and TEM. Catalytic tests were carried out in a bench scale flow reactor at 400 ^oC and 650 psi. The results showed that the Al-SBA-15 supported catalyst has the highest surface area of 711.5 $m^2\ g^{-1}$ and largest pore volume of 0.71 $cm^{-3}\ g^{-1}$. Among the five catalysts, hydrotreating on NiMoC/Al-SBA-15 gave the highest yield of organic liquid product (96%) and highest selectivity (97%) to hydrocarbons in the boiling range of the diesel fraction. For the three zeolite supported catalysts, hydrotreating of soybean oil produced more hydrocarbon products in the boiling range of green gasoline (about 15-40%).

(322a) Magnetic Separation of Algal for Biofuel Production

Jeffrey J. Chalmers, William G. Lowrie Department of Chemical and Biomolecular Engineering, The Ohio State University, Columbus, OH, Wei Xue, Chemical and Biomedical Engineering, The Ohio State University, Columbus, Jie Xu, Chemical and Biomolecular Engineering, The Ohio State University, Columbus, OH, Maciej Zborowski, Department of Biomedical Engineering, Cleveland Clinic, Cleveland, OH and Brad Postier, Phycal, St. Louis, MO

Phycal is an algal biofuels company focusing on a hybrid photoautotrophic/heterotrophic open pond system using freshwater algal species. To **potentially reduce costs associated with algal biofuels production and/or improve the quality of intermediate steps through improved efficiencies downstream, Phycal is d**eveloping strains that can transport and store significantly more iron than wild-type strains. The potential reward is that they can out-compete contaminating species (*e.g*., algae, bacteria, protozoa, and *etc*.) for the available iron. Once stored, the iron is in the form of magnetite inside ferritin complexes that are paramagnetic. Thus, in addition to the advantage of iron scavenging, these strains can be manipulated magnetically with the potential of magnetic separation.

In this presentation we will present progress on our Phase II SBIR work to 1) characterize the magnetic susceptibility of genetically engineered and intrinsic magnetic susceptibility as a result of culture conditions, and 2) pilot scale separation systems to "dewater" algae cultures.

(322b) A Novel Taylor-Couette Photobioreactor for Energy Efficient Micro Algae Cultivation

Bo Kong, Department of Chemical and Biological Engineering, Iowa State University, Ames, IA and **R. Dennis Vigil**, Iowa State University, Ames, IA

Numerous studies have demonstrated that algal growth rate and light utilization efficiency is dramatically impacted by the cycling in the intensity of light exposure as cells move between well-illuminated photic regions near a photobioreactor external surface and the darker regions of the reactor. In particular, when the characteristic frequency of the light/dark cycles is greater than approximately 1 Hz, the "flashing light effect" can boost cell growth rates 50~100% compared to growth rates obtained without light/dark cycling. However, high frequency light/dark cycles are not easy to achieve in large-scale reactors and may be one of the reasons for low yields. In commonly used flat panel bubble column and airlift reactors, the random nature of the agitation provided by the gas bubbles produces a wide spectrum of random light/dark frequencies. In the work presented here, we make use of robust and coherent hydrodynamic structures known as Taylor vortices that arise in the annular region of a Taylor-Couette flow cell (concentric cylinders with the inner cylinder rotating) in order to rapidly and periodically shuttle algal cells between photic and non-photic regions of the reactor. Because the fluid agitation in a Taylor vortex algal reactor is not dependent on bubble-induced turbulence, much lower carbon dioxide flow rates can be used and the carbon capture efficiency is significantly higher than in conventional algal reactors. Here, we present experimental results (using Chlorella vulgaris) and analyses based upon computational fluid dynamics simulations to demonstrate that even at relatively low light-dark cycling frequencies, the Taylor vortex algal reactor offers higher biomass yield than can be achieved in corresponding bubble column experiments. Furthermore, we consider important factors for design and scale up of Taylor vortex algal reactors.

(322c) Effect of Growing Conditions On Algal Carbohydrate to Butanol Production

Alice C. Jernigan, Chemical Engineering, University of Arkansas, Fayetteville, AR and Christa N. Hestekin, Ralph E. Martin Department of Chemical Engineering, University of Arkansas, Fayetteville, AR

There are several environmentally sound reasons to use algae as a feedstock for biofuels. The fact that they are readily abundant, can absorb unwanted elements out of waste water, and can use carbon dioxide as a carbon source are just a few of the reasons that algae are being examined as one of the most likely medium to replace petroleum fuels. An issue of some concern is that microalgae have a high rate of growth in the summer and lower rates in the winter, which can cause a depletion of feedstock for a bioreactor. We are using a species of *Clostridium* bacteria to convert algal sugars to butanol in a bioreactor. In order to facilitate localized bioreactors that can operate year-round, we examined dried and stored stocks of algae taken during a six month harvesting period and examined the effects of climate, location, and processing on the amount of carbohydrates extracted from the algae and converted into butanol.

(322d) Photoautotrophic Growth and Lipid Production Kinetics of the Microalgae Scenedesmus Dimorphus

Joanne Belovich, Jacob Schwenk, Christopher Hardulak and John Van Blargan, Cleveland State University, Cleveland, OH

The design and optimization of culture systems for microalgae for biofuel production are complicated by the conflicting requirements for high biomass productivity and high oil content. Reactor models that are needed for this optimization rely on models of the algae growth and lipid productivity kinetics. Unlike heterotrophic growth of most microorganisms, photoautotrophic growth of algae can have multiple limiting nutrients, further complicated by potential light limitation. Factorial experimental designs were conducted to identify the nutrients, and their concentration ranges, that are potentially limiting for growth and oil production of the microalgae *Scendesmus dimorphus*. The individual and interactive effects of six groupings of the components of 3N-BB medium on growth rate, lipid content, and total biomass yield of *S. dimorphus* were determined. Trace metal and vitamin concentrations were reduced to 1/6 the level of 3N-BB medium without adversely affecting growth rates and biomass concentration, while concentrations of $CaCl_2$ and K_2HPO_4/KH_2PO_4 were reduced to 1/10 that of 3N-BB without adversely affecting biomass and lipid concentrations. Lipid productivity was maximized at the lowest $NaNO_3$ concentration (1/10 that in 3N-BB) and independent of $MgSO_4$ concentration, while the interaction of these two chemicals enhanced biomass concentration. Initial substrate experiments were then conducted to determine in greater detail the effects of magnesium sulfate, sodium nitrate, and potassium phosphate on biomass growth. Kinetic models for growth rate and substrate consumption were developed from the data and validated. These fundamental kinetic models can then be used in a variety of reactor styles and configurations in order to identify the system and operating parameters

(322e) Use of Sodium Bicarbonate for Efficient Carbon and Water Management for Autotrophic Microalgae Cultivation in Open Pond System

Jinsoo Kim, School of Energy, Environmental, Biological and Medical Engineering, University of Cincinnati, Cincinnati, OH, Joo-Youp Lee, School of Energy, Environmental, Biological, and Medical Engineering, University of Cincinnati, Cincinnati, OH and Ting Lu, Wastewater Treatment Division, Metropolitan Sewer District of Greater Cincinnati, Cincinnati, OH

Dissolved inorganic carbon including carbon dioxide(aq) and bicarbonate ion (HCO_3^-) is an essential carbon source for the autotrophic growth of microalgae. However, the use of CO_2 gas only cannot provide a sufficient amount of dissolved inorganic carbon concentrations in a narrow pH window suitable for the growth. In this study, sodium bicarbonate was used to increase dissolved inorganic carbon concentrations in conjunction with CO_2 gas. Added CO_2 gas was able to suppress the increased pH values as a result of bicarbonate ion uptake and to supply additional dissolved inorganic carbon. The fates of inorganic carbon and sodium ion were studied by taking mass balances for the two components. The effects of sodium ion on CO_2 gas absorption and algal productivity will be presented for sustainable water reuse and autotrophic microalgal culture in a typical open pond system. A preliminary economic analysis of the use of sodium bicarbonate based on the experimental results will also be presented.

(322f) Transgenic Expression of a Bacterial Exo-Acting Intracellular α-Amylase in the Chlamydomonas Reinhardtii Chloroplast

Xiaoqing Wang[1], Barbara Sears[2], Yan Susie) Liu[3] and Wei Liao[1], (1)Biosystems and Agricultural Engineering, Michigan State University, East Lansing, MI, (2)Plant Biology, Michigan State University, East Lansing, MI, (3)Department of Biosystems and Agricultural Engineering, Michigan State University, East Lansing, MI

Microalgae with their relatively high starch and lipid contents are one of the most promising non-food-crop-based feedstocks for biofuels production. To date, most studies have focused on accumulating algal lipids for biodiesel production and some have focused on hydrogen generation. Very little attention has been paid to simultaneously converting algal carbohydrate and lipids into biofuels. Although some algal species are able to accumulate significant amounts of carbohydrates (mainly starch) and lipid, their rigid cell wall retards the extraction of those carbon-rich compounds. In order to overcome this barrier, this study constructed an amylase-producing strain of the unicellular green alga *Chlamydomonas reinhardtii*, with the goal of converting algal starch into soluble sugars *in vivo*, which would facilitate their recovery. An exo-acting intracellular thermophilic α-amylase AmyB from the hyperthermophilic bacterium *Thermotoga neapolitana* was used as the amylase gene of choice. The AmyB gene was synthesized with codon usage optimized for the *C. reinhardtii* chloroplast; it was cloned into chloroplast expression vector p666 to yield plasmid p666AmyB. The vector p666AmyB was inserted into the site between the *psaB* and *atpA* genes on the chloroplast genome of *C. reinhardtii* by micro-particle bombardment, without interruption of chloroplast gene function. PCR analysis indicated that *AmyB* was integrated into the chloroplast genome of the transformants, and western blotting demonstrated the AmyB protein accumulation. The activity of AmyB can reach 17.4 U/g dry biomass. Thus, α-amylase was successfully expressed within the chloroplast, accomplishing an important step towards improving the ability to harness and harvest the carbon-compounds synthesized by micoalgae.

(350a) Treated Waste Water Source Heat Pump Design and Improvement

Huajun Yuan, Anum Raza, Bo Song and Sohail Murad, Chemical Engineering, University of Illinois at Chicago, Chicago, IL

A water source heat pump is a central heating/cooling system that pumps energy to or from water to provide heating or air conditioning. It is an efficient alternative to conventional heat, ventilation and air conditioning (HVAC) systems. Waste water treatment facilities provide easy access to large bodies of water which can be readily harvested. Such system generally lead to an almost 50% reduction in energy cost, and can therefore play an important role in maintaining and improving a sustainable ecosystem. The high flow rate and almost constant water temperature in such facilities make them especially attractive energy sources for heat pumps. We have collaborated with Metropolitan Water Reclamation Department of Greater Chicago (MWRDGC) to design and test different layouts of water source heat pump including open and close loops systems. The test data obtained will be used to increase the efficiency and lower maintenance costs, which will significantly reduce the cost of energy usage at the site. The test system has been installed at the Kirie Waste Water Treatment Facility in Des Plaines, IL. The system will in the future be further improved by using Phase Change Materials (PCMs) to take advantage of large latent heat of (about 250kJ/kg) PCMs to store energy during off-peak hours and release it during peak hours. This will reduce overdesign of the system and lower the initial capital cost. The entire process has also been simulated using the Aspen Plus simulation package and these simulations will allow us to further increasing the efficiency in the future as well. This project is a modest contribution to the development of more sustainable energy systems.

(350b) Self-Heat Recuperation and Its Applications in Chemical Industries

Yasuki Kansha, Yuping Liu, Yui Kotani, Hiroyuki Mizuno, Masanori Ishizuka and Atsushi Tsutsumi, Institute of Industrial Science, The University of Tokyo, Tokyo, Japan

The combustion of fossil fuels for heating produces a large amount of carbon dioxide (CO2), which is the main contributor to the global greenhouse gas effect. Hence, the reduction of carbon dioxide (CO2) emission and the reduction of energy consumption for heating has nowadays become a very important issue in the efforts to suppress global warming. Heat recovery technology such as pinch technology, which exchanges heat between the hot and cold streams in a process, has been applied to thermal processes to reduce energy consumption. Although the net energy input seems to be reduced by using exhausted heat as the additional heat, the heat is also provided by the combustion of fossil fuels, leading to exergy destruction during energy conversion from chemical energy to heat.

Recently, Kuchonthara and Tsutsumi have proposed an energy and exergy recuperative integrated gasification power generation system and a design method for the system. Authors following on this concept and developed self-heat recuperation technology (SHRT) based on exergy recuperation. The most important characteristic of this technology is that the entire process stream heat can be recirculated into a process without any heat addition, leading to marked reduction of exergy and energy savings for the process.

In this research, the process design methodology based on self-heat recuperation is illustrated with the theoretical minimum energy required. Then, the energy required for processes based on self-heat recuperation and that for conventional processes with heat recovery in chemical industries are compared. From these results, the direction of energy saving process designs with fundamental developments is shown.

(350c) Simulation of Self-Heat Recuperation Using Magnetocaloric Effect

Yui Kotani, Yasuki Kansha and Atsushi Tsutsumi, Institute of Industrial Science, The University of Tokyo, Tokyo, Japan

The reduction of carbon dioxide (CO_2) and the fossil fuel consumption has been one of our major concerns over past few decades. In chemical processes that involve heating, heat is mainly provided by fossil fuel combustion associated with great exergy loss. Recently, self-heat recuperation technology based on the exergy recuperative heat utilization principle has been developed, which can recirculate all heat of the process stream by providing work instead of heat generated by fuel combustion. By compressing gas at its highest temperature in the process for heat elevation, the energy consumption for heating can be reduced drastically compared to conventional heat-recovery technology.

A novel magnetic heat circulator for thermal process based on self-heat recuperation technology is proposed. In magnetic heat circulator, process heat is recirculated by magnetocaloric effect of ferromagnetic materials through cyclic magnetization and demagnetization. The ferromagnetic material is adiabatically magnetized/demagnetized at the highest/lowest temperature of the process to create a temperature difference required for heat exchange. All heat is recirculated inside the system without heat addition. The theoretical energy consumption of magnetic heat circulator was calculated in terms of the temperature-entropy diagram. The simulation result shows large potential for energy saving in thermal processes.

In this research, because the magnetic heat circulator poses a complex problem of heat transfer, magnetic field, and physical properties of ferromagnetic material, a numerical 1D model of magnetic heat circulator has been constructed. The magnetocaloric effect (MCE), varying magnetic field, thermal conduction and heat transfer has been implemented in the governing equation.

(350d) A Novel Seawater Desalination Based On Self-Heat Recuperation

Hiroyuki Mizuno, Yasuki Kansha and Atsushi Tsutsumi, Institute of Industrial Science, The University of Tokyo, Tokyo, Japan

The demand for potable water has been increasing recently due to the rise in population and industrialization of developing country. There are two well-known desalination processes, thermal and membrane processes. Thermal process which separates potable water from seawater by evaporation can be divided into several types such as Multi Stage Flash and Multi Effect Desalination. Thermal desalination consumes large energy due to the heat requirement of water evaporation. But there are also merits; the contents of feed water do not strongly affect the desalination operation, and the salinity concentration of the product water is relatively low.

On the other hand, membrane processes such as Reverse Osmosis (RO) which separates portable water from brine water by semipermeable membrane is the most widely used desalination technique. Its reason is merely because it consumes less energy than conventional thermal process. Thus, most newly desalination plant is constructed based on RO process. However, when feed seawater contains large amount of dissolved solids and impurities, they damage the membrane. For this reason RO requires a greater degree of pretreatment than the thermal desalination process.

Summing up, thermal desalination does not need the greater degree of pretreatment but requires large amount of energy.

To overcome the problem, we proposed an energy efficient thermal desalination process based on self-heat recuperation. In a thermal process based on self-heat recuperation, both latent heat and sensible heat are recirculated into the process without any heat addition. As a result, energy requirement is drastically reduced.

In this study, the energy consumption of thermal process based on self-heat recuperation in the steady state has been calculated using a process simulator. We obtained some results; the energy consumption grows with increase of the minimum temperature difference in the heat exchangers, and increased quantities of production flow rate make the specific energy consumption smaller. The desalination based on self-heat recuperation consumes approximately 1/4 of the energy compared to the conventional thermal desalination process when the minimum internal temperature difference in heat exchangers is assumed to be 5 K and the percentage of the product in the feed stream is assumed as 30%.

By applying the self-heat recuperation technology to the thermal desalination process, we proposed an innovative thermal desalination process.

(350e) Identifying Similar Environmental Objectives in Multi-Objective Optimization Via Clustering Methods

Diego Gabriel Oliva[1,2], **Gonzalo Guillén Gosálbez**[1] and Laureano Jiménez Esteller[1], (1)Department of Chemical Engineering, University Rovira i Virgili, Tarragona, Spain, (2)CONICET UTN, INGAR Instituto de Desarrollo y Diseño, Santa Fe, Argentina

Multi-objective optimization has recently emerged as a useful technique in environmental engineering, as it allows treating environmental aspects as additional objectives to be optimized rather than constraints on the system. One major limitation of these models (as applied to environmental engineering) is that they typically include a large number of objectives, which causes difficulties regarding the computation and visualization of the Pareto solutions.

Previous works by the authors focused on reducing the number of objectives in these problems using rigorous mixed-integer linear programs (MILP). In this piece of research, we have addressed a broader problem: given a multi-objective optimization model, we aim at grouping the environmental objectives into clusters, with the property that the minimization of any objective within a cluster will result in the minimization of the rest of objectives within the same cluster. Our final goal is to aid decision-makers by providing information on the structure of the problem, thereby offering them the opportunity of selecting the objectives to be minimized according to their preferences and the similarities between the environmental indicators of concern.

We present herein several theoretical and algorithmic developments in this area. We first provide a rigorous definition of clustering error in multi-objective optimization to quantify deviations from the Pareto structure of the problem that take place when optimizing in a reduced domain of objectives defined according to some clustering. We then introduce a new algorithm to perform such a clustering, and compare it with other traditional clustering methods available in the literature. Our clustering strategy is based on finding structural dominance relations between objectives, which are quantified according to different metrics

Finally, examples are taken from the literature to perform the comparison with other clustering methods. The main advantages of our method are its low running time and quality (i.e., error) of the clustering generated. The clusters identified provide in turn valuable insight for designers in multi-objective optimization.

(350f) Holistic Multi-Phase Processing of Natural Materials

Aydin K Sunol, Chemical & Biomedical Engineering, University of South Florida, Tampa, FL and **Rosanna Ayala**, University of South Florida, Tampa, FL

A novel and green multi-fluid process for extraction and isolation of value added products from natural plants will be discussed. The process aims to remove components with different functional groups such as nutrients and anti-oxidants as well as utilize the residue beyond its heating value or energy conversion options. The solvents used are high pressure warm water and supercritical carbon dioxide while the plant used is acai berry. The extract fractionation beyond functional groups for isolation of chiral and isomers is part of extract removal component of the approach. The typical plant material residue options traditionally include its use as biomass. However, in this application residue is fermented for fruit wine. The pilot plant studies, product characterization, and process modeling using a flowsheet simulator illustrate the potential and flexibility of this dual green solvent process. The use of dual solvents multi-phase simultaneously where one is supercritical is a first in high pressure natural plant refining.

(380a) Multidisciplinary Optimization Model for Region Specific Sustainable Biorefining

Sumesh Sukumara[1], Dr. Jeffrey Seay[1], Dr. Fazleena Badurdeen[2], Joseph Amundson[3] and William Faulkner[3], (1)Chemical and Materials Engineering, University of Kentucky, Lexington, KY, (2)Institute for Sustainable Manufacturing, University of Kentucky, Lexington, KY, (3)University of Kentucky, Lexington, KY

Biofuels are a promising alternative to the fossil based fuels, upon which we have become dependent to supply our energy needs. There are many ways in which biomass can be converted into useful products. Integrated biorefining is a possibility which combines a slate of processes involving various feedstock to produce transportation fuel, electric power and other useful chemical products. Widespread availability, varying feedstock and specific regional conditions makes it challenging to develop an optimization model that can be applied to any region to estimate the overall economics of the process. Based on the literature and our previous experiments it is clear that a multidisciplinary approach is required to deal with the above challenge. This contribution will illustrate a novel approach to develop a region specific optimization tool which links various aspects of the biofuel supply chain like transportation, storage, processing, distribution and environmental effects. This contribution will show how various small-scale optimization models can be combined, resulting in a complete, multivariable economic optimization model. For this model, literature data sources as well as mapping and spatial modeling data are used to locate biomass sources in a given geographic region. Based on the source of biomass and available road network, transportation costs are optimized using ILOG-OPL software. The results can be used to determine a suitable location for potential biorefineries in any given region. Processing and capital costs are other key factors in determining overall economics for biorefining and hence demands a detailed process optimization model.

A multi feedstock process optimization model including both preprocessing and conversion of biomass into biofuels and products will be described in this contribution. The process optimization model to be described consists of two major processes to produce transportation fuel and power as end products. This model has the capability of including various types of biomass such as dedicated energy crops, crop residues and animal wastes as feedstock. Aspen Plus V-7.3 is the software package used for this purpose. Multiple process and transportation optimization models are then interfaced using Visual Basic in order to provide an economic assessment for biorefining in a region. This model will be validated using the Jackson Purchase region of Western Kentucky, which is not only rich in coal, lignocellulosic biomass and crop residue but also has large number of poultry farms, as a case study. Finally an optimum location for the proposed biorefinery, ratio of combined feedstock and profitability can be determined based on the output of this model. The model developed will have an ability to show how multiple parameters in various stages of supply chain effects the overall economic of the process. This innovative linking of various optimization models will help develop a large scale multivariable model for biorefining which will include all possible feedstocks for a given region and provide the option to choose from various available processes in the model based on the cost, demand and environmental regulations in a region. The model can not only be used by investors as a tool to estimate the potential profitability of biorefining in a region but will allow any locale to attain the distant goal of sustainability.

Sustainable Processing of Water Hyacinth through a Distributed Biorefinery System

José Ezequiel Santibañez-Aguilar,[1] J. Betzabe González-Campos,[1] José María Ponce-Ortega,[1] Medardo Serna-González[1], and Mahmoud M. El-Halwagi[2, 3]

[1]*Chemical Engineering Department, Universidad Michocana de San Nicolás de Hidalgo, Morelia, Michoacán, México, 58060*

[2]*Chemical Engineering Department, Texas A&M University, College Station, TX, USA, 77843*

[3]*Adjunct Faculty at the Chemical and Materials Engineering Department, King Abdulaziz University, Jeddah, Saudi Arabia.*

Water hyacinth is a major nuisance of water bodies. In addition to blocking pathways, this plant quickly consumes the available nutrients in the water as well as dissolved oxygen thereby affecting the aquatic life in the water bodies as well as human activities such as fishing. To eliminate this plant from the water bodies, several techniques have been implemented (including chemical, physical and mechanical elimination techniques, many of them with severe environmental consequences). Much more work is needed to address this problem because the very high growth rate of water hyacinth. Recently, several papers have reported the possible use of water hyacinth as feedstock to produce biofuels, compost, chemical products, green manure, and paper, among others. In this regard, this paper presents an economical and environmentally sustainable alternative to address this problem, where water hyacinth is used as a raw material to produce a set of products as well as clean water (see **Figure 1**). Since the water hyacinth is geographically located in a given region, a distributed processing system is required considering the installation of both preprocessing and central facilities to take advantages of the economies of scale. In this work the objective is to present a mathematical programming formulation to address the optimal design of distributed biorefinery systems of water hyacinth according to economic and environmental criteria. The proposed model is based on the superstructure shown in Figure 2 and it takes into account a set of water bodies as sources of water hyacinth for biomass and water production. The biomass can be processed using different technologies to obtain several products, which can be sent to the available markets. On the other hand, the water extracted from water hyacinth can be treated to separate the pollutants (because the roots of water hyacinth absorb several pollutants helping this way to clean the water bodies) and to obtain clean water. The proposed model is applied to a case study in the central part of Mexico, which is a region with several lakes highly congested with water hyacinth. The results show that water hyacinth elimination can be economically attractive when the distributed treatment system proposed in this paper is considered.

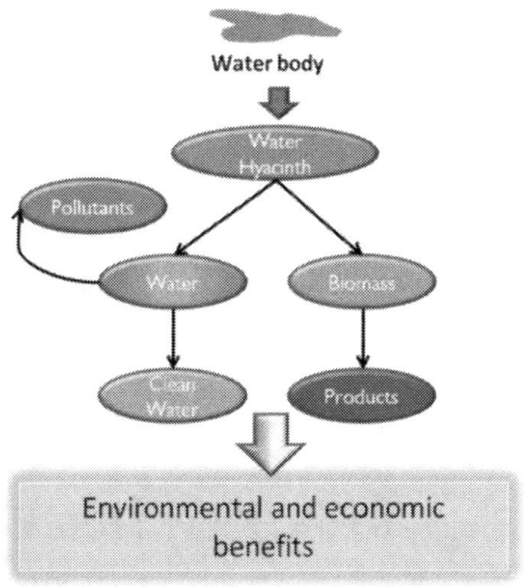

Figure 1. Schematic representation for the addressed problem.

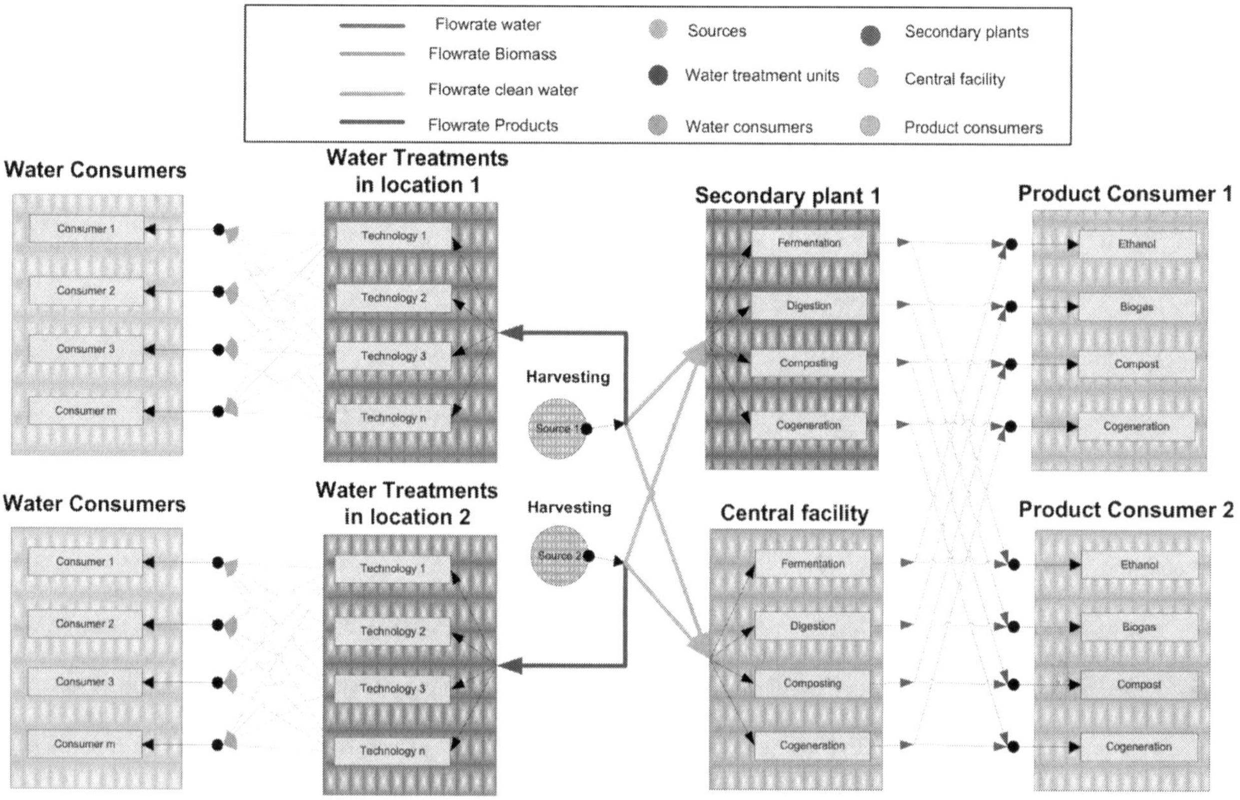

Figure 2. Proposed superstructure for the sustainable use of water hyacinth.

(380c) Optimal Design of County-Level Hydrocarbon Biorefinery Supply Chains Under Uncertainty: A Case Study for the State of Illinois Using Spatially-Explicit Model

Yuan Yao, Berhane H. Gebreslassie and Fengqi You, Department of Chemical and Biological Engineering, Northwestern University, Evanston, IL

Hydrocarbon biorefinery technologies, which convert biomass to hydrocarbon biofuels such as gasoline and diesel, have been considered as promising approaches to replace traditional fossil-fuel-based energy generation technologies. Hydrocarbon biofuels can be potentially used without significant changes of the current fuel distribution and utilization, including pipelines, pumping stations, and vehicles. Compared with their conventional counterparts, hydrocarbon biofuels provide vehicle similar to or better performance.[1] Renewable Fuel Standards (RFS) has put a target of producing 36 billion gallons of biofuels in 2022.[2] In order to achieve this goal in a relatively short time, it is appealing to investigate design of supply chain network for biofuels. Currently, several mathematical models have been developed to address this issue.[3, 4] However, most of them are deterministic approaches that assume all problem parameters can be exactly known in advance. In practice, the hydrocarbon biorefinery supply chain may be affected by a variety of uncertainties, such as seasonal and geographical fluctuation of biomass supplies, population growth and unexpected events.[5] It is necessary to develop a comprehensive model for the optimal design of hydrocarbon biorefinery supply chain under the presence of uncertainties. Another critical issue is the existence of potentially high financial risk when uncertainties are involved in. Risk management should be taken into consideration in the final design of hydrocarbon biorefinery supply chain to reduce the impacts of unexpected outcomes of uncertain parameters.[6]

In this paper, we study the county-level hydrocarbon biofuel supply chain for the State of Illinois. A spatially explicit model is developed to integrate decision making across multiple temporal and spatial scales and simultaneously predict optimal network design, technology selection, capital investment, production operations, and logistics management decisions under hydrocarbon biofuel demand and biomass supply uncertainties. The objective of the model is to minimize the total expected annualized cost and financial risk that is quantified by conditional value-at-risk (CVaR)[7] and downside risk.[8] The State of Illinois consists of 102 counties. Thus, the supply chain network in this case study has 102 harvesting sites, 102 hydrocarbon biorefinery location candidates and 102 demand zones. Three major types of biomass are considered, which are agricultural residues, energy crops, and wood residues. Two types of hydrocarbon biofuels, diesel and gasoline, are produced and shipped to demand zones. We consider two major conversion technologies: gasification followed by FT synthesis and fast pyrolysis followed hydroprocessing. The capacities for both of them have three levels: 1-50 MGY (million gallons per year), 50-100 MGY, and 100-200 MGY.

Case study in the State of Illinois with sample size of 100 scenarios is solved and the minimum total expected annualized cost is $2,822.565MM. It corresponds to a minimum supply chain cost of biofuels is 4.00 $/GEG (gasoline-equivalent gallon). In the optimal solution, 11 biorefineries with capacities ranging from 62 MGY to 200 MGY are built in 11 counties that have high population density. Six of them are in the north part of Illinois, another four are located in central Illinois and the one left is built in southern Illinois. For each biorefinery, biomass used is supplied by local harvesting sites or those in adjacent 1 to 8 counties. The number of counties that are involved in supplying biomass to hydrocarbon biorefineries ranges from 23 to 56. When biomass supply decreases or biofuel demand increases, more counties join biomass supply network, leading to more inter-county transportation. According to production capacity and biofuel demand, hydrocarbon biofuels of each biorefinery are shipped by trucks to satisfy local demand or demand in other counties. The biorefinery with capacity of 80 MGY in Cook County can only meet local demands because of the highest demands of biofuels in Chicago area. On the contrary, as a result of the lowest demands in southern Illinois, the biorefinery in Wayne County with capacity of 125 MGY is able to meet biofuel demands in as many as 28 southern counties. When biomass supply reduces or biofuel demand increases too much, the number of counties whose demand can be met by each biorefinery shrinks a lot. Affected by yields of two technologies using different kinds of biomass that are geographically representative, most of biorefineries choose fast pyrolysis except the plant in La Salle County that adopts gasification. The inventory levels of biofuel products and biomass feedstock in a year always reach peaks in November and decrease gradually in other months, this trend is consistent with the seasonality of typical biomass resources corn stovers in the State of Illinois. Increase of either biofuel demand or biomass supply contributes to a high level of biomass inventory. Biofuel inventories are more sensitive to demand, they fall down sharply when demand grows greatly. In the optimal solution after financial risk management, some of biorefineries shift from counties with high population to counties with rich biomass resources to reduce the risk of supply disruption. Based on the optimal solutions of hydrocarbon biorefinery supply chain, counties with both high population density and abundant biomass resources are perfect locations of biorefineries.

References

[1] F. Q. You and B. Wang, "Life Cycle Optimization of Biomass-to-Liquid Supply Chains with Distributed-Centralized Processing Networks," *Industrial & Engineering Chemistry Research,* vol. 50, pp. 10102-10127, 2011.

[2] EIA, "U.S. Energy Information Administration," 2011.

[3] A. Dunnett, C. Adjiman, and N. Shah, "A spatially explicit whole-system model of the lignocellulosic bioethanol supply chain: an assessment of decentralised processing potential," *Biotechnology for Biofuels,* vol. 1, pp. 1-17, 2008.

[4] B. Aksoy, H. Cullinan, D. Webster, K. Gue, S. Sukumaran, M. Eden, and N. Sammons, "Woody biomass and mill waste utilization opportunities in Alabama: Transportation cost minimization, optimum facility location, economic feasibility, and impact," *Environmental Progress & Sustainable Energy,* vol. 30, pp. 720-732, 2011.

[5] B. H. Gebreslassie, Y. Yao, and F. Q. You, "Design under Uncertainty of Hydrocarbon Biorefinery Supply Chains: Multiobjective Stochastic Programming Models, Decomposition Algorithm and A Comparison between CVaR and Downside Risk," *Submited to AIChE Journal* 2012.

[6] F. Q. You, J. M. Wassick, and I. E. Grossmann, "Risk Management for a Global Supply Chain Planning Under Uncertainty: Models and Algorithms," *AIChE Journal,* vol. 55, pp. 931-946, 2009.

[7] S. Uryasev and R. T. Rockafellar, "Conditional Value-at-Risk: Optimization approach," in *Stochastic Optimization: Algorithms and Applications*. vol. 54, S. Uryasev and P. M. Pardalos, Eds., ed, 2001, pp. 411-435.

[8] G. Eppen, and and R. Martin, "A scenario approach to capacity planning," *Oper Res,* vol. 37, pp. 517–527, 1989.

(380d) Life-Cycle Assessments of Forest Biomass Supply in Michigan

Robert M. Handler, Michigan Technological University, Houghton, MI, Dalia Abbas, Tennessee State University, Nashville, TN, TN, Ajit Srivastava, Michigan State University, East Lansing, MI and David R. Shonnard, Chemical Engineering, Michigan Technological University, Houghton, MI

In this study, we developed measures of the environmental footprint for primary forest products supply chain activities (harvesting and hauling wood) within the state of Michigan. A life-cycle assessment procedure is described, relying on a combination of peer-reviewed literature, national databases, and primary data collected from loggers and truckers within the Michigan study area. Several different equipment configurations and operating scenarios for roundwood harvesting have been considered, with greenhouse gas emissions and fossil energy demand per unit of delivered wood utilized as ultimate environmental indicators. Results indicated that a full processor / forwarder is the best choice of harvesting equipment configuration due to relatively low inputs and high reported productivity, although the burdens of harvesting depend strongly on the intensity of harvest being conducted. Multimodal truck + rail transport had roughly 2X lower environmental burdens than typical log truck transport, which was directly related to the increased fuel efficiency of rail transport. Aggregated results for roundwood supply within the state of Michigan are comparable to similar studies in other regions, although the mechanization of the harvesting industry and large size of MI log trucks are factors in the relatively smaller environmental burden.

(380e) Distributed Small Scale Marine Biorefineries: Scale Analyses and Design for Developing Countries

Alexander Golberg[1], Gregory Linshiz[2], Nathan Hillson[2], Jay D. Keasling[3], Alexander Chemodanov[4] and Mark Koudritsky[5], (1)UC Berkeley, Berkeley, CA, (2)Joint BioEnergy Institute, Lawrence Berkeley National Laboratory, Berkeley, CA, (3)Berkeley Center for Synthetic Biology, UC, Berkeley, Berkeley, CA, (4)Independent, Hedera, Israel, (5)Independent, Toronto, Canada

Abstract

Distributed Small Scale Marine Biorefineries: scale analyses and design for developing countries.

Alexander Golberg[1], Gregory Linshiz[2], Nathan J. Hillson[2], J. D. Keasling, M. Koudritsky

Alexander Chemodanov[3].

[1] Department of Mechanical Engineering, *Etcheverry* Hall, 6124, University of California

at Berkeley, Berkeley, CA 94720 USA. and Center for Engineering in Medicine and Department of Surgery, Massachusetts General Hospital, Harvard Medical School, Boston, MA, USA

Email: agolberg@gmail.com.

[2] Joint BioEnergy Institute ,Lawrence Berkeley National Labs, 1 Cyclotron Road, Berkeley, CA 94720.

[3] Independent researcher, 2-236 Duke st. W. Kitchener ON, Canada.

[4] Independent researcher, Milner 16b Hadera P.B.20995, Israel.

The major growth in liquid fuel demand over the next 20 years will be predominantly due to developing countries. While Gross Domestic Product (GDP) and primary energy production in Organization for Economic Co-operation and Development (OECD) countries is not linearly correlated, GDP growth in developing countries requires increases in primary energy production. The World Energy Council predicts that India and China will overtake developed countries in transportation fuel consumption by 2025. Due to climatic, economic, and fossil fuel resource constraints, there is an urgent need for the sustainable, cost-effective production of carbon-neutral transportation fuels.

Biofuels present an alternative to fossil fuels. First generation biofuel technologies utilize established processes and currently produce biofuels on a commercial scale. First generation feedstocks include sugar beet, starch-bearing grains, and conventional vegetable oil crops, and first generation fuel products include ethanol and biodiesel. Second and third generation biofuel technologies, currently in research and development, utilize animal fat, lignocellulosic biomass, and algae feedstocks, and produce hydrotreated vegetable oil, cellulosic-ethanol, biomass-to-liquids (BtL)-diesel, bio-butanol, and advanced drop-in replacement fuels such as fatty-acid ethyl esters, bisabolene, pinene, n-alkanes, n-alkenes, and methyl ketones.

Although biofuels may collectively supply a portion of future transportation fuel demand, competition between "energy crops" and "food crops" for land and water resources is a growing concern. Furthermore, the extents to which land erosion, potable water consumption, fertilizers, pesticides, and climate change impact biofuel sustainability have yet to be evaluated .

While significant efforts are being directed towards developing feedstocks and conversion technologies, biorefinery design remains in its infancy. The optimization of biorefinery size, feedstock, technology, and serviced area will be required to reduce the costs of the resulting biofuel products.

Recent research in constructal design shows that the optimal distribution of flows of products and services across a populated area depends on a balance between the size of the production centers with sizes of distribution networks that connect these centers to end users. Although larger systems are thermodynamically more efficient in production, they serve larger areas; thus, the collection and distribution logistical costs also increase with size. Therefore, a balance exists between efficiencies of scale, and system distribution losses. This balance has been investigated both for thermodynamics of energy sources and economics of agricultural/biofuel systems. Applications of the balance principle have been demonstrated for the thermodynamical optimization of hot water flow and heating, refrigeration, combined solar power and desalination, and agricultural product processing economics. The balance principle has also lead to proposals for "distributed energy systems" to optimize energy production size and service area.

The goal of this work is to show that the balance between thermodynamic efficiency of system size, collection and distribution is valid for biorefineries. This work also aims to demonstrate that population characteristics, such as density and per capita liquid fuel consumption, play a critical role in the design of biorefinery size, technology choice and serviced area.

We report a model macro algae biorefinery design for midsize towns in low to medium income countries with low liquid fuel consumption per capita. We targeted these populations since the majority of new fuel systems built in future years will serve their growing demands. We focus our model on macro algae, a promising biofuel crop feedstock that does not compete with food crops for arable land or potable water. Furthermore, macro algae, which do not contain lignin, are convenient candidates for cost effective processing with current technology. We analyze biorefinery production of a transportation biofuel from the green macro algae *Ulva sp*, and demonstrate the integration of this model biorefinery into the distributed energy systems. For example, we analyze a system that would supply all the liquid fuel needs of a 20,000 person coastal town in a middle income country, where the annual per capita demand is 26 gallons of liquid fuel per year. The designed system consists of an open controllable photobioreactor and a consolidated bioprocessing unit, which yields renewable liquid fuel. We propose to use a green seaweed *Ulva* as a biomass source. The analyses of prior studies shows that *Ulva* yields of 15-22 kg dry weight$\cdot$ m^{-2} $\cdot$year $^{-1}$, with a heating value of about 19 MJ$\cdot$kg^{-1}, can be achieved with 200 W$\cdot$m^{-2} solar radiation with a zero pesticide footprint. With current ethanol/seaweed conversion rates, a 30 ha cultivation farm, along with 8 ha of a photo-voltaic (PV) facility to provide operational energy, is required to satisfy the liquid fuel demands of a middle income town in a developing country. The system is easy to operate and maintain, and can be manufactured and installed in multiple locations. Finally, in the context of a single biorefinery, we compare macro algae with corn grain and cassava feedstocks.

IDENTIFICATION OF THE MOST APPROPRIATE METHOD TO USE SEWAGE SLUDGE AS A SOURCE OF ALTERNATIVE ENERGY

Yolanda Pliego-Bravo[1], **Marlene Vergara-Hernández**[1], Galo Rafael Urrea-García[1] and María Elena García-Reyes[1],

[1]Instituto Tecnológico de Orizaba, Veracruz, México.

Key words: **SEWAGE SLUDGE, ALTERNATIVE ENERGY**

Abstract

Currently, Mexican law requires the safe use and the recovery of waste, as in the case of sludge from a wastewater treatment plant that can be used as fertilizer, however, the presence of metals can be a limiting. Therefore another option is energy recovery from sludge. In this paper have been analyzed the sewage sludge from a treatment wastewater plant to meet its potential use as alternative energy source. To determine the physicochemical and microbiological characteristics have been made the analysis of moisture, ash, organic matter, volatile matter, total solids, volatile solids, fixed carbon, density, pH, total nitrogen, organic nitrogen and ammonia nitrogen as well as fecal coliform and Salmonella *spp*. Elemental analysis (C, H, O, N and S). The heating value (higher heating value and lower) for recovery as an energy source. Just as the heavy metal content to estimate the waste generated during incineration.

Introduction

Water is the most useful and important resource of life. When wastewater is treated to return this resource to the environment, a semisolid, nutrient-rich by-product called wastewater sludge is produced. Of the amount of water treated in treatment plants, is generated 1% of sludge. The cost of sludge treatment is higher than the treatment of wastewater due to time spent and equipment needed to perform each treatment. Therefore, the processing of waste water sludge must be done in a cost-effective way , (Turovskiy, 2006).

Wastewater sludge generation has increased considerably in recent years due to the amount of treated water, those sludge represent an important source of pollution. Research has concluded that sludge could be converted to a useful source energy in order to remove all its negative effects and to contribute to the power generation, (Midilli, 2000).

Renewable energy is of growing importance in satisfying environmental concerns over fossil fuel usage. The recovery of biosolids alternatives are intended to recover the chemical value of these materials and find ways direct target for processing or as raw material for second-generation processes, basically focused on the recovery of organic content minimization risks of spontaneous combustion, pests, odors and leachate.

There are many ways to dispose the sludge but it is necessary to choose the correct one, in order to have a benefit and minimize the pollution generated by them. There are mainly four commercial means of sludge disposal; fertilizer, land filling, dumping in the sea, and incineration. However, these means are not designed for waste-to-clean energy conversion. On the other hand, pyrolysis and gasification are considered waste-to-clean energy technologies. Additionally, these means can reduce flue gases volume and fix heavy metal in the solid residues, while producing valuable products of oil and syngas. (Nimit, 2010).

Thermochemical processing of biomass uses heat and catalyst to transform plant polymers into fuels, chemicals, or electric power. This contrasts with biochemical processing of biomass, which uses enzymes and microorganisms for the same purpose.

Gasification is the conversion of carbonaceous solids at elevated temperatures and under oxygen-starved conditions into syngas, a flammable gas mixture of carbon monoxide, hydrogen, methane, nitrogen, carbon dioxide, and smaller quantities of hydrocarbons. Gasification can be used to convert any carbonaceous solid or liquid to low molecular weight gas mixtures. In fact, the high volatile matter content of biomass allows it to be gasified more readily than coal.

Pyrolysis is by definition thermal decomposition of organic compounds occurring in the absence of oxygen to produce liquids, gases and char. It is always also the first step in combustion and gasification processes where it is followed by total or partial oxidation of the primary products.

Methodology

The sewage sludge samples were collected from a waste water treatment plant and analyzed by 2500 and 2540 Standard Methods, NOM-21-RECNAT-2000 and NOM-004-SEMARNAT-2002.

Proximate analysis and the ultimate analysis

Moisture was determined, ash, volatile matter and fixed carbon by the proximate analysis, as well as, organic matter, total and volatile solids and the content of heavy metals. The ultimate analysis provided the content of carbon, hydrogen, nitrogen, sulfur and oxygen.

Microbiological analysis

By microbiological analysis were determined fecal coliform and Salmonella *spp*, based on the technical requirements of NOM-004-SEMARNAT-2002.

Heating Value

The higher and lower heating value were obtained by calorimetery, and the thermochemical characterization was done by thermogravimetry and differential scanning calorimetry.

Results and discussion

The purpose of studies of this nature was to identify the thermochemical properties of sewage sludge and thereby take advantage of these properties in a process of alternative energy generation, which contributes to solve the worldwide energy crisis.

Table 1 shows the proximate analysis, which consist of the content of moisture, ash, volatile matter and fixed carbon. These parameters are important in the characterization, since these will define the quality of the fuel in the combustion to determine the content of combustible material present, and produce smoke when burned and the amount of coke or "char" (intermediate combustion products) that can occur, and the content of unburned carbon as fly ash.

Due to the percentage of moisture in sludge samples, would be required a drying thereof prior to its energetic use by the pyrolysis and gasification technologies, reaching a moisture content of about 10-15%. On other hand, there are technologies as hydrothermal processing in which the high moisture content is favorable for the production of hydrogen-rich fuel gas, (Brown, 2011). The results agree with the values displayed by Mejía, (2011), mentions that the sludge that is treated by filtration and centrifugation, such as sludge studied, has a moisture content of 85 to 50%.

Table 1, presents the values of proximate analysis of sludge with different moisture content, presented by Dominguez, (2006). On a dry basis, the volatile matter content of a fuel varies less than 5% for anthracite over 75% in the case of wood. The reactivity of a solid fuel and its conversion to char within the gasifier depends on its content of volatile matter. Fuels with higher volatile matter content are more reactive and therefore can be converted into gas, more easily producing less char, (Basu, 2006).

Table 1. Proximate analysis*

Parameter	*This worK*	*Domínguez (2006)*	
Moisture	83	71	81
Ash	29.91	31.2	38.1
Volatile matter	67.91	62.3	54.7
Fixed carbon	2.18	6.5	7.2

*Dry basis.

Table 2 shows the results of ultimate analysis of the sludge studied and a comparison with sludge studied by other authors that have characterized sludge, for use as an alternative energy source. It is remarkable the low content of nitrogen and sulfur, which are undesirable elements in thermochemical processes, since they increase the production of compounds that generate environmental problems as NOX and SOX.

Table 2. Ultimate analysis

Element	*This work*	*Lou (2011)*	*Zhang (2011)*
N	5.77	2.97	8.3
C	35.26	28.56	42.3
H	7.23	3.29	3.7
S	0.80	0.11	19.1
O*	50.94	65.07	26.6

*By diference

Table 3 shows the average of some heavy metals analyzed from the sample of sewage sludge, and presents the values set by the standard.

Table 3. Heavy metals content in sewage sludge

Metal	*Content (ppm)*	*Reference NOM-002-SEMARNAT-2002*
Antimonio	22.90	---
Arsenico	< 0.30	41-75
Talio	29.02	---
Cadmio	< 1.50	39-85
Cromo	46.76	1200-3000
Mercurio	< 0.30	17-57
Plomo	< 10.00	300-840
Selenio	< 15.00	---
Vanadio	< 15.00	---
Zinc	54.54	2800-7500
Niquel	< 15.00	420
Cobalto	1.91	---
Manganeso	20.78	---
Bario	15.92	---

Figure 1 shows the results of the count of fecal coliforms and Salmonella *spp*, as expected, the NMP is high, indicating that the sludge has to be stabilized for safe and acomplish with regulations. It has been documented that Salmonella, as well as fecal coliforms die when exposed to temperature of 55 ° C for one hour or between 15 and 20 min at 60 ° C (Cota, 2008), therefore, the use of sludge studied, using a thermochemical process is favorable to the elimination of pathogenic microorganisms.

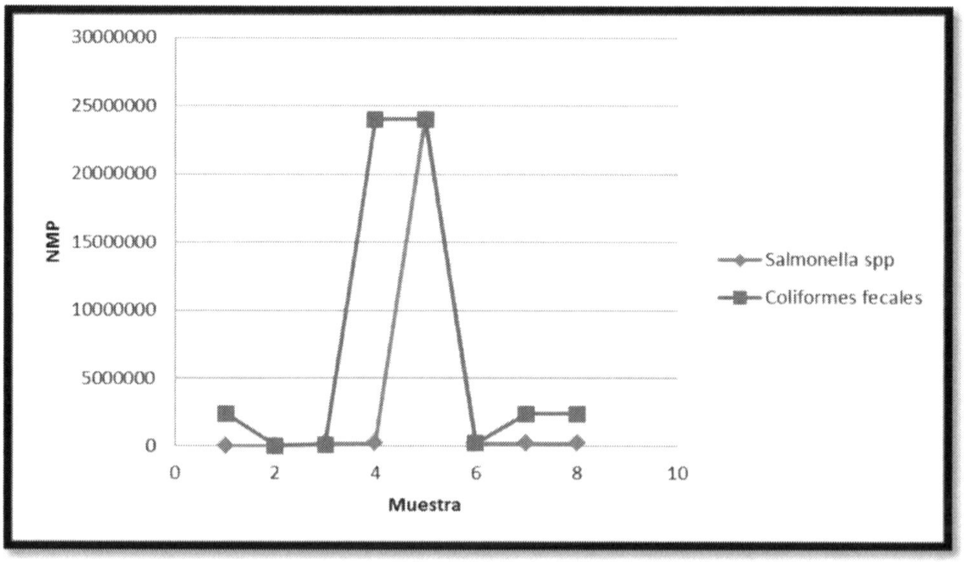

Figure 1. Fecal coliforms and Salmonella *spp*

Table 4 shows higher and lower heating value, obtained by calorimetry

Table 4. Higher and lower heating value

Heating value	*Value (Mj/kg)*
Inferior	1.655
Superior	1.7825

*wet basis

Figure 2 shows the thermogravimetric curve obtained from TGA analysis to which the sludge sample was subjected, it can be appreciated the mass losses when, it was subjected in TGA at heating in a range of 0 to 900 ° C at a rate of 10 ° C / min in a nitrogen atmosphere at a flow of 20 mL / min.

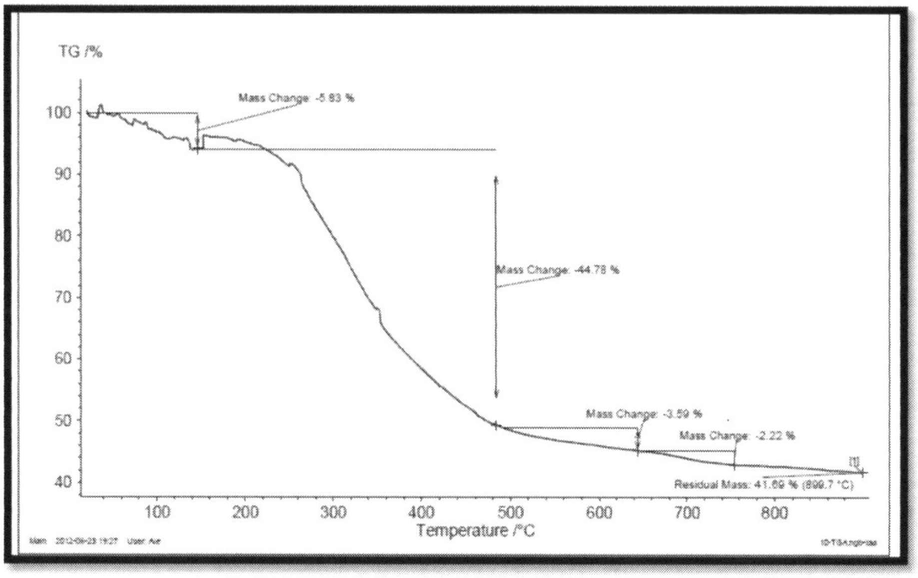

Figure 2. Thermogravimetry curve

Figure 3 shows the results of the curve of DSC performed to the sample of sludge and it was carried out under the same conditions as it was conducted thermogravimetric analysis, in this case was observed endothermic curve because the transformation involves energy absorption, and can observe the ΔH obtained at different temperatures.

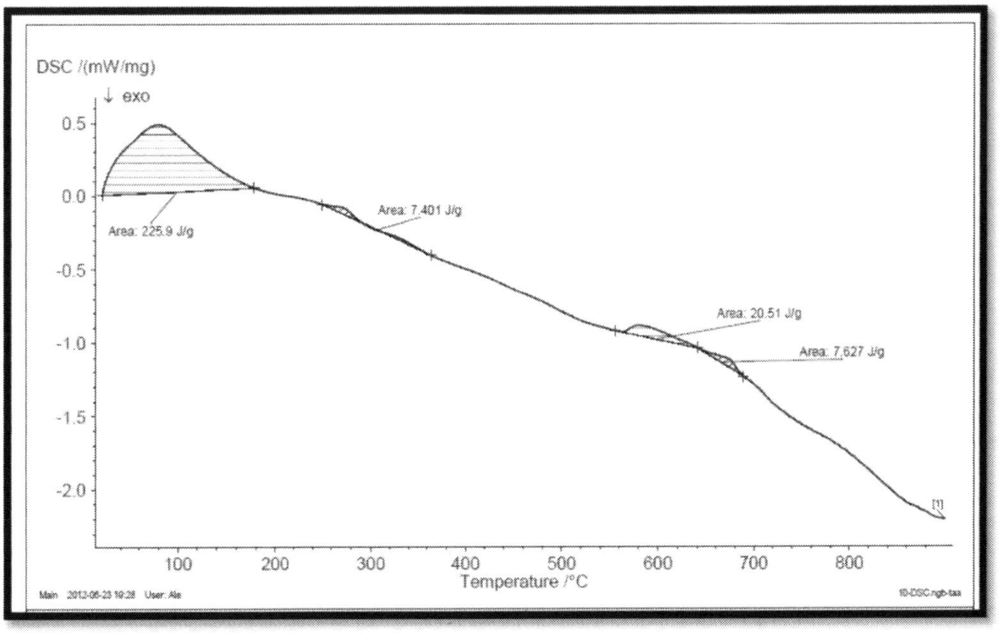

Figure 3. Differential scanning calorimetry (DSC)

Conclusions

There were obtained the general characteristics of the sewage sludge by performing ultimate and proximate analysis, thus fulfilling the objectives set for the characterization thereof. With the characteristics observed in the sludge, it can be concluded that they are feasible for use by a thermochemical process as alternative energy source.

References

1. Beiping, Z., Sijiang, X., Bo, X., Dongke, Y., Xiaoyuan, J. (2010). Mechanism of wet sewage sludge pyrolysis in a tubular furnace. School of Environmental Science and Engineering. International Journal of Hydrogen energy 36(2011)355-363.
2. Brown, R. C. (2011). Thermochemical processing of biomass. Conversion into fuels, chemicals and power. Wiley.
3. Domínguez, A., Menéndez, J. A., Pis, J.J. (2006). Hydrogen rich fuel gas production from the pyrolysis of wet sewage sludge at high temperatura. Instituto Nacional del Carbón, CDIC, apartado 73 Oviedo, España.
4. Girovich, M. J. (1996). Biosolids treatment and management processes for beneficial use. Marcel Dekker, Inc.

5. Grammelis, P., Basinas, P., Malliopoulou, A., Sakellaropoulos, G. (2009). Pyrolysis kinetics and combustion characteristics of waste recovered fuels. Fuel 88 (1), 195–205.
6. Mejía, M. E. (2011). Aspectos técnicos y legales del manejo de lodos en México. academiadeingenieriademexico.mx/...lodos.
7. Nimit, N., Islam, I. A., Somrat, K., Ashwani, K. G. (2010). Hydrogen and syngas production from sewage sludge via steam gasification. The Combustion Laboratory, Dept. of Mechanical Engineering, University of Maryland, College Park, MD, USA. The Waste Incineration Research Center, Dept. of Mechanical and Aerospace Engineering, King Mongkut's University of Technology, North Bangkok, Thailand.
8. Oropeza García Norma. (2006). Lodos residuales: estabilización y manejo. Departamento de ingeniería. Universidad de Quintana Roo.
9. Otero, M., L.F. Calvo, M.V. Gil, A.I. García y A. Morán. (2007). Co-combustion of different sewage sludge and coal: A non-isothermal thermogravimetric kinetic analysis. Bioresource Technology.
10. Principi, P.; Villa, F.; Bernasconi, M. y Zanardini, E. (2006). Metal toxicity in municipal wastewater activated sludge investigated by multivariate analysis and in situ hybridization, Water Research 40, pp 99-106.
11. Ramírez, M. C, Larrubia, M. A, Herrera, M. C., Guerrero, P. y Malpartida, I. (2007). Valorización energética de biosólidos. Algunos aspectos económicos y ambientales en la EDAR Guadalhorce (Málaga). Universidad de Málaga. Facultad de Ciencias. Departamento de Ingeniería Química.
12. Rui, L., Shubin, W., Gaojin, L., Qing, Y. (2011). Energy and resource utilization of deinking sludge pyrolysis. Applied Energy.
13. Turovskiy, I. S., Mathai, P. K. (2006). Wastewater sludge processing. Wiley-Interscience.
14. Vallina, D., Rodrigo, M., Centeno, L. M., Muruais, J., Maílo, A., Ramírez, J., Tolosa, E. (2010). Guía de valorización energética de residuos. Madrid, España.

(381a) Improvements in Bio-Butanol Separation by Adsorption: Adsorbent Screening, Kinetics and Equilibrium

Niloofar Abdehagh[1], F. Handan Tezel[2] and Jules Thibault[2], (1)Chemical and Biological Engineering Department, University of Ottawa, Ottawa, ON, Canada, (2)Department of Chemical Engineering, University of Ottawa, Ottawa, ON, Canada

The rising cost of crude oil combined with depletion of the resources, political instability in oil producing countries and environmental concerns are some of the reasons that have motivated different waves of interest for renewable and sustainable fuels. To achieve a greater use of biofuels, it is paramount to develop innovative techniques to convert biomass to biofuel as efficiently as possible in order to make these production processes more economically viable. Recent developments in biotechnology offer an avenue for bio-production of fuels from renewable and sustainable resources instead of petroleum-based fuels.

Bio-butanol offers several strong attributes and has advantages over other biofuels such as lower volatility, flammability and corrosiveness. The net heat of combustion (NHOC) for butanol is 29.2 MJ/L, which is 83% of gasoline NHOC (which is 32.5 MJ/L) whereas values for ethanol and methanol are 65 and 45% of gasoline NHOC, respectively. Butanol is a very hydrophobic compound with a solubility of 7.7 g/100 mL (at 20°C) in water,whereas lower alcohols (methanol, ethanol and n-propanol) are miscible in water. Also, butanol can be used in car engines without any modification. Butanol was initially produced from fermentation, but in the 1950s, butanol produced from petroleum via the oxo alcohol process became much cheaper and the chemical route remains the only economic source of butanol today. However, the resurgence of interest in the production of butanol from fermentation is being witnessed from research institutions and major players, including large multinational petroleum companies. However, the production of biobutanol via fermentation is facing significant engineering challenges due the very low final concentration, low yield and severe butanol toxicity to microorganisms. It is therefore important to find an efficient separation technique to recover butanol at the end of fermentation or during the fermentation to reduce the level of toxicity and prolong the fermentation in addition to making it an economically feasible process.

Adsorption is an energy-efficient process that can be used to achieve this objective. In selecting a suitable adsorbent for the adsorption process, many factors need to be considered such as : adsorption rate, adsorption capacity, ease of desorption, selectivity for the desired product and cost of the adsorbent. Activated carbon and zeolites are the most common adsorbents used in this process since they are hydrophobic materials having high adsorption capacity for butanol. In this study, two activated carbons (AC F600 and AC F400) with different particles sizes (0.55-0.75 and 1 mm) and two zeolites (ZSM-5 and NaY) with different SiO_2/Al_2O_3 ratios (80 and 1.8) were tested. Adsorbent screening was done based on the kinetic and equilibrium experiment results. In kinetic experiments, AC F400 and AC F600 showed much faster adsorption kinetics in comparison to zeolites., The results obtained from kinetic experiments for both activated carbons showed that the adsorption reaches equilibrium in approximately 100 minutes while for zeolites, the system did not yet reach equilibrium even after 240 min.

Results for equilibrium experiments showed that the highest adsorption capacity for butanol was obtained with AC F400. With an equilibrium butanol concentration of 15 g/L, the butanol adsorption capacity for AC 400 was 283 mg/g while butanol adsorption capacities for AC F600, ZSM-5 and NaY were 150, 155 and 116 mg/g, respectively. Based on these results, AC F400 was selected as the adsorbent to carry out further experiments. Due to the presence of other components such as acetone and ethanol in the fermentation broth in the butanol production process, it is essential to assess the impact of other components o[J1] n butanol adsorption capacity. To investigate the competitive effect of these components on butanol adsorption, ternary solutions of butanol-acetone-water and butanol-ethanol-water with different initial concentrations of acetone and ethanol (1, 5 and 10 g/L) were tested. In this series of experiments, results showed that the presence of ethanol has a negligible effect on butanol adsorption whereas the addition of acetone decreased the butanol adsorption capacity. The negative effect of acetone was more pronounced at higher acetone concentrations. For the 10 g/L butanol equilibrium concentration, in the presence of 1, 5 and 10 g/L initial acetone concentration, butanol adsorption capacity decreased from 258 mg/g to 239, 212 and 206 mg/g, respectively.

In this study, the selectivity of AC F400 for the three main fermentation species (butanol, acetone and ethanol) was investigated using binary solutions of butanol-water, acetone-water and ethanol-water separately. The results showed that AC F400 is more selective to butanol in comparison to the other components. At an equilibrium concentration of 10 g/L, adsorption capacity of AC F-400 for butanol was 258 mg/g, while for acetone and ethanol, the adsorption capacities at the same equilibrium concentration were 115 and 55 mg/g, respectively.

From all the results obtained in this investigation, activated carbon F400 has been chosen to be the best adsorbent for butanol separation due to its high affinity, fast kinetics and higher adsorption capacity.

[J1]Note that acetone is not an alcohol.

(381b) 100% Conversion of Waste Brown Grease to Biodiesel and Syngas As Valuable Products for Heat and Power Cogeneration

Shoucheng Du[1], Fabio Poretti[2], Iman Noshadi[1], Richard Parnas[1] and George M. Bollas[1], (1)Department of Chemical, Materials & Biomolecular Engineering, University of Connecticut, Storrs, CT, (2)Department of Mechanical Engineering, University of Connecticut, Storrs, CT

Hydrocarbon-based biofuels derived from renewable, non-food carbon sources serve as a promising alternative to petroleum-derived fuels and are capable of mitigating greenhouse gas emissions. A sustainable biofuel strategy should utilize widely renewable biomass feedstocks, especially wastes, for which industries carry the high cost of disposal. Fuel production and power cogeneration from waste recovery is a sustainable and potentially profitable option to explore, addressing today's economic challenges of energy independence and waste management. A potentially interesting candidate in this scenario is brown grease. Brown grease is a mixture of high-value hydrocarbons, such as waste vegetable oil, animal fats and grease that is collected in water treatment plants and requires further treatment before disposal. However, brown grease has a higher level of contamination, compared to yellow grease, and cannot be used as animal feedstock or fertilizer. Additionally, brown grease is responsible for about 40% of all sewer overflows, causing back-ups and damage to pipe lines and for about 20,000,000 illnesses each year in the USA [1]. Therefore, special attention must be dedicated to its disposal and the cost associated. The main goal of this project is to explore the possibility of transforming 100% of the raw brown grease into biodiesel and synthesis gas that can be used for power generation. This work comprises a multi-disciplinary study of experimentation and simulation of optimal process options for the conversion of brown grease to energy and energy carriers.

In this study, heat treatment of the raw brown grease results in phase separation of the organic mixture. The light product, mainly characterized by a high free fatty acid (FFA) content (approximately 90%), is extracted and processed to obtain biodiesel. A two-step esterification process, using 5% (v/v) HCL and with a 9 to 1 methanol/oil molar ratio, produces liquid products with an acid value smaller than 0.5% FFA. The triglyceride is then transesterified using 2wt% KOH and a 6 to 1 of methanol/fats molar ratio, reaching higher ester content in the biodiesel. Gas Chromatography (Figure 1) shows that the biodiesel from the extracted brown grease oil not only satisfies the total and free glycerin percentage requirements imposed by the ASTM specifications, but also presents similar value in comparison with the biodiesel derived from Yellow Grease (Table 1), which makes it more competitive as brown grease is more abundant.

Table 1. Comparison of biodiesel produced from brown grease, yellow grease and ASTM standards in total and free glycerin

Sample	Free glycerin	Total glycerin
Biodiesel from brown grease	0.015	0.209
Biodiesel from yellow grease	0.022	0.158
ASTM standard	<0.020%	<0.240%

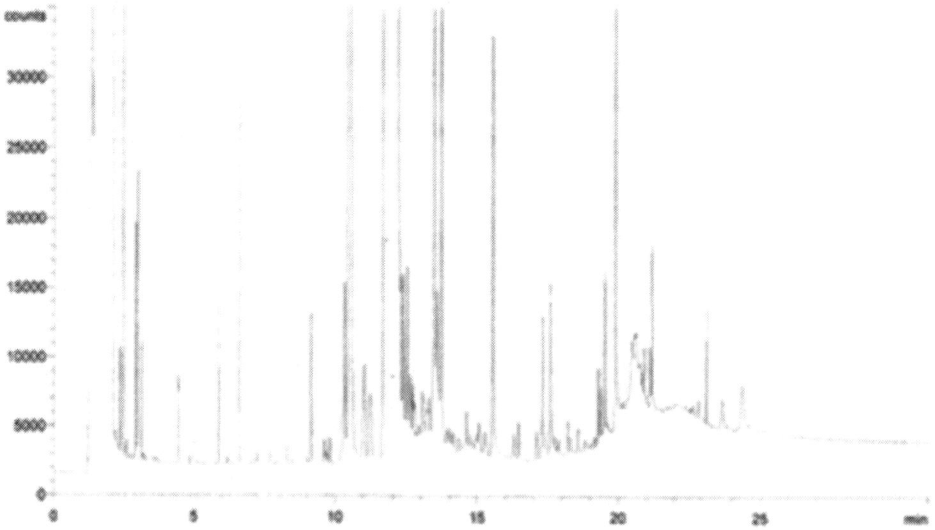

Figure 1. GC analysis of biodiesel produced from brown grease (Signal Count versus time)

The heavy product consists of a solid mixture, which is extracted and gasified to obtain synthesis gas. In Figure 2 thermogravimetric analysis (TGA) of the heavy brown grease product shows that this biosolid is gasifiable in its entirety. However the minor, almost negligible, residue of gasification contain inorganic compounds, of undefined environmental impact. Further analysis of the inorganics and their fate after gasification will be presented and discussed.

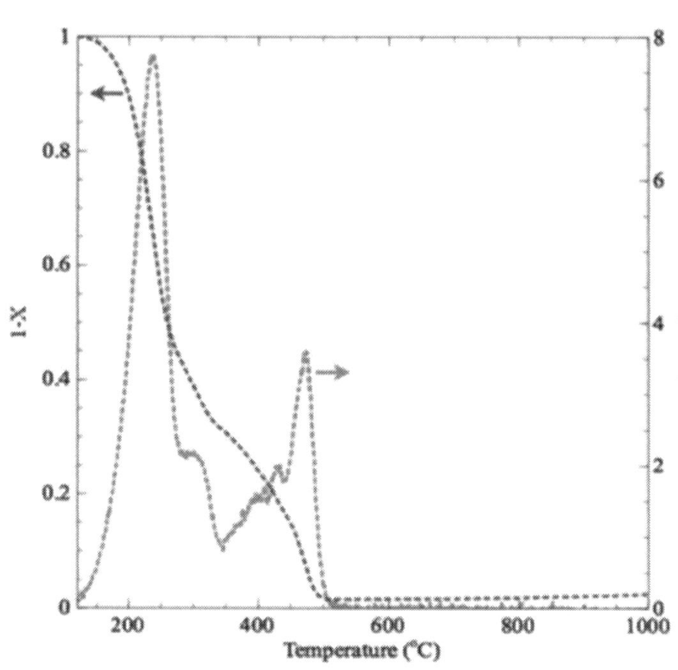

Figure 2. Gasification of biosolid residue from biodiesel production

TGA and gas product analysis (performed in gas chromatography/mass spectroscopy/Fourier transform Infrared, not shown here) show that the heavy product of brown grease is rich in hydrogen; thus, a valuable feedstock. Elemental analysis (Table 2) shows that the carbon to hydrogen ratio of the brown grease heavy fraction biosolid is lower than pine sawdust, which confirms its value as a gasification feedstock.

Table 2. Elemental analyses of biosolid from biodiesel production, biomass and coal (mol%)

Feedstock	N	C	H	S	C:H
Biosolid	0.0263	3.2483	6.4692	0.0157	0.5021
Sawdust	0.0261	4.4880	6.0236	0.0201	0.7451
Coal	0.1007	5.9767	5.0600	0.0881	1.1812

Due to the relatively low available quantity of the brown grease heavy fraction (as opposed to the efficiency increase of large scale gasification), co-gasification of the brown grease residue with fossil fuels, such as coal, is investigated. Co-gasification of coal and brown grease residues is capable of increasing the power generation plant throughput; thus, increasing its overall efficiency [2]. On the other hand, co-gasification reduces the consumption of coal in power generation plants resulting in economic and environmental benefits. Experimental results of co-gasification of biosolids and coal will be compared with the more widely studied co-gasification of woody biomass and coal [3].

On the basis of experimental results, process simulation of an integrated brown grease biorefinery is performed in Aspen Plus. The objective of the simulation study is to identify the cost and profitability of an integrated brown grease refinery with power co-generation, focusing also on the benefits of capturing CO2 and the overall reduction of greenhouse gas emissions. The energy penalties for drying, 02 separation, H2S, SOx, NOx, and CO2 separation are estimated for various co-gasification scenarios. A brief description of the process flow sheet is as follows. Raw brown grease is fed to the system and the first operation consists of heating the mixture in order to evaporate all the containing water and to phase separate the substance in the two layers. The oil at the top layer is extracted and sent to the biodiesel production hierarchy, while the heavier biosolids are taken to the syngas production hierarchy for gasification. In the biodiesel production hierarchy the liquid animal fat enters the pretreatment unit where impurities are filtered and removed. Cleaned animal fat enters an esterification unit where the oil is homogenously mixed at a constant temperature of 60°C with Methanol (MeOH) and Hydrochloric acid (HCl), which serves as a catalyst. The reaction reduces the concentration of the FFA. Then the stream is washed and dried. A second esterification process follows, to convert all the FFA remaining into Methyl ester. The stream is washed and dried again to remove remaining methanol and acid catalyst. The processed flow then enters the Transesterification unit where a base catalyst, Potassium hydroxide (KOH), is used to convert the triglycerides into biodiesel. This reaction also produces glycerol, which can be separated and sold as a chemical. Finally, crude biodiesel enters the post-processing unit where it is refined and tested for quality standards.

In the syngas production hierarchy the dewatered biosolids enter the preparation unit where they are mixed with coal, which has previously passed a process of sizing (crushing and screening) and drying. Air is separated mainly in Nitrogen and Oxygen in the Air Separation Unit (ASU) through cryogenic distillations. The mixed feedstock, together with the stream of almost pure Oxygen, enters the gasifier. The melted ash is collected at the bottom of the reactor and disposed. Then the syngas is cleaned from H2S and CO2 in the Acid Gas Cleaning Unit (Rectisol Process). Captured CO2 is sent to compression and storage, and the H2S is converted to solid S in the Claus Unit [3]. Finally the syngas is burned in a combined cycle, where heat and electricity are produced. The hot gas, after expansion in the gas turbine, pass through a Heat Recovery Steam Generator (HRSG) where superheated, high pressure steam is produced which can be used for district heating and/or expanded in a steam turbine for additional power production.

The experimental analysis integrated with the ASPEN model allows to examine the feasibility of recovering 100% of the raw brown grease that would otherwise have to be disposed at a high cost, due to its contamination as a water treatment waste. The opportunity of transforming it into valuable energy products opens access to new markets, while the sustainability of the overall process and the improved carbon life cycles of this process, lead to its reassessment from an unwanted waste to a precious resource.

Acknowledgment

This work was supported in part by the Faculty Large Grant of the University of Connecticut and in part by the EAGLES (Engineers as Global Leaders for Energy Sustainability) Atlantis Fellowship.

References:

[1] Tyson, K. S. Brown Grease Feedstocks for Biodiesel, National Renewable Energy Laboratory, 2002.

[2] Saw, W.; McKinnon, H.; Gilmour, I.; Pang, S. Fuel. 2012, 93, 473-478.

[3] Valla, L.; Du, S.; Bollas, G. M. 2012 AICHE Annual Meeting, October 2012, Pittsburgh, PA, USA.

(381c) Techno-Economic Analysis of Mixed Alcohol Production Through Hydrogenation of Volatile Fatty Acids Derived From Seaweed

Peyman Fasahati, Chemical Engineering, Pukyong National University, Busan, South Korea and J. Jay Liu, Department of Chemical Engineering, Pukyong National University, Busan, South Korea

Recently, there has been growing interest regarding use of seaweeds as biomass feedstocks. On bioethanol production from seaweeds, there are two main obstacles that make it hardly applicable in industrial scale. First, high seaweed price greatly effects the economy of bioethanol production from seaweeds. Developing mass cultivation sites and enhancing seaweeds cultivation procedures would be necessary steps to address this problem. Secondly, ethanol production from seaweeds has relatively low yield. Research on finding new enzyme, bacteria, or yeast, which are capable of fermenting a wider range of sugars produced by seaweed saccharification, can be the only solution. An alternative way to produce bioethanol would be finding new substitute processes that produce ethanol with higher conversion and lower final ethanol price.

Volatile fatty acids (VFAs) can be easily produced from seaweeds. They have higher conversion yields in comparison with those of direct ethanol production. VFAs can easily be converted to ethanol by hydrogenation. For these reasons, this study develops a techno-economic model to survey economy of VFAs production from seaweeds followed by hydrogenation to produce bioethanol.

Our previous study on direct bioethanol production from seaweed showed that the maximum dry seaweed price (MDSP) should be lower than 160 US$ per ton to allow ethanol production economically viable. This study also calculates the MDSP for this process and survey possible opportunities to improve the economy of ethanol production.

The whole process was designed and simulated in Aspen plus V7.3 and then sent to aspen process economic analyzer for economical evaluations. Plant scale-up effects on the economy of plant were also examined in detail.

Flex Fuel Polygeneration: Optimizing Cost, Sustainability, and Resiliency

Matthew Kieffer, Tristan Brown, Robert C. Brown

Iowa State University
Center for Sustainable Environmental Technologies

Flex Fuel Polygeneration (FFPG) is a new technology that employs multiple energy sources and produces multiple energy carriers to construct optimal energy plant designs in terms of cost, sustainability, and resiliency. This technology allows for adjustments to be made to market fluctuations and improvements to the energy efficiency and environmental performance of power plants. The diversification of energy pathways also increases resiliency to contingencies such as earth quakes, flooding, reduced winds, or other events that may reduce the availability of energy sources. In this study we develop two FFPG plants that employ combinations of natural gas, biomass, wind, and the use of anaerobic digestion, fuel cells, different types of turbines, and Fischer-Tropsch synthesis. To obtain FFPG plant performance results, we combine techno-economic analyses available in literature of existing energy plants to reflect the FFPG plant designs. By doing high level analysis we can compare costs, products, and emissions of theoretical plants to plants that are currently in operation. What we have found are preliminary results regarding the internal rates of return. If the internal rates of return prove to be positive, we will continue to design and perform high level analysis on additional FFPG plants.

(381e) Feed Flexible Process Optimization Model for Region Specific Sustainable Biorefining

Sumesh Sukumara[1], Jeffrey R. Seay[2], Dr. Fazleena Badurdeen[3], William Faulkner[4] and Joseph Amundson[4], (1)Chemical and Materials Engineering, University of Kentucky, Lexington, KY, (2)Department of Chemical and Materials Engineering, University of Kentucky, Paducah, KY, (3)Institute for Sustainable Manufacturing, University of Kentucky, Lexington, KY, (4)University of Kentucky, Lexington, KY

Biofuels in recent years has emerged as one of the most promising alternatives to replace fossil fuel. Researchers around the world are working on developing new processes and improve existing ones to convert biomass into various marketable products. Integrated biorefining is a possibility which combines a slate of processes involving various feedstock to produce transportation fuel, electric power and other useful chemical products. Also, due to the seasonal and regional variability of feedstock, the overall supply chain has to be looked at carefully. Thus, a multidisciplinary economic model needs to be developed which incorporates various small models and allows each to exchange information. According to the existing literature and preliminary experiments performed its evident that transportation and processing cost share a significant portion of the overall cost of an integrated biorefinery. This novel contribution aims to show how a feed flexible process optimization model can be developed and interact with the overall region specific optimization model.

Processing cost in a biorefinery is mainly due to the utilities and raw materials involved. The model developed applies various concepts of chemical engineering to minimize the utilities and maximize the profitability for any given process in a region. Various modules of the Aspen Engineering Suite are combined to achieve the required objective. The process optimization model to be described consists of two major processes to produce transportation fuel and power as end products. These models are first developed considering all the possible mass and heat streams involved to get various products. A thermal pinch analysis is performed to find the best possible heat integration practically possible for a process. Based on the optimum heat exchanger network the process simulations are changed accordingly and finally the capital costs are estimated for the process. The process optimization model developed transfers its data with a Visual Basic interface to the other models which eventually estimates the overall profitability of the process. This model will be validated using the Jackson Purchase region of Western Kentucky, which is not only rich in coal, lignocellulosic biomass and crop residue but also has large number of poultry farms, as a case study. The robust model developed can be run for various combinations of feedstock possible. These process simulation results coupled with the overall supply chain model can predict some critical conclusions like profitability, optimum feed ratio, utility requirement and also some sensitive process variable that effect the overall supply chain of the biorefining process. This innovative linking of various optimization models will help develop a large scale multivariable model for biorefining which will include all possible feedstocks for a given region and provide the option to choose from various available processes in the model based on the cost, demand and environmental regulations in a region. Hence the multiple feedstock process optimization model developed supports the overall supply chain model based on which investors can check the economics of a process and decide if an investment is worthy in a given region.

(381f) Sustainable Design of Energy Supply Chains From P-Graphs: Cost, Ecological Footprint, and Emergy

Leisha Vance[1], Istvan Heckl[2], Botond Bertok[3], **Heriberto Cabezas**[1] and Ferenc Friedler[4], (1)Office of Research and Development, U.S. Environmental Protection Agency, Cincinnati, OH, (2)Department Computer Science and Systems Technology, University of Pannonia, Veszprem, Hungary, (3)Department of Computer Science, University of Veszprém, Veszprém, Hungary, (4)Department of Computer Science and Systems Technology, University of Pannonia, Veszprém, Hungary

A collaboration consisting of the Office of Research and Development (ORD) of the U.S. EPA and the research group led by the founders of the P-graph framework at the University of Pannoniahas developed a methodology for designing sustainable supply chains based on an optimization using the p-graph framework constrained by integrated sustainability indicators and engineering costs. The result was a powerful methodology for designing cost-effective and environmentally sustainable supply chains. We illustrated the methodology with a prototype supply chain designed to produce heat and electricity for a generic district in Hungary. Possible sources of heat and electricity included electricity from the Hungarian grid, and heat and electricity generated from natural gas, corn, corn silage, grass silage, or wood or some combination of these sources. Twenty-one different supply chains, each capable of producing 18 TJ per year of heat and 7.2 TJ per year of electricity were found. Each supply chain was ranked according to cost, and assessed environmental impacts using the ecological footprint (representing land use burden), and emergy (representing energy resource burden). Decisively feasible supply chains were found with cost variations of +2% to -17% compared to "business as usual" scenarios i.e. using only natural gas and electricity from the Hungarian grid. For these supply chains, the sustainability profile as represented by the ecological footprint varied from +8% to -78%, and the emergy results ranged from -54% to -93%. Both comparisons were done in contrast to the conventional natural gas/electricity from the Hungarian grid. Most importantly, it appeared feasibly possible to design supply chains for heat and electricity generation which were both cheaper and more sustainable than the supply chain currently in use.

(464a) Energy & Sustainability – What It Means for Europe

Richard Darton, University of Oxford and European Federation of Chemical Engineers, Oxford, United Kingdom

The countries of Europe are varied in geographical location, political system, state of industrialization and energy needs. Even amongst the 27 countries of the European Union (population ~500 million) there are significant differences. Energy policy within the EU27 is therefore a good illustration of how a group of nations can plan to work together to address the key issues of energy and sustainability. The EU's *Energy 2020 strategy* promotes energy saving (20% by 2020), free market movement of energy, and security of supply. It also plans a significant technological shift in the direction of decarbonisation, with development and demonstration projects such as second generation biofuels, CO2 capture and storage, electricity storage and electro-mobility, and next-generation nuclear. Mechanisms to support this strategy are a mixture of carrot and stick: funded research and innovation projects such as FP7, the world's largest research project, and directives leading to regulation. The strategy will undoubtedly book successes, though whether these will be fast enough or go far enough to meet external pressures, is an open question.

(464b) The Energy-Water Nexus from an Industrial Perspective

Henry T. Kohlbrand, AIChE and Consultant, Rapid City, SD

Water is required to produce energy and energy is required to make use of water. This energy water nexus is playing an increasingly important role in site selection and process synthesis as well as impacting existing facilities. As people's need for water increases, for agricultural and personal use, more and more industrial water use will be under pressure. New energy options including carbon capture systems, fracking systems for natural gas production and some biofuel systems are increasing our water demand. New and creative process solutions to address the energy water nexus will need to be developed and chemical engineers will play a critical role.

(464c) IChemE Roadmap - Framing the Issues and Addressing the Challenges

Ed Daniels, Shell Projects and Technology, London, United Kingdom

What does society need; what are the desirable outcomes and how can chemical engineers work in partnership with others to make it happen? These were the key questions addressed in the *Roadmap for 21st Century Chemical Engineering* published by the Institution of Chemical Engineers (IChemE) on the 50th anniversary of its Royal Charter in 2007.

Five years on, IChemE has conducted a detailed review of the key challenges facing humanity. In an updated report, *Chemical Engineering Matters*, due for publication in November 2012, the Institution aassesses the role of the chemical engineer in delivering process and product solutions that maintain and improve quality of life. The challenges are well recognised: energy, food & nutrition, water, health and wellbeing, but the review also stresses that several key issues and concerns must to be taken into account. These include safety and sustainability as well as the need for international and multidisciplinary collaboration. Education, training and research needs are also assessed. The review clearly identifies the essential role of the fundamental aspects of chemical engineering science and the systems thinking upon which the discipline is founded. It also asks the profession to directly confront the opportunities presented by advances in biotechnology.

In his presentation, Ed Daniels will offer an overview of the report's key findings and recommendations. He will also assert that the capability of the professional chemical engineer to design, build, operate and manage process solution that will maintain or improve quality of life is shaped by external factors that lie beyond science and technology. Drawing on nearly 25 years experience in the global energy business Ed will make the case for stronger engagement by chemical engineers and their professional bodies in wider public policy debate.

(464d) The Role of Chemical Engineering Societies in Facilitation of Carbon Management

Jeffrey J. Siirola, Purdue University and Carnegie Mellon University, Kingsport, TN; Purdue University and Carnegie Mellon University, Kingsport, TN

The role of chemical engineering professional societies in maintaining active promotion and facilitation of carbon management advances are discussed. This is in the context of an environment of increasingly available and affordable fossil gas, global economic stress, and lack of political and economic drivers for greenhouse gas mitigation.

(464e) Decarbonization Options: Policies and Technologies

Kamel Bennaceur, Schlumberger, Rio de Janeiro, Brazil

Without major changes in energy-related policies, the world's energy demand will grow by more than 40 % over the next two decades, and will continue to be dominated by fossil fuels. The share of coal in the global energy mix will continue to grow, leading to an increase in CO2 and other greenhouse gas emissions. Projections for 2050 of those emissions would lead to temperature increases in the range of 4-7 degrees C., according to the 2007 Intergovernmental Panel on Climate Change Assessment Report.

Options for an "energy revolution" in both the energy demand and the energy supply sectors are presented. Scenarios for the deployment of a new energy portfolio are discussed, including ones that would lead to a reduction of emissions by 50 percent in 2050. Among technologies that would mitigate emissions from the supply side, CO2 Capture and Storage (CCS) is a major emission abatement option. The cost of implementing these scenarios is given, along with the implications on the use of fossil-fuels.

(482a) Effects of Sodium Sulfide Supplementation On Alkaline Pretreatment of Herbaceous Biomass

Venkata Ramesh Pallapolu, Chemical Engineering, Auburn University, Auburn University, AL and Y. Y. Lee, Department of Chemical Engineering, Auburn University, Auburn University, AL

In alkaline pretreatments of biomass, it is feasible to retain most of carbohydrates while removing a large fraction of lignin. The enhancement of digestibility is caused primarily by lignin removal. Retention of hemicellulose after pretreatment gives a significant economic benefit since it eliminates the need of detoxifying hemicellulose sugars. Sulfur containing reagents are highly reactive with lignin, thus can serve as delignification enhancer. We have investigated the effects of adding small amount of sodium sulfide in alkaline pretreatment of herbaceous biomass. The alkaline reagents tested were either recoverable ones (ammonia, sodium carbonate) or inexpensive one (lime). Pretreatments were applied to switchgrass, corn stover, and palm residue with and without supplementation of sodium sulfide to see the isolated effects of sodium sulfide. Pretreatments were done in batch mode under low severity conditions: 90 - 120 °C, 12 - 24 hr reaction time, 5 - 15 % alkali, 1 - 3% sodium sulfide, and 5~10:1 of L:S ratio. The treated substrates were put through enzymatic digestibility tests using commercial cellulase (Novozyme CTec 2). Addition of sodium sulfide did not change carbohydrates retention, but increased delignification by 20 - 25 %. Addition of sodium sulfide has significantly increased digestibility for all alkali treatments: 15 – 45 % in glucan digestibility and 10 – 40 % in xylan digestibility depending on type of substrate and pretreatment severity. The results were further assessed to determine the optimum set of pretreatment conditions considering carbohydrate retention and digestibility as criteria. Details of the performance data are also presented.

(482b) Impacts of Cell Wall Hydrophilicity On the Enzymatic Deconstruction of Diverse Biomass Feedstocks and Pretreatments

Daniel L. Williams and David Hodge, Chemical Engineering and Material Science, Michigan State University, East Lansing, MI

This work will present research results relating the environment of the water in the cell wall to the potential for enzyme-catalyzed depolymerization of plant cell wall polysaccharides from diverse plant cell walls treated with a range of pretreatments. These biomass feedstocks will include grasses using alkaline hydrogen peroxide (AHP) as a chemical pretreatment or as a post-treatment following liquid hot water pretreatment (LHW) as well as hardwoods delignified by either soda pulping or organosolv pulping. The effect of pretreatment on the carbohydrate and lignin composition after pretreatment and fiber swelling behavior will be shown. We hypothesize that the digestibility improvement resulting from may be attributed to mild oxidation or solubilization of the lignin remaining in the cell wall and, for grasses, destruction of ferulate crosslinks between cell wall polymers which would have the net effect of increasing the hydrophilicity to allow improved water and enzyme penetration into the cell wall. To test this hypothesis, we will correlate the hydrophilicity of the cell wall as measured indirectly by lignin and cell wall carboxylic acid content and total lignin content to water swelling capacity and digestibility. Additionally the effects of shear-induced fibrillation and particle size reduction on water will be quantified as well.

(482c) Demonstration Scale Alkali Passivation for the Pyrolysis of Biomass

Lysle Whitmer, Center for Sustainable Environmental Technologies, Iowa State University, Boone, IA and Robert C. Brown, Mechanical Engineering, Iowa State University, Ames, IA

Demonstration Scale Alkali Passivation for the Pyrolysis of Biomass

Lysle Whitmer and Robert C. Brown

Center for Sustainable Environmental Technologies

Department of Mechanical Engineering

Iowa State University

Ames Iowa – 50010 U.S.A.

Abstract

The objective of this research is to evaluate the ability to passivate alkali laden biomass feedstock for use in thermochemical conversion processing at demonstration scale. Laboratory Pyrolysis-GC/MS results indicate that naturally occurring mineral matter found in all biomass catalyzes the decomposition of cellulose during pyrolysis to yield light oxygenated compounds in bio-oil such as formic acid and acetaldehyde, rather than uncatalyzed pyrolysis which tends to depolymerize the cellulose to produce anhydrosugars[1]. Acid infusion of the biomass has been shown to form stable salts with the naturally occurring alkali and alkaline earth metals (AAEM), rendering the AAEM less catalytically active and producing bio-oil that is rich in anhydrosugars.

This research focuses on the scale up of AAEM passivation technology to produce consistently pretreated biomass feedstock in large quantities suitable for demonstration scale reactors. A number of independent variables have been investigated and a response surface methodology (RSM) developed for the pretreatment technology. Micro-pyrolysis and the subsequent analysis of the products produced from the pretreated biomass samples served as a proxy for effective passivation.

1. Kuzhiyil, N., et al., *Pyrolytic Sugars from Cellulosic Biomass.* Submitted.

(482d) Examination of Biomass Composition, Structure, and Enzymatic Digestibility As a Function of Ionic Liquid Pretreatment Temperature

Christopher J. Barr[1], Kevin Click[2], Paul Langan[3], B. Leif Hanson[4] and Constance Schall[1], (1)Chemical and Environmental Engineering, University of Toledo, Toledo, OH, (2)Chemistry, Kent State University, Kent, OH, (3)Biosciences Division - Los Alamos National Laboratory, Los Alamos, NM, (4)Instrumentation Center, The University of Toledo, Toledo, OH

Deconstruction of lignocellulosic biomass into simple sugars constitutes a core barrier for producing value added products from the sugar platform. Lignocellulosic biomass is primarily comprised of cellulose (a glucose polymer), hemicellulose (a xylose polymer), and lignin. Enzymatic hydrolysis of lignocellulosic biomass is one of the crucial steps in biochemical conversion to monomeric sugars. The native cellulosic portion of biomass has a crystalline and oriented fibrillar structure in the cellulose I crystal form. Cellulose I can be a major impediment for enzymatic hydrolysis. Ionic Liquid (IL) pretreatment can disorder the crystalline cellulosic component to an amorphous material or result in structural changes in the cellulose fibers to other crystal forms, such as cellulose II. In these studies, the effects of IL pretreatment temperature on various sources of lignocellulosic biomass were examined using X-ray powder and fiber diffraction, compositional analysis and comparative enzymatic hydrolysis. Polysaccharide digestibility was related to the observed structural and compositional changes in biomass. The improvement in hydrolysis is hypothesized to be primarily due to structural changes in the cellulose constituent.

(482e) Flash Hydrolysis for Producing Algal Hydrolysate and Its Application in Enzymatic Saccharification of Lignocellulosic Biomass

Jose Garcia, Old Dominion University, Norfolk, VA, Rui Chen, Department of Biosystems & Agricultural Engineering, Michigan State University, East Lansing, MI, Benjamin Thomas, Department of Biosystems & Agricultural Engineering, Michigan State University, East Lansing, VA, Sandeep Kumar, Civil & Environmental Engineering, Old Dominion University, Norfolk, VA and Wei Liao, Biosystems and Agricultural Engineering, Michigan State University, East Lansing, MI

Flash hydrolysis using subcritical water is a novel and versatile process for selectively hydrolyzing polymeric components of algae biomass. In this study, our objective is to use high water content of algal biomass advantageously for extracting proteins and carbohydrates in a short residence time (few seconds). The present study was focused on the extraction of proteins from *Scenedesmus spp.* using subcritical water in a continuous flow reactor. The effect of temperature and residence time on the yield of peptides and free amino acids in liquid hydrolyzate are studied. About 60-80 wt% of proteins in microalgae could be extracted within few seconds in the temperature range of 200-325°C under subcritical water conditions. The algal samples were first characterized for biochemical compositions and then optimal conditions for protein extraction using subcritical water were determined.

Algal hydrolysate obtained via flash hydrolysis at different temperatures was investigated for its use as reaction medium during saccharification of lignocellulosic biomass. The objective was to replace conventionally used buffer solutions and optimize the enzymatic hydrolysis of biomass in the presence of nutrient rich algal hydrolyzate. It is understood that algal hydrolyzate is an additional protein source to the reaction medium that are irreversibly absorbed by lignin present in lignocellulosic biomass. This helps in increasing the hydrolysis rate by reducing the interference in cellulase activity.

(482f) Effects of Dilute Acid Pretreatment On Cellulose DP and Factors Related to Cellulose Digestibility

Wei Wang, Xiaowen Chen, Ashutosh Mittal, Rui Katahira, Melvin P Tucker and David K. Johnson, National Renewable Energy Laboratory, Golden, CO

Several chemical and physical factors have been proposed to be related to biomass recalcitrance. The degree of polymerization (DP) of cellulose has been considered to be one of the most important properties affecting the enzymatic hydrolysis of cellulose. However, there are few studies investigating the effect of pretreatment on cellulose DP in biomass and the relationship between cellulose DP and enzymatic hydrolysis. This study of the effect of dilute acid treatment on cellulose DP was first started with various pure cellulose materials. A substantial reduction in DP was found for all cellulosic materials studied even at conditions that would be considered relatively mild for the pretreatment. Different cellulosic materials decreased to a similar DP upon treatment with acid regardless of the significant differences in the starting DP of the celluloses. The effect of dilute acid pretreatment on cellulose DP in biomass samples was also investigated. The cellulose DP in corn stover was also found to decrease greatly on acid pretreatment, however not to the extent observed with pure cellulosic samples. Acid pretreated corn stover was found to contain cellulose with DPw of 1600 - 3500, significantly higher than the DPw found with acid treated pure celluloses (150 – 300). The digestibilities of various acid-treated and untreated cellulosic and biomass samples with differing cellulose DP were tested with a cellulase complex (Genencor GC 220). Other cellulose properties, such as particle size and crystallinity were also monitored during digestions. There was little difference in the cellulose conversion for the treated and untreated celluloses despite their very large differences in DP. From this study it does not appear that cellulose DP is a main factor affecting cellulose saccharification. Other properties including particle size, xylan removal, pore volume etc., were also measured for the pretreated biomass samples. A statistical analysis was performed on all these factors to determine which were the most significant in determining cellulose digestibility.

(498a) Hydrolysate Fermentation Characterization for Xylose-Fermenting Saccharomyces Cerevisiae Strains Derived From Directed Evolution for Improved Xylose Utilization and Tolerance to Inhibitors

Tongjun Liu, GLBRC, East Lansing, MI; Chemical Engineering and Materials Science, Michigan State University, E. Lansing, MI, Trey Sato, University of Wisconsin, Great Lakes Bioenergy Research Center, Madison, WI and **David Hodge**, Chemical Engineering and Material Science, Michigan State University, East Lansing, MI

The goal of this work is to quantify the spectrum of pretreatment-derived compounds from alkaline hydrogen peroxide (AHP) pretreatment of grasses, to quantify their impact on fermentation of high sugar content hydrolysates on *Saccharomyces cerevisiae* strains engineered to ferment xylose by introduction of xylose fermentation capabilities by introduction of either xylose reductase and xylitol dehydrogenase or the xylose isomerase followed by evolution for improved xylose utilization and AHP inhibitor tolerance, and to demonstrate improved performance and successful co-fermentation of undetoxified hydrolysates. Specifically, we were able to determine that the classes of inhibitors included inorganics (Na^+ from NaOH and sulfate used for pH adjustment) and solubilized organics (aromatics, acetate, and oxidation prodcucts including formate). It was found that inorganics were not significantly inhibitory at concentrations of 0.5 M Na^+ which would be present in hydrolysates from high-solids pretreatment and hydrolysis. We were able to identify the soluble aromatics and especially the hydroxycinnamic acids as the major contributor to fermentation inhibition. We are able to demonstrated complete sugar utilization from undetoxified AHP corn stover hydrolysates from high-solids pretreatment and enzymatic hydrolysis containing more than 120 g/L of glucose and xylose with conversion to ethanol at high yields with minimal byproduct formation.

(498b) Conversion of E-AFEX(TM) Pretreated Corn Stover to Ethanol by Bcrl Fast SHF Process

Ming Jie Jin, Yanping Liu, Nirmal Uppugundla, Cory Sarks, Leonardo da costa Sousa, Bruce E. Dale and **Venkatesh Balan**, Chemical Engineering and Material Science, Great Lakes Bioenergy Center, Michigan State University, Lansing, MI

Extractive Ammonia Fiber Expansion (E-AFEX^{TM}) pretreatment convents cellulose I to cellulose III and removes part of the lignin in the biomass. Hence, E-AFEX^{TM} pretreated biomass typically has a higher enzyme digestibility compared to AFEX^{TM}. So far, E-AFEX^{TM} has not been fully studied in terms of fermentability and ethanol production potential.

Here we present the performance of BCRL fast SHF process (*Jin et.al, 2012, Energy Environ. Sci., 2012, 5(5), 7168-7175.*) on E-AFEX^{TM} pretreated corn stover for ethanol production. We are aiming to achieve 2-3 times higher productivity with an enzyme loading as low as 8 mg/g glucan and similar or higher ethanol yield when compared to current technology. BCRL fast SHF process has been studied on regular AFEX^{TM} pretreated corn stover and showed high ethanol productivity and reduction of more than 1/3 enzyme loading. In the BCRL fast SHF process, the easily hydrolyzed part of biomass is converted to ethanol first in a 24 h enzymatic hydrolysis and 24 h fermentation process. This way the slow rate period of enzymatic hydrolysis is avoided. The liquid hydrolysate after fermentation is subjected to distillation. The residue solids (difficultly hydrolyzed part of biomass (DHPB)) with some enzymes adsorbed are transferred to the next cycle enzymatic hydrolysis for further hydrolysis. By this approach, parts of the enzymes are reused. After five cycles, the accumulated recalcitrant biomass is further hydrolyzed in the last step. With less degradation products and low initial sugar concentration, the hydrolysis of DHPB is easier to be hydrolyzed. For fermentation, we use high inoculation size (OD_{600} =20) and recycle the yeast cells after fermentation. Details about the condition used and process optimization will be discussed during the presentation.

(498c) High Titer Ethanol Production From SPORL Pretreated Lodgepole Pine by Simultaneous Enzymatic Saccharification and Combined Fermentation (SSCombF) of Enzymatic and Pretreatment Hydrolysates

Tian-Qing Lan, School of Light Industry and Food Sciences, South China University of Technology, Guangzhou, China, Bruce S. Dien, USDA-ARS NCAUR, Peoria, IL, Ronald Hector, NCAUR, USDA ARS, Peoria, IL, Rolland Gleisner, US Forest Service, Madison, WI and J.Y. Zhu, USDA Forest Service, Forest Products Laboratory, Madison, WI

We will present high titer ethanol production of 47 g/L from softwood lodgepole pine by SPORL pretreatment using S. cerevisiae YRH400. SPORL pretreatment was conducted at 180C for 20 min with solids loading of 25%. Sulfuric acid and sodium bisulfite charge on oven dry wood were 2.2 and 8 (wt/wt)%, respectively. Simultaneous enzymatic saccharification of SPORL pretretaed solids and combined fermentation with pretretament liquor were conducted at total soldis of 20% to achieve ethanol yield of 285 L/ton. The SPORL pretreated solids were not washed and combined with pretretament liquor directly in enzymatic saccharification. The CTec2 loading was only 10 FPU/g solids. The liquor was conditioned using lime to adjust pH and evaporated to achieve total solids of 20% in processing. This liquor conditioning process reduced acetic acid and furfural concentration, but did not significantly affect HMF, the major inhibitor derived from sofwood. No sugar loss was observed through evaporation. This study demonstrated the fisrt step process integration by eliminating substrate washing, very important to reduce water usage, without using high enzyme loadings.

Acetone-Butanol-Ethanol fermentation in a continuous and closed-circulating fermentation system with PDMS membrane bioreactor

paper number:254905

Chunyan Chen, Zeyi Xiao, Haidi Cui, Junqing Zhang, Weijia Li, Chao Ying

Introduction

The production of n-butanol with solvent-producing strains of *clostridium*, commonly known as ABE (acetone-butanol-ethanol) fermentation is one of the largest biotechnological processes ever developed [1]. Butanol and acetone are excellent organic solvents and important chemical substances, which are widely used in chemical industry, such as pharmaceuticals, agrochemicals and organic syntheses. In recent years, butanol is regarded as genuine biofuel due to its considerably higher combustion value than ethanol and blending in fuels more readily than ethanol [2,3].

Many researches on ABE fermentation have been indexed. Qureshi et al investigated butanol production from wheat straw by simultaneous saccharification and fermentation in five different batch fermentation processes [4]. Cheng et al studied bio-butanol production by solvent-producing bacterial microflora and the maximum butanol concentration of 12.4 gL^{-1} was obtained in batch fermentation [5]. However, as known, the economics of the ABE bio-fermentation are hampered by the high cost of substrates and butanol toxicity leading to cell inhibition, low product accumulation, low space-time-yields and high purification costs [6,7].

To relieve butanol inhibition and improve the fermentation efficiency, several separation technologies have been investigated for extracting products and controlling their concentration in fermentation broth, such as distillation [8], gas stripping [9], extraction [6], adsorption [10], perstraction [11] and pervaporation [12]. Based on energy requirements of the separation steps, adsorption and pervaporation were indentified as most promising unit operations for in situ product recovery of butanol. The use of pervaporation for product recovery from biomass fermentation processes has been comprehensively reviewed by Vane [13].

In the earlier work, a system integrating fermentation with in situ pervaporation, named as continuous and closed-circulating fermentation (CCCF) system was developed for aerobic ethanol production [14,15]. The results showed the feasibility and prospect of the long term CCCF system in ethanol production with improved fermentation performance. However, it was another challenging and interesting subject whether the CCCF system is suitable for the anaerobic butanol fermentation. In this work, two cycles of CCCF experiment with different coupling operation modes were conducted, then both the overall kinetic parameters and time course profiles of fermentation were explored, intending to find the process principles of ABE fermentation in this specific system.

Methods

Culture and inoculum preparation

C.acetobutylicum strain CICC 8012 was obtained from China Center of Industrial Culture Collection (Beijing, China). Initially, sporulated cells were activated by heat shock at 100? for 1 min. The activated spore culture of 5 ml was inoculated to 100 ml sterile medium in a glass bottle and cultivated anaerobically for 24 h at 37□. The medium used for seed culture and fermentation was identical and contained (for 1L) glucose 60 g, yeast extraction 3 g, phosphate buffer (KH_2PO_4 0.5 g, K_2HPO_4 0.5 g, ammonium acetate 2.2g), vitamins (para-amino-benzoic acid 1mg, thiamin 1mg, biotin 0.01mg), mineral salts ($MgSO_4 \cdot 7H_2O$ 0.2 g, $MnSO_4 \cdot H_2O$ 0.01 g, $FeSO_4 \cdot 7H_2O$ 0.01 g, NaCl 0.01 g). Glucose and yeast extraction were sterilized by autoclaving at 121 □ for 20 min. Phosphate buffer, vitamins and mineral salts were filter-sterilized through 0.45 μm sterile membrane.

CCCF integrated with pervaporation

The CCCF set-up was adopted from the previous study with some modifications [14]. Experiments were carried out in a 10 L jacketed glass bioreactor with the fermentation volume of 8 L. After sparging with nitrogen, the bioreactor was inoculated with 10% inoculums of highly motile cells of *C.acetobutylicum* CICC 8012. Two cycles of fermentation with different operation modes were conducted in the experiments. For the first cycle, the operation mode was of fermentation intermittent coupling with pervaporation, which means fermentation was running for 24 h every day, while pervaporation was only operated from 8:30 to 20:30 (12 h) each day. For the second cycle, the operation mode for the first 192 h was continuous coupling, which means both fermentation and pervaporation were running for 24 h every day. Then, the operation mode was shifted to intermittent coupling which was the same as the first cycle. During the two cycles of fermentation, the temperature of pervaporation and fermentation was the same 37□ controlled by a thermostat, and the pH value was adjusted to no lower than 4 with ammonia water of 25 wt%. Glucose and distilled water were added every 2 h in order to maintain the glucose concentration of about 30 gL^{-1} and broth volume of 8 L. The permeation vapor released at the membrane downstream (with an absolute pressure of 2000 Pa) was collected in cold traps immersed in a -30□ refrigerator.

Pervaporation membrane and module

The pervaporation membrane used here was in-house developed composite polydimethylsiloxane (PDMS) membrane, and two kinds of modules were applied in the two cycles of CCCF respectively. A circular flat-plate membrane module providing an effective membrane area of 0.024 m^2 was used in the first cycle of CCCF, and a square flat-plate membrane module providing an effective membrane area of 0.08 m^2 was used in the second cycle. The preparation procedure of the membrane and the structure of the module had been fully described in the previous work [16,17].

Analysis

Biomass concentration was determined in three terms of dry cell weight, cell number and Optical Density (OD), respectively. For dry cell weight, the broth sample was first centrifuged at 12000 rpm for 5 min, then cells at the bottom was washed with distilled water and finally dried in an oven at 80□. For cell number, broth sample was moderately diluted and then counted by a microscopic in a counting chamber after staining with Iodine solution. For OD, the broth sample was directly determined by a spectrophotometer at 540 nm. Solvents (acetone, butanol, ethanol) were quantified by gas chromatography (GC112A, shanghai, China) equipped with a FFAP capillary column (30m×0.32mm×0.5μm) with flame ionization. Injector temperature 180□, detector temperature 200□, column temperature linearly ramping from 60 to 180□ at 20□ per min. Nitrogen was used as the carrier gas. Isobutanol was used as an internal standard, and the injection volume was 1 μL. The concentrations of glucose were determined using the 3,5-dinitrosalicyclic acid (DNSA) method [18].

Results and discussion

Cell growth profiles of ABE fermentation

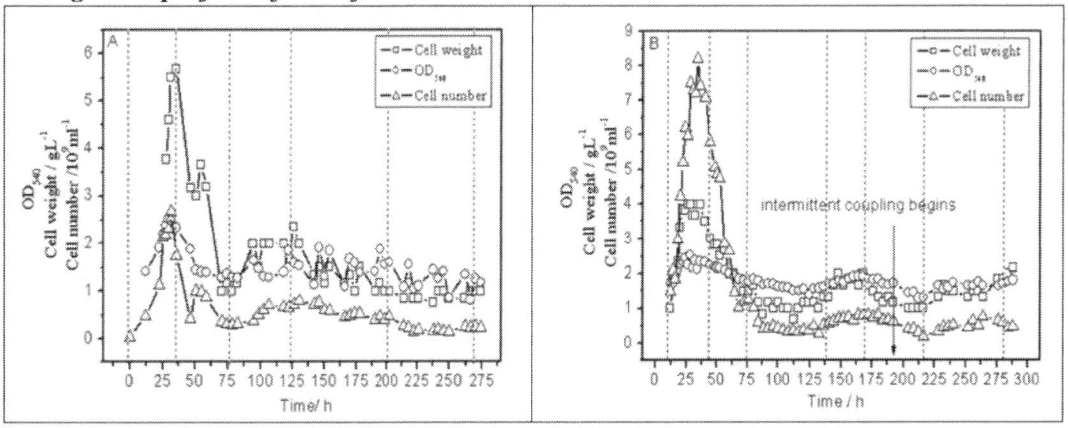

Fig. 1 Time course profiles of dry cell weight, cell number and OD_{540} in two cycles of CCCF. (A) the first cycle; (B) the second cycle.

Three different measurements for microbial concentration/density, dry cell weight, optical density and cell number, were used to describe the cell growth. Time course profiles of those measurements during two cycles of CCCF were shown in Fig. 1. These three measurements were all indicating a similar feature of the cell growth. As described from Fig. 1, the time course profiles of cell growth in the CCCF were very different from those in batch fermentation. For example, the time course profile of cell growth in the first cycle could be roughly divided to 5 phases. From 0 to 30 h (phase 1), cells grew rapidly with dry cell weight increasing to the maximum of 5.6 gL^{-1}, but it dropped sharply to 1.33 gL^{-1} in phase 2 (from 30 to 75 h). Interestingly, the cell concentration increased again from 75 to 125 h (phase 3), and doubled at the maximum of the phase. However, the microorganism decreased again in the phase 4 (from 125 to 225 h). After 225 h (start of phase 5), the cell concentration remained steady, with the average dry cell weight of 1.33 gL^{-1}. For the second cycle, the cell growth profile was similar to the first cycle except for the additionally steady phase 3 between the first decrease phase and the second increase phase (Fig. 1B). However, both cycles had two peaks in the profile of cell growth, which means the microorganism could adapt the CCCF and be surviving in the deteriorated environment in the PDMS membrane bioreactor.

Fermentation kinetic profiles of ABE fermentation

Time course profiles of ABE concentrations in the broth showed two peaks (Fig. 2), at near 40h and at near180h, respectively. This also indicated the adaptable and survival feature of *C.acetobutylicum* CICC 8012 in the long-term CCCF. Besides, compared Fig. 1 to Fig. 2, the cell concentration began to decrease when butanol concentration was over 6.5 gL^{-1} in the first cycle and higher than 8.5 gL^{-1} in the second cycle separately. Obviously, this was the direct evidence that high butanol concentration inhibits the growth of microorganism. Note that the original strain of the second cycle was the descendant of the first cycle, therefore, the higher inhibition concentration of the second cycle may be partly attributed to the adaptive evolution of the microorganism in the long-term fermentation.

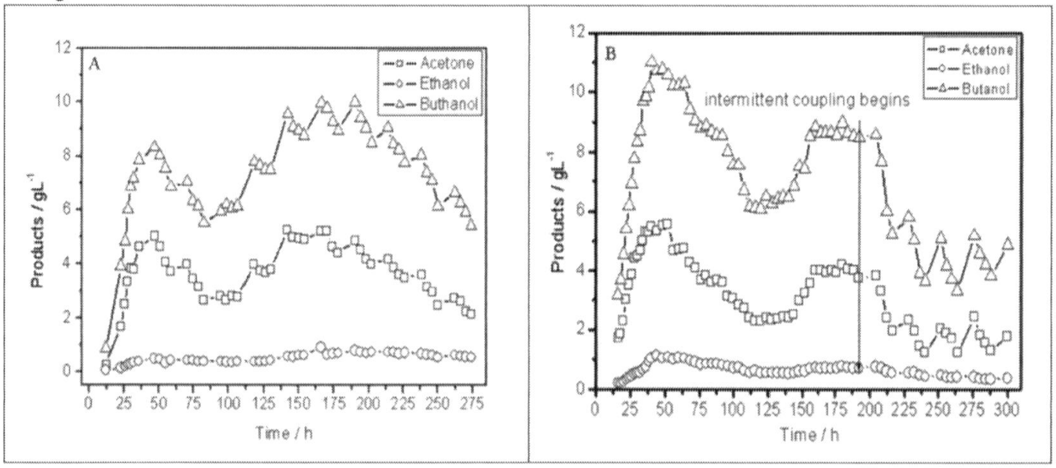

Fig. 2 Time course profiles of ABE concentrations in the broth of two cycles of CCCF. (A) the first cycle; (B) the second cycle.

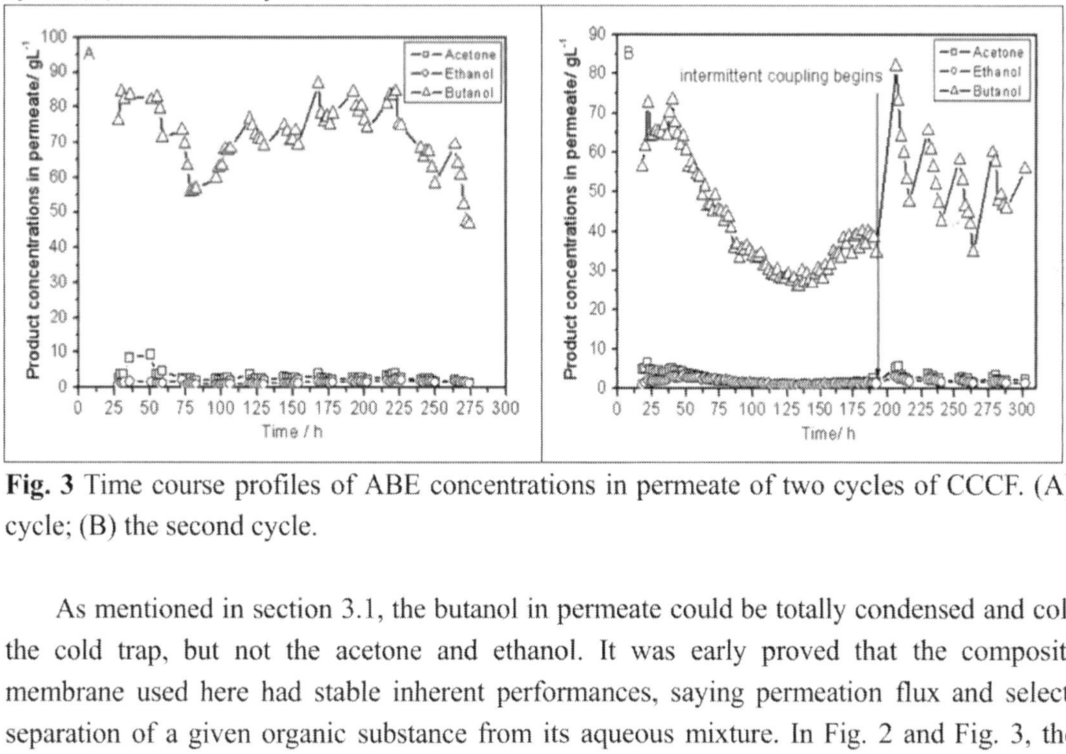

Fig. 3 Time course profiles of ABE concentrations in permeate of two cycles of CCCF. (A) the first cycle; (B) the second cycle.

As mentioned in section 3.1, the butanol in permeate could be totally condensed and collected by the cold trap, but not the acetone and ethanol. It was early proved that the composite PDMS membrane used here had stable inherent performances, saying permeation flux and selectivity, for separation of a given organic substance from its aqueous mixture. In Fig. 2 and Fig. 3, the butanol concentration profiles behaved a similar status, which imply an inherent relationship between the concentrations in broth and in permeate. Nevertheless, it seems that such an inherent relationship for acetone and ethanol could not be found. This supports the observation that the acetone and ethanol in permeate could not be completely condensed and collected. On the other hand, the maximum and average butanol concentration in permeate was 84.41 and 70.96 gL^{-1} in the first cycle, and that in the second cycle was 81.87 and 46.42 gL^{-1} respectively. Compared to the butanol concentration in the broth no higher than 12 gL^{-1}, the highly concentrated production in the permeation could greatly reduce the subsequent cost of recovery. Besides, due to the different membrane modules with different effective membrane areas used in the two cycles, the average butanol concentration in permeate of the first cycle was higher than that in the second cycle as shown in Fig. 3.

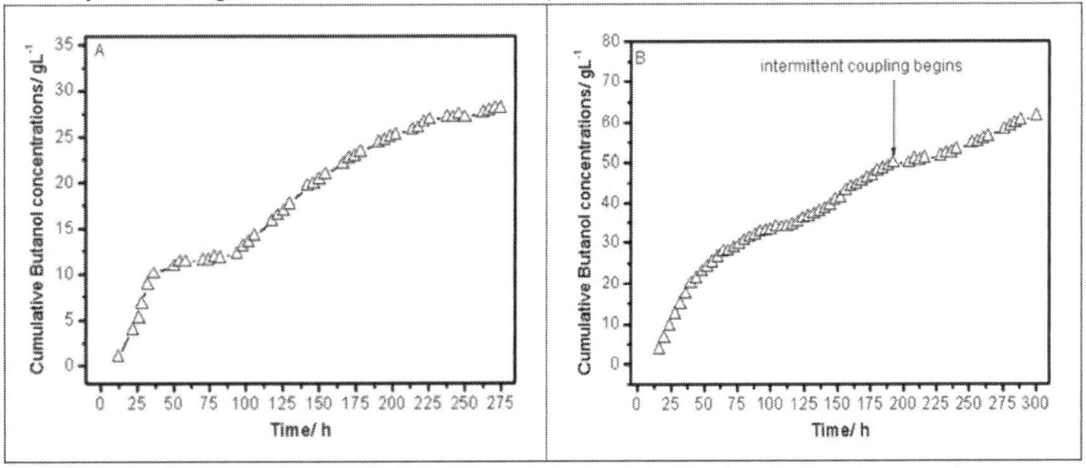

Fig. 4 Time course profiles of cumulative butanol concentrations in two cycles of CCCF. (A) the first cycle; (B) the second cycle.

Figure 4 shows the cumulated butanol amounts in the broth during the fermentation, which was expressed as butanol concentration in the broth. It was reported that butanol concentration of 12.4-21.6 gL^{-1} and butanol productivity of 0.25-0.30 gL^{-1}h^{-1} could be obtained in traditional batch fermentation with duration of about 72 h [4,5]. Correspondingly, during previous 72h, the cumulative butanol concentration of 28.61 gL^{-1} and butanol productivity of 0.40 gL^{-1}h^{-1} had been achieved in the second cycle (Fig. 4B). The higher fermentation efficiency was considered to be attributed to the fermentation-pervaporation coupling in the specific CCCF system. Moreover, the final cumulative butanol concentration of 28.03 and 61.43 gL^{-1} in the two cycles of CCCF far exceeded the butanol inhibition concentration, which showed great superiority over the batch fermentation.

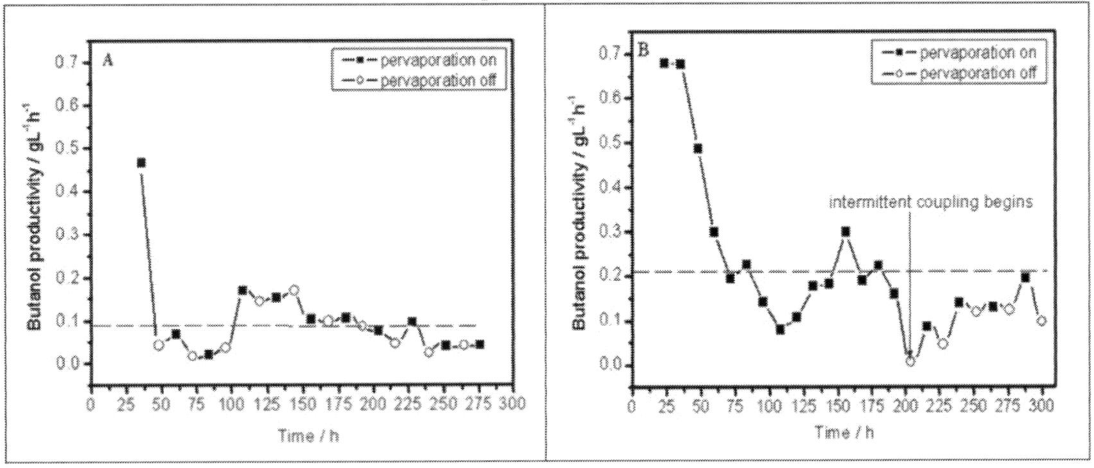

Fig. 5 Time course profiles of butanol productivity in two cycles of CCCF. (A) the first cycle; (B) the second cycle.

Time course profiles of butanol productivity in the two cycles were shown in Fig. 5, the maximum butanol productivity of 0.47 gL^{-1}h^{-1} with an average value of 0.11 gL^{-1}h^{-1} were obtained in the first cycle of CCCF. And the corresponding maximum and average butanol productivity in the second cycle were 0.69 and 0.21 $gL^{-1}h^{-1}$ respectively, which were much higher than those of the first cycle due to the bacterial adaptability and different fermentation-pervaporation coupling modes. The same case was observed for glucose consumption rate in Fig. 6, with the average value of 0.63 and 1.12 $gL^{-1}h^{-1}$ in the two cycles respectively. Similarly, the time course profiles of fermentation kinetic parameters exhibited two peaks, which also indicated the reproducing phenomenon of the clostridium. Besides, the time course profiles of the butanol productivity and glucose consumption rate showed nearly same trend in each cycle, which indicated a constant butanol yield rate and stable fermentation system in the process of CCCF.

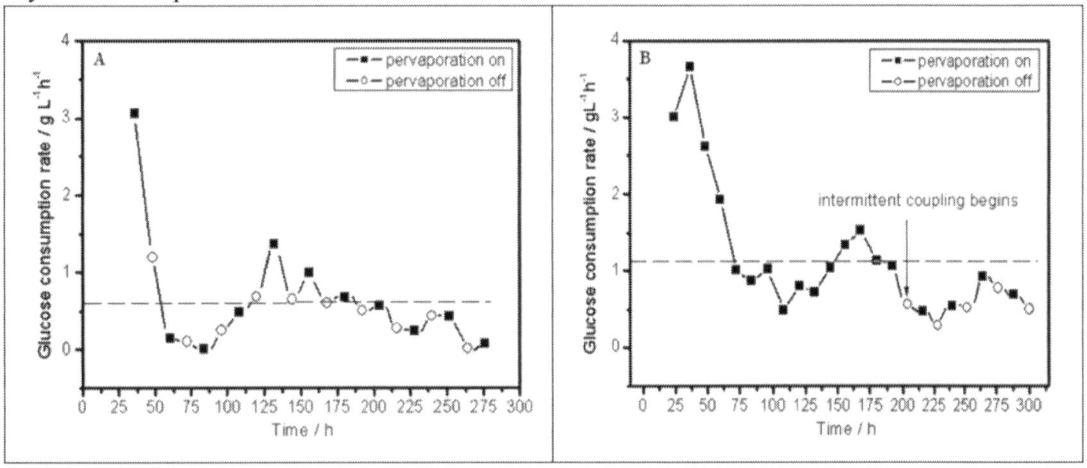

Fig. 6 Time course profiles of glucose consumption rate in two cycles of CCCF. (A) the first cycle; (B) the second cycle.

On the other hand, only glucose was added and ABE removed in the long-term CCCF system. Consequently, in addition to the damaged or dead cells, some metabolites of the microbes which could not permeate the membrane, such as acetic acid, butyric acid, glycerol and lipids, were accumulating in the broth. All these resulted in a gradually worse fermentation condition and caused strong environmental stresses which could be harmful to growth and fermentation of microorganism [5]. Therefore, the fermentation kinetic profiles all demonstrated a generally decreasing trend as fermentation processing (Fig. 5 and Fig. 6).

Conclusions

Continuous and closed-circulating fermentation (CCCF) system integrating fermentation with in situ pervaporation could effectively reduce the butanol inhibition and extend the fermentation duration. The clostridium showed good adaption ability and could be surviving in the deteriorated environment in the PDMS membrane bioreactor. The butanol production in the second cycle was as high as 61.43 gL^{-1}, with the maximum and average butanol productivity of 0.69 and 0.205 $gL^{-1}h^{-1}$. Fermentation performance under continuous coupling mode was superior to that under intermittent coupling mode. The results showed that ABE fermentation in the CCCF system was feasible and promising.

References

1. Van Hecke, W., Vandezande, P., Claes, S., Vangeel, S., Beckers, H., Diels, L., De Wever, H. (2012), "Integrated bioprocess for long-term continuous cultivation of Clostridium acetobutylicum coupled to pervaporation with PDMS composite membranes". Bioresour. Technol. 111, 368-377.
2. Kumar, M., Gayen, K.(2011), "Developments in biobutanol production: New insights". Appl. Energy 88, 1999-2012.
3. Thirmal, C., Dahman, Y.(2012), "Comparisons of existing pretreatment, saccharification, and fermentation processes for butanol production from agricultural residues". Can. J. Chem. Eng. 90, 745-761.
4. Qureshi, N., Sahaa, B. C., Hector, R. E., Hughes, S. R., Cotta, M. A.(2008), "Butanol production from wheat straw by simultaneous saccharification and fermentation using Clostridium beijerinckii: Part I - Batch fermentation". Biomass Bioenerg 32, 168-175.
5. Cheng, C.-L., Che, P.-Y., Chen, B.-Y., Lee, W.-J., Chien, L.-J., Chang, J.-S.(2012), "High yield bio-butanol production by solvent-producing bacterial microflora". Bioresour. Technol. 113, 58-64.
6. Bankar, S. B., Survase, S. A., Singhal, R. S., Granstrom, T.(2012), "Continuous two stage acetone-butanol-ethanol fermentation with integrated solvent removal using Clostridium acetobutylicum B 5313". Bioresour. Technol. 106, 110-116.
7. Garcia, V., Pakkila, J., Ojamo, H., Muurinen, E., Keiski, R. L.(2011), "Challenges in biobutanol production: How to improve the efficiency?" Renew Sust Energ Rev 15, 964-980.
8. Wang, H., Li, C., Li, C.(2012), "RSM optimization of the operating parameters for a butanol distillation column". Asia-Pac. J. Chem. Eng. 7, 117-123.
9. Ezeji, T. C., Qureshi, N., Blaschek, H. P.(2003), "Production of acetone, butanol and ethanol by Clostridium beijerinckii BA101 and in situ recovery by gas stripping". World J Microb Biot 19, 595-603.
10. Qureshi, N., Hughes, S., Maddox, I. S., Cotta, M. A.(2005), "Energy-efficient recovery of butanol from model solutions and fermentation broth by adsorption". Bioprocess. Biosyst. Eng. 27, 215-222.
11. Qureshi, N., Maddox, I. S.(2005), "Reduction in butanol inhibition by perstraction: Utilization of concentrated lactose/whey permeate by Clostridium acetobutylicum to enhance butanol fermentation economics". Food Bioprod. Process. 83, 43-52.
12. Yen, H.-W., Chen, Z.-H., Yang, I. K.(2012), "Use of the composite membrane of poly(ether-block-amide) and carbon nanotubes (CNTs) in a pervaporation system incorporated with fermentation for butanol production by Clostridium acetobutylicum". Bioresour. Technol. 109, 105-109.
13. Vane , L.M.(2005), "a review of pervaporation for product recovery from biomass fermentation processed". J.Chem.Technol.Biotechnol.80,603-629.
14. Chen, C., Tang, X., Xiao, Z., Zhou, Y., Jiang, Y., Fu, S.(2012), "Ethanol fermentation kinetics in a continuous and closed-circulating fermentation system with a pervaporation membrane bioreactor". Bioresour. Technol. 114, 707-710.
15. Ding, W. W., Wu, Y. T., Tang, X. Y., Yuan, L., Xiao, Z. Y.(2011), "Continuous ethanol fermentation in a closed-circulating system using an immobilized cell coupled with PDMS membrane pervaporation". J. Chem. Technol. Biotechnol. 86, 82-87.

16. Shi, E., Huang, W. X., Xiao, Z. Y., Li, D. H., Tang, M.(2007), "Influence of binding interface between active and support layers in composite PDMS membranes on permeation performance". J. Appl. Polym. Sci. 104, 2468-2477.
17. Tang, X. Y., Wang, R., Xiao, Z. Y., Shi, E., Yang, J.(2007), "Preparation and pervaporation performances of fumed-silica-filled polydimethylsiloxane-polyamide (PA) composite membranes". J. Appl. Polym. Sci. 105, 3132-3137.
18. Bergmeyer, H.U., Grassel, M., 1983. Reagents for enzymatic analysis: enzymes—-amylase, third ed. In: Bergmeyer, H.U. (Ed.), Methods of Enzymatic Analysis, vol. 2. Verlag Chemie, Weinhiem, Germany, pp. 151–152.

Adsorption of Galacturonic Acid onto Anion-exchange Resins to enhance the Enzymatic Hydrolysis in a Proposed Citrus Processing Waste Biorefinery

Raul C. Rivas-Cantu[1], Patrick L. Mills[2], Kim D. Jones[1], and Shad Nelson[3]

[1] Department of Environmental Engineering, [2] Department of Chemical and Natural Gas Engineering, [3] Department of Agriculture, Agribusiness & Environmental Sciences. Texas A&M University-Kingsville, Kingsville, Texas.

Extended Abstract

Citrus processing waste (CPW) is considered as a renewable feedstock for biorefining applications. The main structural polysaccharides of CPW as cellulosic biomass are cellulose and pectin. Pectin is one of the main polysaccharides forming the structure of citrus fruits and is composed of monomers of galacturonic acid (GA) forming a polymeric chain. Considering that in USA are generated around 3million metric tons per year of wet CPW (NASS, 2011), and assuming a content of 5-6 wt% of pectin, a considerable amount of 120,000 to 150,000 tons can be converted into galacturonic acid assuming a 80 wt% of the pectin. As part of the development for a CPW-based biorefinery, the enzymatic hydrolysis of grapefruit processing waste (GPW) brake down the structural polysaccharides of the peels and pulp to produce sugars such as glucose (from cellulose and hemicellulose) and galacturonic acid (from pectin) in relatively high concentrations (the total sums approximately 12-14 wt% of the total wet weight of the waste). Currently there is not industrial technology to recover GA from any CPW. One proposed technology for recovering poly-hydroxyl-carboxylic acids is the adsorption-desorption process utilizing anion-exchange resins. Studies in the adsorption of lactic, glutaric and glyoxylic acids utilizing weakly basic anion-exchange resins (WBAER) have been done during the last years by Uslu et al., 2009; Bayazit et al., 2010; and Bayazit, et al., 2011; . Resins such as Amberlite IRA-67 and IRA 96 (Dow Chem. 2008a,b) have been utilized for this purpose. To define the process feasibility and conditions for its application in CPW-Biorefinery is needed to develop a series of tests and experimental procedures.

A proprietary technology for the integrated pretreatment and enzymatic hydrolysis steps is being developed to enhance the hydrolysis reaction rate, and increase the yield of fermentable sugars. In this process, galacturonic acid is produced as monosaccharide during the enzymatic hydrolysis of pectin resulting from the synergistic action of pectinases, cellulases, and β-glucosidase. Issues related to the inhibition of enzymatic hydrolysis due to product and pH take place from the galacturonic acid production. Galacturonic acid cannot be fermented by traditional yeasts and bacteria, and genetically modified microorganisms (i.e. E. Coli K011) have to be utilized for its ethanol fermentation (Stwart et al., 2006; Ingram et al., 1999) making this process more complicated and expensive. Then GA has to be separated from enzymatic hydrolysis liquor to avoid these process inhibitions, and be recovered for its utilization and conversion in value-added products.

The aim of this research is the development of a process for the simultaneous separation of GA during the enzymatic hydrolysis of GPW as model biomass. This process is being performed using anion-exchange resins for the adsorption of GA in batch and semi-continuous systems. Laboratory experiments for the adsorption and recovery of GA from pure solutions and pectin hydrolysis liquor have shown the potential for the application of this technology to enhance the enzymatic hydrolysis rate and the yield of fermentable sugars. Furthermore, this innovative process could alleviate the environmental and economical issues related to wastewater treatment resulting from the biorefining applications of GPW.

Preliminary Results

Initial studies have been completed to investigate the feasibility, and efficiencies of WBAER on the adsorption of model aqueous solutions of GA to simulate its extraction from the enzymatic hydrolysis liquor resulting from GPW conversion. Adsorption experiments were performed in a batch reaction. The aim of these experiments is to determine the effect of important variables on the adsorption of GA onto selected anion-exchange resins, such as quantity of adsorbent (anion-exchange resin), equilibrium time, initial galacturonic acid concentration, and adsorption temperature.

Preliminary results are shown in Figures 1and 2 of the Langmuir adsorption isotherms for galacturonic acid onto a selected WBAER. Figure 3 shows the equilibrium time for the adsorption of GA on WBAER.

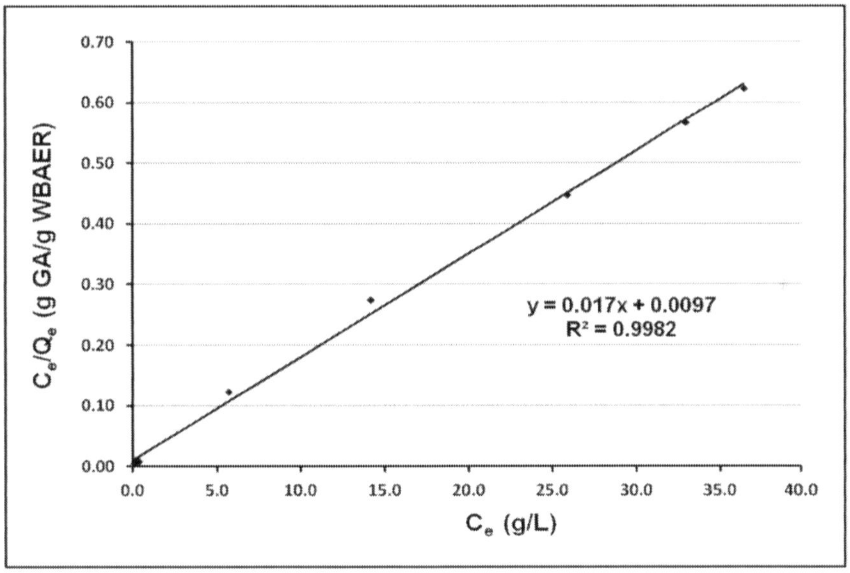

Figure 1. Linear Langmuir adsorption isotherm of galacturonic acid using the weakly basic anion exchange resin Amberlite IRA-67

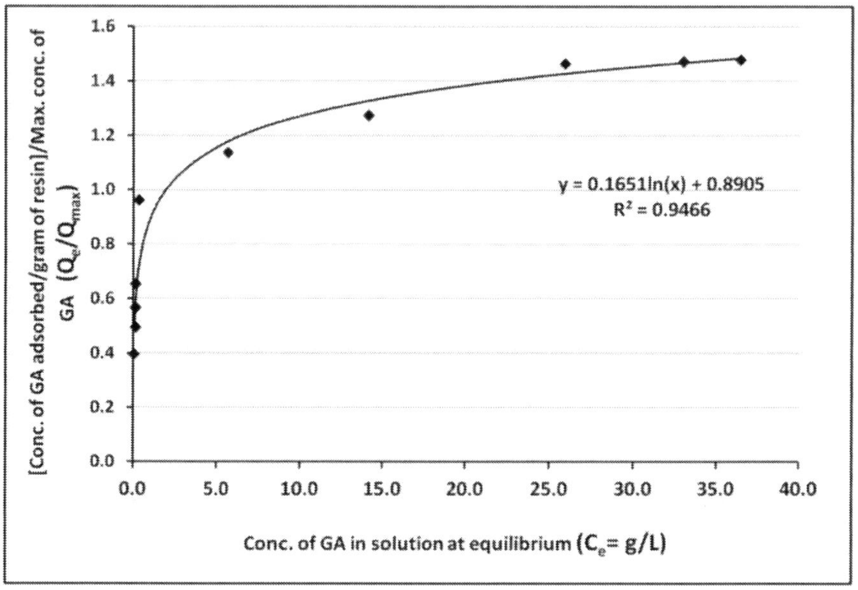

Figure 2. Langmuir adsorption isotherm of galacturonic acid using the weakly basic anion exchange resin Amberlite IRA-67.

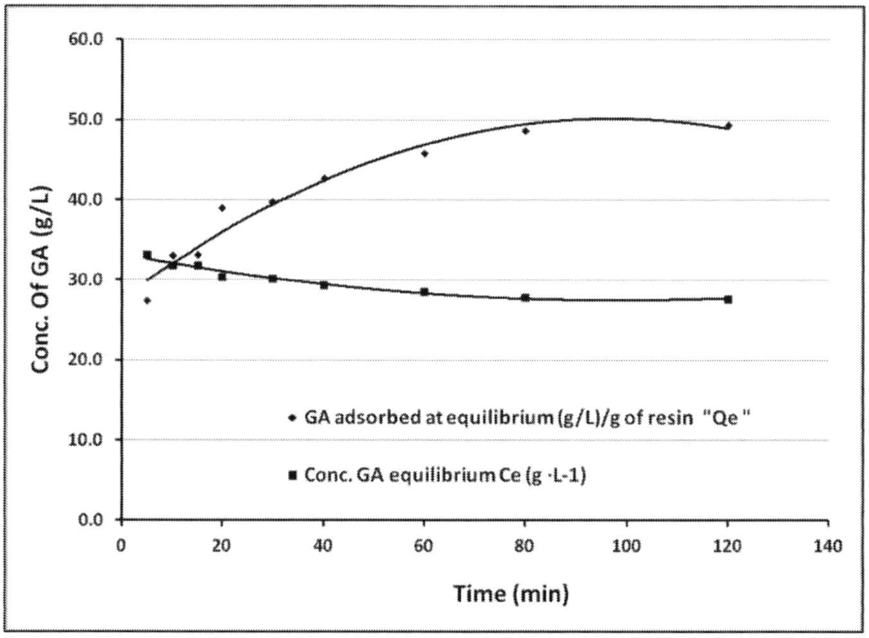

Figure 3. Equilibrium time for the adsorption of galacturonic acid using the weakly basic anion exchange resin Amberlite IRA-67.

Bibliographic References

1. Bayazit, S. S., Inci I., and Uslu, H. (2010). Adsorption of Glutaric Acid and Glyoxylic Acid onto Weakly Basic Ion-Exchange Resin: Equilibrium and Kinetics. J. Chem. Eng. Data 2010, 55, 679–684
2. Bayazit S. S., Inci, I., and Uslu H (2011). Adsorption of Lactic Acid from Model Fermentation Broth onto Activated Carbon and Amberlite IRA-67. J. Chem. Eng. Data 2011, 56, 1751–1754
3. Ingram, L.O., Aldrich, H.C., Borges, A.C., Causey T.B., Martinez, A., Morales, F., Saleh, A., Underwood, S.A., Yomano, L.P., York, S.W., Zaldivar, J., and Zhou, S., (1999), Enteric bacterial catalyst for fuel ethanol production, *Biotechnology Progress,* vol. 15, pp 855-866, 1999.
4. National Agricultural Statistics Service (NASS), (2011), *Citrus Fruit 2011 Summary*. United States Dept. of Agriculture (USDA). Last date accessed on 11.14.2011. http://usda01.library.cornell.edu/usda/ current/CitrFrui/CitrFrui-09-22-2011.pdf.
5. Stewart, D.A., Widmer, W.W., Grohmann, K., and Wilkins, M.R., (2006). Ethanol production from citrus processing wastes. *US Patent Application Publication*, No.US/2006/0177916-A1, 2006.
6. The Dow Chemical Company (2008). AMBERLITE™ IRA-67 Industrial Grade Weak Base Anion Exchanger, Product Data sheet. http://www.dow.com/assets/attachments/business/ier/ier_ for_industrial_water_ treatment/amberlite_ira67/tds/amberlite_ira67.pdf
7. The Dow Chemical Company (2008). AMBERLITE™ IRA-96 Industrial Grade Weak Base Anion Exchanger, Product Data sheet http://www.dow.com/assets/attachments/business/ier/ier _for_industrial_water_ treatment/amberlite_ira96/tds/amberlite_ira96.pdf
8. Uslu H, Inci I, Bayazit S, and Demir G.(2009). Comparison of Solid-Liquid Equilibrium Data for the Adsorption of Propionic Acid and Tartaric Acid from Aqueous Solution onto Amberlite IRA-67. Ind. Eng. Chem. Res., Vol. 48, No. 16, 2009.

(498f) Possibility of Multi-Fuel Production From Trap Grease by Combination of Heat-Driven Upgrading and Methane Fermentation

Hidetoshi Kuramochi[1], Kouji Maeda[2], Takuro Kobayashi[1], Yoshitaka Ebie[1], Keisuke Fukui[2] and Masahiro Osako[1], (1)Center for Material Cycles and Waste Management Research, National Institute for Environmental Studies, Tsukuba, Japan, (2)School of Mechanical and Systems Engineering, University of Hyogo, Himeji, Japan

Trap grease, which is a free-floating aggregate on the surface of the tank between restaurant and sewer, is either an oil or sludge-like grease including food scraps and water. It is a low-quality mixture of fats and oils, and is not recycled as a fuel in Japan. However, oily components such as glycerides and fatty acids in the grease a re considered to be a feedstock for alternatives to fossil fuels such as diesel and heavy oil. Furthermore, we consider that residues discharged during extraction of such oily content from the grease may be converted into biogas via methane fermentation. In the present study, first, we proposed a heating-driven liquefaction and phase separation for recovery of the oily content, and investigated the possibility of the obtained oil to use as an alternative to heavy oil in terms of impurity and property. From the investigation, the oily content was a clean fuel, especially a low-sulfur fuel. Hence, the heat-driven upgrading was useful for recovering a high-quality oil content from trap grease. In addition, we showed this upgrading can be successfully scaled up to a small demonstration phase. Unfortunately, the recovered oil could not meet some of fuel properties for replacement of high-grade heavy oil. Therefore, some improvements were proposed, and then their effectiveness was experimentally evaluated. Second, to evaluate the possibility of the residues to produce biogas, several types of residues from different restaurants were characterized in terms of volatile solid (VS), total nitrogen, etc, and then biogas productivity of the individual residues was examined using a batch-type methane fermentation. The results showed that the residues were applicable for methane fermentation and the biogas productivity ranged from 0.5 to 0.9 L/g-VS. These residues are considered to be good feedstocks for methane fermentation. From a series of investigation, the combination of the heat-driven upgrading of trap grease and methane fermentation of the residue from the grease upgrading can produce not only an alternative to heavy oil but also biogas. We expect both fuels may be used in a dual-fuel engine to generate electricity.

(519a) Advanced Computational Modeling of Photosynthetically Active Radiation (PAR) Distribution in Algae Photobioreactors

Bo Kong, Department of Chemical and Biological Engineering, Iowa State University, Ames, IA and R. Dennis Vigil, Iowa State University, Ames, IA

The development of highly efficient scalable continuous-flow algae photobioreactors (PBRs) could be a major advance towards enabling algae-based biofuel and biorenewable chemicals technology.Conversion of solar energy into chemical energy by photosynthesis is strongly depends on the quantity and quality of light received by microorganisms. Therefore, it is crucial to understand the availability and distribution of light in order to optimize the design and operation of a photobioreactor. In particular, it is important to understand the distribution of photosynthetically active radiation (PAR) for phototrophic microorganisms, which includes wavelengths from 400-700 nm. This radiation is absorbed by various pigments in the organisms, such as chlorophyll, carotenes and xanthophylls, which have different PAR absorption spectra. Despite the fact that only a narrow band of wavelengths are useful for photosynthesis, the most frequently used method for simulating light distribution in PBRs is to employ the Beer-Lambert law to calculate the overall light intensity distribution in one dimension. However, there exist several more advanced CFD techniques to solve the radiation transfer equation, such as the discrete ordinate (DO) and Monte Carlo (MC) methods, which are commonly used in other engineering fields, for example in combustion and HVAC. Although most of these models are used to compute total radiative transfer, in principle they can be extended to include spectral dependencies. Consequently, it is necessary, desirable, and feasible to develop a multidimensional spectral radiation model capable of predicting light distribution in photobioreactors.

In this work we describe a detailed and general numerical method for simulating the spatial distribution of PAR in photobioreactors using the discrete ordinate (DO) method for solving the radiative transfer equation (RTE), which was implemented in a computer program developed using the open source CFD library openFOAM. This non-gray DO implementation divides the radiation spectrum into N wavelength bands, in each of which the treatment of the light is the same as that for the gray DO model.Optical properties of the medium and boundary conditions for the non-gray DO model are applied on a band basis. Thus, the PAR transfer can be modeled by solving RTEs in small wavelength bands in the 400-700 nm wavelength range with a non-gray DO model. In algal cultures, the liquid medium is usually transparent to visible light, and therefore the transport of visible wavelengths are determined mainly by algae cells. The optical properties of the algae cells and the radiative properties of the turbid culture medium were calculated at each wavelength band range using the method developed by Pottier, et.al. 2005, which is based on Mie theory and requires information concerning algal pigmentation, shape, and size distribution. The model was validated in a simple rectangular rector configuration using the experimental data provided in Pottier, et.al. 2005. Subsequently, simulations were carried out in tubular and annular photobioreactor configurations, and implications of the simulation predictions are discussed.

(519b) Mechanistic Modeling of the Enzymatic Saccharification of Cellulose-Iβ and Cellulose-III: Parameter Determination and Model Validation

Andrew J. Griggs[1], Ambarish Nag[2], James J. Lischeske[1], Jonathan J. Stickel[1], Ashutosh Mittal[3], Wei Wang[3] and David K. Johnson[3], (1)National Bioenergy Center, National Renewable Energy Laboratory, Golden, CO, (2)Computational Science Center, National Renewable Energy Laboratory, Golden, CO, (3)Biosciences Center, National Renewable Energy Laboratory, Golden, CO

Cost-effective production of fuels and chemicals from lignocellulosic biomass often involves enzymatic saccharification, which has been the subject of intense research and development. Recently, a mechanistic model for the enzymatic saccharification of cellulose has been developed that accounts for distribution of cellulose-chain lengths, the accessibility of insoluble cellulose to enzymes, and the distinct modes of action of the component cellulases [Griggs *et al.* (2012) *Biotechnol. Bioeng.*, **109(3)**:665–675; Griggs *et al.* (2012) *Biotechnol. Bioeng.*, **109(3)**:676–685]. However, determining the appropriate values for enzyme adsorption, inhibition, and rate parameters required further experimental investigation. In this work, we performed several sets of experiments to aid in parameter estimation and to quantitatively validate the model. Several cellulosic materials having varied degrees of polymerization, crystallinity, and crystalline allomorph (cellulose-I_{β} and cellulose-III_{I}) were digested by component enzymes (EGI and CBHI) and by cocktails of these enzymes. Based on information from the literature and the results from these experiments, a single set of model parameters was determined and used to produce model simulation results. The agreement between simulation results and experimental data will be demonstrated. Cases of discrepancy point the way for further model development.

(519c) A Pore-Hindered Diffusion and Reaction Model Can Help Explain the Importance of Pore Size Distribution in Enzymatic Hydrolysis of Biomass

Jeremy S. Luterbacher, Department of Chemical and Biomolecular Engineering, Cornell University, Ithaca, NY and Larry P. Walker, Department of Biological and Environmental Engineering, Cornell University, Ithaca, NY

Until now, most efforts to improve carbohydrate production from biomass through pretreatment and enzymatic hydrolysis have used random optimization, trend extrapolation and/or educated guessing. These efforts could employ a rational design process if they were guided by a modeling framework that captured the key mechanisms governing the relationship between pretreatment and enzymatic hydrolysis. In this study, we propose a pore hindered diffusion and kinetic model for enzymatic hydrolysis of biomass. When compared to data available in the literature, this model accurately predicts the well-known dependence of initial cellulose hydrolysis rates on surface area available to a cellulase-size molecule. Modeling results suggest that, for particles smaller than $5x10^{-3}$ cm, a key rate-limiting step is the exposure of previously unexposed cellulose occurring after cellulose on the surface has hydrolyzed, rather than binding or diffusion. However, for larger particles, according to the model, diffusion plays a more significant role. Therefore, the proposed model can be used to design experiments that produce results that are either affected or unaffected by diffusion. Finally, by using pore size distribution data to predict the biomass fraction that is accessible to degradation, this model can be used to predict cellulose hydrolysis with time using only pore size distribution and initial composition data.

(519d) Anaerobic Digestion of Waste Materials for the Production of Biogas

Ramalingam Subramaniam, Stephen Dufreche, Rakesh Bajpai and Mark Zappi, Chemical Engineering, University of Louisiana at Lafayette, Lafayette, LA

Anaerobic digestion of waste materials for the production of biogas

Ramalingam Subramaniam, Stephen Dufreche, RakeshBajpai, and Mark E. Zappi

Bioprocessing Research Laboratory

Department of Chemical Engineering

University of Louisiana at Lafayette, Lafayette, LA

Biogas energy can be produced from various sources of biomass which include agriculture crops and residues, municipal solid wastes, and animal waste. In this work, anaerobic biogas was produced from waste materials such as shrimp waste, landfill leachate, and crab waste. The biogas produced from these waste materials in an anaerobic digester contains 65-70% methane and 25-30% carbon dioxide. Biogas with a methane composition of 66% has a heating value of 668 Btu/ft^3. Batch experiments were conducted in 500 ml three neck rectors as well as New Brunswick, microfirm MF114 reactor. Kinetic model was developed from the batch experiments. Based on the batch kinetic results, continuous process was investigated in 14 L bioreactor with a working volume of 10L. Batch kinetics, fermentor design and scale up will be discussed in this presentation.

(519e) Mass Transfer in High Solids Biomass Slurries

Samin Rezania[1], Jonathan Michael D. Thomas[2], Amlan Chakraborty[2] and **R. Eric Berson**[2], (1)University of Louisville, Louisville, KY, (2)Chemical Engineering, University of Louisville, Louisville, KY

Enzymatic saccharification rates and yields are hindered in high solids slurries, which is generally attributed to mass transfer limitations resulting from the high viscosity of the slurries. However, other possible mechanisms exist and no evidence has yet been presented directly linking the lower rates and yields to the mass transfer limitations. The objective of this work was to compare saccharification results for identical high solids slurries in a conventional Erlenmeyer flask with another system capable of overcoming mass transfer limitations in high viscosity slurries.

In order to achieve high mass transfer rates in the high viscosity slurries, a resonating acoustic mixer was employed for the saccharification reaction. Dispersion coefficients in the mixing vessel were determined as a function of viscosity in order to quantify the mass transfer improvements and determine if the system is a valid choice for studying reactions in the absence of mass transfer limitations. Since conventional means of measuring dispersion could not be performed due to the slurry characteristics, a computational fluid dynamics model was developed that simulated motion of the slurries in the mixer. The model was validated by comparing experimentally determined mixing times using an electrolytic tracer with mixing times from the simulations. Computational and experimental mixing times differed by 2% or less in all cases. Dispersion values were achieved on the order of values for water flow in a pipe, indicating the system was a valid choice for studying reactions in the absence of mass transfer limitations.

At low initial solids where good mixing could be achieved in the shake flask, the glucose release in the mixer was just 4% higher than in the shake flask, while for 30% initial total solids, the glucose release in the mixer was approximately 40% higher than in the shake flask. These results provide a direct link between low rates and yields of saccharification of viscous biomass slurries and mass transfer limitations.

(519f) Determining Dispersion in High-Solids Biomass Slurries

James J. Lischeske[1], Michael A. Sprague[2] and Jonathan J. Stickel[1], (1)National Bioenergy Center, National Renewable Energy Laboratory, Golden, CO, (2)Computational Science Center, National Renewable Energy Laboratory, Golden, CO

Conversion of lignocellulosic biomass to fuels can be made more cost-effective by operating at high-solids conditions. However, little is known about the transport properties in these systems. A thorough understanding of convection and diffusion in high-solids biomass slurries is necessary in order to properly develop predictive models for the degradation of biomass in reactor systems. In this work, packed-beds of pretreated biomass were prepared in preparative chromatography columns. Step inputs of tracer molecule solutions of varying size and type were passed through the packed beds. Initial residence-time-distribution analyses were performed to determine the extra-particle porosity of the packed beds and to determine apparent pore volume accessibility of the biomass. Secondly, an advection-diffusion-adsorption model was used to distinguish the effects of dispersion and adsorption of tracer molecules, and to measure parameters for the adsorption and diffusion. The relationship between dispersion and flow rate was examined, and diffusion parameters for a given bed were estimated. These diffusion parameters were then related to tortuosity factors, and the relationship between tortuosity and packing density was examined.

(547a) Fractionation and Characterization of Lignin Extractives From E-AFEX(TM) Pretreatment Process

Leonardo da costa Sousa[1], Shishir Chundawat[1], Vijay Bokade[2], Marcus Foston[3], Ali Azarpira[4], Bruce E. Dale[1] and **Venkatesh Balan**[1], (1)Chemical Engineering and Material Science, Great Lakes Bioenergy Center, Michigan State University, Lansing, MI, (2)Catalysis Division, National Chemical Laboratory (NCL), Pune, India, (3)Department of Chemistry and Biochemistry, Georgia Institute of Technology, Atlanta, GA, (4)Great Lakes Bioenergy Center, University of Wisconsin, Madison, WI

Extractive AFEX^{TM} (E-AFEX^{TM}) is a novel pretreatment technology that is able to simultaneously convert cellulose I to cellulose III and partially extract lignin from lignocellulosic biomass. As a result, pretreated feedstocks contain highly digestible carbohydrate content and significantly reduced lignin content. Depending on the pretreatment conditions, it is possible to obtain variations in the selectivity towards lignin solubilization. Temperature, ammonia concentration and co-solvent type are the major factors that contribute to a selective lignin solubilization and subsequent fractionation. Some pretreatment conditions tend to solubilize ash and hemicellulose residues along with lignin. However, since most applications for lignin require low ash and carbohydrate contents, these extractives have been further fractionated to obtain varying purity lignin fractions. For this purpose, solvent precipitation and filtration based methods weredeveloped to fractionate lignin. In this work, multiple fractions generated from the E-AFEX^{TM} extractives generated from pretreated corn stover were characterized by several techniques, which include GPC, TGA, ^{13}C-NMR, ^{1}H-NMR and 2D NMR. A comparison of E-AFEX^{TM} lignin with other commercial lignin sources is also presented here. The level of lignin extraction and cellulose III conversion was correlated with cellulose digestibility for corn stover. Furthermore the benefits of this new pretreatment methodology to downstream processing are presented in comparison with conventional AFEX^{TM}.

(547b) Effects of Mechanical Refining On the Enzymatic Digestibility of Acid Pretreated Corn Stover

Xiaowen Chen[1], Erik Kuhn[1], Wei Wang[1], Todd B. Vinzant[2], Sunkyu Park[3], Keith Flanegan[4] and Melvin Tucker[5], (1)National Renewable Energy Laboratory, Golden, CO, (2)Biosciences Center, National Renewable Energy Laboratory, Golden, CO, (3)Department of Forest Biomaterials, North Carolina state university, Raleigh, NC, (4)KL Energy Corporation, Rapid City, SD, (5)National Biology Center, National Renewable Energy Lab, Golden, CO

Mechanical refining has been applied in the pulp and paper industry for decades. It has been shown that mechanical refining provides better pulp quality through three different mechanisms: 1) shortening the fiber; 2) crushing and fibrillating the fibers internally; and 3) abrasion and fibrillation of the fibers on the exterior surfaces. However, mechanical refining has not been widely studied for the purpose of application to the concept of the bio-refinery. Recent studies by Koo et al. showed that mechanical refining using a laboratory PFI mill could potentially increase glucose yields in enzymatic saccharifications up to 20% using green liquor pretreated hardwood. The refining reduced the enzyme usage by ~50%. However, studies on disk refining and blending reported by Zhu et al and Dibble et al, respectively, showed minor effects on digestibility in terms of increased glucose yields. In this research, we explored the effects of five different refining methods including PFI milling, disk refining, extrusion, Szego milling and blending. The effects of the various refining methods on enzymatic hydrolysis using low severity pretreated corn stover substrates were investigated. The effects of different methods of refining on pretreated biomass are characterized by measurements of particle size distributions, degree of polymerization (DP), and crystallinity of cellulose, ultrastructural changes using SEM imaging, and enzymatic digestibility are reported.

(547c) Isoconversional Methods to Model Oxidative Alkaline Pretreatment of Lignocellulose

Mayra del Pilar Quiroz Galvan[1], Christian Andres Nova Salazar[2], Jonatan Garcia Mape[2], Juan Manuel Rivera Aragon[2] and Rocio Sierra[3], (1)Chemical Engineering Department, Universidad de los Andes, Bogotá, Colombia, (2)Chemical Engineering, Universidad de los Andes, Bogotá, Colombia, (3)Chemical Engineering, Universidad de los Andes, Bogota, Colombia

Oxidative alkaline pretreatment has proven successful to increase lignocellulose digestibility. In order to properly assess its effects, it is necessary to kinetically model the process. Most efforts to model pretreatment obtain kinetic parameters (i.e., activation energy and frequency factors) from model fitting methods using rate laws. The Isoconversional method is an alternative that in addition to being more accurate because does not depend on a so called constant temperature, is also usually simpler. In spite of these facts, isconversional methods have not been applied to kinetically model pretreatment.

The present work applies isoconversional methods to model oxidative alkaline pretreatment of sugar cane bagasse addressing a main difficulty inherent to the method: the need to provide large amount of data. This has been accomplished by implementing a novel procedure to estimate lignin and carbohydrates contents in lignocellulose. It uses the partial least square (PLS) method combined with Near Infrarred (NIR) Spectroscopy Analysis. The method was proven succesful to obtain kinetic parameters that were shown to be temperature dependent.

(547d) Quantitative Predictions of Bioconversion of Aspen by Dilute Acid and SPORL Pretreatments Using a Unified Combined Hydrolysis Factor (CHF)

Wenyuan Zhu[1], Carl J. Houtman[2], J.Y. Zhu[2] and Rolland Gleisner[3], (1)Nanjing Forestry University, Nanjing, China, (2)USDA Forest Service, Forest Products Laboratory, Madison, WI, (3)US Forest Service, Madison, WI

A combined hydrolysis factor (CHF) was developed to predict xylan hydrolysis during pretreatments of native aspen (*populus tremuloides*) wood chips. A natural extension of previously developed kinetic models allowed us to account for the effect of catalysts by dilute acid and two sulfite pretreatments at different pH values. When xylan is modeled as two fractions with different hydrolysis rates as identified as fast and slow xylan, the model closely matches the experimental data. Extent of xylan hydrolysis is strongly correlated with pretreatment solids yield, energy consumption for size reduction, and substrate enzymatic digestibility (SED). Composition of the pretreatment hydrolysate was less correlated with extent of hydrolysis due to carbohydrate decomposition reactions. Substrate cellulose enzymatic conversion and enzymatic hydrolysis glucose yield can be predicted to approximately 10% accuracy using *CHF* alone. The conventional combined severity fcator is a special and simplied case of CHF. The CHF developed has significant practical importance for scale up experiments which cannot afford to numerous experiments for optimization.

(547e) Study of A Novel Biomass Pretreatment Process Utilizing A Modified Taylor-Couette Mixer and the Effect of Cationic Polyelectrolyte On Alkaline Pretreatment and Enzymatic Hydrolysis

Shaowen Ji, Wei Wang and Ilsoon Lee, Department of Chemical Engineering and Materials Science, Michigan State University, East Lansing, MI

Lignocellulosic biomass has been considered as one of the primary candidates as alternative renewable resources for biofuel production. The presence of lignin is widely recognized as the barrier to cost-effective production, which inhibits enzyme accessibility and decreases sugar yields. The costly pretreatments and large consumption of enzymes are needed to reach high cellulose conversion. Therefore, the removal or effective modification of lignin is required for efficient pretreatment process. Here we report an efficient biomass pretreatment conducted via a novel modified Taylor-Couette mixing system, aiming to overcome the natural recalcitrance of lignocellulosic biomass in a short processing time and facilitate better enzyme accessibility to polysaccharides. Both lignin and hemicellulose were significantly removed after 2 minutes processing, leaving rich glucan content in the residue. Studies on the experimental parameters, such as mixing speed, substrate loading, and enzyme digestibility, were systematically performed. We further investigated the effect of cationic polyelectrolyte on the alkaline pretreated biomass. Structural studies revealed the morphology change of the cell wall in the residue, showing the retention of lignin-containing nanoparticles. The cellulose and hemicellulose conversion were enhanced after pretreatment, suggesting the possibly positive effect of polyelectrolyte on significantly modifying biomass morphological properties, such as increased porosity and improved accessibility of cellulose surface.

(547f) Applying Partial Least Square (PLS) to Quantify Lignin and Carbohydrates Content of Lignocellulose

Rocio Sierra, Chemical Engineering, Universidad de los Andes, Bogota, Colombia

Analytical chemistry is often employed to measure lignin and carbohydrates content of lignocellulose. Most traditionally used methods are long, comprise a large number of steps, are intensive in reactants use, require highly sophisticated equipment such as HPLC and most of the time result in a large standar deviation for replicates. If a large set of data is required, for example when determining kinetic parameters of pretreatment, this method is an important limitation.

Partial Least Square (PLS) combined with Near Infrared Spectroscopy (NIR) was used to determine lignin and carbohydrates content of lignocellulose in solid phase. A calibration curve was obtained using about 100 samples of lignocellulosic material with different lignin and carbohydrates content. An ample spectrum of lignin and carbohydrates contents was attained by using different degrees of maturity and plant parts, as well as by some modifications on the lignocellulose compositional structure through the application of different types of pretreatment. The method was shown succesful to quantify lignin and carbohydrates and much more faster than other traditionally used methods. However, an important limitation is the need to obtain a new calibration curve for each type of biomass used.

(571a) Engineering Yeast for Advanced Biofuels

Jose L. Avalos[1,2], Gerald Fink[1] and Gregory Stephanopoulos[2], (1)Whitehead Institute, Cambridge, MA, (2)Department of Chemical Engineering, Massachusetts Institute of Technology, Cambridge, MA

The development of renewable liquid transportation fuels is a key component of the national strategy to reduce CO_2 emissions, slow the progression of global climate change, and achieve energy self-sufficiency. Although bioethanol is the most commercially successful biofuel to date, its physicochemical properties and current methods of production make it an inadequate long-term solution to these problems. We are developing yeast strains that produce advanced biofuels that overcome ethanol's limitations. In order to accelerate strain development, we have developed a set of molecular tools and strategies that facilitate the cloning, assembly, integration, and optimization of engineered pathways. We demonstrate the use of this molecular toolkit to overcome roadblocks in the molecular and cellular biology of yeast that present bottlenecks in metabolic engineering, such as the screening of isoenzymes, vectors, promoters and strains. These molecular tools also facilitate new strategies for the assembly of multigenic pathways, which we used in the combinatorial assembly of multiple "isopathways" for the production of advanced biofuels. Using these tools we have engineered yeast strains that produce three different advanced biofuels with better fuel properties than ethanol, and better compatibility with prevalent infrastructures.

(571b) Phosphate Optimization for Economical Lipid Production From Lipomyces Starkeyi Grown On Starch

Sharif M. Rahman, Ramalingam Subramaniam, Stephen Dufreche, Mark Zappi and Rakesh Bajpai, Chemical Engineering, University of Louisiana at Lafayette, Lafayette, LA

Phosphate optimization for economical lipid production from *Lipomyces starkeyi* grown on starch

Sharif M. Rahman ,Ramalingam Subramaniam, Stephen Dufreche, Mark Zappi and Rakesh Bajpai, Chemical Engineering, University of Louisiana at Lafayette, Lafayette, LA

Microbial lipids can be obtained by the oleaginous yeast *Lipomyces starkeyi* using sweet potato starch as carbon source which is available largely from processors during cutting and canning process. But high cost of production is the major barrier for industrial production of microbial lipids. This cost can be reduced by optimizing nutrient components. Phosphorus is one of the highest cost contributing elements for microbial culture; and it is mostly incorporated with phospholipids and nucleic acids into the cells. It is under investigation that reduction of phosphate concentration affects cell growth as well as lipid accumulation. However, it is predicted that there is an optimum point of phosphate level where lipid accumulation does not change significantly. This work mainly focuses on optimizing phosphate concentration of the medium since it is a vital factor for economical microbial lipid production using waste materials.

(571c) Platform Pathway for the Synthesis of 3-Hydroxyacids As Value Added Products Derived From Biomass

Himanshu H. Dhamankar and Kristala L.J. Prather, Chemical Engineering, Massachusetts Institute of Technology, Cambridge, MA

The dwindling nature of petroleum and other fossil reserves has provided an impetus to microbial synthesis of fuels and value added chemicals from biomass as a renewable resource. The use of biomass as a feedstock requires the design and engineering of novel pathways connecting biomass derived simple sugars to such value added chemicals and fuels. 3-Hydroxyacids (3-HAs) are a versatile class of molecules that can serve as value added products derived from biomass. 3HAs such as 3-hydroxybutyrate and 3-hydroxyvalerate serve as monomers for biodegradable polyhydroxyalkanoates (PHAs) and chiral building blocks for pharmaceuticals and chemicals. We have established a novel platform pathway in *E. coli* that allows stereospecific synthesis of straight and branched chain 3HAs using glucose and small acid molecules as starting materials. Using this pathway we have been successful in establishing the first biosynthetic route towards the valuable pharmaceutical building block 3-hydroxy-γ-butyrolactone (3-HBL) and demonstrated synthesis of two novel PHA monomers: 3-hydroxyhexanoic acid and 4-methyl-3-hydroxyvaleric acid. Exploiting the full potential of this pathway requires exploring alternative pathway enzymes and engineering them to enhance pathway productivity and improve versatility for the synthesis of other valuable 3HAs. In particular, we are interested in enhancing pathway productivity for synthesis of 3-HBL which features as one of the top ten value added chemicals from biomass in a 2004 Department of Energy report and serves as a versatile chiral building block for the synthesis of drugs like Zetia®, cholesterol reducing synthetic statins such as atorvastatin (Lipitor ®) and rosuvastatin (Crestor ®), antibiotics such as Zyvox® and the nutritional supplement L-Carnitine. Ongoing efforts to identify alternative pathway enzymes for extending the 3HA pathway towards additional value added products as also enhancing enzyme activity and specificity through protein engineering towards specific products will be discussed.

(571d) Engineering a Yeast Conversion System for the Novel Production of Alkanes

John Blazeck, Department of Chemical Engineering, University of Texas at Austin, Austin, TX and Hal Alper, Chemical Engineering, The University of Texas at Austin, Austin, TX

Bio-derived alkanes are attractive second generation biofuels that circumvent some of the disadvantages associated with traditional biofuels. Engineering alkane production in yeasts relies upon the metabolic engineering of fatty acid metabolism. Fatty acids consist of a long aliphatic carbon chain decorated with only one carboxylic acid group, and thus serve as excellent substrates for alkane pathway engineering. An attractive host for alkane production is *Yarrowia lipolytica*, an oleaginous yeast adept at producing high levels of fatty acids. Here, we report the novel heterologous production of pentane in a *Y. lipolytica* fungal system by assembling a plant-derived pentane-production pathway. Pentane production was optimized using environmental and genetic techniques including increasing precursor availability through genetic knockouts and media formulation and by employing promoter engineering and codon optimization for high enzyme expression levels. Here, we will present the results of this optimization and discuss strategies for improved pentane production and for long chain alkane production in this yeast.

(571e) Engineering of Fatty Acid Pathway and Hydrocarbon Production in E. Coli

Fengming Lin, chemical engineering, University of michigan, ann arbor, MI, Yu Chen, Department of Chemical Engineering, University of Michigan, Ann Arbor, MI, Robert Levine, Chemical Engineering, University of Michigan, Ann Arbor, MI, Neil Marsh, chemistry, University of michigan, ann arbor, MI and Xiaoxia Nina Lin, Chemical Engineering, University of Michigan - Ann Arbor, Ann Arbor, MI

To improve the efficiency of producing fatty acid derived hydrocarbons, such as alkane, alkene, and alkyl ester, in model microorganism *Escherichia coli,* we are engineering pathways related to fatty acid biosynthesis using several strategies. First, a metabolic network bi-level optimization model was developed and applied to strain design for maximum fatty acid production. The gene manipulations suggested by this model were performed and led to an increase of total fatty acid from 58.5 mg/g dry cell weight in the wild type to 84.9 mg/g dry cell weight in a multiple knockout strain. Second, the triacylglycerol biosynthesis pathway was successfully introduced into *E. coli* by heterogeneous expression of wax ester synthase/acyl-coenzyme A: diacylglycerol acyltransferase (WS/DGAT) enzymes together with the knockout of dgkA, which not only increases the total amount of fatty acids in *E. coli*, but also improves the fuel quality of fatty acids. Third, an alkane biosynthesis pathway of cyanobacterial origin is being explored by utilization of synthetic protein scaffolds which spatially organize the enzymes involved. All combined, these strategies should lead to the improvement of titer and yield of producing fatty acid derived hydrocarbons in microorganisms.

(571f) Metabolic Engineering of Clostridium Tyrobutyricum for Isopropanol Production

Wenyan Jiang, Chemical Engineering, The Ohio State University, Columbus, OH, Mingrui Yu, Department of Chemical and Biomolecular Engineering, The Ohio State University, Columbus, OH and Shang-Tian Yang, The Ohio State University, Columbus, OH

Isopropanol is one of the most widely used solvents in the world. It is also one of the secondary alcohols that can be used as a direct or partial replacement for gasoline. Isopropanol can be produced by anaerobic microorganisms such as *Clostridium beijerinckii*. However, the native metabolic pathways are limited by tight regulation and the isopropanol production is low, less than 30 mM, in these strains. *Clostridium tyrobutyricum* is an anaerobic acidogen which can convert glucose and xylose to acetate and butyrate. Previously, our group had engineered this bacterium to produce n-butanol at a high rate, yield, and titer that are even higher than most of those reported native solventogenic *Clostridia*. In this work, we studied the feasibility of enabling C. *tyrobutyricum* to produce isopropanol. A plasmid was constructed with *C. acetobutylicum ctf* (acetoacetyl-CoA transferase), *adc* (acetoacetate decarboxylase) and *C. beijerinckii adh* (secondary alcohol dehydrogenase) genes under *C. tyrobutyricum* ATCC 25755 *thl* (acetoacetyl-CoA transferase) promoter. The plasmid was transformed into both *C. tyrobutyricum* wild type and *ack* (acetate kinase) knockout mutant strains. Isopropanol production by these mutants were confirmed in batch fermentation with glucose as the carbon source. The effects of isopropanol production on cell growth and fermentation kinetics were studied and will be presented in this paper.

(571g) Renewable Production of 5-Carbon Polyamide Building Blocks Using Engineered E. Coli

Jake Adkins, Chemical Engineering, Arizona State University and **David R. Nielsen**, Chemical Engineering, Arizona State University, Tempe, AZ

Polyamides are a class of homopolymers and copolymers with properties that balance excellent mechanical strength with durability and chemical resistance. Common examples include nylons which are produced from amino acid or diamine and diacid monomer building blocks. Today, such monomers are principally derived from petroleum resources. In recent years, however, the microbial production of 4- through 6-carbon monomers such as succinic acid, adipic acid, putrescine, and cadaverine have been reported. Here we report how, through *de novo* pathway engineering, our lab has been exploring the 'bottom up' development of microbial biocatalysts to produce other remaining polyamide monomer building blocks from renewable feedstocks. Systematic over-expression of the natural lysine degradation pathway enzymes of *Pseudomonas* sp. in a lysine over-producing *Escherichia coli* strain has allowed us to produce, as individual products, both 5-aminovaleric acid and glutaric acid (precursors to nylon-5 and copolymers such as nylon-5,5, for example). To date, genes from the *davBADT* operon of *P. putida* KT2440 has proven to be support the highest productivities in shake flask cultures. Alternative expression strategies been designed to compare the utility of the natural operon versus the individualized expression of pathway enzymes with the latter proving most effective. In shake flask cultures containing glucose in minimal media, our initial strains have already achieved 5-aminovaleric acid and glutaric acid titers as high as at least 350 mg/L. Here we report an overview of our most recent findings in this project. In all cases of strain development, systematic pathway development will be described through the recruitment and screening of candidate isoenzymes from different genetic sources for each step. In addition, to support greater overall productivity, strategies aimed at increasing precursor availability, improving pathway flux, and overcoming product toxicity are also described.

(586a) Comparison of Syngas Fermentation Reactors for Ethanol Production

Hasan K. Atiyeh[1], James J. Orgill[2], Mamatha Devarapalli[1], John R. Phillips[1], Randy S. Lewis[2] and Raymond L. Huhnke[1], (1)Biosystems and Agricultural Engineering, Oklahoma State University, Stillwater, OK, (2)Chemical Engineering, Brigham Young University, Provo, UT

Ethanol can be produced from biomass feedstocks or municipal solid wastes (MSW) using a hybrid thermochemical-biochemical conversion process called gasification-fermentation. In the gasification-fermentation process, biomass is gasified and converted to synthesis gas (syngas; primarily CO, CO_2 and H_2). The advantage of gasification-fermentation over the saccharification-fermentation process is that gasification utilizes cellulose, lignin and hemicellulose components of the biomass, leading to a greater potential for high carbon conversion efficiency and high ethanol yield from the same amount of biomass. Syngas fermentation rates are subject to gas-liquid mass transfer limitations since CO and H_2 have low solubilities in the fermentation broth. Successful syngas fermentation balances the kinetic capacity of the cell inventory with CO and H_2 supply that is set by the mass transfer characteristics of the reactor used. Continuously stirred tank reactor (CSTR), trickle-bed reactor (TBR) and hollow fiber membrane reactor (HFR) were compared for syngas fermentation with regards to gas utilization (including mass transfer assessment), cell growth, and product formation. The volumetric mass transfer coefficient ($K_{tot}A/V_L$) for O_2 in the three reactors was measured. The HFR provided the largest $K_{tot}A/V_L$ followed by the TBR. The mass transfer characteristics in each reactor were affected by agitation speed, and gas and liquid flow rates. Correlations for $K_{tot}A/V_L$ for each reactor were developed to predict mass transfer during syngas fermentation by *Clostridium ragsdalei* to produce ethanol and acetic acid. Although scale-up issues are different for each reactor, the comparison of the three reactors provides guidance as to which reactor(s) should be evaluated for scale-up, especially if one reactor significantly outperforms the others regarding ethanol yield and productivity and syngas utilization efficiency. Further, insight gained from the analysis of the three reactors can be extended to other fermentation processes.

(586b) Electrode Surface Kinetics of Methyl Viologen for Biological Processes

Chang Chen, Chemical Engineering Department, Brigham Young University, Provo, UT and Randy S. Lewis, Chemical Engineering, Brigham Young University, Provo, UT

Electrons are essential for biological processes and are often supplied by reduced species (e.g. glucose, H_2, CO, etc.). However, a recent focus has also been the use of electrodes to supply electrons for biological processes. For electrodes, electrons are either transferred by an extracellular electron mediator or directly transferred to immobilized bacteria on the electrode.

External chemical electron mediators, such as methyl viologen (MV), NADH and neutral red, can support metabolic functions of various bacteria. There is great potential in using electrodes but a more rigorous engineering analysis is needed to assess constraints such as electron efficiency, toxicity, thermodynamics, reaction rates, and availability of species. This work focuses on characterizing the kinetics of methyl viologen (MV) at the surface of an electrode. MV was chosen as a model compound as a result of its reported beneficial effects on various biological processes.

MV has three redox states (MV^{2+}, MV^{+} and MV^{0}). MV^{2+} undergoes a two-step reduction to acquire 2 electrons, forming MV^{0}. MV^{0} is capable of delivering electrons to cells. The reduction by an electrode is an electrochemical reaction and the reaction rate is a function of the applied potential. The relationship can be expressed by the Bulter-Volmer equation. The MV^{2+} reduction is also a surface reaction with three steps of absorption, surface reaction and desorption. The MV^{2+} reduction is demonstrated to be surface reaction limiting. This study combined the surface reaction with the electro-chemical expression which provides a more detailed engineering analysis to improve future design and control for electrode-mediated biological processes. Results of the kinetic study will be presented, along with thermodynamic constraints of utilizing MV. A key component of the presentation will be a discussion of the importance of engineering analysis when applying electrode systems to biological processes.

(586c) Characterization and Kinetic Study of Fast Pyrolysis and Torrefaction Products From Woody Feedstocks Using Direct Py-GC/MS

Jordan Klinger, Chemical Engineering, Michigan Technological University, Houghton, MI, David R. Shonnard, Chemical Engineering, Michigan Technological University & Sustainable Futures Institute, Houghton, MI and Ezra Bar-Ziv, Mechanical Engineering-Engineering Mechanics, Michigan Technological University, Houghton, MI

Pyrolysis and torrefaction are complex thermal degradation pathways that can produce valuable biofuels products – pyrolysis for bio-oils and torrefaction for bio-coals. Due to the complexity of these processes, decomposition mechanisms and product formations can vary based on the starting feedstock. Yet, detailed mechanistic understanding of the native material degradation is essential for process/product formation control. Current kinetic models and speciation as based around global phenomenological and empirical models, with little to no direct relation to actual decomposition mechanisms. This presentation will present chemical speciation and models based on the product identification that provides new insights for model development. Product speciation includes major gaseous and vapor products formed during both torrefaction (200-300°C) and pyrolysis (500-700°C). The preliminary models are formed around observed products/product formation rates during torrefaction at fast heating rate (1000°C/s). The models were developed around temporal histories of two primary observed product groups during torrefaction: (1) CO, CO_2, H_2O and (2) organic acids (acetic acid, formic acid). Group 1 products are likely formed through dehydration and deoxygenation reactions, followed by Group 2 formation through ring scission reactions, depolymerization, and other complex hemicelluloses degradation reactions. The direct observation of these chemical groups leads to the proposed degradation mechanisms, and kinetic parameters. This work was accomplished with the use of a Pyroprobe 5200 micropyrolysis reactor (CDS Analytical, Inc.) in-line with a Trace Gas Chromatograph Ultra (Restek Rxi-5ms capillary column) and Trace DSQ II Mass Spectrometer (Thermo-Fisher Scientific).

(586d) Kinetic Study of Autothermal Reforming of Glycerol Using a Dual Layer Monolith Catalyst

Yujia Liu[1], Bi Chen[2], Bob Farrauto[3] and Adeniyi Lawal[2], (1)Dept of Chemical Engineering and Material Science, Stevens Institute of Technology, Hoboken, NJ, (2)Dept. of Chemical Engineering and Materials Science, New Jersey Center for MicroChemical Systems, Stevens Institute of Technology, Hoboken, NJ, (3)BASF Catalysts LLC, Iselin, NJ

Kinetic study of autothermal reforming of glycerol using dual layer monolith catalyst

Yujia Liu[a], Bi Chen[a], Robert Farrauto[b] Adeniyi Lawal[a]*

[a] New Jersey Center for MicroChemical Systems, Dept. of Chemical Engineering and Materials Science, Stevens Institute of Technology, Castle Point on Hudson, Hoboken, NJ 07030, USA

[b] BASF Catalysts, 25 Middlesex Turnpike, Iselin, NJ 08830, USA

Crude glycerol is a by-product during biodiesel production via transesterification of triglycerides. The conversion of glycerol to synthesis gas and then to methanol represents an important niche for crude glycerol processing that could relieve small biodiesel producers from high glycerol disposal cost while reducing their dependence on methanol from fossil fuels. The autothermal reforming (ATR) of glycerol into synthesis gas (syngas) was studied using the BASF dual layer monolith catalyst.

This paper presents an approach for developing the kinetic model for the ATR of glycerol. The determination of reaction pathways/mechanisms on one hand and chemical kinetics and parameter estimations on the other make up the two vital aspects of kinetic modeling of catalytic processes. A LHHW type rate model is developed and subjected to parameter estimation and model discrimination. The kinetic parameters in conjunction with the kinetic equations can then be used as a design tool for process scale-up.

(586e) Predicting Biomass Fuel Bound Nitrogen (FBN) Conversion in an Oxygen/Steam-Blown Fluidized Bed Gasifier

Karl Broer, Center for Sustainable Environmental Technologies, Iowa State University, Ames, IA and Robert C. Brown, Mechanical Engineering, Iowa State University, Ames, IA

Predicting Biomass Fuel Bound Nitrogen (FBN) Conversion in an Oxygen/Steam-Blown Fluidized Bed Gasifier

Karl M. Broer (kbroer@iastate.edu)

Robert C. Brown (rcbrown@iastate.edu)

When biomass is processed in a fluidized bed (FB) gasifier, its nitrogen content, also known as the fuel bound nitrogen (FBN), is converted primarily into ammonia (NH_3) and nitrogen gas (N_2). Smaller amounts are also present in the raw syngas as hydrogen cyanide (HCN), char-bound nitrogen, and tar-bound nitrogen. NH_3 is of the most interest because it is typically found in the syngas in high enough concentrations to form problematic amounts of NO_X when the syngas is combusted for integrated gasification-combined cycle applications. This NH_3 can also be responsible for poisoning catalysts in biomass-to-liquid applications. Chemical equilibrium calculations indicate that for typical FB biomass gasification temperatures, pressures, and equivalence ratios, nearly all FBN should be converted to harmless N_2. These results contradict experimental results, which show significant NH_3 concentrations. More sophisticated biomass gasification models based on fundamental gasification kinetics have been developed that are capable of predicting NH_3 concentrations. The model developed by de Jong et al. [1] exhibits modest ability to accurately predict NH_3 concentration, but relatively poor ability to predict the concentrations of other important syngas constituents, such as CO and H_2. Additionally, the de Jong model was designed for steam-air gasification, whereas steam-oxygen blown gasification presents a new range of operating conditions that are characterized by much higher initial oxygen concentrations at the bottom of the fluidized bed, and different relative proportions of permanent gases in the resulting syngas.

A revised kinetic model involving fundamental homo- and heterogeneous combustion reactions has been generated to predict the final concentration of the nitrogen compounds in syngas, along with the other major constituents, such as H_2, CO, CO_2, and CH_4, for steam/oxygen-blown gasification. In addition to predicting the syngas composition, the model suggests which reaction pathways of nitrogen evolution are most important under FB gasification conditions. Knowledge of these major mechanisms allows insight into which methods may be worth further investigation for *in situ* reduction of NH_3 concentration in syngas in favor of N_2 and char-bound nitrogen, which are less problematic for downstream syngas applications.

1. de Jong, W., Unal, O., Andries, J., Hein, K., & Spliethoff, H. (2003). Biomass and fossil fuel conversion by pressurized fluidised bed gasification using hot gas ceramic filters as gas cleaning. *Biomass and Bioenergy*, *25*, 59-83.

(586f) An Experimental Investigation of Solid Carbon Conversion to Fuels in the Presence of CO2

Marco J. Castaldi, Department of Earth and Environmental Engineering (HKSM), Columbia University in the City of New York, New York City, NY

To meet the challenges of increased energy demand and waste management we must move to an atom economy where every atom is utilized in the best possible manner. To achieve that goal, a fundamental understanding of the underlying mechanisms and processes of energy and waste generation is necessary. Furthermore, if we consider where the world stands today in terms of energy use and where it will be in 100 years, there is a significant challenge to meet the expected demand in a more CO_2-neutral way. Therefore, a primary objective for new energy technology development must be to produce carbon-free energy through the advancement of technologies that use both conventional and alternative sources.

We have recently reported on the impact of CO_2 on biomass gasification and its ability to more effectively gasify biomass than steam. Continuing this investigation has led to understanding the impact of heating rates and different reaction environments. This paper will present the results from the gasification of various biomass feedstocks. The heating rates were varied from 1°C min^{-1} to 100 °C min^{-1} to ballistic rates (~ 9000°C min^{-1}). The gasification mediums investigated include H_20/N_2, $CO_2/N_2/H_20$, and O_2/N_2. Global activation energies for pyrolysis were found to be significantly higher than for gasification while those for the grasses were significantly lower than the woods, possibly indicating a catalytic effect during pyrolysis of the high mineral content herbaceous feedstocks. CO_2 pyrolysis (110-450°C) activation energy values for lignin, cellulose and biomass averaged 36, 216, and 50 kJ mol^{-1} and CO_2 gasification (500-700°C) values for lignin and biomass were 25 and 33 kJ mol^{-1}, though cellulose did not exhibit significant gasification behavior. Using a least squares fit on the rate of mass loss fraction, the global decomposition reaction during pyrolysis for lignin in either medium was found to be third order while that for cellulose was first order and for the various biomass samples either first or second order. In addition, the impact of pretreatment on biomass will be presented.

In addition, rapid heating rates (greater than or equal to 700°C min^{-1}) were utilized to gasify Clean Wood and two Refuse Derived Fuel (RDF) samples using thermogravimetric analysis coupled to gas chromatography (TGA/GC). Reaction atmospheres included Air, 5% O_2/95% Ar, 10% O_2/90% Ar, 100% Ar and steam and were used to produce gas evolution profiles for hydrocarbons ranging from H_2 to C_4H_{10}. While expected results were obtained using Air reaction atmospheres, some interesting results were observed using steam and Ar. Different concentration profiles and production rates of C_2H_6 compared to C_2H_4 and C_2H_2 enabled some understanding of the reaction sequence occurring during gasification under rapid heating conditions. Kinetic analysis showed pre-exponential factors of $8.00x10^{27}$, $2.02x10^{29}$ and $3.71x10^{23}$ (sec^{-1} K$^{1/2}$) for samples Clean Wood, RDF C (Industrial Solid Waste basis) and A (Municipal Solid Waste basis), respectively. Furthermore the apparent activation energy was determined to be 22, 71, and 185 (kJ mol^{-1}) for Clean Wood, RDF C and A respectively indicating that the Clean Wood is slightly more reactive than RDF C and more reactive than RDF A.

Finally, it will be shown how CO_2 can adjust the products of gasification and fast pyrolysis to a desired output. This process allows the decoupling of the biomass feedstock from the final product. Data will be presented focusing on the relation between gasification medium, feedstock selection and major chemical species evolution.

(602a) Axial Dispersion in a Horizontally Rotating Scraped Surface Bioreactor

Mahyar Ghorbanian, University of Louisville, Louisville, KY and **R. Eric Berson**, Chemical Engineering, University of Louisville, Louisville, KY

Horizontally rotating bioreactors offer many advantages for high-solids enzymatic saccharification. By operating horizontally: (1) particle settling and dead zone formation are avoided by horizontal rotation of the blades, which scrape against the surface of the reactor. (2) Impellers in conventional stirred tank reactors need to be run above the just suspended speed to maintain an off-bottom/uniform suspension, which demands significant power. The horizontal scraping action of the impellers in the SSBR requires very low rotation rates and, hence, much less power while more effectively preventing particle settling. (3) The scraping action of the blades maintains a clear reactor surface and, thereby, improves heat transfer.

Despite these advantages, all the transport motion occurs only in the angular direction. Multi-directional transport is important for dispersion of enzyme and carrying away product from the reaction sites. The purpose of this study was to investigate the degree of axial dispersion, if any, in a horizontally rotating scraped surface bioreactor (SSBR) during high solids enzymatic saccharification.

Computational fluid dynamic modeling qualitatively showed clear evidence of axial motion within the slurries. Axial dispersion coefficients were then quantified as a function of the solids concentrations and fluid viscosity. The dispersion coefficients were obtained experimentally by measuring the transient concentration of an injected electrolytic tracer and then fitting the concentration/time data to an axial dispersion model. Axial dispersion was strongly dependent on viscosity and there were order of magnitude differences between coefficient values for water and biomass slurries approaching 20% solids content. The degree of axial transport was still significant enough to consider horizontally rotating reactors as a viable option for processing high solids biomass slurries.

Cycle Operation of Laboratory-Scaled Adsorption Heat Pump for Regenerating Steam from Waste Water

Yoshiho Iwama, Kyushu University, Fukuoka, Japan
Yuuki Tanaka, Kyushu University, Fukuoka, Japan
Kazuya Nakashima, Kyushu University, Fukuoka, Japan
Bing Xue, Kyushu University, Fukuoka, Japan
Agung Tri Wijayanta, Kyushu University, Fukuoka, Japan
Koichi Nakaso, Kyushu University, Fukuoka, Japan
Jun Fukai, Kyushu University, Fukuoka, Japan

Introduction

Steam is demanded and generally provided from large quantities of fossil fuel while large amount of waste water with the range of 60-90 °C is released from these industrial processes [1]. The system to generate steam from waste heat is therefore required for the effective utilization of energy. In this study, a novel steam generation process using a direct heat exchange system with adsorbent-water pair is proposed. Contacting water and adsorbent makes excess water evaporate due to the release of adsorption heat from adsorbent. Because this system does not require any heat exchangers, increment in packing density of adsorbent particles in the reactor is expected. The purpose of this study is to investigate the novel steam generation system using water-adsorbent pair. As a basic study, cycle operation consists of steam generation process and regeneration process of adsorbent is experimentally studied under atmospheric pressure. Steam generation at high pressure is also studied.

Experimental

Apparatus

Fig. 1 illustrates schematically the experimental setup for generation in an adsorption heat pump (AHP) cycle. The setup consists of a thermostatic bath, a cylindrical reactor, a heater, a drier, a compressor, a condenser, a reservoir an electronic balance. The connecting metallic tube with diameter of 6.35 mm (0.25 inch) and outer wall of the reactor are insulated with 12.5 mm thickness of ceramic fiber. The reactor is made of stainless steel with height of 100 mm, inner diameter of 71 mm and outer diameter of 80 mm. Cylindrical zeolite 13X particles with diameter of 1.6 mm and lengths 2–6 mm are used. Mass of zeolite is 0.25 kg.

The high-pressure experimental apparatus includes a relief valve, an evaporator and a vacuum pump in addition to the cycle experimental apparatus. A relief valve is installed for high-pressure steam generation and set 0.2 or 0.4 MPa. The relief valve, which is initially closed, opens when the gas (steam plus air) pressure reaches a pre-set value. The evaporator and the vacuum pump are installed for the preheat process. The evaporator temperature is 80°C (saturated water vapour pressure is 0.047 MPa).

Procedure

The cycle operation begins with generation and ends with regeneration. In the steam generation process, water at 80 °C was introduced to the reactor by the feeding pump. Steam was then generated by adsorption heat. The mass of steam was measured by an electric balance as condensed water. The steam generation process was terminated when water interface reached the top of packed bed. After the steam generation process, remained water was drained from the bottom of the reactor. Dried air at 120 °C and 0.62 Nm/s was then introduced to the reactor for a given period of time in the regeneration process. The temperature in the packed bed was measured by thermocouples inserted in the packed bed. Both steam generation process and regeneration process was cycled 5 times. As the basic study, the cycle experiment was carried out under atmospheric pressure.

In the high-pressure experiment, saturated water vapor at 80 °C is introduced as preheating. After that generation process follows the same procedure as steam generation process of cycle experiment.

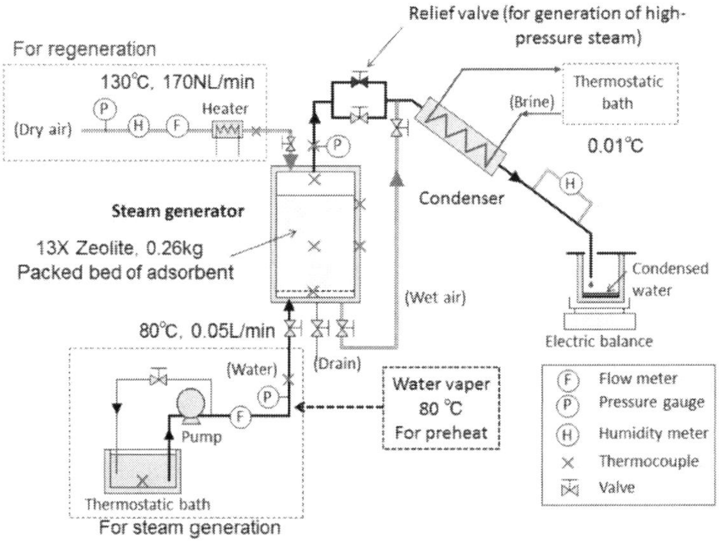

Fig. 1 Schematic illustration of experimental setup

Results

Cycle Operation at 0.1 MPa

Fig. 2 describes transient temperatures for all five cycles based on t_{regen}=3600 s. The thermal response is distinguished in the first cycle from subsequent cycles due to the difference in initial temperatures inside the reactor and mass ratio of water to dry zeolite (x). Results from the second to the fifth cycles depict good agreement in thermal response.

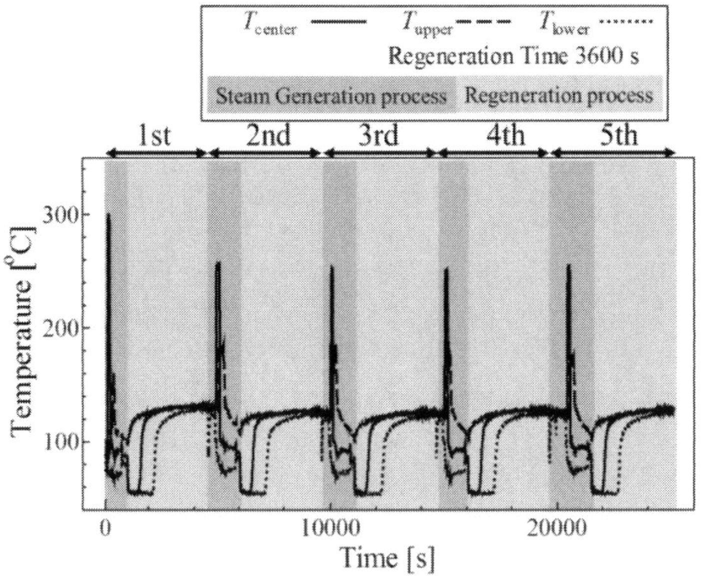

Fig. 2 Transient temperatures over five cycles based on regeneration for 3600 s

Fig. 3 shows the results for the fifth cycle. Fig. 3(a) shows the time variation in temperatures and steam production in the generation process. The feeding water temperature (T_{lower}) initially reaches 100 °C (boiling point under atmospheric condition) due to adsorption heat. Then T_{lower} gradually decreases to 7 °C. T_{center} dramatically increases from 130 °C to 260 °C due to adsorption of steam on zeolite. The maximum is retained for 100 s because adsorption is paused under a local equilibrium condition. As water passes through this part, the temperature rapidly drops to 100 °C. A slow decline to 90 °C follows uniform with water temperature until the end of the process.

Fig. 3(b) shows the transient temperatures during regeneration process. The feeding dry gas temperature (T_{upper}) on the top remains stable at 130 °C during the whole process. The temperature (T_{center}) in the center quickly declines to 50 °C largely from free water evaporating away as a flow of dry gas. T_{center} then quickly rises to 110 °C because of the decreasing desorption rate, followed by a slow increase to 130 °C at the feeding gas temperature (T_{upper}). A similar response to the temperature in the center (T_{center}) is then observed. Finally, regeneration stops as the feeding (T_{upper}) and outlet (T_{lower}) temperatures reach almost the same value.

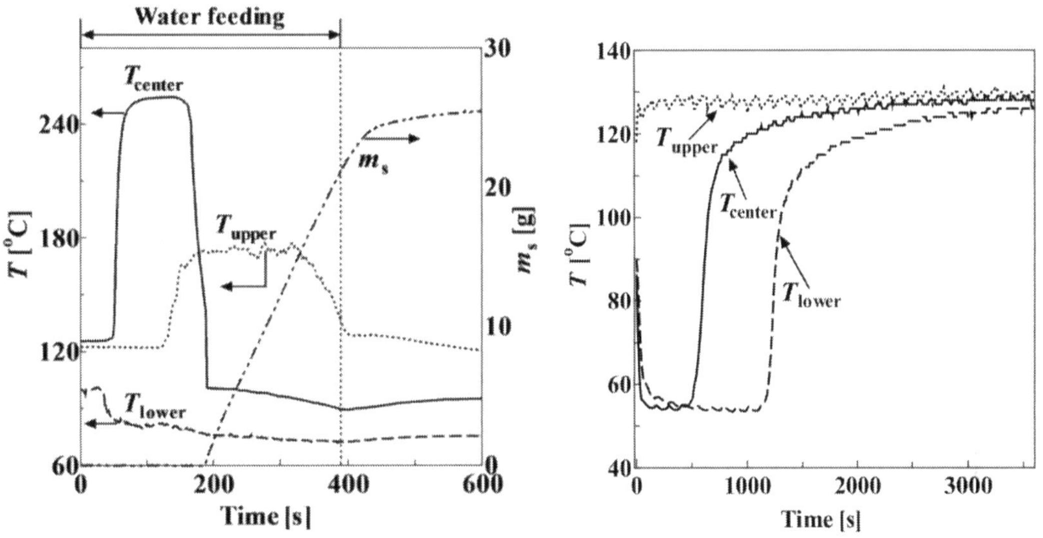

(a) Generation (b)Regeneration

Fig. 3 Transient behavior in the fifth cycle

Table 1 summarizes the cyclic operation conditions and results based on t_{regen}=3600 s. The generation rate per zeolite mass ($\dot{m}_{cycle}/m_z$) from the second to fifth cycle is obviously larger than that at the first cycle, in spite of a decrease in adsorption heat ($\Delta H_{ads,zw}$) at 1126 kJ/kg compared with that at 1350 kJ/kg for the first cycle. $\dot{m}_{cycle}$ is defined as mass of steam generated per the cycle time including generation (t_{gen}) and regeneration (t_{regen}).

$$\dot{m}_{cycle} = m_s /(t_{gen} + t_{regen}) \quad (1)$$

The rise on $\dot{m}_{cycle}/m_z$ is mainly due to a higher initial zeolite ($T_{z,ini}$) and reactor ($T_{rea,ini}$) temperatures. $T_{z,ini}$ increases from 83.3°C at the first cycle to an average at 126°C in the following cycles, while $T_{rea,ini}$ rises from 83 to 121°C. $\dot{m}_{cycle}/m_z$ varies in a range from 2.08 to 2.48×10^{-5} kg-s/kg-z·s at different operation conditions. Table 1 also summarizes cycle operation conditions and results based on t_{regen}=1200 s. Compared with t_{regen}=3600 s, the mass of generated steam (m_s) is expected to decline due to lower $T_{z,ini}$, $T_{rea,ini}$ and a decline in total heat of adsorption ($Q_{ads} = m_z(x_{eq}-x_{ini})\Delta H_{ads,zw}$). Average $T_{z,ini}$ and $T_{rea,ini}$ decrease from 125 to 110 °C and 121 to 94.2°C, respectively. Declines in both Δx (from 0.195 to 0.135) and $\Delta H_{ads,zw}$ (from 1126 to 1080 kJ/kg) with a larger x_{ini} result in a decrease in Q_{ads}. However, average $\dot{m}_{cycle}/m_z$ is improved by 10% from 2.22×10^{-5} 1/s (t_{regen}=3600 s) to 2.44×10^{-5} kg-s/ kg-z s (t_{regen}=1200 s). This is because m_s drop to only a half when the regeneration time is decreased to one-third.

Table 1 Cycle operation conditions and results based on regeneration for 3600 s and 1200 s

		Regeneration for 3600s					Regeneration for 1200s					
		1st	2nd	3rd	4th	5th	1st	2nd	3rd	4th	5th	6th
Generation												
$m_{w,in}$	[g]	327	325	314	306	305	319	216	255	247	245	252
$u_w\times10^4$	[m/s]	1.98	2.02	2.48	2.27	2.10	2.02	3.70	2.19	2.19	2.06	2.23
$T_{w,feed}$	[°C]	69.4	72.4	68.1	68.5	71.5	69.8	71.4	71.1	71.9	71.0	72.0
$T_{rea,ini}$	[°C]	83.0	125	120	119	119	79.7	92.9	94.1	94.7	94.9	94.2
$T_{z,ini}$	[°C]	83.3	130	124	124	124	81	104	110	112	112	113
$x_{ini}\times10^2$	[-]	1.7	6.5	6.5	6.5	6.5	1.7	12.5	12.5	12.5	12.5	12.5
$\Delta H_{ads,zw}$	[kJ/kg-w]	1350	1126	1126	1126	1126	1350	1085	1085	1085	1085	1085
$T_{rea,fin}$	[°C]	85.7	93.8	90.6	89.9	92.5	84.2	86.1	87.0	87.3	86.8	86.9
$T_{z,fin}$	[°C]	87.8	90.2	88.4	88.6	90.3	88.5	88.2	88.2	88.2	87.9	88.0
T_s	[°C]	140	180	180	178	178	184	175	183	184	186	186
m_s	[g]	22.0	31.7	28.8	26.7	26.6	18.9	12.6	13.7	14.3	13.7	13.8
$\dot{m}_{cycle}/m_z\times10^5$	[kg-s/kg-z s]	1.72	2.48	2.25	2.08	2.08	3.39	2.26	2.45	2.56	2.46	2.44
Regeneration												
$T_{g,feed}$	[°C]	131	131	131	131	131	131	131	131	131	131	131
u_g	[Nm/s]	0.64	0.64	0.64	0.64	0.64	0.64	0.64	0.63	0.64	0.64	0.64
$w_{s,feed}\times10^6$	[kg-s/kg-a]	4.77	4.77	4.77	4.77	4.77	4.77	4.77	4.77	4.77	4.77	4.77

Steam Generation at High Pressures

Table 2 shows experimental conditions for high-pressure steam generation without and with preheat. Fig. 4(a) shows the results for $u_w = 2.57\times10^{-4}$ m/s and P_{preset} = 0.4 MPa. There is a difference of 50 degrees in the maximum temperature of T_{z1} and T_{z2} because partial pressure of steam at beginning T_{z2} rising is higher than that at beginning T_{z2} rising. Superheated steam is generated at 220°C and 0.45 MPa. Fig. 4(b) shows the time variations in T_{upper} and m_s for P_{preset} = 0.4 MPa. The effective time for steam generation decreases to 50 s, which is only 14.2% of the water feeding time. At the end of water feeding, m_s has reached 1.4 g and continues to increase to 4.28 g due to the delay of condensed water.

Table 2 Experimental conditions for steam generation without and with preheat

		No preheat	Preheat	No preheat	Preheat
P_{preset}	[MPa]	0.2	0.2	0.4	0.4
$P_{g,ini}$	[kPa]	101.3	≤2.5	101.3	≤2.5
$T_{z,ini}$	[°C]	78.8	78.5	79.2	77.7
x_{ini}	[-]	0.017	0.017	0.017	0.017
$T_{w,feed}$	[°C]	78.8	79.1	79.1	81.3
$u_w\times10^4$	[m/s]	2.44	2.78	2.53	2.61

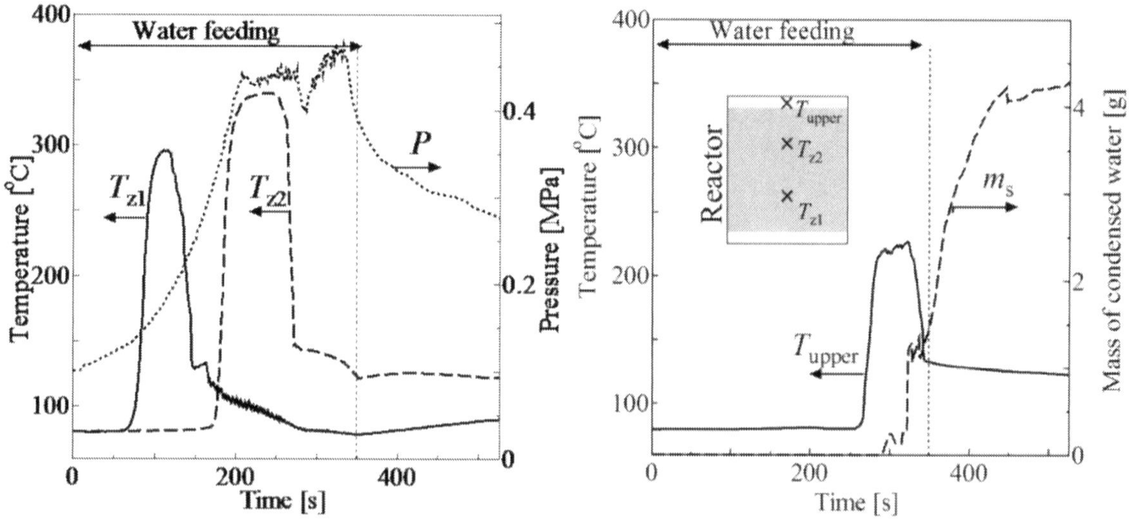

(a)Time variation in temperature and total pressure (b) Time variation in temperature and mass of steam

Fig. 4 Transient behavior in steam generation process at P_{preset} = 0.4MPa

Fig. 5(a) shows the time variations of temperatures (T_{z1}, T_{z2} and T_{upper}) and total pressure (P_g) at P_{preset} = 0.4 MPa. During preheat, both T_{z1} and T_{z2} increase as a result of steam adsorption on zeolite. T_{z1} and T_{z2} rise to their maximum temperatures of 270 and 230°C in 55 s, respectively. During water feeding, T_{z1}, T_{z2} and T_{upper} rise in sequence with increases in position height, because feed water flow originates at the bottom of the reactor.

Fig. 5(b) shows the effect of preheat on normalized time variations in steam temperature. The time ratio of effective steam generation to water feeding ($t_{s,eff}/t_{w,feed}$) is 45%, which is much larger than the 25 % without preheat. The increase is significant because most of the zeolite-gas region achieves local equilibrium at the start of water feeding. A local equilibrium for zeolite and gas is reached in most of the packed bed, and also in the upper part where steam escapes. Therefore, generated steam passes through the packed bed more easily with preheat.

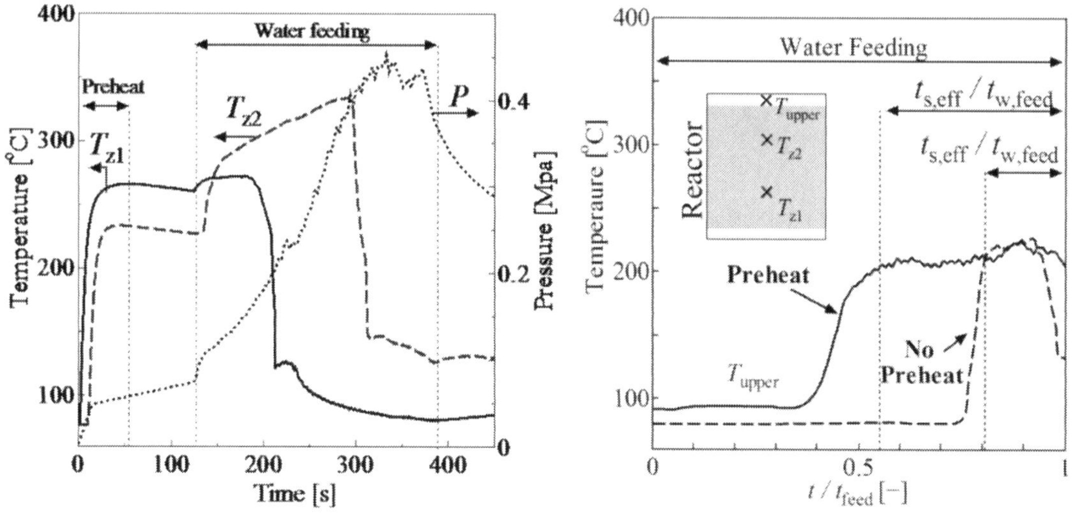

(a)Time variation in temperatures and total pressure with preheat (b) Effect of preheat on normalized time variation of steam temperature

Fig.5 Transient behavior in steam generation process with preheat at P_{preset} = 0.4MPa

Fig. 6 shows the mass of steam (m_s) without and with preheat at two preset pressures (P_{preset}). The increases in m_s of 20.5 and 54.8% are provided with preheat at P_{preset} = 0.2 and 0.4 MPa, respectively. The reason why increase the mass of steam generated with preheat is explained from the following heat balance. During experiment, initial conditions for both with and without preheat are same. According to the heat balance in Eq. (2), Q_{in} (heat flow in) is zero. Q_{acc} (sensible heat change in zeolite, water and reactor) is almost the same for various P_{preset}. Q_{out} (heat of steam flowing out) is determined by Q_{gen} (overall adsorption heat).

$$Q_{in} - Q_{out} + Q_{gen} = Q_{acc} \quad (2)$$

Q_{gen} for the experiment without preheat is given in Eq. (3), and that with preheat is shown in Eq. (4), respectively:

$$Q_{gen} = m_{z,dry}(x_{eq} - x_{ini})\Delta H_{ads,zw} \quad (3)$$

$$\begin{aligned} Q_{gen} &= m_{z,dry}(x_{eq} - x_2)(\Delta H_{ads,zw} + L) + m_{z,dry}(x_2 - x_{ini})\Delta H_{ads,zw} \\ &= m_{z,dry}(x_{eq} - x_{ini})\Delta H_{ads,zw} + m_{z,dry}(x_{eq} - x_2)L \end{aligned} \quad (4)$$

where x_2 is the mass ratio of water to dry zeolite at the end of the preheat process. Although x is measured before the start and after the finish of an experiment, the change in x during the process could not be measured during the experiment. As described above, the reason for the increase in m_s is that Q_{gen} increases by the latent heat of vaporization of steam from preheat.

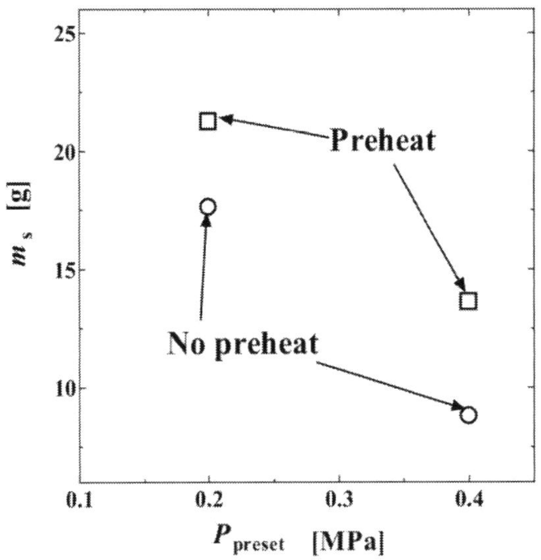

Fig. 6 Steam generation with and without preheat at two preset pressures

Summary

Cycle operation for generation steam using an AHP has been studied experimentally. Experiments have been conducted with and without a preheat process at various pre-set pressures. Concluding remarks in this study are summarized as follows:

1. Steam generation in cycle operation in a zeolite-water AHP using low-grade waste heat has been experimentally demonstrated.
2. Reproducibility in temperature response and mass of generated over cycles proves the durability in practical applications.
3. Experiments demonstrate a significant increase in high-pressure steam generation with a preheat process using low-pressure steam in the zeolite-water AHP system, because the latent heat of vaporization in water is transported to the zeolite before the steam generation process.

Nomenclature

L	= latent heat of vaporization	[kJ/kg]
m	= mass	[kg]
$\dot{m}_{cycle}$	= steam generation rate in cycle operation	[kg-s/s]
P	= total pressure	[Pa]
p	= partial pressure	[Pa]
Q	= heat	[J]
T	= temperature	[K]
t	= time	[s]
u_g	= superficial velocity of feed gas	[m/s]
u_w	= superficial velocity of feed water	[m/s]
V	= volume	[m^3]
w	= mass ratio of steam to dry air	[–]
x	= mass ratio of water to dry zeolite	[kg-w/kg-z]
ΔH_{ads}	= adsorption heat	[kJ/kg]

<Subscripts>

acc = accumulation
center = center of packed bed height
dry = dry
eff = effective
eq = equilibrium
feed = feed
fin = final
g = gas
gen = generation
in = inlet
ini = initial stage
lower = bottom of the packed bed
out = outlet
preset = preset value of relief valve
rea = reactor
regen = regeneration
s = vaporized water (steam)
upper = top of the
w = liquid water
z = zeolite
z1 = one third of the packed bed height
z2 = two thirds of the packed bed height

Acknowledgements

This work was partially supported by the NEDO project under Research and Development Program for Innovative Energy Efficiency Technology, and by a grant from the Global Centre of Excellence in Novel Carbon Resource Sciences, Kyushu University.

Reference

1. Japan Machinery Federation; "Research on new business development through the use of energy technologies", pp. 12-15, Tokyo, Japan (2007) [In Japanese]

(602ab) Investigation of Application for Star Block Copolymer As a Material for Photocatalytic Degradation of Chlorophenols

Seung-Gun Chung[1,2], **Yong-Soo Choi**[1], Jae-Woo Choi[1] and Sang-hyup Lee[1], (1)Water Resource Cycle Research, Korea Institute of Science and Technology, Seoul, South Korea, (2)Earth and Environmental Science, Korea University, Seoul, South Korea

This study investigated the photocatalytic degradation of chlorophenols using two different types of porphyrin core star block copolymer (P-PSD and P-PD) under visible light irradiation. Both P-PSD and P-PD catalysts show similar tendency in degradation of chlorophenols. Pentachlorophenol was degraded faster than other compounds (Pentachlorophenol > 2,4,6-Trichlorophenol > 2,4-Dichlorophenol). The removal rate of chlorophenols by P-PSD was higher than P-PD, which could be attributed by the existence of hydrophobicity on P-PSD. The degradation intermediates and by-products of chlorophenols were also identified by HPLC and LC-MS analysis. The analysis results revealed that the degradation of highly-chlorinated phenols were rapid than the less-chlorinated phenols as confirmed by residual chlorinated compounds and released chloride ions. The toxicity of PCP was greatly reduced after P-PSD catalyzed photocatalytic treatment. The process of photocatalytic degradation using porphyrin core star block copolymer is found to have potentially advantageous on dissolved chlorophenol pollutants in water.

This study investigated the photocatalytic degradation of chlorophenols using two different types of porphyrin core star block copolymer (P-PSD and P-PD) under visible light irradiation. Both P-PSD and P-PD catalysts show similar tendency in degradation of chlorophenols. Pentachlorophenol was degraded faster than other compounds (Pentachlorophenol > 2,4,6-Trichlorophenol > 2,4-Dichlorophenol). The removal rate of chlorophenols by P-PSD was higher than P-PD, which could be attributed by the existence of hydrophobicity on P-PSD. The degradation intermediates and by-products of chlorophenols were also identified by HPLC and LC-MS analysis. The analysis results revealed that the degradation of highly-chlorinated phenols were rapid than the less-chlorinated phenols as confirmed by residual chlorinated compounds and released chloride ions. The toxicity of PCP was greatly reduced after P-PSD catalyzed photocatalytic treatment. The process of photocatalytic degradation using porphyrin core star block copolymer is found to have potentially advantageous on dissolved chlorophenol pollutants in water.

(602ac) Adsorption of Fluoroquinolone Antibiotics Onto Activated Sludge: Implications for Membrane Bioreactor Operation

Ke He, Chemical, Biochemical and Environmental Engineering, University of Maryland, Baltimore County, Halethorpe, MD and Lee Blaney, Chemical, Biochemical and Environmental Engineering, University of Maryland, Baltimore County, Baltimore, MD

Abstract

Given the increasing detection of pharmaceuticals in treated wastewater, the need to optimize wastewater treatment processes for pharmaceutical removal is considerable. In conventional wastewater treatment plants, the activated sludge process contains the greatest potential for pharmaceutical treatment. That potential can be harvested using novel treatment processes, such as membrane bioreactors (MBRs), which allow the operator to uncouple the solids retention time (SRT) and the hydraulic residence time. By increasing the SRT, greater degrees of pharmaceutical removal can be attained through adsorption to biosolids and biological transformation mechanisms. This research focuses on two aspects of such systems: (1) development of analytical methods for detecting trace concentrations of fluoroquinolone antibiotics and (2) description of fluoroquinolone adsorption onto activated sludge solids.

A rapid, accurate, and sensitive method has been developed for the quantitative determination of six fluoroquinolone antibacterial agents (*i.e.*, ciprofloxacin, norfloxacin, levofloxacin, sparfloxacin, sarafloxacin, and orbifloxacin) by online solid-phase-extraction high performance liquid chromatography with fluorescence detection (oSPE-HPLC-FLD). That method was used as a template for development of an ultrahigh performance liquid chromatography tandem mass spectrometry (UHPLC-MS/MS) detection method. The UHPLC-MS/MS method allows quantitation of fluoroquinolone antibiotics to the ng/L level. With this method, the six fluoroquinolone antibiotics of concern were measured in raw wastewater, the activated sludge reactor, and treated wastewater from the Little Patuxent Water Reclamation Plant (LPWRP; Savage, MD). These concentrations of fluoroquinolones were employed in laboratory-based investigations of the adsorption of fluoroquinolones onto activated sludge.

Activated sludge was collected from LPWRP; the specific biomass concentration was determined by weighing the dry solids. Biological degradation of fluoroquinolones was minimized through the addition of 0.1% w/v of sodium azide. Batch adsorption experiments were designed to investigate the impact of pH and ionic strength on fluoroquinolone partitioning between the aqueous and solid phase. Specific masses of biosolids were added to batch reactors containing synthetic wastewater containing environmentally-relevant (from LPWRP) concentrations of fluoroquinolones. At specific times, samples were collected and analyzed using UHPLC-MS/MS to determine the fluoroquinolone concentrations in the aqueous phase; solid phase concentrations were determined via mass balance. The equilibrium data (contact time = 36 hours) was fit to typical isotherms to fully describe adsorption of fluoroquinolones onto activated sludge. These data will further understanding of how SRT affects adsorption of antibiotics in activated sludge / MBR operation. Ultimately, the results of this research will be used to design and operate a laboratory-scale MBR to investigate fluoroquinolone removal through adsorption and biological transformation mechanisms.

(602ad) Ozone Treatment of Tetracycline Antibiotics: Transformation Kinetics and Removal in Water/Wastewater Matrices

Zachary Hopkins, Chemical, Biochemical, and Environmental Engineering, University of Maryland, Baltimore County, Baltimore, MD and Lee Blaney, Chemical, Biochemical and Environmental Engineering, University of Maryland, Baltimore County, Baltimore, MD

Ozone treatment of tetracycline antibiotics: Transformation kinetics and removal in water/wastewater matrices

Abstract by Zachary Hopkins researching under Dr. Lee Blaney

The fate of pharmaceuticals in the environment is of increasing concern. Antibiotics represent unique threats to human and ecological health due to the potential for development of antibiotic resistance, inhibition of select microorganism populations, and various sub-lethal effects stemming from the ability of antibiotics to act as intermicrobial signaling agents. To preclude lasting effects to environmental and human health, advanced water and wastewater treatment processes are required to successfully remove trace concentrations of antibiotics from water supplies.

In this research, three tetracycline antibiotics (*i.e.*, tetracycline, chlortetracycline, and rolitetracycline) were treated using concentrated aqueous ozone solutions (~16 mg/L as O_3), which were generated by bubbling gaseous ozone into DI water. The baseline structure of tetracycline antibiotics contains four potential sites for ozone interaction. All three tetracyclines have pK_a values of approximately 3.3, 7.6, and 9.3; above $pK_{a,2}$, ozone reacts rapidly with the free tertiary amine functionality (~4×10^6 $M^{-1}s^{-1}$). Below pH 7.6, ozone reacts with the tetracycline ring structure. Regardless of pH, ozone was extremely reactive with the three tetracyclines; for example, the apparent second order rate constants ranged from 5×10^4 to 5×10^6 $M^{-1}s^{-1}$ across the pH range of interest (*i.e.*, pH 5–9). Tetracyclines were measured using ultrahigh performance liquid chromatography tandem mass spectrometry (UHPLC-MS/MS).

While pH affects the reaction kinetics of tetracycline transformation by ozone, a variety of water quality parameters, including alkalinity and natural organic matter (NOM), influence the ultimate transformation of tetracyclines in water/wastewater treatment. To draw general conclusions about the impacts of the background water quality on ozonation of the three tetracyclines of concern, three different water matrices, namely tap water, surface water from Herbert Run (Baltimore, MD), and tertiary effluent from the Little Patuxent Water Reclamation Plant (Savage, MD), were employed. These three experimental solutions were chosen to resemble treatment of finished drinking water, raw drinking water, and treated wastewater. Tetracyclines (approximately, 2×10^{-9} to 2×10^{-7} M) were spiked into these background matrices and dosed with ozone (approximately, 3×10^{-4} M) to result in molar ratios of applied ozone to initial (total) tetracycline antibiotics of 0–5 mol/mol. The ultimate transformation of the three tetracycline antibiotics in these matrices will be presented as a function of pH, NOM content (measured as total organic carbon), and alkalinity. This data will be beneficial to the design of ozonation processes aimed at treating trace concentrations of pharmaceuticals in water/wastewater treatment processes.

(602ae) Kinetics Study of Calcium Carbonate Decomposition At High CO2 Environment

Yong-Hua Duan[1], De-Long Xu[2], Hui Li[3], Lele Zhang[4], Xiao Fan[1], Wen-bin Yang[1] and Yong Min[3], (1)Material Institute, Xi'an University of Architecture and Technology,, Xi'an, China, (2)Xi'an University of Architecture and Technology, Xi'an, China, (3)Xi'an University of Architecture &Technology, Institute of Powder Engineering, Xi'an, China, (4)Xi'an University of Architecture &Technology, Institute of Powder Engineering, Xi'an, China, xian, China

Kinetics Study of Calcium Carbonate Decomposition at High CO2 Environment

Yong-Hua Duan[1], Delong Xu[2], Hui Li[3], Le-le Zhang[1], Xiao Fan[1], Wen-bin Yang[1], and Yong Min[3],

(1)Material Institute, Xi'an University of Architecture and Technology, Xi'an, China, (2)Institute for Powder Eng, Univ. of Archi. & Tech., Xi'an, China,

(3)Xi'an University of Architecture &Technology, Institute of Powder Engineering, Xi'an, China

With the concerns of "global warming", CO2 sequestration from cement industry are becoming significant important for the "green processing" for the environments. This is because the cement processing released about 36% of CO2 comparing to the 12% from regular coal-power plant. CaO/CaCO3, a major component and raw material from cement, was investigated in order to design suitable processing to sequestrate the CO2 during the cement production. It was found that the raw material particle size, surface area as well as chemical components were important factors, which will effect on the CO2 absorption efficiency, cycle times, particle activities, and so on. Various analytical equipments such as SEM, XRD, FTIR, TGA, were used to monitor the processing and analysis materials during the processing.

It was found that there are two competitive mechanisms involved during the processing i.e., chemical reaction and physical diffusion. In order to investigate the dominated effect, we selected the different particles as model compounds for the analysis. As a result, we found the small particle (particle size less than 5 micron) is dominated by chemical reaction mechanism, i.e., CO2 can completely diffusion into the core of the particle and result in a complete chemical reaction to form CaCO3; and the larger particle is dominated by the physical diffusion mechanism, i.e., only partial CO2 can diffuse into the CaO particles and formed a "core-shell" particles, where the centre core is CaO and out-shell is CaCO3. In addition, the small particle always accomplished with a high cycle times and more complete chemical reaction, which was recommended to use in the industry.

Based on the cement processing, various chemical looping processes were used to demonstrate the CaCO3 Calcinations / CaO Carbonation Cycle such as, vacuum processing, solids flow processing, fluidization processing, and cocking gas recycling processing. The energy efficiency, raw material activities, and corresponding thermodynamic date will be presented during the presentation.

(602af) Looking for Optimal Conditions to Manufacture Activated Carbon From Barley Husk by Single- or Dual-Optimization

Margarita Loredo Cancino, Eduardo Soto Regalado, Maria Teresa Garza González, Refugio Bernardo García Reyes, Felipe de Jesús Cerino Córdova and Azucena García León, Facultad de Ciencias Químicas, Universidad Autónoma de Nuevo León, San Nicolás de los Garza, NL, Mexico

Activated carbon has a strong affinity for binding organic or inorganic pollutants from aqueous solutions due to its high specific surface area and the possibility of incorporating functional groups on surface. For these reasons, activated carbons are commonly used for wastewater treatment. Despite of the above-mentioned advantages, the use of commercial activated carbon as adsorbent is often limited due to economical reasons, and many researchers have paid attention in the use of alternative precursors, such as agro-waste materials for activated carbon production. In this research, the manufacturing of activated carbon from barley husk (BH) by chemical activation with zinc chloride was optimized using a factorial design with two responses (yield and iodine number) and three factors (activation temperature, activation time and impregnation ratio). Optimal conditions for manufacturing barley husk activated carbon (BHAC) were obtained using a 2^3 factorial design followed by a central composite design by adding axial points to the initial design to explore the space curvature. Single or dual optimization was performed to determine the optimal conditions which improve surface area and yield.

Findings demonstrate that a linear model was not appropriate to represent the design space based on the ANOVA (at a confidence level of 99%) and, therefore, a central composite design was used to fit a quadratic model to the experimental data. According to the ANOVA for the quadratic model of iodine number, the most significant individual factors were temperature and impregnation ratio, double interaction between temperature and impregnation ratio, and the quadratic term of temperature (similar significant factors were determined for yield response except that the quadratic term was the impregnation ratio instead of temperature). It is important to mention that activation time was not significant and, for this reason, it was not considered in the quadratic model but for economical reasons time was set equal to 20 minutes.

Single optimization using the quadratic model of each response gives different optimal conditions depending on the response, for instance: maximal iodine number (1046.26±105.52 mg/g) was predicted at 530°C and impregnation ratio of 1.5 g $ZnCl_2$/g BH, while for maximal yield (61.58±3.32%) were found at 300°C and impregnation ratio of 0.76 g $ZnCl_2$/g BH. Taking into account that maximal responses were found at different conditions and both responses play an important role in manufacturing activated carbon from lignocelullosic precursors, both responses were optimized simultaneously by using the desirability functions approach included in Design expert software.

After simultaneous dual optimization, the maximal desirability function value (0.779) was obtained at 436°C, activation time of 20 minutes, and impregnation ratio of 1.1 g $ZnCl_2$/g BH. At these conditions, predicted responses for iodine number and yield were 829.58±78.30 mg/g and 46.82±2.64%, respectively, whereas experimental tests achieved 901.86 mg/g and 48.48% (values within the predicted interval). Obtained iodine number (901.86 mg/g) is similar or greater than those reported by other researchers dealing with lignocelullosic precursors. Correspondingly, BHAC produced at the optimal conditions (assessed by simultaneous dual optimization) reached a surface area (811.44 m^2/g) similar to commercial activated carbons and lignocelullosic-based activated carbons.

In summary, simultaneous dual optimization seems to be better than single optimization due to the fact that both yield and surface area are very important properties when manufacturing activated carbon from lignocelullosic precursors.

FOULING RATES OF SYNTHETIC THIN STILLAGE

Ravi Challa[1], David Johnston[2], Vijay Singh[1], Mike Tumbleson[1], Nicki Engeseth[3] and Kent Rausch[1,*]

[1]Agricultural and Biological Engineering, University of Illinois at Urbana-Champaign, 1304 West Pennsylvania Avenue, Urbana, IL 61801, USA
[2]Eastern Regional Research Center, Agricultural Research Service, USDA, 600 East Mermaid Lane, Wyndmoor, PA 19038, USA
[3]Food Science and Human Nutrition, University of Illinois at Urbana-Champaign, 1201 West Gregory Drive, Urbana, IL 61801, USA

INTRODUCTION

In the US, fuel ethanol is produced by two processes: dry grind (86%) and wet milling (14%) (Mueller 2010). Ethanol production became more energy efficient with innovative energy saving methods, which reduced the energy requirements per liter ethanol from 33.4 MJ in 1981 to 7.2 MJ in 2008 (Mueller 2010). Following fermentation, whole stillage is separated into thin stillage (5 to 10% soluble solids, w/w) and wet grains (suspended solids) by centrifugation. Thin stillage is concentrated to form condensed distillers solubles (25 to 30% dry solids, w/w) in multiple effect evaporators (Singh et al. 1999).

All dry grind plants use evaporators to concentrate thin stillage to syrup or condensed distillers solubles. For a 40 mmgy ethanol plant, 240 to 280 mmgy thin stillage is produced (Rausch et al 2003). Thin stillage contains 90% water and a portion is concentrated in evaporators and mixed with wet grains to produce distillers dried grains with solubles (DDGS). During evaporation, undesirable deposits form on heat transfer surfaces, increasing resistance to heat transmission and decreasing energy efficiency (Taborek et al. 1972). Fouling deposits increase over time and require periodic evaporator shutdown for cleaning (Wilkins et al. 2006a). Costs associated with fouling include labor and equipment to clean heat transfer surfaces, increased capital, antifoulant chemical costs, production losses and increased environmental impact from cleaning chemical disposal (Epstein 1981).

Process steps involved in converting starch to sucrose are gelatinization, liquefaction and saccharification. Starch is an insoluble granule and resistant to enzymatic breakdown. During ethanol production, first of several steps in starch gelatinization is cooking where the starch granules adsorb water and swell to lose the crystalline structure. Gelatinization is followed by starch enzymatic hydrolysis. Starch hydrolysis is usually expressed in terms of dextrose equivalent (DE) and it is inversely proportional to degree of polymerization. Enzymatic hydrolysis results in starch breakdown into shorter chain oligosaccharides with DE ranging from 0 to 20 (Kenyon 1988). Complete starch hydrolysis to glucose can be represented from the number of reducing ends (DE =100). However, in corn ethanol production by dry grind process, thin stillage composed of unconverted starch, yeast, protein, oil and ash (Singh et al. 1999). Enzymatic hydrolysis results in starch breakdown into shorter oligosaccharides but even under strict reaction conditions, it is difficult to achieve consistent composition at industrial scale and therefore changes occur in the composition of dextrins, maltose and glucose from batch to batch. These compositional changes in process streams could contribute to fouling problems on the evaporator surfaces. Earlier research on thin stillage fouling include the effect of Reynolds number (Wilkins et al. 2006a), pH (Wilkins et al. 2006b), membrane filtration (Arora et al. 2010) and fat content (Singh et al. 1999) but did not identify thin stillage components that accelerate evaporator fouling. In the current study, the effect of sucrose:starch ratio on evaporator fouling was investigated by preparing synthetic thin stillage solutions. Starch is insoluble in water whereas sucrose is soluble at room temperature. The objective is to determine the effect of soluble and insoluble solids on evaporator fouling by varying the starch:sucrose (STA:SUC) in synthetic thin stillage.

MATERIALS AND METHODS

An annular fouling probe was used to perform fouling experiments (Fig. 1). Annular probe design and operation were described previously (Agbisit et al. 2003; Arora et al. 2010; Fischer et al. 1975; Singh et al. 1999; Wilkins et al. 2006a). Fouling experiments were performed using synthetic thin stillage (STS) solutions for 1, 3, 5 and 10% total solids in water. STS components selected were starch and sucrose to demonstrate the individual and interaction effects in evaporator fouling. STS was starch suspended in water, sucrose dissolved in water or sucrose and starch mixed as per the starch to sucrose ratio. Each solution was transferred into a stainless steel tank and tap water was added to make 50L batches. Recycling STS through the fouling system provided for uniform distribution of model components (starch and sucrose) in water.

STS was pumped through the test loop at a flow rate of 15 L/min (Fig. 1). Bulk temperature (T_b) was maintained at 50±2°C for all tests. STS bulk temperature was attained by using water as the heating fluid in a shell and tube heat exchanger (1.5 m, Graver Technologies, DE). The hot water temperature (T_{HW}) was maintained at 55±2°C. A 100 ml sample was collected before start of the experiment to measure test fluid viscosity. Power (410 ± 10W) was supplied to the resistance heater within the probe to an average initial temperature of 100°C after T_b reached 50°C. Probe began to foul immediately, which was reflected in the increased thermocouple temperatures.

The overall heat transfer coefficient (U) (kW/m2.K) was found from the following equation:

$$U = \frac{Q/A}{T_s - T_b}$$

Where T_s is the probe surface temperature (°C) and T_b is the STS bulk temperature (°C).

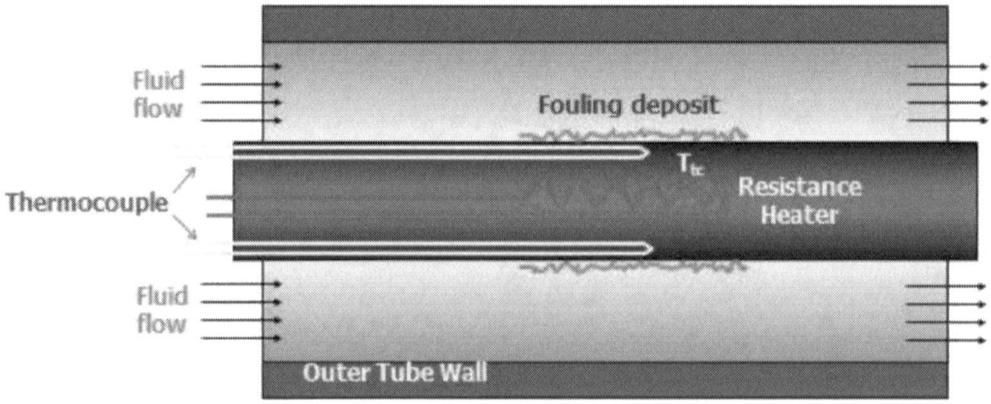

Fig. 1. Annular fouling probe.

Fouling resistance (R_f) was found by the difference of the reciprocals of the overall heat transfer coefficient of the fouled and clean surface.

$$R_f = \frac{1}{U_f} - \frac{1}{U_i}$$

Where U_f is the overall heat transfer coefficients of a fouled probe at time *t* and U_i is the initial overall heat transfer coefficient for a clean probe at time t=0. The temperature inside the probe increased as the fouling deposit accumulates on the probe surface.

A schematic of the fouling apparatus is shown in Fig 2. All the tests were terminated when the interior probe temperature reached 170°C or at the end of 10 hr. A 100 ml sample was collected after each test and stored at 4°C for viscosity measurement. Fouling apparatus and annular probe were cleaned following each test. The annular probe along with fouling deposit was soaked overnight in 5% (w/v) NaOH solution. Fouling deposits were removed with a wet sponge; fouling apparatus (batch tank, tubing, heat exchanger, pump and rod housing) was cleaned by recirculating 1% (w/v) hot detergent solution (Alconox, New York, NY), followed by a hot water rinse.

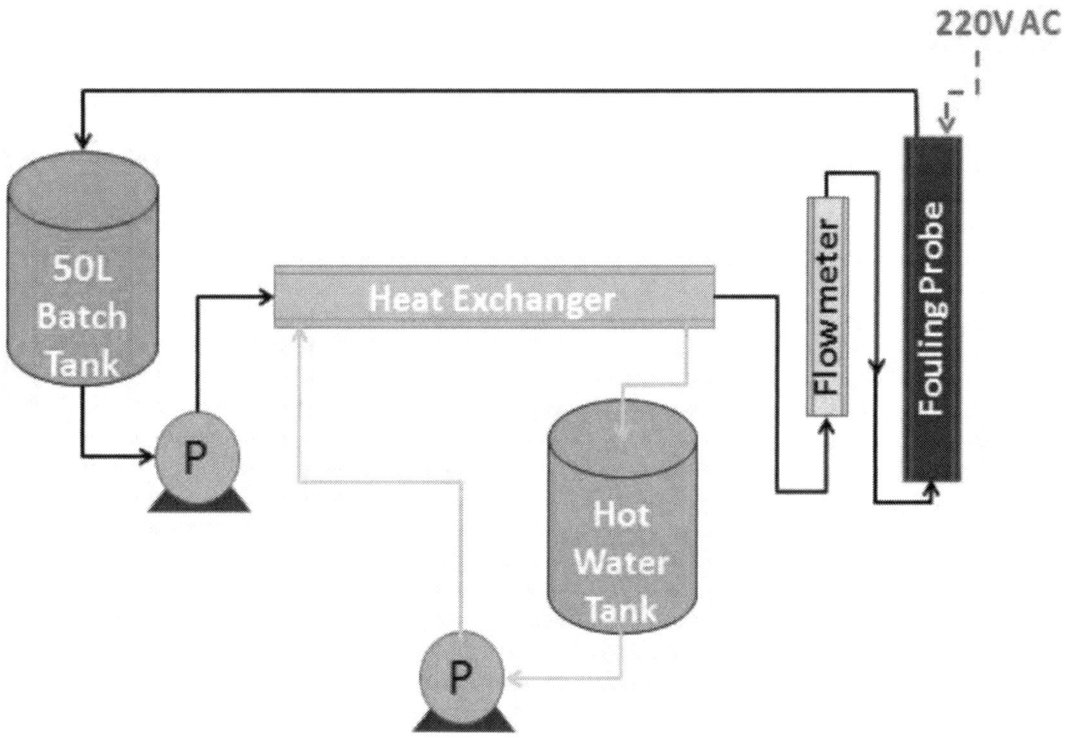

Fig. 2. Fouling apparatus.

RESULTS AND DISCUSSION

1% Total Solids

No induction period was observed for any STA:SUC concentration in synthetic thin stillage (Fig.3). Initial probe temperature (100°C) was too high to get an induction period since the probe did not need any surface conditioning for the onset of fouling. R_f increased rapidly at the and the rate slightly increased with time representing asymptotic fouling. For all STA:SUC concentrations, fouling profiles were similar, increased starch contents increased the fouling resistances except for the STA:SUC of 9:1. Thermocouple readings were recorded for 10 hours and the probe did not reach the maximum temperature, 170°C for STA:SUC treatments at 1% total solids. Fouling profiles followed the trend of increased fouling resistance with increased starch concentration. Fouling profiles had a "saw-tooth" shape which is attributed to signal noise generated from thermocouple recordings in the data logger.

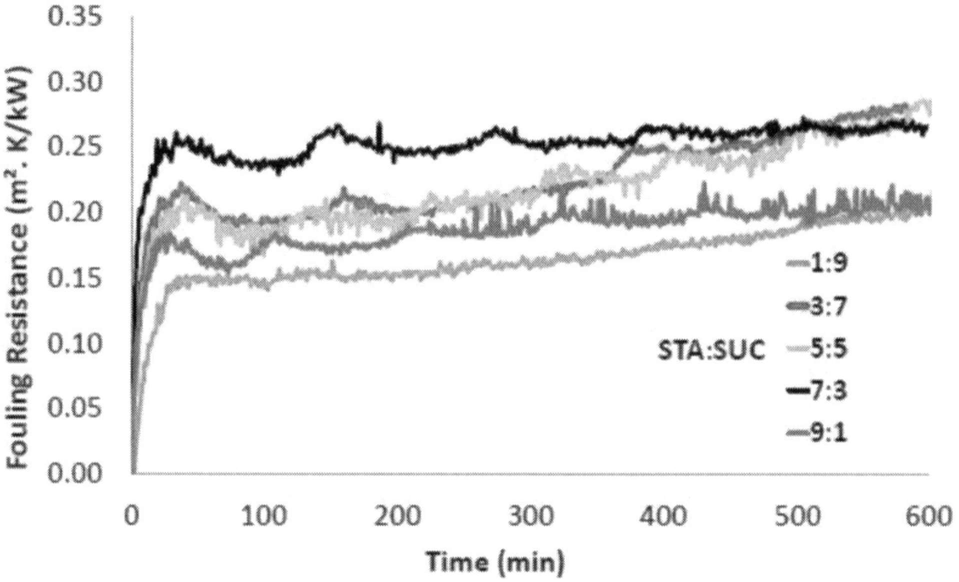

Fig. 3. Effect of STA:SUC on fouling resistance at 1% TS.

10% Total Solids

Fouling profiles for 10% total solids were different from 1% total solids in terms of the slopes (Fig. 4). For STA:SUC of 1:9, $R_f = 0.163$ m^{2}K/kW at the end of 10 hr but did not reach the maximum probe temperature, 170°C. As STA:SUC increased to 3:7, maximum probe temperature was reached in 4.2 hr with final $R_f = 0.264$ m^{2}K/kW. For STA:SUC of 5:5, maximum probe temperature was reached in 0.77 hr with final $R_f = 0.267$ m^{2}K/kW. Treatments having sucrose only (0:10) did not foul measurably by the end of 10 hr. The slopes of the fouling were different for different starch content. Fouling profile for STA:SUC of 5:5 initially had a higher slope and it continued till the probe reached 170°C. The initial slope decreased with decreasing starch content. Sucrose had an effect on fouling in combination with starch but sucrose alone did not foul the probe at the end of 10 hr.

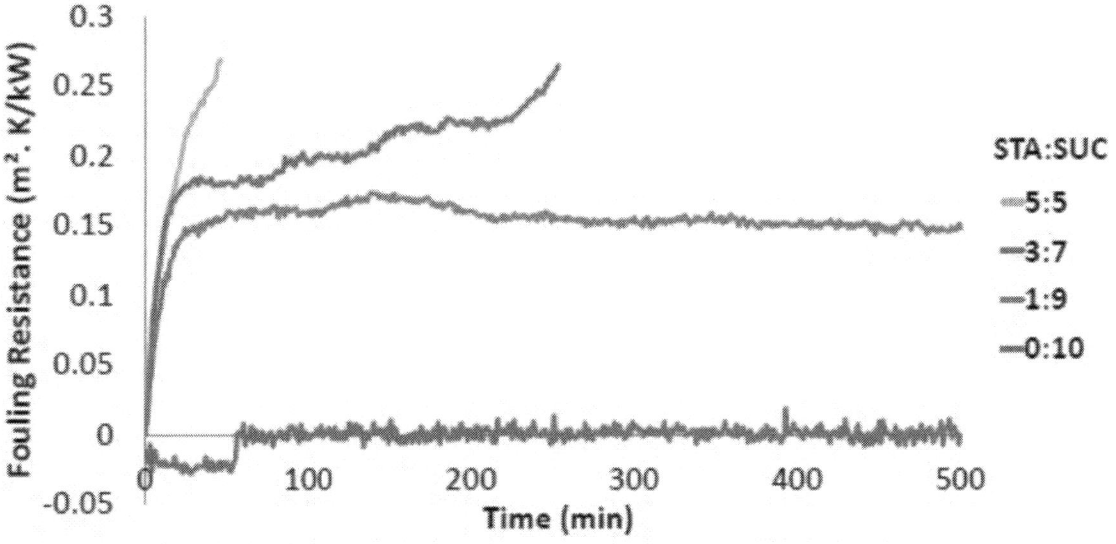

Fig. 4. Effect of STA:SUC on fouling resistance at 10% TS.

Effects of Starch

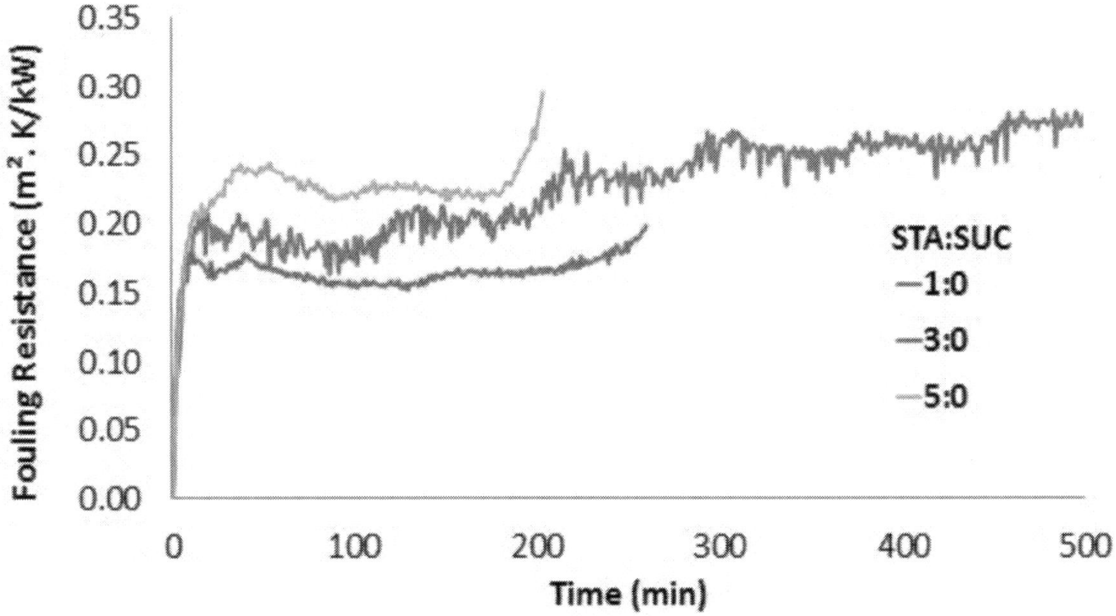

Fig. 5. Effect of STA and % total solids on fouling resistance.

Fouling profiles for 1, 3 and 5% total solids and starch only were obtained (Fig. 5). Synthetic thin stillage with 1% starch had a final R_f of 0.28 m^2K/kW at the end of 10 hr. At 3% starch, probe reached maximum temperature in 4.33 hr with final R_f of 0.19 m^2K/kW and time for 5% starch further decreased to 3.42 hr with final R_f of 0.29 m^2K/kW.

CONCLUSIONS

Higher concentrations of starch in synthetic thin stillage shortened time to reach maximum probe temperature. For 1, 3, 5 and 10% TS, R_f increased mainly because of the starch present in synthetic thin stillage, while glucose had a smaller effect on R_f. Synthetic thin stillage with glucose only did not foul after 10 hr run. Increased starch concentration decreased time to reach maximum probe temperature of 170°C. Glucose had an effect on R_f in combination with starch at 10% TS for STA:GLU of 5:5 compared to STA:GLU of 5:0 in decreasing the time.

REFERENCES

Agbisit, R. M., Singh, V., Valenti, J. J., Kakleas, M., Tumbleson, M. E. and Rausch, K. D. 2003. Technique to measure surface-fouling tendencies of steepwater from corn wet milling. Cereal Chemistry 80:84-86.

Arora, A., Dien, B. S., Belyea, R. L., Singh, V., Tumbleson, M. E. and Rausch, K. D. 2010. Heat transfer fouling characteristics of microfiltered thin stillage from the dry grind process. Bioresource Technology 101:6521-6527.

Epstein, N. 1981. Thinking about heat transfer fouling: a 5 x 5 matrix. Heat Transfer Engineering 4:43-56.

Fischer, P., Suitor, J. W. and Ritter, R. B. 1975. Fouling measurement techniques. Chemical Engineering Progress 71:66-72.

Kenyon. 1988. Maltodextrins and Low-Dextrose-Equivalence Corn Syrup Solids. Pages 7-11 in: Flavor Encapsulation. American Chemical Society.

Mueller, S. 2010. 2008 National dry mill corn ethanol survey. Biotechnology Letters 32:1261-1264.

Singh, V., Panchal, C. B. and Eckhoff, S. R. 1999. Effect of corn oil on thin stillage evaporators. Cereal Chemistry 76:846-849.

Taborek, J., Palen, J. W., Aoki, T., Ritter, R. B. and Knudsen, J. G. 1972. Fouling - major unresolved problem in heat - transfer. Chemical Engineering Progress 68:59-68.

Wilkins, M. R., Belyea, R. L., Singh, V., Buriak, P., Wallig, M. A., Tumbleson, M. E. and Rausch, K. D. 2006a. Analysis of heat transfer fouling by dry-grind maize thin stillage using an annular fouling apparatus. Cereal Chemistry 83:121-126.

Wilkins, M. R., Singh, V., Belyea, R. L., Buriak, P., Wallig, M. A., Tumbleson, M. E. and Rausch, K. D. 2006b. Effect of pH on fouling characteristics and deposit compositions in dry-grind thin stillage. Cereal Chemistry 83:311-314.

(602ai) A Comparative Study of Biodiesel Synthesis From Soybean Oil Via Transesterification Reaction Using Calcium Methoxide in the Absence and Presence of Ultrasound

Vishwanath G Deshmane, Mechanical Engineering, North Carolina Agricultural and Technical State University, Greensboro, NC and Yusuf G. Adewuyi, Chemical and Bioengineering, North Carolina A&T State University, Greensboro, NC

Biodiesel is produced by the reaction of lipid feedstocks with an alcohol such as methanol in the presence of a catalyst (NaOH, KOH or their methoxides) to yield methyl esters (biodiesel) and byproduct glycerol [1,2]. However removal of the base after reaction is a major problem since aqueous quenching results in the formation of stable emulsions and saponification (i.e., soap formation), making separation of the methyl ester difficult and resulting increased production costs. Also, this reaction does not tolerate the presence of water or fatty acids and the use of acids such as H_2SO_4, H_3PO_4, and HCl pose environmental and corrosion problems [3]. On the other hand, use of solid base catalyst offers several process advantages. It can be easily separated from the reaction products and the much more simplified product separation steps result in high yields of methyl esters. Glycerol is also directly produced with high purity (at least 98 %) and is free from any contaminations, thus increasing the profitability of the process. It also gives the opportunity to operate in a continuous process increasing the production capacities and further lowering the production costs. The use of alkaline earth metal oxides as solid base catalysts for transesterification of vegetable oils has been reported in the literature [4]. The basic strength of alkaline earth metal oxides and hydroxides was observed to increase in the order of Mg < Ca < Sr < Ba. From the economical and ecological point of view, calcium derived bases were found to be the most promising as they are inexpensive, exhibit low methanol solubility, and are the least toxic of all. In the present study a commercially available calcium methoxide catalysts has been used in the transesterification of soybean oil to produce biodiesel in a cylindrical jacked glass reactor stirred by mechanical agitation and a 20 kHz ultrasonic, and a novel low-intensity, multi-frequency ultrasonic reactor operated at 611 kHz. In this presentation, the effects of various process parameters such as catalyst concentration, methanol to oil molar ratio and the temperature of the reaction systems on the kinetics of the reaction and the optimization of the process will be discussed. The characterization of the catalytic material (fresh and used) by nitrogen adsorption–desorption (BET) and thermogravimetric and differential scanning calorimetry (TGA-DSC), and the effect of storage time on the activity of the catalyst will also be discussed.

References

1. Mahamuni, N.N.; Adewuyi, Y.G. Optimization of the Synthesis of Biodiesel via Ultrasound-Enhanced Base-Catalyzed Transesterification of Soybean Oil Using a Multifrequency Ultrasonic Reactor". Energy & Fuels. 2009, 23, 2757-2766.

2. Mahamuni, N.N.; Adewuyi, Y.G. Application of Taguchi Method to Investigate the Effects of Process Parameters on the Transesterfication of Soybean Oil Using High Frequency Ultrasound. Energy & Fuels, 2010, 24, 2120-2126.

3. Deshmane, V.G., Gogate, P., Pandit, A. Ultrasound-assisted Synthesis of Biodiesel from Palm Fatty Acid Distillate. Industrial Engineering Chemistry Research. 2008, 48 (17), 7923-7927

4. Watkins, R.S, Lee A.F et al. Li-CaO Catalyzed Triglyceride Transesterification for Biodiesel Application. Green Chemistry. 2004, 6(7), 335-340.

(602aj) Biofuels From Pyrolysis of North-American Grass Species

Shantanu Kelkar, Chemical Engineering, Michigan State University, East Lansing, MI and Christopher M. Saffron, Biosystems & Agricultural Engineering, Department of Forestry, Michigan State University, East Lansing, MI

Native perennial grasses, indigenous to the Midwest, are ideal bioenergy crops due to their potential to be productive on marginal lands. Fast pyrolysis is a simple, flexible process to convert these feedstocks into bio-oil, which is precursor to liquid hydrocarbon fuels. However, organic acids and oxygenates in bio-oil lead to storage instability and limit its early adoption as a transportation fuel. Biomass constituents such as inorganic salts dramatically alter the speciation of pyrolysis products. Therefore, the objective of this study was to prospect amongst several native grasses for cultivars suitable for pyrolysis to biofuels, by investigating the effects of biomass composition on pyrolysis products.

In this work, the composition of native grasses, including big bluestem, coastal panicgrass, deertongue, indiangrass, miscanthus, sandreed, sideoats grama and switchgrass, was determined. Pyrolysis of the grasses was studied using analytical pyrolysis-GC/MS and TGA. The resulting pyrolysis gas contains hundreds of chemical species. Principal component analysis revealed that only six chemical compounds, glycolaldehyde, acetic acid, acetol, methyl glyoxal, 4-vinyl phenol and levoglucosan, were needed to describe more than 90% of the variability associated with pyrolysis products. Correlation analysis revealed that potassium had a significant effect on production of acetic acid, acetol, levoglucosan, char yield and TGA peak reaction rate temperature. Pyrolysis of extractive-free as well as water-washed biomass was studied and compared with untreated biomass. Bio-oil from the native grasses was produced using a screw-conveyor pyrolysis reactor and bio-oil quality was subsequently assessed.

(602ak) Comparing Performance and Re-Acclimation of Two Phase Partitioning Bioreactor and Ordinary Bioreactor Under Starvation Period

Amin Yoosefi Booshehri, Chemical and biomedical engineering, NTU, Singapore, Singapore

Biological treatment plants operated for the removal of pollutants from waste streams are often exposed to periods of no chemical load (starvation periods). The length of such periods can range from a few hours (due to facility operation for a fraction of the day) to a few weeks or months (shut down for factory retooling, equipment maintenance, holiday breaks, process changes) and can negatively affect the elimination capacity of biological processes In the present study, we compare the performance of a two phase partitioning bioreactor(TPPB) and an ordinary bioreactor (OB) after being exposed to hexane starvation period. In OB, after 3 of days starvation period the removal efficiency was reduced by 99%, and about 6 days were needed to recover the initial removal efficiency when the hexane feeding resumed. In contrast, in the TPPB the hexane removal efficiency was not changed significantly and recovered to its initial removal efficiency within 10 hours. The results of this study show that the TPPB system can effectively sustain the biological treatment of waste gas during starvation periods of hydrophobic pollutants.

(602al) Comparison of Biocrude From Micro- and Macroalgae

Griffin W. Roberts[1], Marie-Odile Fortier[2], Belinda S.M. Sturm[2] and Susan M. Stagg-Williams[1], (1)Chemical & Petroleum Engineering, University of Kansas, Lawrence, KS, (2)Civil, Environmental & Architectural Engineering, University of Kansas, Lawrence, KS

Hydrothermal liquefaction (HTL) of algae to a hydrocarbon rich biocrude is growing in popularity primarily because it uses subcritical water as the reaction media eliminating the high energy intensive drying step associated with lipid extraction. HTL efficiently extracts existing lipids in addition to converting other macromolecules such as carbohydrates and proteins to smaller molecules that can be readily upgraded to fuels and other chemicals. Such a process also eliminates the need for a high lipid producing algae and can solely focus on high growth rate algae. Most HTL studies have involved the utilization of microalgae; the few using macroalgae have been marine seaweeds, to our knowledge briocrude has not been produced hydrothermally from fresh water macroalgae.

This study involves pilot scale algae growth tanks utilizing effluent from the second clarifier at the Lawrence, KS wastewater treatment plant. Algae were harvested and hydrothermal treated toward biocrude. Grown under the same conditions both a mixed-culture microalgae and macroalgae identified as *Cladophora sp.* were studied. Hydrothermal treatment was performed with 3g of freeze dried algae and 50mL of water at 350°C for 1 hour. The starting algae and product yields were evaluated on an ash free dry weight basis. High oil yields were obtained for both the micro- and macroalgae (54% and 57%) but had significant chemical and physical differences. Residual solids were determined to have higher heating values of 12.4 and 10.8 MJ/Kg for micro- and macroalgae, respectively. Aqueous co-products had carbon, nitrogen, and phosphorus concentrations of 3100-4000 ppm, 1200-1600 ppm, and 3.5 ppm respectively. The biocrude produced was very similar to that of petroleum crude in terms of energy density (39-42 MJ/Kg) and elemental percentages of carbon, hydrogen, and oxygen.

(602am) Determination of Triacylglyceride Hydrolysis Kinetics in Soybean Oil and Algal Systems

Joshua C. Wissinger[1], Amber L. Bosley[1] and Constance Schall[2], (1)Chemical and Environmental Engineering, University of Toledo, Toledo, OH, (2)Chemical & Environmental Engineering, University of Toledo, Toledo, OH

Determination of triacylglyceride hydrolysis kinetics in soybean oil and algal systems

Joshua C. Wissinger, Amber L. Bosley, Constance A. Schall

The third generation of biofuels, algae, is of growing interest as a renewable feedstock due to its high yield of energy per unit area, use of carbon dioxide and other available resources for growth, and minimal competition with the food industry. Biodiesel and jet fuel can be produced from lipid-rich algae. However, extracting lipids from algae is complicated due to the high water content of harvested and mechanically dewatered algae (60 to 90% water). Thus, there is a need for processes that convert wet algal biomass into fuels or fuel precursors directly in the aqueous phase. Such a process alternative is to use the water in the harvested and partially dewatered algae for hydrolysis of triacylglycerides (TAG) to free fatty acids (FFA), forming aqueous and oil rich phases for ease of extraction and separation of the lipid component. TAG processing techniques include refining FFA by catalytic cracking and hydrotreating for jet fuel production, or the UOP/Eni Ecofining™ process to obtain clean green diesel. In our first step, subcritical water is reacted with TAG producing FFA through hydrolysis eliminating the need for expensive and water sensitive catalysts. In the initial development of an empirical kinetic model, soybean oil, a system much less complex than alga, was used in hydrolysis studies at subcritical water (SCW) conditions. This model was applied to algal TAG hydrolysis. The robustness of our model has been tested by verifying both the autocatalytic behavior of TAG hydrolysis reactions, as well as hydrolysis reactions over a range of temperatures.

(602an) Effect of Particle Size On Enzymatic Hydrolysis of Pretreated Miscanthus

Esha Khullar[1], Bruce S. Dien[2], Kent D. Rausch[1], M. E. Tumbleson[1] and Vijay Singh[1], (1)Agricultural and Biological Engineering, University of Illinois at Urbana-Champaign, Urbana, IL, (2)USDA-ARS NCAUR, Peoria, IL

Effect of Particle Size on Enzymatic Hydrolysis of Pretreated *Miscanthus*

Esha Khullar[1], Bruce S. Dien[2], Kent D. Rausch[1], M. E. Tumbleson[1] and Vijay Singh[1*]

[1]University of Illinois at Urbana-Champaign, Urbana, IL 61801

(217-333-9510) vsingh@uiuc.edu and [2]National Center for Agricultural Utilization Research, ARS, USDA, Peoria, IL 61604

The effect of particle size on enzymatic hydrolysis of pretreated *Miscanthus x giganteus* was determined. *Miscanthus* was ground using a hammer mill equipped with screens having 0.08, 2.0 or 6.0 mm sieve openings. Particle size distribution and geometric mean diameters were determined for all samples. Ground samples were subjected to hot water, dilute acid or dilute ammonium hydroxide pretreatment. Enzyme hydrolysis was conducted on washed pretreated solids; sugar generation was used as a measure for pretreatment efficiency. Glucose and xylose concentrations were monitored using HPLC. Glucose and xylose profiles were generated and hydrolysis rates estimated. Glucan, xylan and total conversion yields were determined by comparing final sugar concentrations obtained to theoretical amounts present in raw biomass.

Geometric mean diameters were the smallest from 0.08 mm sieve screen (56 μm) followed by 2.0 mm (301 μm) and 6.0 mm (695 μm) screens. Across all pretreatments, an increase in total polysaccharide conversion (12 to 26%) was observed when particle size was decreased from 6.0 mm to 0.08 mm. Enzyme hydrolysis of unpretreated biomass samples also resulted in increased total conversions as particle size decreased, although mean conversions (10 to 20%) were much lower than for pretreated biomass samples (40 to 70%), indicating the need for chemical pretreatments in biomass conversion.

(602ao) Fiber Reactor Engineering Design Package for Ultra-High Efficiency Biodiesel Manufacturing

Raul Rivas-Villarreal, Chemical Engineering, Texas A&M University - Kingsville, Kingsville, TX, Patrick L. Mills, Dept of Chemical & Natural Gas Engineering, Texas A&M University-Kingsville, Kingsville, TX and John Massingill Jr., Advanced Materials and Processes, San Marcos, TX

An engineering design package based upon a combination of experimental measurements, data analysis, and reactor modeling tools is described to facilitate development of a new process technology for ultra-high efficiency biodiesel manufacturing. The process design is centered around a new reactor concept called the Fiber Reactor™, which is a novel contactor for conducting reactions involving two partially miscible liquid phases that offers one to two orders-of-magnitude improvement in liquid-liquid mass transfer rates when compared to conventional reactors, such as stirred tanks. This effort is part of a research subproject within the NSF-funded TAMUK CREST on Sustainable Technologies for Environment (STE). The main goal of the STE effort is to develop innovative products by stressing processing of both natural and recycled materials while reducing pollution in a variety of applications through improved process technology. The proposed research on the Fiber Reactor™ is encompassed by the mission and goals of the CREST since the reactor is a key component of sustainable biodiesel processes.

The poster will summarize recent research results whose outcomes enable further understanding of the Fiber Reactor™ by investigating underpinnings of the engineering technology for application to biodiesel transesterification chemistry. Emphasis is placed on developing key reactor scale-up parameters through a combination of experiments, reaction engineering principles, and engineering correlations. These three elements are combined to create a fiber reactor model that is useful for design and scale-up of biodiesel transesterification chemistry from the lab-scale (up to 1-inch OD) to demonstration-scale (ca. 1 m OD) systems. Particular Fiber Reactor™ parameters of interest include fluid pressure drop, fluid flow patterns, fluid flow regimes, liquid-fiber interfacial areas, mass transfer coefficients, heat transfer coefficients, and fiber particulars (material of construction, layout, size, geometry, and surface/wetting properties). Validation of the model-predicted results is also provided using experimental data generated using ½-in OD and 1-inch OD fiber reactor systems.

(602ap) Induction of Microalgal Lipids for Biodiesel Production

Wilbel J. Brewer, Chemical Engineerin, Michigan Technological University, Houghton, MI and Ching-An Peng, Department of Chemical Engineering, Michigan Technological University, Houghton, MI

The production of biofuels such as biodiesel from biomass has gained accelerated interest in recent years as these fuels are becoming increasingly economically-viable, renewable, and carbon-neutral energy sources. One reason for this renewed interest derives from new policy and legislation such as sustainable biofuel targets in the US Energy Policy Act (EPA 2005) and Energy Independence and Security Act (EISA 2007). There have been many attempts in deriving biodiesel fuel from different bioenergy crops such as corn, canola, soybeans, palm, sugar cane and vegetable oil. However there are some significant challenges, these includes: depleting feedstock supplies, land use change impacts, food use competition resulting in high prices and inability to completely displace fossil fuel.

In this study, *Chlorella Vulgaris* microalgae was cultivated under different stressful conditions to produce high-yield biomass with high lipid content which were converted to biodiesel fuel by transesterification. These stressful conditions include the use of high CO_2 concentration, high (25-40 W/m^2) and low (0-10 W/m^2) light intensity. Different operating constraints were investigated to optimize the lipid induction process in *Chlorella Vulgaris*. The approach used in this study was to culture the *Chlorella Vulgaris* microalgae with inoculation 1x10^5 cells/ml in a 250-ml Erlenmeyer flask, irradiated with fluorescence light bulbs and cultured at room temperature of 25°C. Using these conditions we were able to determine which were suitable for cultivating the algae. During the experiment, Nile red dye was used as a hydrophobic fluorescent probe to detect the intracellular lipid deposits in the cell concentrations.

(602aq) Modeling of High Temperature Biochar Gasification and Methane Reforming

Aaron W. Palumbo[1], Erica L. Jorgensen[1] and Alan W. Weimer[2], (1)Chemical & Biological Engineering, University of Colorado at Boulder, Boulder, CO, (2)Chemical Engineering, University of Colorado at Boulder, Boulder, CO

Temperatures above 1300°C may be used to increase conversion and synthesis gas quality and decrease carbon dioxide and hydrocarbon formation in an allothermally heated gasifier. Using methane, or natural gas, as a reactant increases the hydrogen content of the synthesis gas to be compatible with liquid fuel synthesis technologies. Thermodynamic equilibrium results confirm an adjustable H_2:CO given a variable methane-to-biomass ratio as well as superior utilization of hydrogen and carbon monoxide compared to methane reforming. Experimental results show that a complex, kinetic-limited relationship exists between char gasification and methane reforming. Reactions of pyrolysis further complicate experimental controllability of the product gas composition. Char gasification and methane reforming were modeled in a 1-D plug flow reactor configuration assuming a monodisperse particle distribution and an entrained particle flow. Experiments with biochar, methane, and steam were used to obtain global kinetic models of the relevant reactions. Effects of reactant flow rate, particle diameter, and wall temperature were explored.

(602ar) Process Synthesis for an Organic Rankine Cycle Using Low Enthalpy Geothermal Sources

Jesús D. Arrieta[1], Omar J. Guerra[2,3] and **Ariel Uribe Rodriguez**[2], (1)Chemical Engineering, Industrial University of Santander, Bucaramanga, Colombia, (2)Colombian Petroleum Institute (ICP), ECOPETROL S.A., Piedecuesta, Colombia, (3)Chemical Engineering, Industrial University of Santander, Bucaramanga, Colombia

Energy sustainability is one of the most important issues in the 21th century because of the many problems (global warming, atmospheric pollution, sustainability, etc.) associated with the use of fossil energy sources. In this context, renewable energy sources, like geothermal water, are promising energy sources basically in term of sustainability. In addition, those energy sources can provide energy without negative impacts in our environment. In recent years, many works have been focused on the development of efficient energy conversion processes using renewable sources.

In this work a Mixed Integer Nonlinear Programming (MINLP) model for process synthesis in an Organic Rankine Cycle using a low enthalpy geothermal source is presented. The proposed model was applied to a case study in order to evaluate many design option for the process involved. The results showed that the proposed model is adequate for the optimization of process configuration and operation in Organic Rankine Cycles, resulting in the generation of electricity with a competitive production cost using a low enthalpy geothermal source.

(602as) Rapid Lipid Quantification of Algal Biomass Using Thermo-Gravimetric Analysis

Balakrishna Maddi[1], Patrick Bollin[2], **Agasteswar Vadlamani**[3], Sridhar Viamajala[1] and Sasidhar Varanasi[1], (1)Chemical & Environmental Engineering, University of Toledo, Toledo, OH, (2)Chemical Engineering, Kansas State University, Manhattan, KS, (3)Bioengineering, The University of Toledo, Toledo, OH

Algae, especially green algae and diatoms, have the potential to accumulate high amounts of lipid (upto 50 wt% of dry weight) through chemical and physical stimuli, when grown in laboratory conditions. However, most of these algae cannot withstand the harsh and non-sterile environmental conditions encountered in actual production systems. At present, intense research efforts are devoted to screening for algal species that can survive and accumulate lipids in the open-pond non-sterile production environments. Simple, rapid and accurate lipid quantification techniques are of great value for rapidly screening a large variety of strains. In this regard, traditional laboratory techniques for lipid quantification, such as Bligh and Dyer, Folch, acid/base catalyzed trans-esterification methods are not convenient. They are time-consuming, require volatile chemicals and expensive analytical facilities, and involve several process-steps. In this paper, we will introduce a simple, rapid and accurate lipid quantification technique based on thermo-gravimetric analysis (TGA).

The main components of oleaginous algae are protein, starch and lipids. Our technique is based on the observation that the protein, starch and lipid portions undergo thermal degradation over distinct non-overlapping temperature intervals. Proteins are most thermally labile followed by starch and lipids. Lipids undergo *complete* volatilization over a narrow temperature interval of 370-450 °C, without leaving behind *any* biochar. Therefore, lipid content in the algal sample can easily be correlated to the weight loss of the sample during TGA over this narrow temperature range. Lipid content estimates, for several oleaginous algae and oil seeds, from this approach will be reviewed relative to the results from traditional lipids quantification techniques.

(602at) Scalable Biosynthesis of Quantum Dots for Bioenergy Applications

Victora F. Berard[1], Chris K. Kiely[2], Steven McIntosh[1] and Bryan W. Berger[3], (1)Chemical Engineering, Lehigh University, Bethlehem, PA, (2)Center for Advanced Materials and Nanotechnology, Lehigh University, Bethlehem, PA, (3)Chemical Engineering and Program in Bioengineering, Lehigh University, Bethlehem, PA

The cost-efficient and energy- efficient generation of renewable fuels will be one of the most important factors in the future economic prosperity of the United States. Solar energy is one of the most promising routes to develop a new, domestic, source of energy. While current solar cells generate electrical power, we need a chemical route to store large amounts of energy. The direct conversion of solar energy to a chemical fuel (hydrogen) by splitting water would enable large scale energy storage and create a clean burning fuel for transportation. The possible commercial applications are as large as the current power and fuel markets. The activity of the nanoparticle catalysts utilized for this water-splitting reaction is dictated by their physical size, shape, structure, and their chemical composition. Thus it is imperative that we develop synthesize routes that enable control over all of these parameters. We have created a unique approach to environmentally benign, *in situ* semiconductor nanoparticle biosynthesis from the gram-negative bacteria *Stenotrophomonasmaltophilia*. *S. maltophilia* is able to generate novel nanoparticles of the relevant materials unable to by synthesized by conventional methods, but little is currently known regarding how to control the chemical and physical attributes of the resulting particles from biosynthesis. We will describe our recent efforts to synthesize novel nanostructured materials from *S. maltophilia,* including a detailed description of their physical properties determined using electron microscopy, fluorescence spectroscopy and UV-visible spectroscopy. Applications of the synthesized nanoparticles to photocatalysis will also be discussed.

(602au) Treated Effluent Water Use in Cellulosic Ethanol Production

Divya Ramchandran, Environmental Engineering, University of Illinois, Urbana-Champaign, Urbana, IL, Kishore Rajagopalan, Illinois Sustainability Technology Center, Champaign, IL, Timothy J. Strathmann, Department of Civil and Environmental Engineering, University of Illinois at Urbana-Champaign, Urbana, IL and Vijay Singh, Agricultural and Biological Engineering, University of Illinois at Urbana-Champaign, Urbana, IL

The bioethanol industry exerts a significant demand on water supplies. Current water consumption rate in corn dry grind ethanol plants is 3 to 4 gallons of water per gallon of ethanol produced (gal/gal) and 6 to 10 gal/gal for cellulosic ethanol plants. The main goal of this study is to examine the use of treated wastewater effluent in place of potable freshwater for cellulosic ethanol production. The effects of using two different types of filtered treated effluent; Bloomington- Normal, IL (Residential type) and Decatur, IL (Industrial/Residential Mix type); on the rate of fermentation and final ethanol yield from a pure cellulosic substrate were evaluated. Final ethanol concentration with Bloomington- Normal and Decatur effluent and our control study using de-ionized water were similar, resulting in 4.57 ± 0.22 % v/v (0.36 g/g, db), 4.74 ± 0.13 % v/v (0.37 g/g, db) and 4.55 ± 0.28 % v/v (0.36 g/g, db), respectively. Residual glucose concentrations were <0.04% w/v at 48 hr in all cases, suggesting complete fermentation.

Further study with Decatur effluent using 0.08 mm finely ground *Miscanthus* as the substrate resulted in a final ethanol concentration of 0.46 ± 0.008 % v/v (0.14 g/g db) which was similar to ethanol concentration of 0.52 ± 0.07 % v/v (0.17 g/g db) obtained with control treatment using de-ionized water. These findings suggest that with proper characterization studies and under appropriate conditions, the use of treated effluent water in cellulosic ethanol production is feasible.

(602av) Waste From Beer Fermentation Broth; A Potent Source of Bioethanol and Hydrolyzing Enzymes

Waleed Ahmad Khattak, Mazhar Ul-Islam, Minkyung Kang and Joong Kon Park, Chemical Engineering, Kyungpook National University, Daegu, South Korea

Waste from beer fermentation broth; A potent source of bioethanol and hydrolyzing enzymes

Waleed Ahmad Khattak, Mazhar Ul-Islam, Minkyung Kang, Joong Kon Park*

Department of Chemical Engineering, Kyungpook National University, Daegu 702-701, Korea

(parkjk@knu.ac.kr)

Abstract

A semi solid discharged residue produced at the end of fermentation process, called as waste from beer fermentation broth (WBFB) showed tremendous potential for bioethanol production without the addition of saccharifying enzyme, microbial cells or carbohydrates. Number of hydrolyzing enzymes produced from malt in beer fermentation broth during brewing, were confirmed via sodium dodecyl sulfate polyacrylamide gel electrophoresis (SDS PAGE) analysis. The enzymes extract from WBFB after partial purification was evaluated for their hydrolyzing activities. Various parameters including pH, temperature, thermal stability and buffer system was investigated for hydrolyzing activities. With gradual increase in temperature (25-70°), the presence of hydrolyzing enzymes in WBFB caused the degradation of yeast cell wall and reduced the live cell density to a greater extent. However, elevated temperature enhanced both the hydrolyzing activities and bioethanol production without additional enzyme, microbial cells and carbohydrates along with shortening the simultaneous saccharification and fermentation time. The enhanced bioethanol production at elevated temperature seems to be partially resulted via cell free enzymes system. Comparing to bare supernatant or sediments of the WBFB, the saccharification and bioethanol production efficacy of original WBFB were superior.

Key words; Waste from beer fermentation broth, Bioethanol, Hydrolyzing enzymes, Cell wall degradation, Cell free enzyme system

(602ax) Effect of Engulfing Fire to Thermal Decomposition of Type III Composite Cylinder

Jae Hun Lee, Safety Research Devision, Institute of Gas safety R&D, Gyeonggi-do, South Korea and **Hyo Kim**, University of Seoul, Seoul, South Korea

The characteristics of the thermal decomposition of composite cylinder has been thoroughly investigated when it is exposed to a engulfing fire. In the computations, one-dimensional heat transfer model and the Arrhenius type kinetic equations are incorporated to construct the robust framework for predicting the surface recession profile of the cylinder wall. An experiment was conducted to measure in surface temperatures and internal pressure of a commercial composite cylinder that was exposed to engulfing fire. The cylinder used in the experiment is a Type ¥² NGV composite cylinder which was manufactured with epoxy/carbon fiber by the hoop and helical winding methods. The cylinder liner is made of aluminum alloy, and glass bulb type TPRD is built in the cylinder valve. An engulfing fire test was conducted with kerosene on a NGV composite cylinder. Surface recession depth profiles were calculated using surface temperature data obtained from the experimental results, and they were compared with the time dependent pressure change.

Figure 1. Photograph of engulfing fire test.

Key words: Composite cylinder, Thermal Decomposition, Surface Recession, Engulfing Fire

(602ay) Mixed Metal Oxide Reduction/Oxidation Cycle for Thermochemical Energy Storage in Solid Particles

Brian D. Ehrhart[1,2], Nathan P. Siegel[3], Eric N. Coker[2] and Alan W. Weimer[1], (1)Chemical and Biological Engineering, University of Colorado at Boulder, Boulder, CO, (2)Sandia National Laboratories, Albuquerque, NM, (3)Mechanical Engineering, Bucknell University, Lewisburg, PA

Solid particle heat transfer and storage media have been shown to be able to operate at temperatures >1000°C in concentrated solar power (CSP) applications, much higher than the operational limit of 600°C for currently state-of-the-art molten nitrate salt heat transfer fluid. Solid particles can be endothermically reduced by direct exposure to concentrated solar energy, thus absorbing and storing thermal energy beyond that possible with sensible heating alone. The particles can then be oxidized exothermically at a later time, releasing the stored chemical heat and effectively augmenting the thermal energy storage capacity of the solid particles. A mixed metal oxide spinel material that reduces in the temperature range of interest (1000°C-1200°C) has been examined for applicability to this thermochemical energy storage concept. A description of this application, prospective materials, and details of the thermochemical cycle are presented. The heats of reduction and oxidation for the thermochemical cycle have been determined for various operating conditions to evaluate the amount of thermal energy that may be stored. Additionally, the extent of reduction possible at various operating conditions has been measured.

(602az) Bioinspired Electrets for Solar Energy Conversion

Srigokul Upadhyayula[1], Robert Bonderer[2], Duoduo Bao[1], Brent Millare[3], Khalid Ashraf[4], Amy Ferreira[3], Roger Lake[5] and Valentine I. Vullev[3], (1)Bioengineering, University of California, Riverside, CA, (2)Chemical and Environmental Engineering, University of California, Riverside, Riverside, CA, (3)Department of Bioengineering, University of California, Riverside, Riverside, CA, (4)University of California, Berkeley, CA, (5)Electrical Engineering, University of California, Riverside, CA

Protein-mediated electron-transfer processes sustain a broad range of redox functions in biological systems, such as respiration and photosynthesis. Therefore, such biological systems are inspirational models for materials for energy-conversion applications. This presentation will focus on bioinspired macromolecular electrets with rectifying long-range charge-transfer properties. We have demonstrated molecular designs based on anthranilamide derivatives, which similar to polypeptide α-helices, contain ordered arrangement of amide and hydrogen bonds, but in contrast to polypeptides and proteins, have an extended Π-electronic conjugation for mediating efficient long-range charge transfer. Ab initio calculations, along with dielectric and NMR measurements, revealed that polyanthranilamides possess intrinsic dipole moments in the order of 3 D per residue. Furthermore, the macromolecular dipoles remove the degeneracy between the frontier orbitals of neighboring residues, causing splitting of ~0.2 eV for 10 residues that can readily result in rectification of vectorial charge hopping. Doping the macromolecules with electron withdrawing and electron donating groups did not only alter the bandgap and the energy levels of the frontier orbitals, but also modulated the intrinsic dipole moments, ranging from 0 to 5 D per residue. The design of the presented bioinspired electrets will prove instrumental for the development of materials with rectifying charge-transduction properties essential for solar-energy-conversion applications.

(602az) Integrated Use of Geographic Informational System and MIXED-Integer Linear Programming for Indentifying Optimal Agricultural Areas for Sewage Sludge Amendment: A Case Study of Catalonia

Pavel Vaskan, Chemical engineering, University Rovira i Virgilii, Tarragona, Spain

INTEGRATED USE OF GEOGRAPHIC INFORMATIONAL SYSTEM AND MIXED-INTEGER LINEAR PROGRAMMING FOR INDENTIFYING OPTIMAL AGRICULTURAL AREAS FOR SEWAGE SLUDGE AMENDMENT: A CASE STUDY OF CATALONIA

P. Vaskan[1], A. Passuello[2], G. Guillén-Gosálbez[3], M. Schuhmacher[4], L. Jiménez[5]

Departament d'Enginyeria Química (EQ), Escola Tècnica Superior d'Enginyeria Química (ETSEQ), Universitat Rovira i Virgili (URV), Campus Sescelades, Avinguda Països Catalans, 26, 43007 Tarragona, Spain

Abstract

Deciding the best agricultural soils for sewage sludge amendment is not a trivial task. Several environmental aspects, such as groundwater contamination by nitrates, open waters and soil protection, as well as human exposure to the contaminants present in the sludge matrix must be considered along with economic aspects, making this task quite challenging. This is because the management of this residue concerns different stakeholders, with different views of the problem and conflicting interests. In addition, they may find that their interests are not reached, fact that leads to a low acceptance of the practice.

Reaching a final solution satisfying all the decision-makers involved is indeed very challenging. Systematic tools based on multi-objective mathematical programming are well suited to tackle this type of problems, as they allow generating and assessing a large number of potential solutions. The final goal is to calculate a set of Pareto optimal alternatives, that is, a set of solutions with the property that it is impossible to improve them in one of the objectives without necessarily worsening at least another criterion.

In this work, we study the use of multi-objective optimization for identifying the best geographical areas for sewage sludge (SS) amendment in terms of economic and environmental criteria. Our approach integrates geographic informational systems (GIS) and multi-objective mixed-integer linear programming (MILP) within a unified framework that allows exploring in a rigorous and systematic manner a large number of alternatives for sewage sludge amendment from which the best ones (according to the decision-makers' preferences) are finally identified.

The capabilities of our approach were tested through its application to a case study based on Catalonia. Given a certain amount of sludge to be treated and a set of available agricultural areas. Numerical results show that it is possible to improve the environmental performance of the final solution by compromising the associated cost. Our methodology is intended to assist decision-makers in such a challenging task.

---

[1] pavel.vaskan@urv.cat

[2] ana.passuelo@urv.cat

[3] gonzalo.guillen@urv.cat

[4] laureano.jimenez@urv.cat

[5] laureano.jimenez@urv.cat

(602b) Algal De-Watering Using Temperature Sensitive Hydrogels: A Preliminary Step in the Harvesting Process

Agasteswar Vadlamani, Bioengineering, The University of Toledo, Toledo, OH, Sridhar Viamajala, Chemical & Environmental Engineering, University of Toledo, Toledo, OH and Sasidhar Varanasi, Chemical and Environmental Engineering, University of Toledo, Toledo, OH

Algal cultures from the growth reactors are usually dilute (~0.1-0.5% (w/w)) and direct use of conventional solid-liquid separation methods (e.g. centrifugation, filtration or flocculation) may be cost prohibitive due to large process volumes. In this study, temperature-sensitive gels made of N-isopropylacrylamide (NIPAAm) were investigated as a re-usable absorbent for increasing the concentration of dilute algal suspensions. We observed that when algal cultures were incubated in the presence of NIPAAm gel (at 25°C), cell-free media was absorbed resulting in an increase in cell concentration. The swollen hydrogels were then recovered and heated to 34°C to deswell and release the absorbed media. Since swelling rates were high initially (t<4h) and decreased subsequently (4h < t <15h), cyclic swelling (for 4h) followed by deswelling was applied to more rapidly concentrate the algal cultures. Our experimental results show that the concentration of algal cultures increased from 0.16% to 4.3% (w/w) in 6 cycles. In addition, slurries initially at 3% (w/w) could further be concentrated to 8% (w/w) by incubation with hydrogels without any discernible change in gel swelling/deswelling kinetics. These results suggest that higher solid concentrations could be achieved by applying more swelling/deswelling cycles. In fact, our studies show that removal of nearly all free water is possible to obtain algal pastes with concentrations of nearly 18% (w/w) solids. However, recovery of gels from such concentrated slurries without incurring significant loss of cells is difficult. Overall, our results demonstrate that temperature-sensitive hydrogels can be used to easily dewater dilute algal cultures to concentrations (8%(w/w) or more) where more traditional solid-liquid separation methods can be easily applied. An economic analysis of the hydrogel based dewatering process will be presented.

(602ba) Verification of Biochar Pyrolysis

Lucas J. Rice, Chemical Engineering, University of Missouri, Columbia, MO

Verification of Biochar Pyrolysis

Lucas J. Rice and Tom R. Marrero

Department of Chemical Engineering

University of Missouri

Columbia, MO 65211

Abstract

This study attempts to verify the ability to generate a carbon-negative cycle from the pyrolysis of biomass materials to supplement soil nutrients.

The pyrolysis process is simulated with ChemCAD software with inputs from new of field and bench scale data.

Field data correspond to a kiln approximately 100 feet long with a nominal cross-sectional area of 100 ft^2. The kiln maintains a nitrogen atmosphere to pyrolysis the biomass to biochar. A series of operating characteristics are measured; temperatures in the chamber, moisture of raw materials and biochar, and process residence time. Furthermore, biochar characteristics such as porosity, elements, and carbon content were determined.

Results indicate the maximum and minimum temperatures required for the process. The maximum temperature of the kiln allows for pyrolysis energy balances to determine other factors required such as the residence time. Furthermore, an ideal temperature is calculated in order to create the most efficient.

Biomass materials are used for the biochar include: corn stover, milo stover, and hay.

Recent publication describes the carbon life cycle of normal energy production. There are three potential cycle types mapped out in this process: carbon positive, carbon neutral, and carbon negative.

Carbon positive represents the burning of fossils and release carbon dioxide into the atmosphere. This process results in a net gain of carbon dioxide in the atmosphere, which is why it's called carbon positive.

The burning of biomass in a similar process as fossil fuels also can produce the same carbon dioxide in the atmosphere. During photosynthesis, plant life takes up the produced carbon dioxide facilitating more green growth, which in turn can be used as biofuels. This process is called carbon neutral since the net gain of carbon in the cycle is zero. However, the half-life of carbon dioxide in the atmosphere is significantly longer than 100 years which results in a loss of carbon.

The third type of process discussed is the focus of this presentation, the carbon negative cycle. Biomass is subjected to slow burning pyrolysis which results in two usable products. Similarly to the previous two cycles, carbon dioxide is generated as a bi-product and can be used as energy or recycled back through green life. The biomass is converted to biochar and can be used as soil amendments thereby a portion is sequestered or captured into the soil due to its refractory properties. This sustainable process is called the carbon negative cycle.

To obtain the nutrient rich qualities of biochar for soil enhancement, the biomass is subjected to slow pyrolysis. During pyrolysis, nitrogen and/or nitrogen containing compounds are adsorbed by biochar.

The amount of soil amendment obtained from the biomass is quantified by measurements of the dry weight and the burn-down of the biomass to biochar.

The production of biochar provides fertilizers with a variety of benefits for the environment. The use of biochar instead of traditional fertilizers (ex. ammonia) can potentially provide a more efficient alternative to soil supplements in theory and in practice. By returning the biomass materials to the soil in this manner, the nutrients are recycled and already in a form useable by vegetation. Furthermore, the biochar in the soil helps the ground to maintain its strength preventing the fast erosion due to wind or rain. Water has a higher rate of retention in soil which contains biochar. Biochar is also able to revitalize grounds that have had all the nutrients drained by over cultivation or other methods. This would allow for a larger area base for farmable land. This also leads to the ability forgo such strategies as crop rotations. Lastly, as the carbon negative cycle utilizes more energy from the pyrolysis of the biomass, the efficiency helps to convert energy.

The final results of the study are a new simulation of a sustainable biomass pyrolysis using the ChemCAD software. This simulation is verified and calibrated by field and bench scale data.

(602bb) Microwave-High Temperature Pretreatment and Enzymatic Hydrolysis of Sugarcane Bagasse

Milad R.Esfahani, Chemical Engineering, Tennessee Technological University, Cookeville, TN and Mehrdad Azin, IROST, Tehran, Iran

Sugarcane bagasse is a promising lingocellulosic biomass for production of bioethanol. High-temperature treatment is one of the common prerequisites for almost all of the different kinds of pretreatment methods. Using microwave power, besides of creating high temperature, has physical effects on degradation of cellulosic polymeric structure of lingocellulosic biomass. To study the effect of microwave power for creating high temperature needed for pretreatment of sugarcane bagasse, the degree of hydrolysis was measured by assaying both the released total reducing sugars and the glucose, specifically. Different particle size, temperature and duration of exposures to the heat, were the main studied factors. Afterwards, the enzymatic hydrolysis was studied at different levels of enzyme load and pH. The optimal conditions for microwave pretreatment and enzymatic hydrolysis of sugarcane bagasse were determined by using Taguchi method. The result indicates that while particle size of 35-80 mesh and 1 minute duration time for microwave exposure are common optimal condition for production of both reducing sugars and glucose, other factors like pH, enzyme Load and temperature varies for optimal production of each of them.

(602bc) Life Cycle Assessment of Plasticizers: DEHP Versus "Green" Alternatives

Yuan Li, Department of Chemical and Petroleum Engineering, University of Pittsburgh, Pittsburgh, PA and Eric J. Beckman, University of Pittsburgh, Pittsburgh, PA

Phthalate plasticizers are added to polyvinyl chloride (PVC) during manufacturing to increase resilience, low temperature performance, softness, and ease of processing. Phthalates are widely applied in toys, construction materials, medical devices, food packaging, flooring and other product categories. After several decades of commercial use, recent investigations have prompted concern that phthalate exposure can cause human health issues. At the same time, a number of new plasticizers have been introduced as potentially greener alternatives to the most widely used phthalates. We have employed process life cycle analysis (process LCA) to examine the life cycle impacts of di(2-ethylhexyl)phthalate (DEHP) to those arising from the use of 1,2 cyclohexane dicarboxylic acid diisononyl ester (DINCH) and citrate esters. Initial work has focused on the cradle to gate stage of the life cycle, comparing the raw material acquisition and manufacturing phases. Energy consumption is also an important concern for the environmental impacts. A complete LCA study to get a much "greener" plasticizer substitute is expected.

(602bd) Improved Reproducibility in HPLC Analytical Techniques and Use of MALDI-MS in Biofuel Related Research

Samira Vasheghani Farahani, Chemical and Environmental Engineering, University of Toledo, Toledo, OH and Michael Mayer, University of Toledo, Toledo, OH

Quantitative biomass analyses and molecular weight determination are essential in assessing lignocellulosic feedstocks in biofuel related research. High-performance Liquid Chromatography (HPLC) equipped with a suitable carbohydrate resin column is commonly used for quantitative analysis of polysaccharide constituents in biomass. When sample preparation includes steps such as acid neutralization, introduction of soluble salts can lead to an unstable chromatographic baseline. The contribution of this work is in introducing HPLC sample pretreatment which results in improved baseline stability of resultant chromatograms and therefore reproducible data. Characterizing cellulose, especially its degree of polymerization (DP), is another important step in evaluating its deconstruction to monomeric sugars or in its use for materials. Matrix Assisted Laser Desorption Ionization Time of Flight Mass Spectroscopy (MALDI-TOF) has been used to determine the degree of polymerization for crystalline cellulose with a DP in an excess of 200. A new technique has been developed for MALDI-TOF, utilizing an ionic liquid matrix and solvent system. The technique has potentially wide application in analysis of lignocellulosic substrates.

(602be) Modification of Calcium Based CO2 Sorbents Using Modified Metal Oxides Powders

Wen-bin Yang[1], Hui Li[2], Yong-Hua Duan[1], Xiaodan Fan[3] and Yong Min[2], (1)Material Institute, Xi'an University of Architecture and Technology,, Xi'an, China, (2)Xi'an University of Architecture &Technology, Institute of Powder Engineering, Xi'an, China, (3)Xi'an University of Architecture and Technology, Xi'an 710055, Shaanxi, P.R. China

Modification of Calcium Based CO2 Sorbents Using Modified Metal Oxides Powders

Wenbin Yang, Hui Li, Yonghua Duan, Xiaodan Fan, Yong Min

1Xi'an University of Architecture and Technology, Xi'an 710055, Shaanxi, P.R. China

Abstract:

In order to solve the "sintering phenomena" by using calcium-based sorbents for CO2 sequestration, we has modified calcium based sorbents (MCS) using metal oxides and metal carbonates dry powders, such as MgCO3, Fe2O3, Al2O3, SiO2, ZrO2, etc. The modified calcium based sorbents had significantly improved the cycle times, CO2 sequestration efficiency, particle size distribution, as well as operation temperatures.

Several analytic equipments, such as TGA, DMA, SEM, XRD, particle size analyzer, mercury analyzer, were used to evaluate the particle composite, structure, and material activities. Our preliminary results indicated that the modified calcium based sorbents (MCS) have not only improved the cycle times, but also improved the CO2 sequestration efficiency. SEM morphology studies indicated that the surface of modified calcium based sorbents had formed an stable highly porous surface structure, which has significantly improved the CO2 sequestration and releasing. Comprehensive studies of material composition and scale up test are on the way.

Pretreatment of Wheat Straw Using the H_2O-SO_2 System at Elevated Temperatures

Weina Liu,[a] Yucui Hou,[b] Weize Wu,[a,*] Muge Niu,[a] Wenhua Wang[a]

[a] *State Key Laboratory of Chemical Resource Engineering, Beijing University of Chemical Technology, Beijing 100029, China;* [b] *Department of Chemistry, Taiyuan Normal University, Taiyuan 030031, China*

Extended Abstract

To improve the use value of lignocellulosic materials, which is abundant, renewable and low-cost, intensive research on their conversion is currently carried out[1]. However, Ligmocellulose have an extremely complex components, so a pretreatment process is needed in most cases to get rid of the hemicellulose fraction or the lignin fraction[2, 3].

Many pretreatments, including physical treatments[4, 5]and chemical treatments[6, 7], have been suggested. Among them, sulfuric acid hydrolysis is efficient, but has the disadvantages of pollution, difficulty in recovery, and hard control of products. SO_2 has been validated to be an efficient catalyst for the conversion of lignocellulose[8, 9]. Its greatest advantage lies in its ability to promote the steps of cellulose into glucose rather than further degradation. Moreover, SO_2 can be totally swept out by steam from an aqueous solution, and the recovered SO_2 can be reused after collection. However, no report is available to validate that the pretreatment process of lignocellulose in H_2O-SO_2 system is prior to the formation of glucose rather than furfural.

This work deals with the pretreatment of wheat straw in H_2O-SO_2 system. Kinetics models of the hemicellulose hydrolysis and xylose degradation with time are established, and the reaction rate constants and activation energy are calculated according to Saeman model.

Wheat straw with an hemicellulose content of 30.1 % was dried and milled (less than 1 mm) for use. SO_2 aqueous solution with a mass fraction of 6 % was purchased from Aladdin Reagent Inc. (Shanghai, China). D-glucose anhydrous, xylose, arabinose, furfural and levulinic acid (LA) were all analytical reagent.

* Corresponding author. E-mail: wzwu@mail.buct.edu.cn. Tel./Fax: +86 10 6442 7603.

In a typical procedure, 0.3 g of wheat straw, 6 mL of SO_2 aqueous solution and 9 mL of H_2O were loaded into the reactor. The reactor was sealed and purged with N_2. Then, the reactor was submerged into the furnace under an desired temperature for a desired time. After reaction, the reactor was placed into ice water to quench the reaction immediately. At last, the obtained reaction solution was purged by steam produced at 120 °C. The solution was filtrated, washed and analyzed by HPLC and the solid material was dried and weighed to calculate the conversion of reactant.

Treatments were performed in H_2O-SO_2 system with a SO_2 concentration of 0.024 g/mL under the temperature range of 130 °C to 160 °C, and the samples were withdrawn at several reaction times in the range of 0 to 90 min. The yields of products based on hemicellulose are shown in Table 1.

It is clear from Table 1 that temperature influences greatly the distribution of products at a given SO_2 concentration. At 130 °C, the main products are monomers of hemicellulose, such as xylose, arabinose, glucose and acetic acid. As the temperature increased from 130 °C to 160 °C, the monomer sugar is more inclined to further degrade to furfural. The reason is that increasing temperature favors not only the hydrolysis of hemicellulose and also the further degradation of xylose.

At a given temperature, with the prolonging of reaction time, the yield of xylose and arabinose first increase and then decrease for the further degradation. The yield of glucose gradually increased when the temperature was lower than 150 °C. It reached 17.2%, and then changed little at 150 °C from 45 min to 90 min. When the temperature was increased to 160 °C, the glucose yield began to increase again, likely due to the hydrolysis of cellulose.

In this work, the experimental value of hemicellulose hydrolysis in H_2O-SO_2 system is modeled by Saeman model (equation 1).

$$w_X = w_{X,0} e^{-k_2 t} + w_{H,0} \frac{k_1}{k_2 - k_1} (e^{-k_1 t} - e^{-k_2 t}) \tag{1}$$

Where k_1 and k_2 are reaction rate constant of hemicellulose hydrolysis and xylose degradation, respectively, min^{-1}; w_H is the mass fraction of hemicellulose; w_x is the mass fraction of xylose; $w_{X,0}$ and $w_{H,0}$ is the initial value for xylose and hemicellulose, which is designated as 0 and 100 %, respectively in our system. Therefore, the equation becomes as follows in our work:

$$Y = \frac{k_1}{k_2 - k_1} (e^{-k_1 t} - e^{-k_2 t}) \tag{2}$$

Where Y is the yield of xylose at certain reaction time. The fitting curves and the results of reaction rate constants are shown in Table 2.

The kinetics equation was obtained by fitting lnk with 1/T according to the Arrhenius equation. The activations for hemicellulose hydrolysis and xylose dehydration were 7.8 and 9.0 KJ/mol, respectively. It is noted that the activation energy for hemicellulose hydrolysis in our work is lower than that for xylose dehydration. Therefore, the xylose formation proceeded faster than xylose degradation, making it possible to decrease the yield of furfural. The reason is that SO_2 in water not only can react with water to form sulfurous acid to generate H^+ as a Brønsted acid, also can exist in the form of molecular as a Lewis acid, which is benefit for the breakage of glycosidic bond.

In this work, wheat straw is efficiently pretreated in H_2O-SO_2 system with a high xylose yield of 61.1% and a high total products yield of 93.9%. The Kinetic study based on Saeman model and experimental data suggested that SO_2 in water as catalyst favors the conversion of hemicellulose to monosaccharide rather than further degradation.

This work was supported by the National Natural Science Foundation of China (21176020), the Beijing Natural Science Foundation (2082017), and the Program for New Century Excellent Talents in University (NCET-08-0710)

Key Words: Pretreatment; Wheat straw, Hemicellulose, H_2O-SO_2 system, Xylose

Reference

(1) Zhang, Z. H.; Zhao, Z. B. K., Microwave-assisted conversion of lignocellulosic biomass into furans in ionic liquid. *Bioresour. Technol.* **2010,** 101, 1111–1114.

(2) Dadi, A. P.; Varanasi, S.; Schall, C. A., Enhancement of cellulose saccharification kinetics using an ionic liquid pretreatment step. *Biotechnol. Bioeng.* **2006,** 95, 904 – 910.

(3) Kumar, P.; Barrett, D. M.; Delwiche, M. J.; Stroeve, P., Methods for pretreatment of lignocellulosic biomass for efficient hydrolysis and biofuel production. *Ind. Eng. Chem. Res.* **2009,** 48, 3713–3729.

(4) Yang, F.; Li, L. Z.; Li, Q.; Tan, W. Q.; Liu, W.; Xian, M., Enhancement of enzymatic in situ saccharification of cellulose in aqueous-ionic liquid media by ultrasonic intensification. *Carbohydr. Polymers* **2010,** 81, 311–316.

(5) Ha, S. H.; Mai, N. L.; An, G.; Koo, Y. M., Microwave-assisted pretreatment of cellulose in ionic liquid for accelerated enzymatic hydrolysis. *Bioresour. Technol.* **2011,** 102, 1214–1219.

(6) Lu, X. B.; Zhang, Y. M.; Yang, J.; Liang, Y., Enzymatic hydrolysis of corn stover after pretreatment with dilute sulfuric acid. *Chem. Eng. Technol.* **2007,** 30, 938–944.

(7) Guo, G. L.; Chen, W. H.; Chen, W. H.; C., M. L.; Hwang, W. S., Characterization of dilute acid pretreatment of silvergrass for ethanol production. *Bioresour. Technol.* **2008,** 99, 6046–6053.

(8) Shi, J.; Ebrik, M. A.; Wyman, C. E., Sugar yields from dilute sulfuric acid and sulfur dioxide pretreatments and subsequent enzymatic hydrolysis of switchgrass. *Bioresour. Technol.* **2011,** 102, 8930–8938.

(9) Ewanick, S.; Bura, R., The effect of biomass moisture content on bioethanol yields from steam pretreated switchgrass and sugarcane bagasse. *Bioresour. Technol.* **2011,** 102, 2651–2658.

Table 1 The yield of products (Y) on time (t) under different temperatures

	130 °C			140 °C			150 °C			160 °C		
t (min)	Y_x	Y_g	Y_f	Y_x	Y_g	Y_f	Y_x	Y_g	Y_f	Y_x	Y_g	Y_f
15	14.8	1.8	0.0	32.8	2.7	0.0	39.5	7.6	0.0	46.6	8.9	0.0
30	32.8	4.9	0.0	50.7	9.6	0.0	61.0	14.2	0.8	53.5	14.4	2.4
45	45.6	12.7	0.0	54.3	12.6	0.3	59.5	17.2	2.8	48.2	18.7	5.6
60	54.7	13.5	0.0	59.2	14.0	0.7	57.5	17.6	3.9	37.0	21.3	11.9
75	57.5	13.4	0.1	57.2	15.8	1.2	55.8	17.8	4.6	20.6	20.0	15.8
90	54.5	14.9	0.7	55.3	17.5	2.3	49.7	17.7	5.6	13.1	18.3	20.2

Note: Y_x is yield of xylose, %; Y_g is yield of glucose, %; Y_f is yield of furfural, %.

Table 2 The fitting equation of xylose yield (*Y*) on time (*t*)

T (°C)	Equation	R^2
130	Y=143.39×[exp(−0.0152t)− exp(−0.0047t)]	0.9493
140	Y=144.08×[exp(−0.0276t)− exp(−0.0084t)]	0.9587
150	Y=136.87×[exp(−0.0392t)− exp(−0.0106t)]	0.9014
160	Y=170.28×[exp(−0.0605t)− exp(−0.0250t)]	0.9405

$$Y = \frac{k_1}{k_2 - k_1}(e^{-k_1 t} - e^{-k_2 t})$$

(602bg) Evaluating the Potentials of Paper Mill Sludge (PMS) As Feedstock for Biofuels Production

Joseph Reckamp[1], Rene Garrido[2], Nydia Ruiz-Felix[2] and Justinus A. Satrio[2], (1)Chemical Engineering, Villanova University, Villanova, PA, (2)Chemical Engineering Department, Villanova University, Villanova, PA

Heightened concerns over escalating greenhouse gas emissions as well as the depletion of petroleum oil and the continuing rise of its prices have resulted in focus on obtaining energy, traditionally produced from the combustion of fossil fuels, from renewable sources. One such renewable source of energy is biomass, in particular lignocellulosic biomass which is rich in lignin, cellulose, and hemicellulose. These carbon-containing components can be converted to liquid fuels by biochemical approaches, thermochemical approaches, or the combination of biochemical and thermochemical approaches.

The ideal biomass materials for energy production are those that are readily available and do not require a lot of energy to produce and collect. More importantly, the biomass materials should not be produced on fertile lands used for growing food crops. Such biomass materials are commonly readily available in the forms of what are considered wastes. These materials are produced from agriculture, forestry, household, or manufacturing activities. Examples of the materials include wheat straws, wood chips, spent mushroom substrate, municipal solid waste (MSW) and paper mill sludge (PMS). This study is focused on PMS materials.

As byproduct of paper making, PMS is produced in the order of 15 to 50 tons per day for a typical paper mill. It is estimated that five million tons of PMS are produced annually in the United States alone. Currently there are not many uses PMS, that most of PMS is disposed to landfills, a financial burden for paper mills. The decline of the landfill space availability has caused the cost of disposing PMS to increase rapidly to the point where the cost of PMS disposal now is about 60% of paper mill's operating costs. Therefore, there is growing interest in the paper industry to research possible uses of PMS, instead of just disposing them.

Thus, the purpose of this study is to show the viability of PMS as a feedstock for the production of liquid biofuels. Exploring the viability first requires characterization of the PMS samples. Characterization by way of Induced Coupled Plasma (ICP), Thermogravimetric Analysis (TGA), Differential Scanning Calorimeter (DSC), Fourier Transform Infrared Spectroscopy (FTIR), and Fiber Analysis will provide information on the physicochemical properties of PMS. The ash content and the mineral composition of the ash material in PMS will be determined. Functional groups and heat capacities will also be determined as well as the amount of lignin, hemicellulose, and cellulose in PMS. The characterization will determine whether PMS is within the optimal specifications as feedstock for biofuel production.

In addition to characterization study, a study on converting PMS to fuels will also be conducted. In order to convert the PMS into a liquid fuel, this study will utilize a method called fast pyrolysis. Fast pyrolysis is a thermochemical conversion technology that rapidly heats organic material to temperatures of 450°C to 600°C in the absence of air which results in rapid decomposition of the material to produce a liquid product typically called bio-crude oil which can be used as a heat source or an intermediate feedstock for producing gasoline or diesel fuels. The fast pyrolysis experiments will be performed by using a micropyrolyzer/GC-MS system. To be reported in this presentation are the results of these characterization and fast pyrolysis studies.

(602bh) Characterization and Fast Pyrolysis Study of Phragmites Australis to Evaluate Its Potential As Feedstock for Biofuels Production

Lindsay Peterson[1], Rene Garrido[2], Nydia Ruiz-Felix[2] and Justinus A. Satrio[2], (1)Chemical Engineering, Villanova University , Villanova, PA, (2)Chemical Engineering Department, Villanova University, Villanova, PA

The use of lignocellulosic biomass as an alternative carbon source for energy production, particularly liquid fuels, is promising due to its widespread availability. A potential biomass source is *Phragmites australis*, commonly known as phragmites or common reeds. It is an invasive wetland weed that is one of the most abundant plant species on the planet, covering approximately ten million hectares, and it can be found on every continent excluding Antarctica. In addition to being abundant, phragmites can produce biomass in high yields. It is reported that it can provide up to 18-28 tons of biomass per acre per year. The phragmites grow easily, producing lignocellulosic biomass in high yield, on land areas that are not used for growing food crops, which potentially makes the plant species an excellent feedstock for biofuel production. It has been reported that *Phragmites australis* is rich in cellulose and lignin, the main components in lignocellulosic biomass that are particularly important for biomass conversions to biofuels.

The purpose of this research is to understand the physical traits of *Phragmites australis* and to understand the feasibility of *Phragmites australis* as a biomass feedstock via a thermochemical conversion process called fast pyrolysis. Fast pyrolysis is a process in which organic materials are rapidly heated to 450 - 600 °C in absence of air. Under these conditions, organic vapors, permanent gases (CO, H2, CO2, CH4), and charcoal are produced. The vapors are condensed to a liquid called bio-crude oil (BCO). The chemical and physical properties of BCO are dependent on the biomass materials used as feedstock. The advantage of pyrolysis over biochemical-based process is that it can transform difficult-to-handle biomass of different natures into clean and uniform liquid (BCO), which then can be used as a heating source or further processed to produce transport fuels, such as gasoline and diesel.

The phragmites samples used for this study were collected from a wetland area located within Villanova University's campus. The material is prepared for physicochemical characterization by drying and grinding. The characterization study includes fiber analysis, proximate analysis (using Thermogravimetric Analyzer (TGA)), mineral content composition (using Induced Coupled Plasma (ICP)), and heating value (by using differential scanning calorimeter (DSC)). DSC is also used to study the thermal behavior of the biomass material when it undergoes fast pyrolysis reaction. Another instrument, Fourier Transform Infrared spectroscopy (FTIR), is used to identify chemical functionality of the biomass material. A limited number of fast pyrolysis experiments by using a micropyrolyzer/GC-MS system were performed to evaluate the yields and chemical product distribution from pyrolyzing phragmites. To be reported in this presentation are the results of these characterization and fast pyrolysis studies.

(602bi) Development of Biodiesel Production Facility for Utilizing Waste Cooking Oil

Brian Delancy[1], Brooks Hopkins[1] and Justinus A. Satrio[2], (1)Chemical Engineering, Villanova University, Villanova, PA, (2)Chemical Engineering Department, Villanova University, Villanova, PA

Development of Biodiesel Production Facility for Utilizing Waste Cooking Oil

Brian Delancy, Brook Hopkins and Justinus Satrio

Department of Chemical Engineering
Villanova University, Villanova, PA 19085

Three years ago, a team of chemical engineering students at Villanova University developed, constructed, and began operating a biodiesel production facility that reuses the waste cooking oil produced from Villanova University dining centers. The production facility is designed to process an annual maximum of 3500 gallons of waste cooking oil. Over the past two years, additional process systems, including a methanol recovery system and soap-production process have been developed as a way to utilize the reaction's glycerol by-product. The focus of our current research is multifaceted, but focuses primarily on the production optimization and output maximization of Villanova's Biodiesel Production Facility. Several quantitative analysis techniques, including Calorimetry and Fourier Transform Infrared Spectroscopy were introduced to measure and maintain quality control of the biodiesel product. Improved methanol recycling techniques and natural soap production minimize any waste, thus making the process more economically efficient and environmentally friendly. Finally, the project increased its relations with Villanova Grounds and Maintenance Department, providing the University with a biodiesel power option in accordance with the University's "Green Energy" initiative. The Villanova Biodiesel Team also worked to expand the program into the local community through a marketing plan, in partnership with the Engineering Advisory Board.

(602bj) Enhanced Lignocellulose Conversion to Biofuels with Thermophilic Bacteria and Thermostable Enzymes

Aditya Bhalla, Chemical and Biological Engineering, South Dakota School of Mines & Technology, Rapid City, SD, Kenneth Bischoff, Bioproducts and Biocatalysis Research Unit, USDA-ARS, NCAUR, Peoria, IL and Rajesh K. Sani, Chemical & Biological Engineering, South Dakota School of Mines & Technology, Rapid City, SD

Enhanced Lignocellulose Conversion to Biofuels with Thermophilic Bacteria and Thermostable Enzymes

Aditya Bhalla[1], Dr. Kenneth Bischoff[2] and Rajesh Sani[1]*

1. Department of Chemical and Biological Engineering, South Dakota School of Mines and Technology, Rapid City, SD.
2. Renewable Product Technology Research Unit, Agricultural Research Service, National Center for Agricultural Utilization Research, U.S. Department of Agriculture, Peoria, IL 61604

Second generation feedstock, especially nonfood lignocellulosic biomass, has been seen as a potential source for biofuel production. Lignocellulose conversion technologies, which are carried out at ≤50°C, have several limitations. Therefore, this study focuses on the importance of thermophiles and thermostable enzymes to overcome the limitations of existing lignocellulosic biomass conversion processes. We have isolated several thermophilic cellulose- and xylan-degrading pure cultures belonging to the genera *Bacillus* and *Geobacillus* using soil samples collected from the Homestake Gold Mine, SD and a local compost facility. Xylanases produced by one of our isolate are found to be highly thermostable as compared to the reported literature. Greater hydrolysis rates of xylan were observed with xylanase from our isolate as compared to commercially available xylanases at high temperatures. In order to produce these enzymes in bulk, we overexpressed thermostable xylanases in a heterologous expression system. Purified xylanase was characterized in terms of pH, temperature, thermostability, and molecular weight. The purified thermostable recombinant enzymes, in the near future, will be used to prepare a thermostable mixed enzyme system for lignocellulose hydrolysis to fermentable sugars.

A Green Carbon Supported Au Catalyst for Acetylene Hydrochlorination Into Vinyl-Chloride-Monomers for Polyvinyl Chloride Production

Kai Zhou, Chmical Engineering, Beijing Key Laboratory of Green Chemical Reaction Engineering and Technology Tsinghua University, Beijing, China

Introduction

Catalysis has been widely used in chemical industries, selecting precious metals or metal oxides as the active components in the majority of the processes. However, the preparation and post-processes of the conventional metal-based catalysts exhaust great resources and environmental cost, which are always toxic and energy-consuming and cannot well meet the demanding of sustainable development for whole society. Therefore, searching promising materials possessing environmental acceptability and low emission is a topic of increased interest and importance for the green chemical processes. Hydrochlorination of C_2H_2 is the dominant method for the synthesis of VCM (vinyl chloride monomer) and further polymerize into poly vinyl chloride. The process takes place in fixed bed reactor in tubes packed with activated carbon loading around 10 ~ 15 wt% $HgCl_2$. The exothermic reaction releases 124.8 $kJ \cdot mol^{-1}$ heat during adding HCl onto C_2H_2. The poor heat transfer of fixed bed reactor resulted in heat spots and hence increasing volatility of $HgCl_2$. The $HgCl_2$ vapor releasing into the environment easily accumulates in the creatures and causes chronic poisoning, which is severely harmful to human beings and environment.

Carbon materials, playing the role as the catalytic substance, have significant advantages over the conventional metal-supported systems. For instance, nanostructured carbon and nitrogen-doped carbon composites have been proved to be the effective alternatives to conventional metal-containing catalysts for heterogeneous reaction processes, such as oxidative dehydrogenation (ODH) of ethyl benzene, Friedel–Crafts Reaction, oxygen reduction reaction (ORR).[1-3]

Therefore, developing new cost-effective and non-mercury catalysts and related process is quite crucial for a sustainable society. In this contribution, a novel carbon based Au catalyst for acetylene hydrochlorination was developed, which showed advanced activity and life time. Active carbon (AC) with hierarchical pore structure was selected as catalyst support. To increase the interaction between the support and active phase, AC was pre-treated by a mixture of H_2SO_4 and HNO_3 to facilitate the anchoring of active metal nanoparticles. The catalyst was prepared by simple co-impregnation in the solution containing $HAuCl_4$ (active phase) and chlorides of metal promoters, following by drying under 130 °C for 8 hrs. The morphology and catalytic performances were characterized by SEM, TEM, XPS, TPR, SEM and XPS results indicated that the existence of promoter metals facilitated the dispersion of Au nanoparticles on AC, which increased the initial reaction activity. Besides, the activity of the main active substance Au^{3+} was also improved. Au-only-based catalyst showed the highest activity with 1wt.% Au content (Au atomic weight to AC weight), while the optimized Au weight content can be reduced to below 0.5 wt.% with the addition of metal promoters. Under the condition of acetylene hydrochlorination of 200 °C and 1200 h^{-1} (Space Velocity, Volume based), the acetylene conversion over 70 % can be achieved and maintained for 10 hrs, which indicated that both the activity and the stability of the catalyst were enhanced. This demonstrated that the carbon based Au catalyst can be a very promising green catalyst for polyvinyl chloride industry.

Table 1. The characterizations of the catalyst based on the different substrates

Catalysts	S_A, ($m^2 \cdot g^{-1}$)	X_A^*, (%)	S_{VC}^*, (%)
Au/MWNT	938	91.25	>99.5
Au/N-CNTs	196	3.64	>98
Au/Celite	178	57.5	>98
Au/ZrO_2	46	-	-

*The Au loading was 1 % in each catalyst and the conversion and selectivity were analyzed under 360 h^{-1}(GHSV(C_2H_2)) and 180 °C

References

1. M. F. R. Pereira, J. L. Figueiredo, J. J. M. Orfao, P. Serp, P. Kalck, Y. Kihn, *Carbon* 2004, 42, 2807-2813;
2. N. Keller, N. I. Maksimova, V. V. Roddatis, M. Schur, G. Mestl, Y. V. Butenko, V. L. Kuznetsov, R. Schlogl, *Angewandte Chemie-International Edition* 2002, 41, 1885;
3. F. Goettmann, A. Fischer, M. Antonietti, A. Thomas, *Angewandte Chemie-International Edition* 2006, 45, 4467-4471;

(602bl) Preparation of MgO Nanoparticles and Its Utilization for CO2 Chemisorption

Vishwanath G Deshmane, Mechanical Engineering, North Carolina Agricultural and Technical State University, Greensboro, NC and Yusuf G. Adewuyi, Chemical and Bioengineering, North Carolina A&T State University, Greensboro, NC

The increased CO_2 concentration in the atmosphere due to emissions of CO_2 from the utilization of fossil fuels including coal has caused serious concerns for global warming. Of the available CO_2 emission reduction options, CO_2 capture and sequestration has been identified to be the most effective one. In regards to this, development of solid adsorbents to capture the CO_2 at elevated temperature-pressure, as well as atmospheric temperature-pressure, from the natural gas and flue gas effluent has got considerable attention. In the present study, MgO nanoparticles were synthesized and utilized for the adsorption of CO_2 at three different temperatures of 30 °C, 100 °C and 250 °C. The effect of preparation conditions such as surfactant concentration, effect of presence of ultrasound, type of precipitating agent and the precursor concentration on the crystalline sizes and the surface area of the MgO nanoparticles and their subsequent effect on the CO_2 adsorption capacity were studied. Preliminary results showed that the prepared nanoparticles have significant CO_2 adsorption capacities. The prepared material was characterized by nitrogen adsorption–desorption (BET), X-ray diffraction (XRD), scanning electron microscopy (SEM), thermogravimetric and differential scanning calorimetry (TGA-DSC), energy dispersive X-ray spectroscopy (EDX) and Fourier transform infrared spectroscopy (FTIR).

(602bm) Utilization of Process Performance Criteria At the Conceptual Design Stage in Green Production of Organic Chemicals

Sencan Haydaroglu, Chemical Engineering, Northeastern University, Boston, MA and Turker Gurkan, Department of Chemical Engineering, Middle East Technical University, Ankara, Turkey

Biochemical and petrochemical production routes for the derivatives of 12 most promising sugar-based building blocks proposed in the U.S. Department of Energy Top Value Added Chemicals from Biomass Report were evaluated at the conceptual design stage utilizing criteria. Sugar-based building blocks were compared to their petrochemical surrogates consisting of persistence, bioaccumulation, aquatic toxicity, atom and mass efficiency and their economical potential. Raw materials and products do not show a significant difference with respect to bioaccumulation criteria. In contrast, persistence of bio-based raw materials and products in air and their aquatic toxicity results indicate that bio-based routes are more environmentally-benign. The relationship between the economical potential and the standard Gibbs free energy change of major reactions is examined to make a further comparison at this stage. A direct relationship was not observed. However, most of the utilized reactions have a negative standard Gibbs free energy of change indicating that chemical industry exploits the spontaneity of reactions.

(602c) Situ Detection of Fermentative Activity by Redox Potential Measure and Its Applications During Ethanol Fermentation

Chen-Guang Liu, School of Life Science and Biotechnology, Dalian University of Technology, Dalian, China

Redox potential reflects the complex physiological information in cells that can be detected for optimizing the fermentation processes. However, control redox potential by delicate approaches is not easy due to lack of mathematic relationships between redox potential profiles and fermentation parameters. In this work, three important parameters including temperature, pH and dissolved oxygen are chosen to buildup a dynamic model according to Nerst equation. Fermentative activity defined as the specific consummation rate of substrate is connected with the redox potential profiles slope, but not the redox potential value. Base on this model, a feeding scheme was developed to maintain the glucose concentration at a low level in order to eliminate the substrate inhibition, and an aeration scheme was proposed to prolong the duration with rapid ethanol production under high cell viability. Besides, when utilizing the self-flocculating yeast, redox potential controlled repeated batch fermentation realizes the cells recycle and then higher ethanol productivity is achieved because of very high density yeast cells.

Abstract: CFD Modeling of a Gravity Settler for Algae Harvesting

By Scott A. Hug, Karl K. Scott, Jorge E. Gatica, and Joanne M. Belovich

Future world energy needs necessitate the search for alternative and sustainable energy sources. Biodiesel is one such source with many upsides, but current production methods are flawed. Existing large scale biodiesel production facilities rely primarily on soybean and canola oil as the lipid source for the biodiesel. The production of soybean and canola oil is not viable in the long term as a transportation fuel lipid source.

Algae are the future of lipid sources for biodiesel production. While soybean and canola oil can only be grown in lands and climates that support agriculture taking away land that could otherwise be used for food crops, algae can be grown in large open pools on non-arable lands in which they will not compete for space with food crops. Algae can also be grown in seawater, which is more abundant than the fresh water required for other sources of biofuel. Algae cultivation can produce up to 20,000 gallons per acre of algae, whereas canola or soybean based biodiesel cultivation can only produce 38-100 gallons per acre. The cells use carbon dioxide as a carbon source and thus can be used to offset carbon dioxide production of facilities, namely power plants.

The problem with Algae as a lipid source for biodiesel is the cost of dewatering of the cells. Algae concentrations in the best conditions in bioreactors will not surpass 20 g/L due to light limitations. The water must be removed before the lipids can be used in the transesterification process that produces the biodiesel. Typical methods for dewatering the algae involve using settling ponds to concentrate the algae a little and then using centrifugation to do most of the dewatering. The costs of this centrifugation step limit the commercial viability of algae based biodiesel.

A patented novel gravity settler design at Cleveland State University is analyzed in this project. This gravity settler design is displayed in Figure 1 below. Preliminary results obtained with this design show promise to help the commercial viability of the algae biodiesel process by lowering the dewatering costs. A major difference between this and traditional gravity settlers is that the inlet of this gravity settler is at the top, whereas traditional gravity settlers have inlets at the bottom. Two prototypes have been tested and experiments indicate that the technique is effective at dewatering algae.

Figure 1: Novel Algae Settler

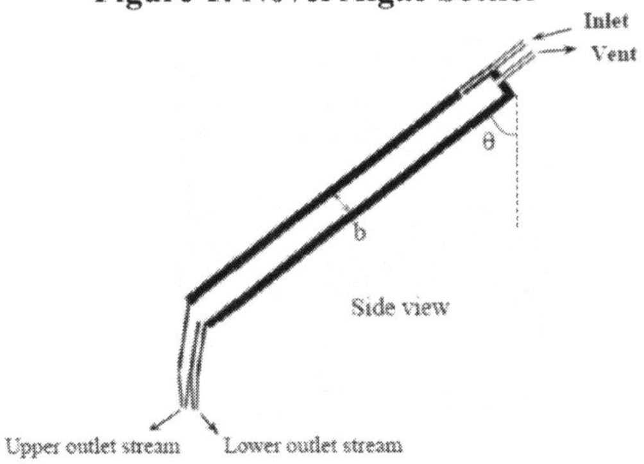

In order to benefit the design process, a computational fluid mechanics model for the system has been developed using COMSOL™ (formerly known as FEMLab), a finite element based multi-physics modeling environment. The model formulation is based on the mixed model that considers a system where there is a dispersed phase of either solids or liquid droplets or bubbles in a continuous liquid phase. This study makes use of COMSOL's particle tracing module, which allows for trajectories of specific particles to be examined over a period of time.

The geometrical model has been developed based on the simplified geometry of the gravity settler shown in Figure 1. After validation experiments are completed and results with simplified relations for settling velocities are evaluated, the simulation environments is to be updated to the more advanced geometry of the prototypes by means of Solidworks® and COMSOL™ LiveLink™ for Solidworks®. This LiveLink™ enables to import complex geometries into COMSOL™ as well as formulation of parametric studies where the CAD software is used to interactively modify geometric parameters that may show relevance in the performance of the settler. Solidworks® parameters can also be used in COMSOL™ to perform parametric sweeps.

Using this LiveLink™ functionality, parametric sweeps will be demonstrated as tools to optimize the settler's design and operation by analyzing geometric (such as length, angle of inclination, and height for dewatering) as well as operating variables (such as inlet and outlet flow rates and velocities). COMSOL™ ant the LiveLink application will be used to optimize these variables in order to create the most efficient model for algae dewatering. The optimized design will be physically prototyped and experimented on to validate the computational fluid dynamics (CFD) modeling.

This study shows how the powerful environments of a detailed CFD modeling interfaced with a commercial computer-aided design (CAD) software can be used to optimize the design of gravity settlers. Comparative studies carried out with alternative designs and results from validation experiments are presented.

The model formulation is based on the mixed model. The mixed model considers a system where there is a dispersed phase of either solids or liquid droplets or bubbles in a continuous liquid phase. It was developed for hard spherical molecules, so some adaption is required to correctly model algae cells which can clump and deform, these modifications will be discussed.

The complex geometry of the gravity settler was mimicked using SolidWorks®. A model of the gravity settler, shown in Figure 2 on the next page, is divided into two parts: The inlet region, and the outlet region. The inlet region is a rectangular prism measuring 215 cm long, 10 cm wide, and 5 cm high. The algae particles enter through the top of the inlet stream. The inlets are placed on the top of the inlet region, and there is a vent on the side of the region. The outlet region is consists of upper and lower outlet streams. The objective of this study to determine the settings in which the water and algae achieve the greatest level of separation in this region based on gravity's influence on the algae particles and water in the inlet region.

The early stages of the project consist of working on each part of the model separately. This allows the simulations to be run on the less complex geometry of each part (i.e, a simple rectangular prism for the inlet region), rather than the complex geometry of the entire gravity settler. These simple geometries allow for the simulation to be run on COMSOL™ much quicker because a more coarse mesh may be used to accurately measure the particle trajectories and other properties in a simulation. For complex geometries, a more fine mesh is needed to accurately measure properties, especially in regions with many edges and corners.

Figure 2: SolidWorks design of the gravity settler. The fluid and particles enter at upper-right corner of the settler, travel through the inlet region, and exit in either the upper or lower outlet stream.

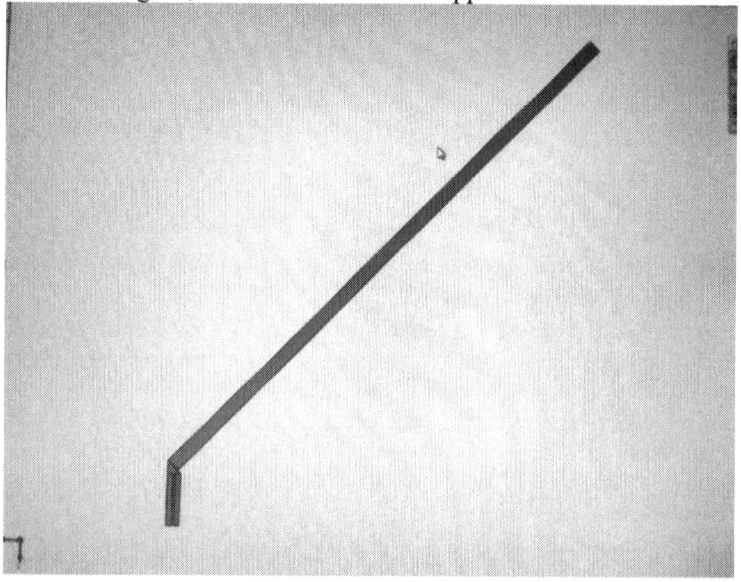

After these runs are made on the simple geometry of the inlet region, the most optimal flow rates, velocities, and length and height of dewatering will be examined throughout the more complex geometry of the entire gravity settler. It is here in which the angle of inclination will be measured

A few modifications must be made from the COMSOL model. The model was developed for hard spherical molecules, so some adaption is required to correctly model algae cells which can clump and deform. These modifications will be discussed.

(602e) Algal Genotyping in an Open System and a Bioreactor

Alice C. Jernigan, Chemical Engineering, University of Arkansas, Fayetteville, AR and Christa N. Hestekin, Ralph E. Martin Department of Chemical Engineering, University of Arkansas, Fayetteville, AR

Algae are a high growth source for biomass with a great potential to be a significantly eco-friendly bio-fuel source. Since different species of algae have different oil and lipid concentrations it is important to genotype the algae in an open environment to determine the best species for biofuel production. Alternatively, in a bioreactor it is important to maintain the dominance of the best-fuel producing species. For genotyping, we examine the 16S and the 18S genes, which are highly conserved among prokaryotes and eukaryotes (respectively), but both have unique small variable regions that can be used to differentiate between species. We examined 2 methods of DNA analysis, capillary electrophoresis single strand conformation polymorphism (CE-SSCP) and single base extension (SBE). For the open system, we were able to use CE-SSCP to identify different algal species, which were confirmed by sequencing. Alternatively, for the bioreactor, we were able to detect and monitor 2 different algal types (*Spirulina* and *Chlorella*).

(602f) Enhancement of H2 and CO2 Uptake for the Production of Biodiesel in Ralstonia Eutropha

Ryan Sullivan, Pin-Ching Maness, Carrie Eckert and Jianping Yu, Biosciences Center, National Renewable Energy Laboratory, Golden, CO

Ralstonia eutropha fixes CO_2 through the Calvin-Benson-Bassham (CBB) cycle using electrons and energy obtained from the oxidation of H_2. Producing biodiesel-equivalent electrofuel from renewable CO_2 and H_2 has immense potential, especially if the fuel is compatible with the existing fuel infrastructure. This research addressed enhanced H_2 and CO_2 substrate utilization by focusing on two strategies: (1) optimizing transcriptional regulations to afford over-expression of Ribulose-1,5-bisphosphate carboxylase/oxygenase (RubisCO), the enzyme responsible for assimilation of CO_2 into the CBB cycle; and (2) hydrogenase over-expression by introduction of additional copies of genes encoding a membrane-bound hydrogenase (MBH), a soluble hydrogenase (SH), and their maturation machinery to enhance oxidation of H_2 to generate NAD(P)H and ATP, both are required for CO_2 fixation. Incorporation of these strategies into a single production *R. eutropha* strain had resulted in 6-fold improvement in CO_2 fixation and 3-fold improvement in H_2 uptake, *in vitro*, with the overarching goal of providing abundant reducing equivalents towards the economic production of biodiesel in *R. eutropha*.

(602g) Biodiesel Production From Cynara Cardunculus Crop: Economic, Environmental and Social Assessment

Carmen M Torres[1], Sergio D. Ríos[1,2], Carles Torras[1], Joan Salvadó[1,2] and **Laureano Jiménez**[1], (1)Departament d'Enginyeria Química, Universitat Rovira i Virgili, Tarragona, Spain, (2)Catalonia Institute for Energy Research (IREC), Tarragona, Spain

The generation, transport and consumption of conventional sources of energy, particularly in the transport sector, have raised many concerns about air pollution, environmental degradation, and petroleum consumption. In this sense, the development of alternative technologies is acquiring importance, and many efforts are being done in the gradual replacement of fossil fuels. Within the renewable sources of energy, the production of liquid biofuels from organic feedstocks sources represents a feasible alternative that avoids important modifications of vehicle engines, maintaining most of the infrastructures for the supply chain and can replace partially the petroleum-based fuels.

Biofuels have many advantages over fossil fuels: as well as being more environmentally friendly and a renewable source of resources, every region in the world has a raw material that can be used, to a greater or lesser extent, to produce some sort of biofuel, thus providing work for the population and improving the economies of the countries that so far have been dependent on petroleum imports.

On the other hand, there is the ethical question of the direct competition for land that is required for producing food. Several groups (Green Peace, Intermon Oxfam, etc) claim that the growing production of biofuels is responsible for the increasing food prices, and the destruction of forests. Second-generation biofuels (those that do not threaten food supplies) were developed to overcome this obstacle.

Of particular interest are the residual parts of agricultural and forest vegetation that contain cellulose for producing bioalcohols, recycled oils, and other plants that contain oils and which do not threaten food supplies (*Jatropha curcas, Cynara cardunculus*, microalgae, etc).

The aim of this work is to analyse the feasibility of biodiesel production obtained by transesterification of vegetable oil extracted from *Cynara Cardunculus* (cardoon), a wild robust perennial plant, native from the Mediterranean basin. The performance assessment of a process with a capacity of 5000 t/year biodiesel production is carried out from the triple bottom line perspective. The characterization of the process includes its rigorous simulation in AspenHysys® from which an automated evaluation tool programmed in Matlab® retrieves the inventory of relevant energy and material inputs, as well as the environmental releases. In the evaluation tool the calculations are grouped into different modules: acquisition of the simulation data, environmental, economic and social assessment, and results analysis.

The simulation performs the alkali-catalyzed transesterification with methanol and a pre-treatment section for the free fatty acids removal by esterification. The operational conditions are based on the cardoon oil characterization and the kinetic data extracted from published experimental studies (Pasqualino, 2006). The simulated process includes the methanol recovery, the catalyst neutralization and the glycerol separation and purification as valuable by-product.

In the environmental module, the Eco-indicator 99 methodology is used for the impact assessment, achieving similar results to other vegetable sources of biodiesel. With respect to the economic criterion, a profitability analysis is performed calculating the net present value, thus considering the initial investment, manufacturing costs and products revenue. As expected, the economic metric points out the necessity of government subsidies to produce positive rates of return. From a social perspective, quantitative indicators (*e. g.*, provision of employment) and qualitative indicators (*e. g.*, nuisance) were selected among the social metrics available in the literature. The three indicators are then aggregated in a single metric using a normalization and weighting procedure based on the geometric mean of the metrics' ratios for pair-wise comparison of alternative processes (Torres et al., 2011).

The performance of the plant is optimized considering the potential alternatives. With this aim, the use of two different catalysts and an additional unit for the recuperation of glycerol in the pre-treatment section are considered, allowing a higher reusing of materials and reducing the waste treatment requirements. Furthermore, a sensitivity analysis is carried out to quantify the robustness of the procedure and identify the critical variables.

References

Pasqualino, J.C., 2006. Cynara cardunculus as an alternative crop for biodiesel production. Ph.D. dissertation. Universitat Rovira i Vigili, Tarragona, Spain.

Torres, C.M., Gadalla, M.A., Mateo-Sanz, J., Jiménez-Esteller, L., 2011. Evaluation Tool for the Environmental Design of Chemical Processes. Ind. Eng. Chem. Res. 50 (23), 13466-13474.

Torres, C.M., Gadalla, M.A, Mateo-Sanz, J.M., Jiménez-Esteller, L., 2012. An automated environmental and economic evaluation methodology for the optimization of a sour water stripping plant. J. Clean. Prod. (in press).

(602h) Reducing the Energy Consumption of Biofuel Production Using a Solar Assisted Steam Generation System with Heat Storage

Robert Brunet[1], Gonzalo Guillén[2] and Laureano Jiménez[1], (1)Departament d'Enginyeria Química, Universitat Rovira i Virgili, Tarragona, Spain, (2)Chem. Enginering, Universitat Politécnica de Catalunya, Barcelona, Spain

In this work, we address the problem of reducing the energy consumption of biofuel plants through their integration with a solar thermal energy system that generates steam. A mathematical model of the solar energy system that includes energy storage is coupled with a rigorous simulation model of the biofuel facility developed in a commercial process simulator. The solar energy system model takes the form of a bi-criteria nonlinear programming (NLP) formulation that accounts for the simultaneous minimization of cost and energy consumption.

We demonstrate the capabilities of this strategy with two case studies in which we address the design of a 12,000 ton/year alkali-catalyzed biodiesel process using vegetable oil modeled in Aspen Plus® and a 120,000 tones/year dry-grind corn-to-ethanol production plant modeled in SuperPro Designer®. In both cases we considered weather data of Tarragona (Spain). A set of Pareto solutions were generated using the epsilon constraint methodology. The results obtained show that is possible to achieve reductions in energy consumption of up to 24.90 % for the biodiesel and 17.24% for the bioethanol with respect to the minimum cost design.

(602i) Biomass-Solvent Catalytic Reactions: Assessing Sugars and Intermediates by HPLC and GCMS

Sipho C. Ndlela, Iowa Energy Center, Iowa State University, Nevada, IA

The first phase of this project evaluated the reaction pathways for biomass feedstock in the presence of various cosolvents including acetone, methanol, CO_2, and water using a 1 liter (7.6cm x 23.1cm, ID x L) high pressure batch reactor. Additionally, the effect of temperature on the extent of the reaction was assessed at 200 to 350°C. In both instances biomass conversion rates compared to the starting material ranged from 70 to 90%, with the remaining portion being the unreacted fraction and lignin. The resulting products consisted of complex chemical fractions: organic acids, organic ester derivatives, sugars, sugar-alcohols, thermal degradation products and more. Although the impact of using various cosolvents was noticeable at lower temperatures, at higher temperatures the extent thermal degradation products increased due to longer heating times to reach the target temperature (2 to 8 h). Despite the challenges with longer residence time when using the batch reactor, results from this study gave an important understanding of the reaction behavior and interaction of the solvent and biomass.

To have better control of the reaction extent at pilot scale, a flow reactor was assembled. The system consisted of a 250 ml/min slurry pump with reactor chamber that is 212 cm longer by 3 cm diameter, a maximum system working pressure up to 69 Mpa, electrical heating jackets and a condenser. Biomass slurry of 5 to 10% was fed into the reactor at varying temperatures from 200 to 350°C and at residence times of 5 to 30 min. Analysis results from the flow reactor study will be main focus for this discussion.

(602j) High Pressure, Oxygen Blown Entrained-Flow Gasification of Bio-Oil

Nicholas Creager, Iowa State University, Ames, IA and Robert C. Brown, Mechanical Engineering, Iowa State University, Ames, IA

The goal of this research is to investigate the gasification of bio-oil in a pressurized entrained flow gasifier. Gasification at high pressure is essential for integrated gasification combined cycle power (IGCC) and efficient syngas upgrading to fuels and chemicals. Feeding solid fuels into high pressure reactors is one of the most difficult challenges of gasification. Dual chamber lock hoppers are an expensive and sometimes unreliable option. Among solid fuels, coal is sufficiently hydrophobic to be mixed with water and pumped at high pressure as high-solids slurry. Biomass is too hydroscopic to be pumped in this fashion. Bio-oil from fast pyrolysis of biomass is essentially "liquid biomass" that can be pumped at high pressures.

This gasifier operates with active electric insulation at temperatures of 700°C to 900°C and under pressures up to 50 bar to simulate the operation of industrial-scale gasifiers. The gasifier consists of two sections: the reactor and the quench vessel. The gasification section is constructed of a 1.0 m long, 3.8 cm dia. silicon carbide tube surrounded by several layers of insulation and encased inside a 15.2 cm dia. stainless steel pressure vessel. Hot syngas entering the quench vessel is sprayed with water to reduce gas temperature to below 100°C followed by pressure let-down and exhaust. Product gases are analyzed by gas chromatography. The system can gasify up to 2.5 kg/hr of bio-oil in oxygen and steam flows of 1.0 kg/hr.

Tests were performed on two bio-oils from the pyrolysis of switchgrass and cornstover using design of experiment protocols. This allowed the construction of statistical models to predict the effect of biomass composition, gasification temperature and pressure, and biomass/oxygen equivalence ratio on gas composition.

OPTIMAL DESIGN OF ALGAE-BASED BIOREFINERIES USING CO_2 FROM DIFFERENT INDUSTRIAL FACILITIES

Rubio-Castro E.[1], *Ponce-Ortega J.M.*[2,*], *Serna-González M.*[2], *Hernández-Calderón O.M.*[1], *Ortiz-del-Castillo J.R.*[1], *El-Halwagi, M.M.*[3]

[1]*Chemical and Biological Sciences Department, Universidad Autónoma de Sinaloa, Culiacán, Sinaloa, México, 80000*

[2]*Chemical Engineering Department, Universidad Michoacana de San Nicolás de Hidalgo, Morelia, Michoacán, México, 58060*

[2]*Chemical Engineering Department, Texas A&M University, College Station, TX, USA, 77843*

* Corresponding author. *E-mail*: jmponce@umich.mx

Abstract

This work proposes an optimization approach for the carbon dioxide sequestration from different industrial facilities to yield an algae-based biorefinery. The proposed approach is based on a distributed system to account for the economies of scale and it involves the site selection for the processing facilities as well as the transportation cost. Additionally, the model considers the optimization for the technologies used in each facility for the CO_2 sequestration, for the algae growth, harvesting, lipid extraction, and different technologies to yield several products like biodiesel, methane, ethanol, between others, depending on the customer demand. The proposed approach considers the mass exchange between different industrial facilities, because the algae from one plant can be used to produce products used in the same or in a different plant. The optimization model considers the maximization for the total net profit, which includes the sales of products plus the tax credit for reduction of CO_2 emissions minus the capital and operational costs for the different processing stages minus the overall transportation costs associated to the system. The constraints for the model include the mass balances for the system, design and operational constraints for the processing units and constraints for the demands for the products. The proposed model is a mixed integer non linear programming model and this was coded in the software GAMS. Finally, two case studies from Mexico were analyzed and the results show several economic and environmental benefits simultaneously.

Keywords: CO_2 sequestration, Optimization, Distributed system, Bio-refineries.

INTRODUCTION

A biorefinery is a system where several processes are integrated to transform the biomass into fuels, energy, chemical products and foods (Ohara, 2003). This concept is analogous to an oil refinery but in a biorefinery the raw material is biomass instead of oil. The biomass can be waste biomass from an industrial or natural process or it can be obtained from the processing of vegetables and animals wastes (for example algae, corn and jatropha (Pokoo-Aikins, et al., 2010)); therefore, the difference between the biomass and the oil is that the former one is a renewable raw material. In the industrial processes there are several CO_2 emissions, and this can be captured for the algae growth; and from this can be obtained oil and biomass that can be used to produce several additional products that can be used in the same industrial facility or sent to a different place where they are required. Therefore, this work proposes an optimization approach for the design of distributed systems for the CO_2 sequestration considering the integration of emissions from different industrial facilities and determining where and what product obtained and the optimal distribution associated through an algae based biorefinery system.

METHODOLOGY

The overall system to obtain alga oil is composed by three process modules (i.e., algae growth, algae harvesting and algae oil separation), like it is shown in **Figure 1a**. In the first module, for the algae growth can be used an open system (raceway pond) or a closed system (photobioreactor). Notice that the growth of algae requires water, sunlight, carbon dioxide and nutrients (phosphorous and nitrogen). Regarding to the carbon dioxide (see **Figure 1b**) the algae growth is an opportunity to sequester this gas by using flue gas emissions from industrial sources. After the algae growth process, a separation process is required. Many separation processes are available for the harvesting method. In the harvesting module there are two outlet streams. One of them is rich in algae oil and low water content and the other one (poor stream) presents a low algae oil content but a high content of water. And the poor stream can be segregated and sent to the waste that is discharged to the environment and/or to the growth module to reduce the fresh water usage while the rich stream is sent to the extraction module for extracting the algae oil (see **Figure 1b**). In this stage of the process, there are three outlet streams: the algae oil, water and the cell paste. Respect to the water and cell paste streams, these ones can be sent to the waste that is discharged to the environment or it can be reused in the algae growth stage (water stream) like it is shown in **Figure 1b**. Finally, like shows the **Figure 1a**, the algae oil can be processed with different technologies to obtain different products (biodiesel, methane, ethanol, etc.) and byproducts (heat, steam, fertilizers, etc.). Additionally, after the oil extraction is important to remove the undesirable components depending of the desirable product.

In this regard, the production (biodiesel, ethanol, methane, etc.) depends on the demand by the specific customers, selling price, process costs and plant capacity. Hence, in this work is proposed an optimization model based on **Figure 1b** to determine the optimal pathway for getting different products from algae oil via the sequestration of carbon dioxide of the flue gas coming from different industries. Then, a novel superstructure is developed to consider multiple technologies on each stage of the algae oil processing. These technologies have specific efficiencies and costs (operating and capital costs). The proposed superstructure is presented in **Figure 2,** which is an extended representation of **Figure 1b**.

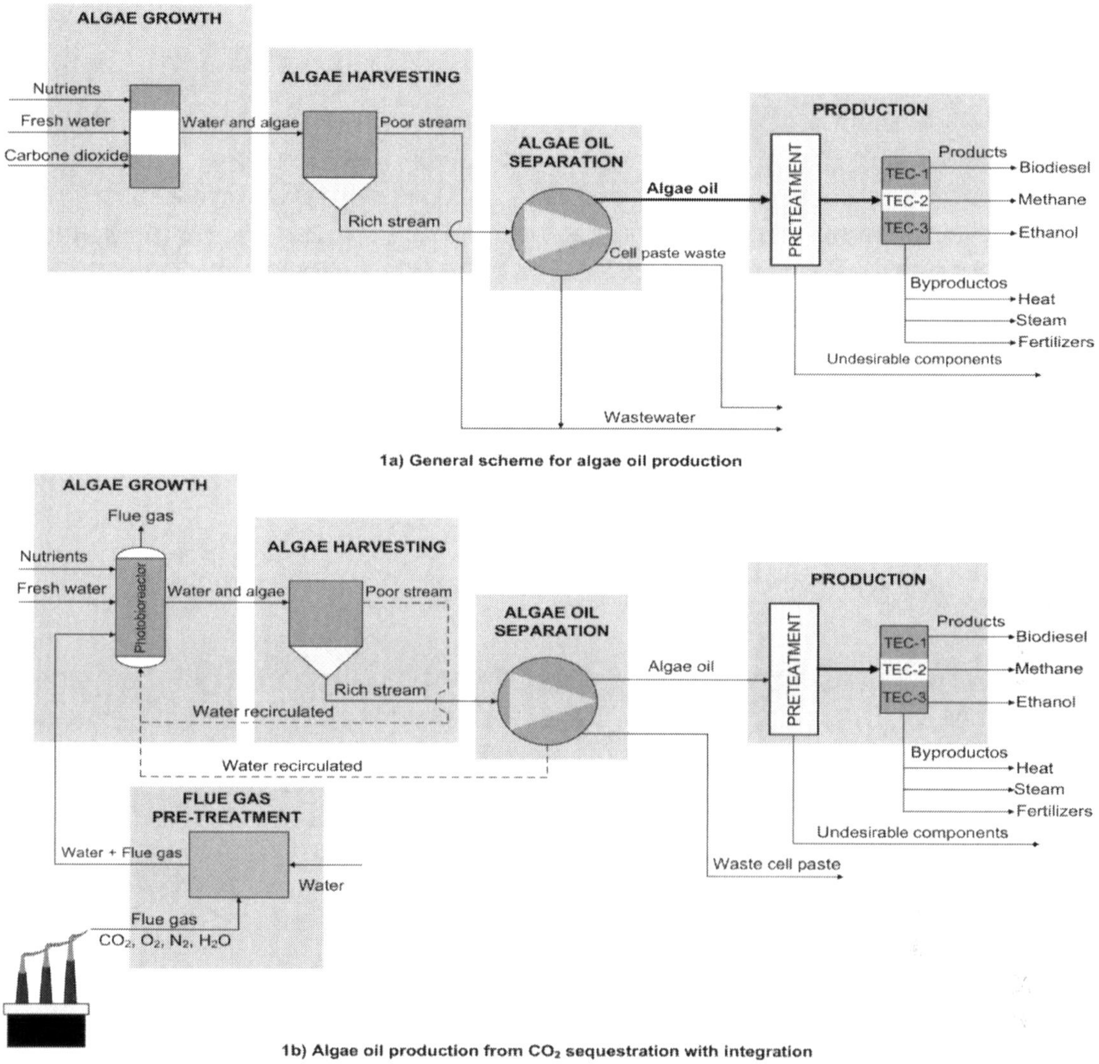

1b) Algae oil production from CO_2 sequestration with integration

Figure 1. General production system for the algae oil processing.

Figure 2 is a new scheme to determine the optimal process of the algae oil for obtaining several products to satisfy specific demands of several consumers as well as to determine the feasibility of algae oil processing. Furthermore, there is the opportunity to sequester flue gas to use it like nutrients in the algae growth.

The proposed superstructure considers the optimal allocation of facilities for the algae oil processing through a distributed system. It is because the processing of the algae oil can take place in a *central processing facility* and/or a *hub processing facility*, which could be installed inside the participating plants.

The objective function is the maximization of the overall process profit (*Profit*):

$$Profit = Spc + Gse - Cap - Cop - Cta - Ctp$$

where *Cap* and *Cop* are the capital and operating costs in the system, *Ctp* is the cost to transport the algae oil to the processing facility and to transport the products to the consumers, *Spc* is the profit owing to the product sales and *Gse* is the tax credit for reduction of CO_2 emissions.

RESULTS

The proposed methodology is applied to a case study from Mexico, in this case two industrial facilities located in different geographical places are considered, also transesterification, anaerobic digestion, pyrolisys, gasification and fermentation processing alternatives are considered and the product involved are biodiesel, methane and ethanol which can be sent to two consumers located in different places. Both consumers demand biodiesel and ethanol and just one require methane. The problem then is coded in the GAMS software and the results indicate that the distributed system is able to capture the industrial emissions to obtain economic and environmental benefits. It is because of part of ethanol and all the biodiesel is generated in the central processing facility. And like the methane is demanded just by one consumer this product is generated in the closer processing facility which represents a less transport cost.

CONCLUSIONS

This paper proposes a systematic approach for the optimal design of an algae based distributed biorefinery system to capture the industrial emissions to obtain simultaneously economic and environmental benefits; the proposed approach determines simultaneously the type and location of the processing facilities as well as the distribution of the emissions and products in the distributed system. The application of the proposed methodology to a case study from Mexico shows the potential benefits that can be obtained.

REFERENCES

1. Ohara, H. (2003), "Biorefinery", *Applied Microbiology and Biotechnology*, 62 (5-6) 474-477.
2. Pokoo-Aikins, G., Nadim, A., El-Halwagi, M.M., Mahalec, V. (2010), "Design and analysis of biodiesel production from algae grown through carbon sequestration". *Clean Technologies and Environmental Policy*, 12 (3) 239–254.

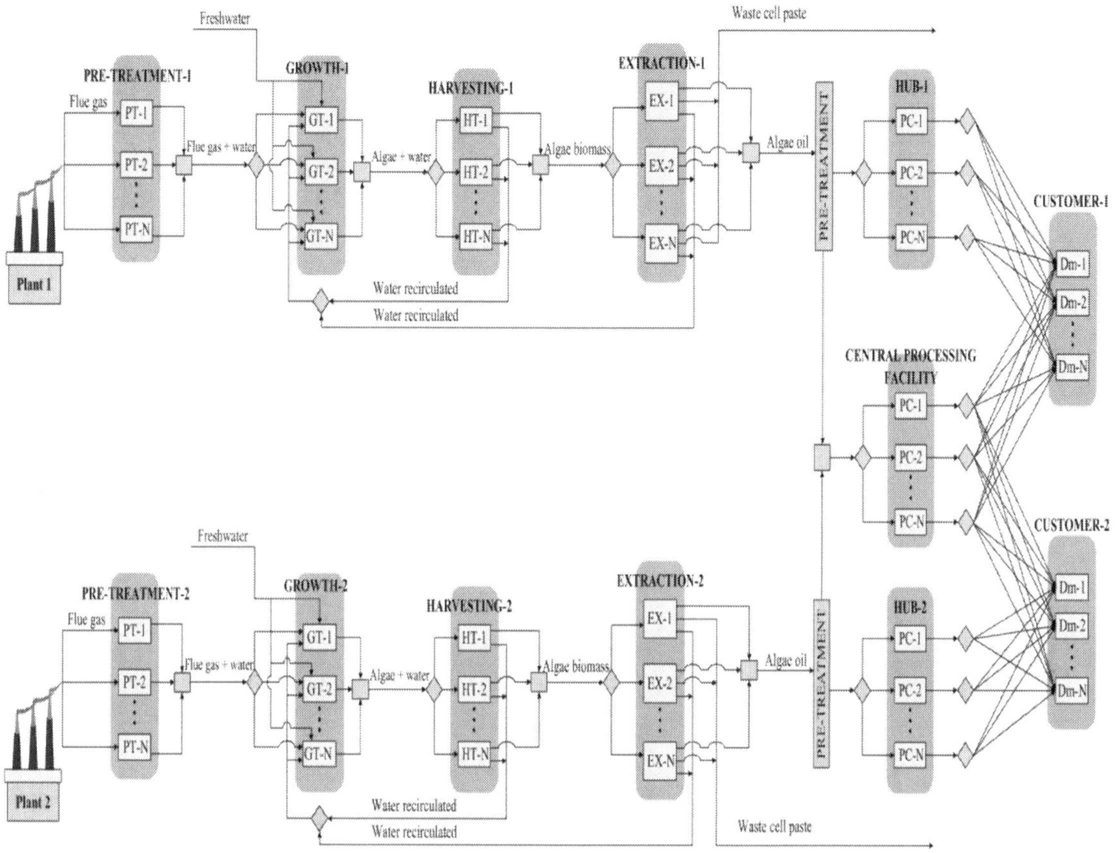

Figure 2. Proposed superstructure for the algae oil processing.

(602l) Production of the Recombinant Trichoderma Reesei Endogluconase Protein Cel7B by Using the Kluyveromyces Lactis Yeast

Zainab Alshoug[1], Michael J. Brodeur-Campbell[2] and David R. Shonnard[2], (1)Chemical Engineering, MTU, Houghton, MI, (2)Chemical Engineering, Michigan Technological University, Houghton, MI

This research is about producing recombinant *Trichoderma reesei* endogluconase protein Cel7B by using *Kluyveromyces lactis* yeast as a host. Cel7B is one of the glycoside hydrolyses family proteins that are produced by *T. reesei*. Cel7B with other endoglucanases, exoglucanases, and β-glucosidases hydrolyze cellulose to glucose, which can then be fermented to fuels or other products.

Examining favorable fermentation conditions for enzyme production and improved activity were the research objectives. Production of protein on different types of media was examined, and measured the activities of the protein by using different tools or procedures. The first method was using different concentration of galactose as a carbon and energy source. The purpose of this method is to determine the relationship between production of enzyme with increasing sugar concentration. The second method was to use different types of media: a complex media-yeast extract, peptone, galactose (YPGal); a minimal media-*yeast nitrogen base* (YNB); and a minimal media with supplement- yeast nitrogen base with casamino acid (YBC). The third method was using different types of reactor or fermenter: a small reactor (shake flask) and a big reactor (BioFlow 3000 fermenter). The purpose of this method is to determine the quantity of the protein produced by using different environments of production.

Different tools to determine the presence and activity of protein were used: for presence of protein sodium dodecyl sulfate-polyacrylamide gel electrophoresis (SDS-PAGE) was used. Secondly, to detect enzyme activity, the carboxymethyl cellulose- 3,5-dinitrosalicylic acid (CMC- DNS) assay was employed.

SDS-PAGE showed that the enzyme band was at 67,000 Da. CMC-DNS activity assay showed that protein activity was greatest with 1% sugar (Galactose) concentration. For the different type of media used in our fermentation, protein was produced from yeast extract peptone galactose (YPGal), and yeast nitrogen base with casamino acid (YBC), but protein was not produced and no activity was detected from *yeast nitrogen base* (YNB). That means the protein production requires the amino acid resources. The quantity of protein which was produced from the rich media (YPGal) over time of production was equal to the quantity of protein that was produced from the minimum media supplemented with casamino acids (YBC) over time with a concentration of 1% galactose.

(602m) Thermal-Catalytic Cracking of Lignin for Production of Value-Added Chemicals

Tae Hyun Kim, Department of Environmental Engineering, Kongju National University, Cheonan, South Korea, Chang Geun Yoo, Agricultural and Biosystems Engineering, Iowa State University, Ames, IA and Kyong-Hwan Lee, (2) Korea Institute of Energy Research, Daejeon, South Korea

Lignin is the second largest components in the lignocellulosic biomass, which can be the valuable byproduct of biorefinery process. If the lignin would be modified or hydrolyzed, high value-added lignin-derived products can be formed.

In the previous reports, ammonia-based pretreatment methods have been introduced as the one of the most effective delignification methods for lignocellulosic biomass. Approximately 60-70% of lignin was hydrolyzed and this chemically modified lignin hydrolysate has a potential for the production of value-added chemicals through thermal-catalytic cracking of lignin because low pH decreases the reaction temperature and increases the reaction rate.

In this paper, various reaction conditions were explored to convert lignin into value-added products. Reaction temperature and pressure were two main factors in this study. From the thermal cracking reaction in the presence of catalysts, several phenolic compounds and carboxylic acids were produced in the form of liquid. Produced gas and solid were also collected and analyzed.

(602n) Potential Carbonyl Condensation Reaction in Detoxification of Phenolic Model Compounds for Bioethanol Fermentation

Maobing Tu[1], Rui Xie[1] and Yonnie Wu[2], (1)Auburn University, Auburn, AL, (2)Chemistry and Biochemistry, Auburn University, Auburn, AL

We investigated the inhibitory effects of ortho-phthalaldehyde (OPA) as a phenolic model compound on fermentation and growth of *Saccharomyces cerevisiae* and its alkaline detoxification. OPA was a potent inhibitor on fermentation and growth of *S. cerevisiae* compared to vanillin, furfural and hydroxymethyl furfural (HMF) at the same concentration level. The inhibition of OPA on fermentation and growth of *S. cerevisiae* was dose dependent. Ethanol production and growth of *S. cerevisiae* were both completely inhibited in the presence of 1.0 mM OPA. The inhibition on fermentation and growth of the yeast decreased with the decrease of the OPA concentration between 0.02 mM-1.0 mM. OPA at 0.02 mM showed no inhibition on both fermentation and growth of yeast, the ethanol final yield was even increased by 4.6% compared to the control. The inhibition of OPA could be overcome by increasing inoculation size of yeast. Most interestingly, we found OPA inhibition could be detoxified under alkaline condition (pH~10) at 60 °C for 2 h in the presence of a reducing sugar (ketose or aldose), but not non-reducing sugar. Mass spectral analysis of OPA reaction products in negative ion mode revealed a high intensity mass at 313.09 m/z. This mass was predicted to be the aldol reaction product of reducing sugar and OPA under alkaline condition. One of the CHO groups on OPA was converted into hydroxyl group by nucleophilic addition of enolate ion of reducing sugar. Loss of one CHO group of OPA could be the key factor for the removal of OPA inhibition.

(602o) Medium and High Through-Put Characterization of Biofuel Feedstocks

Erica Gjersing[1], Angela Ziebell[2], Robert Sykes[1], Geoffrey Turner[3], Renee Happs[1], Melvin Tucker[1], Stephen R. Decker[4], Mark F. Davis[2] and Brian H. Davison[5], (1)National Bioenergy Center, National Renewable Energy Laboratory, Golden, CO, (2)National Renewable Energy Laboratory, Golden, CO, (3)BioSciences Center, National Renewable Energy Laboratory, golden, CO, (4)Biosciences Center, National Renewable Energy Laboratory, Golden, CO, (5)BioEnergy Science Center and Oak Ridge National Laboratory, Oak Ridge, TN

Abstract: The high through-put pipeline has analyzed tens of thousands of samples using NREL's high through-put (HTP) pipeline which characterizes cell wall chemistry and sugar release on pretreatment and enzymatic saccharification (Sykes et al. 2008; Selig et al. 2011). In the last year we have added two new capabilities and worked increasingly closely with plant groups to analyze their data.

The characterization group works together with plant biologist to understand the chemical modifications occurring when plants are genetically modified and why they might cause changes in recalcitrance to pretreatment and saccharification. The characterization group also studies natural populations to understand how natural populations vary and what we might learn from natural diversity.

Plant cell wall sugar content is being determined by a HTP NMR (Nuclear Magnetic Resonance) adaptation to NREL's traditional wet chemistry analysis. This technique is also being used to assess the tendency of a plant to produce inhibitors e.g. furfural and HMF. The method uses proton NMR and Partial Least Squares analysis in conjunction with a set of standards which have the full wet chemistry method. The current NMR facility houses a HTP NMR sample changer which can house >500 samples at a time and sample analysis time take 1-2 minutes. The original full wet chemistry method has also been down scaled to 1/300th by researchers at University of California Riverside and is available for high through-put analysis (DeMartini et al. 2011).

References:

DeMartini, J. D., M. H. Studer, et al. (2011). "Small-scale and automatable high-throughput compositional analysis of biomass." Biotechnology and Bioengineering **108**(2): 306-312.

Selig, M., M. Tucker, et al. (2011). "High throughput determination of glucan and xylan fractions in lignocelluloses." Biotechnology Letters: 1-7.

Sykes, R., B. Kodrzycki, et al. (2008). "Within tree variability of lignin composition in *Populus*." Wood Science and Technology **42**: 649-661.

(602p) Comparison of Enzymatic Reactivity of Corn Stover Solids Prepared by Dilute Acid, AFEX, and Ionic Liquid Pretreatments

Xiadi Gao, Chemical and Environmental Engineering, University of California, Riverside, RIVERSIDE, CA, Rajeev Kumar, Center for Environmental Research and Technology, University of California, Riverside, Riverside, CA and Charles Wyman, Chemical and Environmental Engineering, University of California Riverside, CE-CERT, Riverside, CA

This study is part of a collaboration among the BioEnergy Science Center (BESC), the Great Lakes Biomass Research Center (GLBRC), and the Joint Biological Energy Institute (JBEI), the goal of which is to understand how biomass pretreatments with much different deconstruction patterns impact the chemical and structural features of biomass and its biological conversion to sugars.

Enzymatic hydrolysis was performed on solids resulting from dilute sulfuric acid (DA), ammonia fiber expansion (AFEX), and ionic liquid (IL) pretreatments of corn stover at low to high protein loadings to define how ratios of cellulase, xylanase, and pectinase affected sugar yields. Avicel cellulose, regenerated amorphous cellulose (RAC), and beechwood xylan were also subjected to enzymatic hydrolysis. Initial hydrolysis rates, sugar release patterns over 120 h of hydrolysis, cellulase accessibility to cellulose, and oligomers concentrations were measured over the course of enzymatic hydrolysis to determine enzyme-substrate interactions. These results were then integrated with compositional changes to identify key features controlling enzymatic hydrolysis of solids from these pretreatments and similarities and differences in their impacts on enzyme effectiveness. Ionic liquid pretreated corn stover displayed the highest initial reactivity at all enzyme loadings and the highest final digestibility for a low enzyme loading of 3 mg protein/g glucan in the raw material. However, increasing the enzyme loading to 12 mg/g glucan resulted in dilute acid and AFEX pretreated corn stover attaining the highest overall cellulose digestions. Hydrolyzate from AFEX pretreated corn stover had the highest proportion of xylooligomers, while ionic liquid pretreated produced the most glucooligomers, but amounts of both oligomers dropped with increasing enzyme loadings and hydrolysis times. Ionic liquid pretreated corn stover had the greatest accessibility to enzymes as measured by the maximum cellulase adsorption capacity.

(602q) Understanding How Cell Wall Differences in Agave, Poplar, and Switchgrass Affect Deconstruction in Pretreatment and Enzymatic Hydrolysis

Hongjia Li, Chemical and Environmental Engineering, University of California, Riverside, RIVERSIDE, CA, Sivakumar Pattathil, Complex Carbohydrate Research Center, University of Georgia, Athens, GA, Marcus Foston, Department of Chemistry and Biochemistry, Georgia Institute of Technology, Atlanta, GA, Reichel Samuel, 3Institute of Paper Science and Technology, Georgia Institute of Technology, Atlanta, GA, Rajeev Kumar, Center for Environmental Research and Technology, University of California, Riverside, Riverside, CA, Arthur J. Ragauskas, School of Chemistry & Biochemistry, Georgia Institute of Technology, Atlanta, GA, Michael G. Hahn, University of Georgia, Athens, GA and Charles E. Wyman, BioEnergy Science Center, Oak Ridge, TN; Chemical and Environmental Engineering, University of California Riverside, CE-CERT, Riverside, CA

Agave has emerged as promising biofuels feedstock mainly because it offers high productivity with low water and nutrient demands on semiarid lands. However, our recent results indicated that agave is also significantly less recalcitrant than typical lignocellulosic feedstocks such as poplar and switchgrass. Systematic understanding of structural differences in agave, poplar, and switchgrass and the consequences on cell wall deconstruction can provide clues on paths to reduce recalcitrance that is the key impediment to large-scale commercialization of processes for making fuels from cellulosic biomass. Thus, our high throughput pretreatment and hydrolysis (HTPH) system was used to evaluate the effect of hydrothermal pretreatment conditions and enzyme formulations on sugar yields for four agave materials along with poplar and switchgrass. At the same time, a series of characterization techniques were applied to compare structural differences in cell walls, including lignin composition, non-cellulosic wall components and extractability, cellulose crystalline structure, and surface accessibility. These studies identified important structural features that contribute to lower recalcitrance of agave species and suggested promising directions for more cost effective deconstruction of cell walls of other plants.

(602r) Bacteria Engineered for Bio-Styrene Production

Rebekah McKenna[1], Shawn Pugh[1], Warinsinee Phusitkanchana[2] and **David R. Nielsen**[1], (1)Chemical Engineering, Arizona State University, Tempe, AZ, (2)Chemical Engineering, Arizona State University

Styrene is an important commodity chemical with versatile commercial applications, predominantly as a monomer building block for the production of many useful polymers and co-polymers. Today, all styrene is produced via chemocatalytic routes from petroleum-derived benzene or ethylbenzene according to energy intensive processes. Through *de novo* metabolic pathway engineering, we have pathway engineered a functional pathway to instead produce styrene from renewable feedstocks. By over-expressing both *PAL2* from *Arabidopsis thaliana* and *FDC1* from *Saccharomyces cerevisiae* in a phenylalanine over-producing *Escherichia coli* host we have engineered the first styrene-producing microbe. Initial shake flask cultures produced ~300 mg/L styrene, approaching the determined toxicity threshold. Subsequent works have since sought to overcome several of critical limitations of this preliminary work. Here we present our progress in this regard on several fronts. First, we have increased and balanced pathway flux through the systematic tuning of the expression of *PAL2*, the previous bottleneck enzyme. This was achieved through the systematic tuning of each of gene copy number, promoter strength, and RBS strength. Next, we have sought to overcome the toxicity issue by the 'bottom up' engineering of an aromatic tolerant *Pseudomonas* sp. to produce styrene from glucose and other renewable substrates. At the same time, the ability to circumvent styrene toxicity through its in situ recovery has been explored with the use of solvent extraction, gas stripping, and membrane pervaporation. At last we have sought to improve the sustainability and economics of bio-styrene production by engineering photosynthetic cyanobacteria to produce styrene from CO_2 and light inputs alone.

(602s) Decomposition of Microcrystalline Cellulose by Organic Solvents At Sub- and Supercritical Conditions

Anton Koriakin, Department of Chemical and Biomolecular Engineering, Yonsei University, Seoul, South Korea, Nguyen Van Hai, Chemical and Biomolecular Engineering, Yonsei University, Seoul, South Korea and Chang-Ha Lee, Chemical and Biomolecular Engineering, Yonsei University, Seoul, South Korea

Renewable biomass sources can be converted to fuel, and they are a logical choice to replace oil. By deriving more energy from renewable feedstocks, many countries might be able to decrease their reliance on fossil fuels significantly. In this study, direct thermo-chemical decomposition of microcrystalline cellulose was investigated in various sub and supercritical organic solvents (m-xylene, dodecane, dioxane and methanol). Since the decomposition reaction was performed at 300°C, 350°C and 400°C, the applied pressure was different among the solvents. And the influence of hydrogen and heterogeneous catalysts on decomposition was evaluated for each supercritical solvent. The composition of liquid product was analyzed by Gas Chromatography with a mass selective detector (GC-MS). The char samples were analyzed by X-ray diffraction (XRD).

The addition of hydrogen to the supercritical reaction increased the conversion ratio by about 4-7% in all solvents except non-polar dodecane. At the temperature of 300°C, the conversion ratio was higher in the aprotic solvents than in methanol. At 350°C and 400°C, the conversion in methanol was the highest, but the operating pressure became very high compared to the other solvents. The addition of surface modified silica by acid groups in the methanol solvent led to a significant increase in the conversion ratio. However, the catalyst did not influence the conversion ratio in the aprotic solvents.

(602t) Transgenic Expression of Thermomyces Lanuginosus Lipase in Plants for the Enzymatic Production of Biodiesel

Michael Gagnon[1], Jonathan Fogel[2], Subhash Minocha[2] and Palligarnai Vasudevan[1], (1)Chemical Engineering, University of New Hampshire, Durham, NH, (2)Biological Sciences, University of New Hampshire, Durham, NH

Diminishing petroleum reserves and increasing environmental concerns have led to an urgent need to develop environmentally friendly fuels, such as biodiesel. A relatively new and promising development in the production of biodiesel is via enzymatic transesterification with lipase as the catalyst. The motivation for this research arises from the need to develop inexpensive biocatalysts to replace environmentally harmful chemical catalysts. We are genetically engineering plants to constitutively express a lipase for biodiesel production from spent oils and non-edible plant oils. We have cloned the gene of a lipase with known transesterification activity from *Thermomyces lanuginosus*, a thermophylic fungus. Cloning involved isolation of total RNA, reverse transcription of the mRNA into cDNA and PCR amplification of the lipase gene using specific primers. The gene was first inserted into a cloning vector (pCR8/GW/TOPO) and sequenced to confirm its identity. The gene has been inserted into a plant destination vector (pGWB408 and pMDC83) via LR clonase reaction. Pieces of *Nicotiana tabacum* (tobacco) leaf was transformed with the lipase gene using *Agrobacterium tumefaciens* (strain GV3101) and transferred onto selection media plates. We are currently growing *Arabidopsis thaliana* from seeds that have been transformed using the floral dip transformation method. The recombinant enzyme will be collected from the genetically engineered plants, purified, and tested for its activity. This approach should provide a relatively cheap and environmentally safe source of lipase for use as a biocatalyst.

(602u) Comparison of Catalysts for Applications in Biomass Fast Pyrolysis

Shantanu Kelkar, Chemical Engineering, Michigan State University, East Lansing, MI and Christopher M. Saffron, Biosystems & Agricultural Engineering, Department of Forestry, Michigan State University, East Lansing, MI

Thermochemical methods such as pyrolysis combined with catalysis may be used to produce a hydrocarbon-rich liquid fuel from renewable biomass. The objective of this work was to explore the use of microporous, mesoporous and bi-functional catalysts for catalytic fast pyrolysis of biomass.

ZSM-5, SO_4^{2-} ZrO_2, Al-MSU-S from zeolite beta seeds and sulfated zirconia on mesoporous support 20% SO_4^{2-} ZrO_2 - MCM-41, were synthesized in MSU laboratories. In the first part of this study, microporous catalysts, ZSM-5 and sulfated zirconia, were compared with mesoporous catalysts Al-MSU-S and meso sulfated zirconia for their potential to upgrade oxygenated pyrolysis products. Catalyst properties such as BET surface area, BJH pore volume and acidity using NH3-TPD were measured. Catalyst preparations for upgrading poplar (DN-34) were examined using analytical pyrolysis connected to a GC/MS. Separate experiments were conducted to analyze C1-C9 volatiles using a GC/FID. The product yield, carbon selectivity and yield of hydrocarbon products was evaluated. Although the mesoporous catalysts in this study had many desirable properties such as high pore volume and high acidity, they showed low selectivity for hydrocarbons when compared with conventional ZSM-5 catalyst. The mesoporous catalysts produced a greater amount of non-condensable gases when compared with the microporous catalysts. The char and coke yields of these catalysts were estimated using a TGA. Significant coke production was observed on all catalysts except ZSM-5.

ZSM-5 catalysts may be synthesized with various silica:alumina ratios (SAR) and the change in aluminum content affects the acidity and the hydrophilicity of the catalyst. ZSM-5 catalysts with five different SAR (23, 30, 50, 80, 280) were obtained and tested for the catalytic pyrolysis reaction with the poplar biomass. The catalyst with the lowest SAR was observed to have the highest acidity and production of hydrocarbon molecules. However, the lowest coke yields were observed with SAR=30. The selectivity for various hydrocarbon molecules varied significantly with change in SAR. Metals provide the Lewis acid functionality on zeolite catalysts and may moderate coke production as well as promote olefin production, thus increasing the activity and on-stream lifetime of the catalyst. Bi-functional ZSM-5 (SAR=23) were prepared by loading various dehydrogenating metals. The addition of metals to ZSM-5 enabled a significant increase in hydrocarbon production and decrease in coke production.

(602v) Gas Emissions From the Storage of Western Red Cedar

Xiao He[1], Anthony K. Lau[1], Shahab Sokhansanj[2], C. Jim Lim[1] and Xiaotao T. Bi[1], (1)Chemical & Biological Engineering, University of British Columbia, Vancouver, BC, Canada, (2)Environmental Sciences Division, Oak Ridge National Laboratory, Oak Ridge, TN

The utilization of woody biomass as alternative fuel to generate heat and power has become an important element of sustainability in today's world. Before direct combustion and/or processing in biorefineries, the biomass is usually stored for an extended time period. Gas emission is one of the major problems with biomass storage due to its impact on the surrounding environment and human health, as well as the dry matter losses which would reduce the heating value of the materials. Western Red Cedar (WRC) is a popular wood species in British Columbia, and its harvesting residues can be used as a sustainable biofuel to replace fossil fuels. Aside from the above-mentioned problems associated with biomass storage, WRC has a characteristics pungent smell which would cause concerns about odor nuisance problems. The objective of this paper is to investigate and quantify gaseous emissions from stored WRC woodchips. Fresh WRC chips with an initial moisture content of approximately 50% (wet basis) were used as raw materials in the study. Experiments were conducted in lab-scale reactors under both aerobic and non-aerobic (airtight) conditions at a range of temperatures (5°C, 20°C, 35°C, 45°C and 50°C), which represents cool to hot climate conditions in different geographic locations. Air was pumped into the aerobic reactors after the daily gas sampling event in order to replenish and maintain a high oxygen level in the reactors. The concentrations of gaseous emissions - carbon dioxide (CO_2), carbon monoxide (CO), and methane (CH_4) along with oxygen (O_2) were analyzed by GC, whereas volatile organic compounds (VOCs) were determined by GC/MS.

Results showed that under non-aerobic conditions, the concentrations of CO_2 and CO increased gradually with time for all temperatures. At higher temperatures, higher CO concentrations were measured. In contrast, we observed lower concentrations of CO_2 along with higher O_2 concentrations at higher temperatures. At 20°C, the concentration of CO_2 was the highest; after three weeks, CO_2 concentration reached a plateau of 16% as oxygen content was depleted to 0%, indicating that the storage environment has turned anaerobic. By comparison, oxygen content was lowered to 10-15% for other temperatures. Results further showed that CO_2 and CO emissions from the aerobic reactors exhibit similar trends as the non-aerobic reactors with respect to the effect of temperature. The concentrations of both CO_2 and CO emitted at each temperature decreased slowly with time despite higher oxygen levels of close to 20% being maintained during the entire test period. Biological reaction could be dominant during the storage of wet biomass, leading to the highest level of microbial activity at 20°C rather than at higher temperatures. Microbiological analysis of the WRC woodchips was performed to verify this phenomenon and the extent of microbial activity in the materials. Results from the qualitative GC/MS analysis indicated that the major VOCs emitted include methanol, terpene, aldehydes, acid, alkane, benzene, indoles, acetone, ethers and esters. A TVOC meter was also available for determining the total concentration of VOCs; it was found to be higher at higher temperatures for both aerobic and non-aerobic storage conditions. Under non-aerobic conditions, the initial VOC concentration was approximately 53 ppm and 2 ppm at 50°C and 5°C, respectively; and it gradually decreased with time. Overall, the gas emission results from this study reaffirm the importance of ensuring safe storage conditions for wet woody biomass.

Title: **Wastewater treatment of fluoroquinolone antibiotics using the UV-H_2O_2 advanced oxidation process**

Authors: Sebastian Snowberger; Lee Blaney

Keywords: Advanced oxidation process; pharmaceuticals and personal care products; sustainability; UV-H_2O_2; chemical reactions

---

With recent advancements in medicine, consumption of pharmaceuticals is steadily increasing and, therefore, unprecedented amounts of pharmaceuticals are entering wastewater streams. Currently, wastewater treatment plants do not specifically treat for pharmaceuticals. For that reason, trace concentrations of pharmaceuticals are not only discharged to the environment, but also introduced to drinking water supplies. Health concerns of long-term ingestion of trace concentrations of pharmaceuticals through drinking water are widespread; however, toxicological studies regarding the human health impacts are ongoing. Regardless, numerous studies have identified ecological impacts associated with pharmaceutical contamination. To address these concerns, we studied the transformation of pharmaceuticals under ultraviolet (UV) irradiation and in the UV-hydrogen peroxide (UV-H_2O_2) advanced oxidation process.

This research examined the ability of a lab-scale UV reactor to transform antibiotics into benign compounds. Specifically, a batch-recycle UV reactor was employed to study the transformation of three prominent fluoroquinolone antibiotics: ciprofloxacin, levofloxacin, and norfloxacin. Transformation of fluoroquinolones was studied in DI water and also tertiary effluent from a local wastewater treatment plant. The experimental solutions were prepared by spiking pharmaceuticals (approximate concentrations of 10^{-5} M) into DI and tertiary effluent; hydrogen peroxide (ranging from 10^{-5} to 10^{-3} M) was added during UV-H_2O_2 testing. The pH in the DI solution was controlled using a 0.1M phosphate buffer. Experimental solutions were continuously pumped through the UV cell and samples were generally collected over ten minutes of operation. Fluoroquinolone concentrations were measured using high performance liquid chromatography with fluorescence detection (HPLC-FLD). Transformation of fluoroquinolones occurred through direct photolysis and indirect photolytic mechanisms involving hydroxyl radicals, which were formed via hydrogen peroxide decomposition.

This research has elucidated the quantum yields for transformation of the three fluoroquinolones as a function of solution pH, the impact of the molar ratio of hydrogen peroxide to the fluoroquinolone concentration on transformation kinetics, and the effect of background water quality (*i.e.*, DI water vs. tertiary effluent) on the transformation of fluoroquinolones in the batch-recycle UV-H_2O_2 reactor. Given the widespread adoption of UV technology for disinfection purposes, the results of this research will prove useful to understanding removal of antibiotics in conventional and advanced wastewater treatment processes.

Adsorption Properties of Functionalized PFA-Derived Carbons

Ali Qajar, Maryam Peer, Ramakrishnan Rajagopalan, Henry C. Foley

PFA-derived carbons are synthesized and activated under controlled conditions. The samples are functionalized with concentrated nitric acid under reflux condition to different extends. Surface and structural properties of the samples are characterized by collecting IR spectra, TGA and gas adsorption. Also the samples are compared to commercial activated and functionalized carbons. Our study shows exceptional adsorption properties of PFA-derived carbons for carbon dioxide and ammonia adsorption. Also, functionalization process drastically changes adsorption affinity of the carbons for the adsorption of super-critical gases resulting in higher selectivity for oxygen adsorption over other gases. While, functionalization of the pristine PFA-derived carbon limits pore accessibility resulting in less gas uptake, for the activated PFA-derived carbon the same process promotes adsorption uptakes and dynamics. Ammonia adsorption on the functionalized activated PFA-derived carbon, hits the best reported adsorption uptakes by COF-10 at ~16 mmol/g at 298 K and 1 bar.

(602z) Surfactant Addition to Alkali Solutions for Increased Carbon Dioxide Absorption Rates

Brett P. Spigarelli, Chemical Engineering, Michigan Technological University, Houghton, MI

To meet the growing need for CO_2 capture and storage technology, Michigan Technological University is studying CO_2 capture. Our current approach to CO_2 capture is to bubble CO_2 through alkali solutions. The objective of the present study was to find a way to increase the absorption rate of CO_2 into the alkali solution without reducing the absorption capacity of the solution. One approach to increasing absorption rate is to use a surfactant to chemically alter the gas bubble size. The surfactant chosen for the study was a neutral charge surfactant called DOWFROTH 200 (DF200). Experiments were conducted to study the absorption rate of CO_2 at varying surfactant concentrations of 0, 0.12, 0.24, 0.36, and 0.48 g/L. Results showed that as the concentration of DF200 increased in solution, the absorption rate also increased. The CO
mol/min CO_2 at 0.48 g/L DF200. This amounted to a 14% increase in the CO_2 absorption rate with only a 3% loss in absorption capacity of the solution.

(608a) Sustainability of Biofuel Production Systems

Douglas J. Reinemann, Biological Systems Engineering, University of Wisconsin-Madison, Madison, WI

The Great Lakes Bioenergy Research Center is one of three DOE funded Bioenergy Research Centers in the US and the only one that has a major effort devoted to the study of the sustainability of biofuels production systems. We have used a Spatially-Explicit Integrative Modeling Framework (SEIMF) to evaluate the productivity and environmental impacts of biofuel production systems. The modeling framework was implemented to simulate feedstock supply to a cellulosic ethanol bio-refinery located at the center of each of two Regionally-Intensive Modeling Areas (RIMA). Feedstock sources modeled included crop residues and perennial species. Each production system was evaluated in terms of its effect on soil erosion, water-use efficiency, soil carbon stocks, nutrient losses, nitrate leaching, and nitrous oxide emissions. Perennial biomass crops grown on marginal lands can potentially contribute to meet cellulosic feedstock targets while avoiding food-vs.-fuel conflicts and maintaining or even enhancing ecosystem services. The SEIMF approach together with long-term environmental data and analysis was used to model potential cellulosic ethanol production from perennial mixed prairie systems grown on marginal lands across the 10-state U.S. North Central Region. These RIMA modeling results used to develop a spatially-explicit bio-economic model for the study of potential cellulosic biomass supply at regional scale.

We have applied ISO 14000 Life Cycle Assessment (LCA) methods to quantify the environmental impacts of a number of complex biofuels production scenarios including: a) four major ligno-cellulosic agricultural feedstocks (corn stover, switchgrass, poplar and mixed-prairie species) for RIMAs in southern Michigan and Wisconsin, b) biomass feedstocks from native forests in Northern Wisconsin, c) use of the primary AFEX pretreatment process for agricultural residues and SPORL and Dilute Acid pretreatment for woody biomass, d) final fermentation for ethanol production, e) logistics for transportation to regional preprocessing centers and final transport to a centralized ethanol production facility, and f) heat and electric power energy generation scenarios from co-products. Further, we have analyzed three potential breakthrough technologies a) creation of oil-rich grasses as dedicated energy crops, combining both the oil platform and cellulosic platform in one crop, b) feedstock pre-processing to potentially allow the use of high-diversity feedstocks (such as mixed prairie grasses) and other species, and c) use of AFEX-treated lignocellulosic biomass for cellulase production.

A paradigm shift to cellulosic biofuels seems unlikely unless several challenges can be met, including reconciling biofuels with food production, improving system environmental performance, avoiding harmful land use changes, and assembling the biomass supply chains. In particular, local economic and environmental sustainability of the supply system is vital to advance the farmers' interests. Appropriately priced, low supply risk feedstocks are needed to insure the growth of the nascent biofuel industry, as is the ability to inexpensively transport and store the biomass. Local economic and environmental concerns must be resolved in the feedstock supply chain, and diversifying the market for cellulosic feedstocks via multiple co-products would be advantageous as well. Unfortunately, few of these properties are currently demonstrated by cellulosic feedstocks. One promising means to address these and other issues is to process the biomass near where it is produced in so-called local biomass processing depots (LBPDs). RBPDs would increase the stability, uniformity, density, and economic value of a potentially wide variety of cellulosic biomass.

(608b) Neutron Reflectometry, QCM-D , and TIRF Study of the Interaction of Endoglucanases with Films of Amorphous Cellulose

Michael Kent, Joint BioEnergy Institute, Sandia National Laboratory, Emeryville, CA

Cellulase enzyme cocktails include exoglucanases that digest cellulose chain ends and endoglucanases that cleave randomly at interior points along the chains. While it is known that these enzymes work synergistically, the details are not fully understood. In addition, cellulose binding domains (CBDs) are known to play an important role in the digestion of crystalline cellulose but much less is known about the benefit of CBDs in the digestion of amorphous cellulose. Amorphous cellulose is of interest as pretreatment of biomass with ionic liquids, a promising next generation technology, results in a combination of amorphous cellulose and cellulose II. Determining the full effects of endoglucanase activity is challenging because these enzymes can alter the structure of insoluble cellulose in addition to releasing soluble oligomers. To unravel the actions of endoglucanases and the role of cellulose binding domains in enhancing activity on amorphous cellulose, we have combined studies of the profile of water through cellulose films during digestion by neutron reflectivity, measurements of changes in mass and film stiffness using a quartz crystal microbalance (QCM), and visualization of the motion of individual enzymes by total internal reflection fluorescence (TIRF) microscopy. A trend has emerged in that the enzymes possessing a CBM penetrate into and digest within the bulk of the films to a much greater extent than those lacking a CBM. In addition, much different results were observed for a classic endoglucanase and a processive endoglucanase that both possess CBMs, suggesting that penetration is driven by different mechanisms in the two cases.

(608c) Understanding and Redesigning Secondary Cell Wall Deposition in Plants

Dominique Loque, Joint BioEnergy Institute, Berkeley, CA; Physical Biosciences Division, Lawrence Berkeley National Laboratory, Berkeley, CA

The plant cell wall represents a large source of polysaccharides that could be used to substitute for sugar derived from starchy grains, which are currently used to feed and produce biofuels. This lignocellulosic biomass, largely under-utilized, is mainly composed of sugar polymers (cellulose and hemicellulose) embedded very strong aromatic polymer called lignin. Recalcitrant to degradation, lignin inhibits efficient extraction and hydrolysis of the cell wall polysaccharide and prevents cost-effective lignocellulosic-biofuel production. Unfortunately, lignin cannot simply be genetically removed without incurring deleterious consequences on plant productivity. The cost effectiveness of the conversion of the lignocellulosic biomass into sugars is still one of the major components to produce cheap biofuels. Therefore, strategies that can be used to reduce the lignin recalcitrance and that can increase polysaccharide deposition into the cell wall without altering plant growth should be developed.

We used synthetic biology to redesign cell wall biosynthesis and deposition without affecting the plant growth. We generated strategies to reduce lignin recalcitrance either by manipulating its spacio-temporal deposition or by manipulating its composition. We also reengineered the control of cell wall biosynthesis and used it to enhancing cell wall polysaccharides deposition and biomass density. Finally, several of these engineering strategies were combined and translated into an improvement in saccharification efficiency without affecting plant development. Developed approaches and preliminary data will be presented.

¶*This work conducted by the Joint BioEnergy Institute was supported by the Office of Science, Office of Biological and Environmental Research, of the U. S. Department of Energy under Contract No. DE-AC02-05CH11231*

(608e) Development of the Extractive-AFEX^{TM} (E-AFEX^{TM}) Pretreatment Process

Shishir Chundawat[1], Leonardo da costa Sousa[1], Venkatesh Balan[1], Vijay Bokade[2], Mingjie Jin[1], Nirmal Uppugundla[1], James Humpula[1] and Bruce E. Dale[1], (1)Chemical Engineering and Material Science, Great Lakes Bioenergy Center, Michigan State University, Lansing, MI, (2)Catalysis Division, National Chemical Laboratory (NCL), Pune, India

The ammonia fiber expansion or AFEX™ process is a leading thermochemical pretreatment process that facilitates the rapid deconstruction of lignocellulosic biomass to fermentable sugars in an economical fashion. Two major limitations of the conventional AFEX™ process are; (i) it is unable to reduce the inherent recalcitrance of crystalline cellulose I to enzyme biocatalysis, and (ii) it is unable to extract any lignin during the process that are potentially inhibitory to biological catalysts. In order to address these limitations, we have recently developed a novel ammonia based pretreatment called Extractive-AFEX™ (or E-AFEX™). One of the driving forces that led to the development of the E-AFEX™ process was based on our recent discoveries related to unique physicochemical properties of cellulose III (a less recalcitrant allomorph of cellulose; *Chundawat et al, 2011, JACS, 133:11163*). Fortuitously, the conditions that favor cellulose III formation also facilitate selective extraction of lignin from plant cell walls during the pretreatment. In this presentation, we will explore the salient features of the E-AFEX™ technology and its impact on the enzymatic and microbial processability of lignocellulosic biomass. The E-AFEX™ process is able to overcome several of the common bottlenecks to cellulosic biofuel production and can be readily integrated within a non-centralized biomass processing framework. Additionally, the lignin extracted during the process can be converted to valuable chemicals or materials. AFEX™ and E-AFEX™ processing technologies open up exciting new research avenues for cellulosic biorefineries, that allow selective conversion of the aromatic and carbohydrate fractions to a whole suite of fuels, materials and chemicals using desired catalytic or biocatalytic routes analogous to the petrochemical refineries.

(608f) ORNL: Initial Binding of a Cellulose Chain Into the Cellulase Catalytic Tunnel

Pavan GhattyVenkataKrishna, Computational Biology Group, ORNL, Oak Ridge, TN, Edward Uberbacher, Computational Biology Group, ORNL and Xiaolin Cheng, Center for Molecular Biophysics, ORNL

Cel7A proceeds along a single cellulose chain in a processive manner, cleaving one cellobiose unit per catalytic event. We have conducted extensive molecular dynamics (MD) simulations to understand the initial binding of a cellulose chain into the catalytic tunnel of the Trichoderma reesei Family 7 cellobiohydrolase (Cel7A). In unrestrained MD simulations, the cellulose chain diffuses into the tunnel by the length of a cellobiose unit. Among those residues lining the tunnel, the four tryptophan residues exhibit small structural fluctuations, and may thus play a role in guiding the substrate sliding, while the more dynamic polar residues are observed to promote the initial diffusion of the substrate into the tunnel. Further MD simulations show no evident preference for the reducing end or the non-reducing end binding of the cellulose chain, suggesting that the spontaneous initial binding is not driven by the dipole-dipole interaction between Cel7A and the cellulose chain. Rather, the shape complementarity of the cellulose chain end to the tunnel vestibule (involving local hydrophobic stacking and hydrogen bond interactions) plays a more dominant role. The potential of mean force (PMF) calculations, however, show a clear difference at -5 subsite between the reducing end and the non-reducing end binding, implicating that the selective hydrolysis of cellulose from the reducing end can be partially achieved through the preferential initial binding of cellulose to the Cel7A tunnel. The comparison of the PMFs for the wild type and the W40A mutant enzymes highlights the importance of the Trp40 residue in the initial threading of the cellulose chain to the catalytic tunnel.

(626a) Production of Drop-in Biofuels Using a Hybrid Biological/Chemical Biomass Conversion Process

Cesar B. Granda, Terrabon, Inc., College Station, TX

Conversion technologies to biofuels can be biochemical in nature, such as fermentations, or thermochemical conversions such as gasification and pyrolysis. Both fermentation and, in particular, pyrolysis show great promise, however, they also have several disadvantages. Fermentation produces well-defined products, which are typically easily recoverable, but often requires the pretreatment and hydrolysis of polymeric sugars prior to the conversion of the resulting soluble sugars into the desired product. Fermentation is also often performed using pure cultures that are prone to contamination and thus require aseptic conditions, which are difficult and expensive to maintain. In addition, the pure cultures are commonly able to uptake only one type of substrate, such as sugars, which limits the types of feedstocks that can be converted. On the other hand, pyrolysis often does away with the need to priorly break down the feedstock into simpler molecules and is able to convert different types of substrates without the need for aseptic conditions. But, by the same token, the product generated by pyrolysis is much undefined and highly corrosive, as a result, it is very difficult to upgrade to an acceptable biofuel quality. In addition, high-moisture feedstocks cannot be efficiently processed through pyrolysis. In an effort to capture the best of both worlds without falling victim of their limitations, Terrabon, Inc. is commercializing a process that can convert any anaerobically biodegradable material into liquid drop-in biofuels by firstly fermenting the feedstock with natural mixed cultures of microorganisms, which are able to degrade many different types of substrates (e.g., proteins, cellulose, hemicellulose, fats, pectin, chitin, glycerol) with little differentiation and that, as a result, do not need aseptic conditions, making the fermentation very robust. The product from the fermentation, which is a mixture of carboxylate salts, is a much more defined feedstock for pyrolysis than raw biomass; therefore, the product from the pyrolysis of these salts is more defined, with its composition being mostly ketones such acetone. The ketones are very easy to upgrade and convert into liquid biofuels. The feedstock flexibility that this hybrid biochemical/thermochemical provides, has allowed Terrabon, Inc. to initially focus on converting the low-hanging fruit in biomass resources, municipal and agricultural solids wastes, into drop-in biofuels, such as bio-gasoline and bio-jet fuel. Terrabon has successfully demonstrated the science of this process and is making great strides towards its commercialization. Terrabon operates demonstration facilities with the capacity to process the equivalent to 5 dry ton/day of biomass and is able to generate enough fermentation products to produce 300 gal/day of bio-gasoline. The demonstration plant monitors different parameters in the fermentation such as acid yield and biogas production. In the downstream chemical processing into hydrocarbons, many parameters are monitored, such as yields and selectivity, as well as the quality (e.g., octane number, IBP, FBP) of the final product. Results from these studies and their implications in the economics will be presented.

Development of a C5 and C6 to Ethanol Consolidated Bioprocess with An "Old New" Bug; Use of the Deinococci Platform

Dr. Michael KREL, Deinove,

Introduction

Deinove is a green technology company specialized in development and commercial exploitation of innovative, environmentally friendly, high-performance processes for production of biofuels and other compounds of industrial or pharmaceutical value. The processes developed by Deinove use bacteria with exceptional natural properties - **the deinococci.**

Deinove was incorporated in late 2006 under the joint impetus of Philippe Pouletty, MD (Managing Partner of the Paris-based private equity firm Truffle Capital), and Miroslav Radman (Professor at Descartes University of Paris, a member of the French Academy of Science and winner of the 2003 Grand Prix *INSERM* science prize). Miroslav Radman discovered the genetic basis of the extraordinary biodiversity and robustness of the deinococci (the bacteria on which Deinove's innovation strategy is founded) and, in particular, the molecular mechanism which enables *Deinococcus radiodurans* to repair its genome after the latter has been shattered by ionizing radiation.

Deinove is the only company in the world with an activity based on systematic exploration of the deinococci - a bacterial genus with exceptional biodiversity but which has rarely been studied and never exploited commercially.

By combining the selection of wild-type strains (with natural properties that are already of industrial value) with bacterial engineering, Deinove is building a unique portfolio of intellectual property and innovative biomanufacturing processes. Deinove is aiming to create breakthrough technologies in second-generation biofuel production and green chemistry.

Technological Approach

Deinove has decided to develop a process based on the potentiality of the Deinococci micro-organisms as we detected that these microorganisms are very well suited for industrial purpose even though there was no existing metabolic engineering tools. In order to do that Deinove had to develop the whole genetic toolbox needed for strain optimization (including insertion, knock out, plasmid, promoters….).

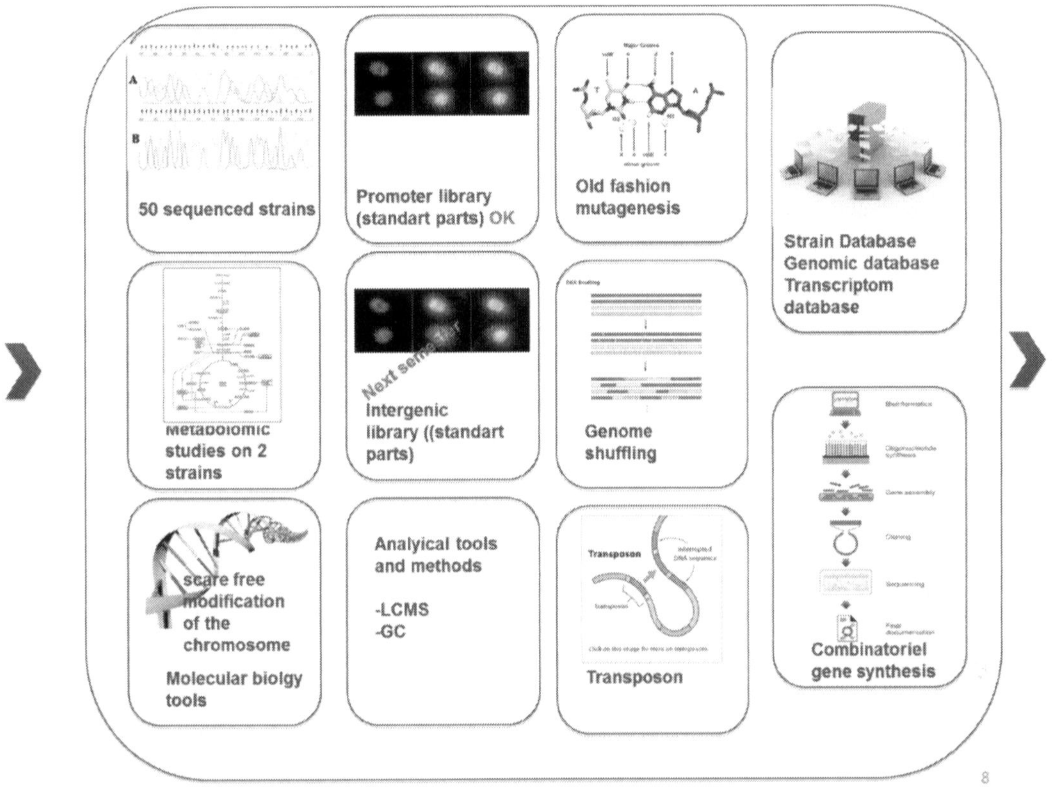

Now that this had been achieved, we are currently working on developing a bacterium that is capable of degrading and digesting both hemicellulose and cellulose (i.e. C5 and C6 sugars) and transforming it into ethanol, in a process known as Consolidated Bio Process (CBP) in order to implement it on industrial scale in the three coming years.

Deinove's research work has confirmed the extreme robustness of the deinococci vis-à-vis various types of stress (ionizing radiation, desiccation, solvents, ethanol, butanol, acid and alkaline milieus, high temperature, etc.) and their ability to integrate large genomic fragments from other micro-organisms. These properties are key success factors for the company's target industrial applications.

The deinococci's properties (robustness, biodiversity and metabolic properties) and Deinove's research results explain these bacteria's huge potential as specialized "cell factories" for the production of ethanol and other components of industrial value in a cost-effective, environmentally friendly way.

Deinove project is a breakthrough scientific challenge as it opened the path to production of cellulosic ethanol from cellulose, hemicellulose and their derivatives (starch, glucose and pentoses) in a simultaneous way. There are no such process running in the world today and this would be of great interest as it is very difficult to get a higher valuation of the C5 stream of the biomass for a mass market (as biofuel). Being able to develop a process that give a higher value to C5 stream (by transforming it into ethanol for instance) and using it in a parallel manner to C6 fermentation is a technical and scientific challenge that has not yet been achieved

Results

- Development of metabolic engineering tools efficient of the Deinococci species that allow insertion on the chromosome or knock out of genes in a non GMO approach (no antibiotic markers, no plasmid and no scars left)
- First ever proof of insertion and/or knock out of selected functionality in Deinococci species
- Development of a bi-functional microorganism able to digest cellulose and hemicellulose material and produce ethanol.

The improvement of the industrial strain regarding the ethanol production can be seen on the following chart:

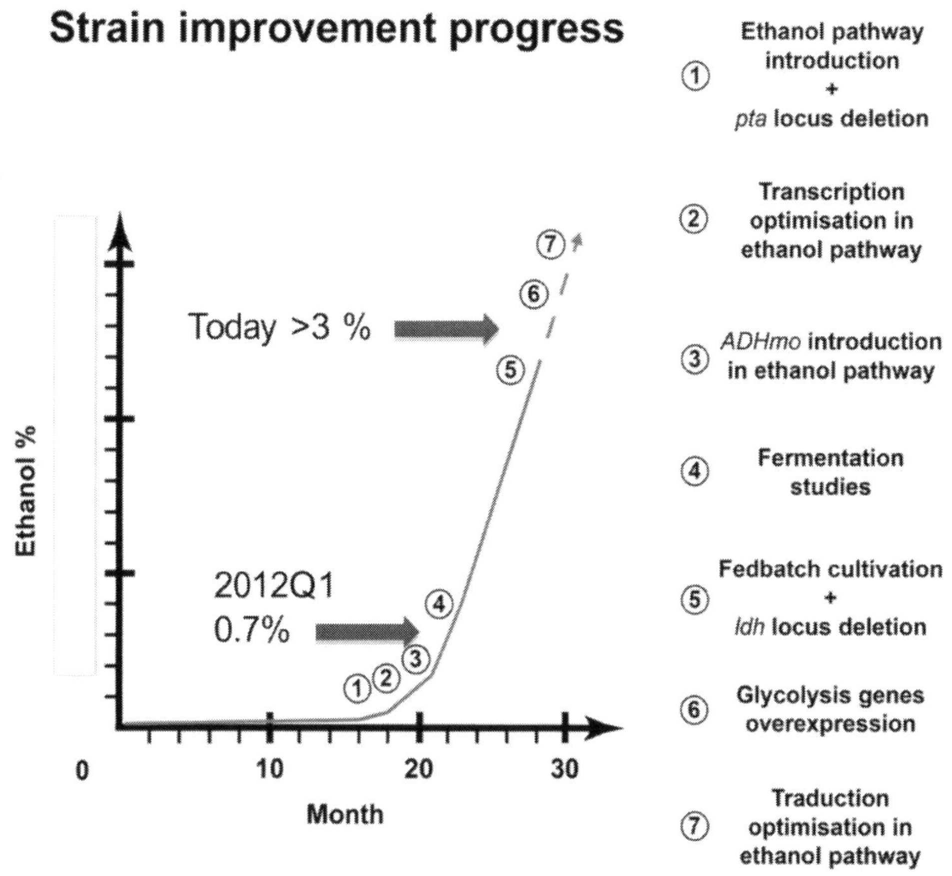

Deinove has already a partnership going on with one of the leading European sugar company, Tereos, on the use of wheat and wheat derivatives for ethanol production as Tereos runs the world biggest wheat-to-ethanol production facility. In the coming months, Deinove will be able to transfer its process to higher scale facilities where the development will be perform with its industrial partners Tereos

(626c) Biomass Particle Size Influences Sugar Conversion

Venkatesh Balan[1], Shuhaida Harun[2], Leonardo da costa Sousa[1], Seema Singh[3], Ming Jie Jin[1] and Bruce Dale[1], (1)Chemical Engineering and Material Science, Great Lakes Bioenergy Center, Michigan State University, Lansing, MI, (2)Department of Chemical Engineering, Universiti Kebangsaan Malaysia, Selangor, LA, Malaysia, (3)Joint BioEnergy Institute, Emeryville, CA

The lignocellulosic biomass (LCB) from crop residues contains about 65-75% of cellulose and hemicellulose that could be converted into fermentable sugars, a renewable source for producing biofuel. It is widely believed that when we reduce the particle size, the surface area increases which results in higher digestibility. We designed experiments to test the effect of particle size (2mm, 5mm, 2cm and 5cm) on rice straw conversion. We pretreated rice straw using two different AFEX™ pretreatment conditions C1 (100°C; 2:1 ammonia to biomass ratio; 80% moisture and 30 minutes residence time) and C2 (140°C; 1:1 ammonia to biomass ratio; 130% moisture and 50 minutes residence time). Subsequently, we hydrolysed the pretreated biomass at different glucan loading (1%, 3% and 6%) using commercial enzymes (Spezyme CP and Novozyme 188) and determined the sugar yield. AFEX™ C2 conditions made the treated rice straw substrate more susceptible to enzymatic hydrolysis than AFEX™ C1 conditions. Interestingly, the largest particle size of rice straw (5 cm) was found to be more digestible when compared to smaller particle sizes (2mm, 5mm and 2cm) under the given enzyme hydrolysis conditions. Viscosity measurements and imaging has been done to explain the phenomenon. To the best of our knowledge, this is the first time such a trend has been observed.

(626d) Efficient Bioconversion of Glucose/Xylose Mixtures for Ethanol Production Using a Novel Co-Culture System

Min Hea Kim[1], Meng Liang[1], Q. Peter He[2] and Jin Wang[1], (1)Chemical Engineering, Auburn University, Auburn, AL, (2)Chemical Engineering, Tuskegee University, Tuskegee, AL

Efficient bioconversion of glucose/xylose mixtures for ethanol production using a novel co-culture system

With the inevitable depletion of the world's conventional oil and gas supply, there has been an increasing worldwide interest in alternative sources of energy. In recent years, growing attention has been devoted to the conversion of lignocellulosic biomass into fuel ethanol, which offers great environmental and potential economic benefits over fossil fuel. The cellulosic biomass is cost-effective compared to low-cost petroleum; the cellulosic biomass is widely available at $40 per dry ton, and this price is equivalent to petroleum at $17.5 per barrel on both a mass and energy basis (Lynd et al., 1999). This potentially abundant lignocellulosic biomass contains fermentable sugars, mainly composed of glucose and xylose. It is worth noting that ethanol production from lignocellulosic hydrolysates in an economically feasible process requires complete utilization of both glucose and xylose. Specifically, the efficient utilization of xylose offers an opportunity to reduce the cost of producing ethanol by 25% (Aristidou and Penttila, 2000). The conventional yeast strain, *Saccharomyces cerevisiae*, is the most frequently used microorganism for ethanol production and it has both high ethanol tolerance and high fermentation yield and rate, but its inability to ferment xylose limits its use in ethanol production with lignocellulosic biomass. Although successful cycles of metabolic engineering have improved xylose utilization in recombinant *S. cerevisiae*, the ethanol production from xylose is still low (Madhavan et al., 2009; Bengtsson et al., 2009; Matsushika et al., 2008). The xylose-fermenting yeast, *Pichia stipitis* shows the highest native capacity for xylose fermentation of any known microbes and can convert both glucose and xylose into ethanol. However, the ethanol productivity of this strain is at least five to ten-folds lower than that obtained with *S. cerevisiae* during glucose fermentation (Alfenore et al., 2004; Agbogbo et al., 2006). Furthermore, *P. stipitis* shows a diauxic growth on the mixed substrates and has low ethanol tolerance, which inhibits its metabolism when ethanol concentration reaches above 30 g/L.

To achieve efficient utilization of glucose/xylose mixtures for ethanol production, most existing research has focused on developing a single recombinant strain which can ferment the mixed sugars derived from lignocelluloses to produce ethanol. In parallel with the ongoing development of the genetic recombinant strategy, co-culture strategy has been recognized as a promising and cost-effective way to co-ferment a mixture of glucose and xylose for ethanol production. However, efficient ethanol production by conventional co-culture systems faces several difficulties and the major ones are the competition for dissolved oxygen between the two microbes, the diauxic behavior of xylose-fermenting strain (i.e., they cannot use xylose in the presence of glucose) and their low ethanol tolerance (Gutierrez-River et al., 2012; Chandel et al., 2011; Li et al., 2011). In addition, very limited research has been done to investigate the dynamic properties of co-culture systems. Since the interactions between two microorganisms (glucose-fermenting and xylose-fermenting strains) are complex due to different assimilation pathways and regulatory mechanisms, it is important to develop new approaches to help understand and explore the dynamic behaviors of the co-culture system.

In this study, we designed and customized a novel bioreactor with an innovative fermentation scheme to offer a promising alternative of overcoming the difficulties of the existing co-culture systems. With this innovative fermentation scheme, the simultaneous complete utilization of both glucose and xylose was achieved. In addition, both diauxic phenomenon and oxygen deficiency of *P. stipitis*, were circumvented by providing the optimal dissolved oxygen required for efficient xylose fermentation. Also, the pseudo-continuous fermentation, i.e. continuous fermentation with cell retention, enhanced ethanol tolerance of both strains significantly by allowing the cells to adapt to a high ethanol concentration and environmental pressure. Specifically, the ethanol tolerance of *P. stipitis* (strain from pseudo-continuous fermentation) had increased 3.5-fold compared with unadapted strain of *P. stipitis* in the presence of 60g/L of ethanol. Furthermore, the dynamic properties of co-culture system were studied under pseudo-continuous fermentation to identify the optimal co-culture conditions by manipulating oxygen transfer rate ($mmolg^{-1}hr^{-1}$), substrate feed rate (ghr^{-1}), and substrate feed ratio (glucose/xylose ratio).

Reference

Agbogbo FK, Coward-Kelly G, Torry-Smith M, Wenger KS. (2006). Fermentation of glucose/xylose mixtures using *Pichia stipitis*. Process Biochem, 41, 2333-2336.

Alfenore S, Cameleyre X, Benbadis L, Bideaux, C, Uribelarrea JL, Goma G, Molina-Jouve C, Guillouet SE. (2004). Aeration strategy: a need for very high ethanol performance in *Saccharomyces cerevisiae* fed-batch process. Appl Microbiol Biotechnol, 63, 537-542.

Aristidou A, Penttila M. (2000). Metabolic engineering applications to renewable resource utilization, Curr Opin Biotechnol, 11, 187-198.

Bengttson O, Hahn-Hagerdal B, Gorwa-Grauslund ME. (2009). Xylose reductase from *Pichia stipitis* with altered coenzyme preference improves ethanolic xylose fermentation by recombinant *Saccharomyces cerevisiae*. Biotechnol Biofuels, 2, 9.

Chandel AK, Singh OV, Narasy ML, and Rao V. (2011). Bioconversion of *Saccharum spontaneum* (wild sugarcane) hemicellulosic hydrolysate into ethanol by mono and co-cultures of *Pichia stipitis* NCIM3498 and thermotolerant Saccharomyces cerevisiae-VS_3, New Biotechnol, 28, 593-599

Gutierrez-Rivera B, Waliszewski-Kubiak K, Carvajal-Zarrabal O, and Aguilar-Uscanba MG. (2012). Conversion efficiency of glucose/xylose mixtures for ethanol production using *Saccharomyces cerevisiae* ITV01 and *Pichia stipitis* NRRL Y-7124. J Chem Technol Biotechnol, 87, 263-270.

Li Y, Park J, Shiroma R, and Tokuyasu K. (2011). Bioethanol production from rice straw by a sequential use of *Saccharomyces cerevisiae* and *Pichia stipitis* with heat inactivation of *Saccharomyces cerevisiae* prior to xylose fermentation, J Biosci Bioeng, 111, 682-686.

Lynd LR, Wyman CE, and Gerngross TU. (1999). Biocommodity Engineering. Biotechnol Prog, 15, 777-793.

Madhavan A, Tamalampudi S, Srivastava A, Fukuda H, Bisaria VS, and Kondo A. (2009). Alcholic fermentation of xylose and mixed sugars using recombinant Saccharomyces cerevisiae engineered for xylose utilization. Appl Microbiol Biotechnol, 82, 1037-1047

Matasushika A and Sawayama S. (2008). Efficient bioethanol production from xylose by recombinant *Saccharomyces cerevisiae* requires high activity of xylose reductase and moderate xylulokinase activity. J Biosci Bioeng, 106, 306-309.

(626e) Upgrading Biomass Derived Lignin for Hydrocarbon Production

Dhrubojyoti D. Laskar, Center for Bioproducts & Bioenergy and Biosystems Engimeering, Washington State University, Richland, WA, Melvin Tucker, National Biology Center, National Renewable Energy Lab, Golden, CO and **Bin Yang**, Center for Bioproducts & Bioenergy and Biosystems Engimeering, Washington State University , Richland, WA

The current vision of biofuels production undervalues lignin's potential to address the nation's high quality liquid fuel requirements. The utilization of waste lignin as a feedstock for conversions to hydrocarbons offers a significant opportunity for enhancing the overall operational efficiency, carbon conversion rate, economic viability, and sustainability of biofuels production. Despite the potential, the conversion of lignin to biofuels has proven to be challenging. Several solubilized and depolymerized lignin fractions with a range of structural features were separated from cellulosic biomass as intermediates for further conversion to hydrocarbons. The detailed release and fate of lignin and lignin mono/oligomers during hydrothermal pretreatment with or without additives (e.g. catalysts) was characterized under various conditions (e.g., temperature, time, flow rate etc.). Various advanced analytical techniques, such as 1D and 2D NMR, FTIR, GC/MS, and GPC, were applied to derive the associated chemical changes in lignin, and lignin-carbohydrate complexes during the various pretreatment and depolymerization processes. The strategies for developing catalysts that can both selectively depolymerize the lignin framework and remove oxygen by hydrodeoxygenation will be addressed in the presentation.

(626f) A New Energy Densification Approach to Convert Lipids to Hydrocarbons

Jonathan Dubinsky, Department of Civil Engineering, University of Coloraod Denver, Denver, CO and **Arunprakash T. Karunanithi**, Department of Civil Engineering, University of Colorado Denver, Denver, CO

We present a thermo chemical process to convert lipids to renewable diesel fuel. Several thermo-chemical processes, such as hydrodeoxygenation (HDO), have been previously proposed for energy densification of fuels from bio feedstocks. However, HDO is an energy intensive process and it requires high pressure hydrogen stream. The use of hydrogen gas in the conversion process has two major drawbacks 1) typical hydrogen production processes are based on fossil fuel and hence are non-renewable 2) significant safety issues related to storage and transportation of hydrogen gas. We present a decarboxylation based approach that does not require the use of gaseous hydrogen and hence is more sustainable, safe and cheap as compared to the current-state of the art thermo-chemical-catalytic processes. We show that this process produces hydrocarbons which can be further refined to produce renewable diesel fuel. A comparative life-cycle assessment of renewable diesel produced from a commercial renewable diesel process and our proposed approach will also be presented.

(665a) Sustainable Design and Operations At the David L. Lawrence Convention Center

Angelica Ciranni, Sports & Exhibition Authority of Pittsburgh and Allegheny County, Pittsburgh, PA

In 2003, the David L. Lawrence Convention Center (DLCC) received a Gold rating under the LEED for New Construction rating system, making it the first convention center to receive LEED certification. At that time, the DLCC was also the world's largest green building at 1.5 million square feet. From the early stages of design through current operations, the DLCC has remained committed to sustainability. In order to demonstrate the sustainable operations of the facility, the DLCC registered under the LEED for Existing Buildings: Operations & Maintenance (O&M) rating system and was certified Platinum in April 2012. The DLCC is the first convention center to receive two LEED ratings.

Concurrent with the LEED for Existing Buildings: O&M documentation process, the DLCC underwent a "building in operation" study that examined building performance, facility management, occupant satisfaction, and organizational sustainability. A thorough set of analyses contributed to the findings, including quantitative and qualitative assessments of energy, natural ventilation, water, site, transportation, waste, purchasing, commissioning, occupant comfort, and indoor air quality.

This session will provide an overview of the case study findings and the LEED for Existing Buildings: O&M process for the host facility of the AIChE Annual Meeting, including specific strategies and lessons learned.

Variations in Temperature in Mini-Roof Structures Employing Different Roofing Materials

Jason T. Kirby[1], Ph.D., Ronald D. Sherrod[2], Matt Winslett[3], and Robert W. Peters[4*], Ph.D., P.E.

[1]Associate Professor of Environmental Engineering
Department of Civil, Construction and Environmental Engineering
University of Alabama at Birmingham
1075 13th Street South
Birmingham, Alabama 35294-4440
Phone: (205)-934-8434; e-mail: rwpeters@uab.edu

[2]Undergraduate Student
Department of Chemical Engineering
Worcester Polytechnic Institute
100 Institute Road
Worcester, Massachusetts 01609-2280

[3]Engineering Manager, Energy Management
Facilities Management Department
University of Alabama at Birmingham
822 Facilities Support Building
Birmingham, Alabama 35294-4564

[4*]Professor of Environmental Engineering
Department of Civil, Construction and Environmental Engineering
University of Alabama at Birmingham
1075 13th Street South
Birmingham, Alabama 35294-4440
Phone: (205)-934-8434; e-mail: rwpeters@uab.edu

ABSTRACT

The overall goal of green roof research at the University of Alabama at Birmingham (UAB) campus is to identify a suitable roofing material combination that reduces energy cooling costs. Since 2008, ambient temperature data has been recorded to better understand how the roof heats and cools on a diurnal, seasonal, and yearly basis. During this study, 15 mini-roof combinations were observed for trends in internal temperatures. Conventional roofing systems included river rock, crushed marble chips, black roofs, reflective (white) roofs, concrete pavers coated with epoxy, and an alternative roofing system involving vegetated roofs (planted with sedum plants). To investigate the thermal properties of the roofing structures, an ambient temperature probe was placed inside of each roof to record data every 10 minutes.

The University of Alabama at Birmingham (UAB) is located in the Southeastern United States. This region is characterized by its hot summers and relatively mild winters. Due to these hot summers during which the temperatures can approach 37.8°C (100°F), along with conditions of high humidity, it is crucial to have a roof which minimizes cooling costs. These hot summers also create an urban heat island effect in the Birmingham area due to the current roofing structures and facility usage. An urban heat island is defined on the basis of a temperature difference between urban and rural area, caused by human modifications to the natural geographic landscape. The human modifications observed here are the buildings and roads added to the natural habitat. These structures retain the heat transferred from the sunlight increasing the temperature. This property results in an increased temperature in the highly modified areas (urban downtown areas) versus a less modified area (rural areas).

Various roofing materials are being studied for their thermal efficiency properties; roofing materials studied include black roofs, white reflective roofs, river rock, crushed marble chips, cement blocks, and vegetated roofs [having sedum plants (*Sedum hispanicum*)]. The temperature behavior both on a daily basis and on an annual basis vary in a sinusoidal fashion. Additionally, a pilot vegetated roof was installed on top of Hulsey Center in July 2008. This pilot green roof occupies ~1388.0 m ^{2} (14,940 ft ^{2}) and contains ~20,000 sedum type plants. Utility bill information has been collected on natural gas usage, water usage, and electricity usage for ~3 years prior to the installation of the vegetative roof, and for more than a year since installation of the green roof. Reductions in the usage of natural gas, water, and electricity has been observed and documented since installation of the green roof on top of Hulsey Center.

Historically, studies on green roofs have explored their energy performance compared with traditional roofs. Thermal performance indicated a significant reduction (~40%) of a building cooling load during the summer period [Spala *et al.*, 2008]. Similar results were achieved for a nursery school, with reductions ranging from 6% to 49%, and reduction ranging from 12% to 87% on the last floor of the nursery school [Santamouris *et al.*, 2007]. Wong *et al.* [2007] note that green roofs tend to experience lower surface temperatures than the original exposed roof, especially in areas well covered by vegetation. When green roofs are well covered by vegetation, the resulting substrate moisture will tend to keep substrate temperature lower than the original exposed bare roof. These studies determined that over 60% of the heat gain was mitigated by vegetated roof systems. Summertime data have indicated significant lower peak roof surface temperature and higher nighttime surface temperature for green roofs as compared to conventional roofs [Sonne, 2006].

After acquiring the necessary raw data from the mini-roof sensors, we noticed that the temperatures cycled in a sinusoidal fashion on yearly and daily time frames, so using the general form sine function was an appropriate model to describe the data. We determined the frequency and amplitude for all fifteen mini-roof systems. The determined functions supported that most roofs are statistically different from one another from an amplitude aspect but the phase angles are statistically the same. We also have discovered that most roofs are significantly different from the standpoint of the average mean roof temperature, but the significance is most prevalent in the summer months and during the other times of the year the roofs behave in a more similar fashion to one another. The model does well in describing the average daily temperature during the year.

Key Words: Roofing materials, temperatures of roofing structures, thermal properties
*Corresponding author

Introduction/Background

Green building design, as defined by the United States Environmental Protection Agency (USEPA), is "the practice of creating structures and using processes that are environmentally responsible and resource-efficient throughout a building's life-cycle from design, to construction, operation, maintenance, renovation and ultimately deconstruction" [EPA, 2010]. However, according to the United States Green Building Council (USGBC), commercial buildings in the US alone account for:

- 65% of electricity consumption,
- 30% of greenhouse gas emissions,
- 30% of raw materials use,
- 30% of waste output (136 million tons annually), and
- 12% of potable water consumption [USGBC, 2010].

The building infrastructure (Residential, Commercial, Institutional, and Industrial) in the United States (US) consumes over two-thirds of the nation's electricity demand and accounts for one-third of all domestic energy consumption [EPA, 2009]. Regrettably, conventional forms of energy production have an adverse impact on natural ecosystems. Collectively, our buildings contribute to 38.9% of the nation's total greenhouse gas emissions. Faced with rising energy costs, diminishing fuel resources and emerging environmental concerns, scientific research has begun to address these challenges by adopting sustainable or green building alternatives.

Common Roofing Systems

The choice and type of urban roofing system is dependent on environmental concerns such as sun exposure and meteorological factors such as temperature, wind and rain. Some roofs may have covenants to determine their usage. Different systems have been developed and designed to perform at its most effective given the exposure conditions of the building location. As a result, many roofing systems exist and are commonly used by the construction community.

There are many roofing systems used today that depend on the building type, whether is for residential or for commercial applications. The most common systems are listed as follows [Reid, 2000].

- Built-up,
- Preformed metal, and
- Shingles.

A brief description of the three roofing system is provided involving advantages, disadvantages and durability characteristics of each.

Built-Up

Built-up roofing systems are most common for flat roof applications. They consist of several layers of material built-up from the internal structural frame of the building. They are most commonly made of alternating layers of plywood, felt, asphalt, tar and gravel [Reid, 2000]. The major advantage of this system is its waterproofing characteristics, and out of all roofing systems, built-up roofs are considered to have the highest degree of waterproofing. However, the system has poor fire resistance, generally requires a professional to install, and is difficult to locate roof leaks. The system typically lasts for 22.5 years [Cash, 1999; Editors of Time Life Books, 1979].

Preformed Metal

Another common roofing system is made of preformed metal panels. The panels are generally made from aluminum, steel and copper [Reid, 2000]. They are most commonly found with contemporary designs. They come in flat, ridged, ribbed or corrugated forms. These systems are easy to install and repair, and can be painted any color. They are considered fire resistant. The panels are easily damaged by wind, falling trees and tree limbs, or any other type of contact. They generally last for 25 years [Cash, 1999; Editors of Time Life Books, 1979].

Shingles

Shingle roofs come in many forms. Shingles are commonly made of asphalt, slate, wood and clay tiles [Reid, 2000]. Asphalt shingles are the most popular, especially for residential construction. They are available in a variety of sizes, weights, and colors. They require little maintenance and are easy to install. However, they are considered to have poor fire resistive qualities. Asphalt shingles generally last for 16–17 years with proper maintenance [Cash, 1999]. Slate and ceramic tiles are highly expensive due to their aesthetic and durability properties. They are fire resistant and have attractive appearances. Yet, they are extremely heavy and require strong structural support. They are also very brittle, and require increased amount of time to install which often entails specialized tools. Repairs are in most cases difficult. Slate and ceramic tiles are considered the most durable, lasting for 20–100 years with proper maintenance. Wood shingles and shakes are also attractive options. They are easy to install and are a natural insulator. However, they are highly flammable and require treatments for weather and insect protections. They generally last for 10 to 20 years with proper maintenance [Cash, 1999; Editors of Time Life Books, 1979].

Urban Heat–Island Reduction and Building Energy Conservation

For millions of Americans living in and around cities, heat islands are of growing concern. This phenomenon describes urban and suburban temperatures that are 2 to 10°F (1 to 6°C) hotter than nearby rural areas. Elevated temperatures can impact communities by increasing peak energy demand, air conditioning costs, air pollution levels, and heat-related illness and mortality. The Environmental Protection Agency (EPA) recommends installing cool or vegetative green roofs, planting trees and vegetation and switching to cool paving materials as a way of reducing the negative effects of urban heat islands. The EPA says green roofs, if installed widely in a city, can contribute to heat island reduction by replacing heat-absorbing surfaces with plants, shrubs, and small trees. The vegetation cools the air through evapotranspiration (or evaporation of water from leaves). Planted rooftops remain significantly cooler than a rooftop constructed from traditional heat-absorbing materials. Further, green roofs reduce summertime air conditioning demand by lowering heat gain to the building.

Energy Modeling

Energy modeling (i.e., energy simulation) is a method for predicting the energy consumption of an occupied structure. Building energy analysis must consider numerous thermal characteristics including: wall and roof materials, the size and orientation of the building, how the building is occupied and operated, as well as influences from the local climate.

The surface temperature of a roof exposed to solar radiation, the resulting heat flow into the building, along with associated indoor temperatures and cooling needs depend on the effect of solar radiation, surface absorptivity, ambient air temperature and wind speeds adjacent to the surface [Givoni, 1994]. When vegetative roofs are considered, because of added thermal mass, it is also important to take into account hourly heat transfer when determining energy consumption; as the heat flux through a vegetated roof can be quite different from conventional roofing.

R-values and *U*-values have been used for many years as a measurement of a building envelope's thermal performance. However, these attributes do not fully take into account the effects of thermal mass, and by themselves, are inadequate in describing the heat transfer properties of construction assemblies with significant amounts of thermal mass [Kosny *et al.*, 2007]. Vegetated roofs are more dependent on the interaction between the roofing systems' unit weight, density, thermal conductivity, moisture content, vegetal coverage and specific heat. Therefore, it is often necessary to utilize computer software, which incorporates these elements into the analysis of high thermal mass roof structures and associated energy consumption. The steady-state *R*-values traditionally used to measure energy performance will not accurately capture the complex, dynamic thermal behavior of vegetated roof systems.

Vegetated (Green) Roof Systems

Green roofs are engineered ecosystems that rely on vegetation to provide benefits such as reduction of roof temperatures and stormwater retention [Wolf and Lundholm, 2008]. Green roofs offer benefits of reducing stormwater runoff, improving air and water quality, and providing habitat and biodiversity for urban centers [Wong *et al.*, 2003][0]. Hydrologic modeling has demonstrated that widespread green roof implementation can significantly reduce peak runoff rates, particularly for small storm events [Carter and Jackson, 2007]. By combining the Green-Ampt method with evapotranspiration of green roofs, Roehr and Kong [2010] estimated the potential runoff reduction achieved by green roofs is 20%. Green roofs provide an excellent option to improve stormwater runoff [Hathaway *et al.*, 2008]. Green roofs are primarily valued based on their increased roof longevity, reduced stormwater runoff, and decreased building energy consumption [Clark *et al.*, 2008]. Carter and Jackson [2007] noted that research studies have primarily been focused on roof-scale processes such as individual roof stormwater retention, plant growth, or growing medium composition. Few studies have examined the impact that widespread green roof application could have on the hydrology of a real-world watershed [Carter and Jackson, 2007]. A major barrier to increasing the prevalence of green roofs is the lack of scientific data available to evaluate their applicability to local conditions [Williams *et al.*, 2010].

Green roofs are typically classified as being either an intensive or extensive roof [Green Roof Plan, undated]. Intensive green roofs are often used on commercial buildings in order to have large green areas that incorporate all sizes and types of plants. These roofs use grasses, ground covers, flowers, shrubs and even trees. They often include paths and walkways that travel between different architectural features to provide space where people can interact with the natural surroundings. Intensive green roofs, sometimes termed "rooftop gardens", utilize planting mediums that have greater depth than extensive green roofs; the deeper soil allows intensive roofs to accommodate large plants and various plant groupings. Intensive green roofs require more maintenance than extensive green roofs because of the plant varieties they will support.

Extensive green roofs have a planting medium that ranges from 1.6 to 6 inches deep. Typically, drought-tolerant sedums (succulent plants) and grasses are used since they are shallow-rooted and use little water. Plant diversity on these roofs is kept low to simplify care and to be sure all plants have similar moisture requirements.

Extensive green roofs can significantly reduce both the timing and magnitude of stormwater runoff relative to a typical impervious roof [Schroll *et al.*, 2011]. They note, however, that regional climatic conditions such as seasonality in rainfall and potential evapotranspiration can strongly alter the stormwater performance of vegetated roofs. Factors such as type of green roof and its geometrical properties (slope), soil moisture characteristics, season, weather and rainfall characteristics, age of the vegetated roof, and vegetation affect the runoff dynamics from green roofs [Berndtsson, 2010]. Fioretti *et al.* [2010] noted that green roofs significantly mitigate storm water runoff generation, as well reducing the daily energy demand. Aitkenhead-Peterson *et al.* [2011] note that most studies on runoff quality from green roofs have been conducted in cooler northern climates. Villarreal and Bengtsson [2004] recommended the use of a combination of best management practices; additionally, they observed that green roofs are effective at lowering the total runoff from Augestenborg (Sweden) and that detention ponds should successfully attenuate storm peal flows. Niu *et al.* [2010] noted that over the lifetime of a green roof (~40 years), the net present value is ~30% to 40% less for a green roofs as compared with conventional roofs (not including green roof maintenance costs). Kirby *et al.* [2011] note that extensive vegetated roof systems offer at least 1 16% enhancement in reducing stormwater runoff as compared to conventional roofs. Clark *et al.* [2008] further note that the additional upfront investment of a green roof is recovered at the time when a conventional roof would be replaced. Rosatto *et al.* [2010] concluded that green roofs contribute positively in reducing runoff, with greater retention with vegetated plots and thicker substrate.

Vegetated roof systems have a number of advantages over that of conventional roof systems. Benefits associated with green roof systems include [Green Roofs for Healthy Cities, undated]:

- Urban greening has long been promoted as an easy and effective strategy for beautifying the built environment and increasing investment opportunity.
- With green roofs, water is stored by the substrate and then taken up by the plants from where it is returned to the atmosphere through transpiration and evaporation.
- Depending on the plants and depth of growing medium, during the summer, green roofs retain 70% to 90% of the precipitation that falls on them; in winter they retain between 25% to 40%.

- Green roofs not only retain stormwater, but also moderate the temperature of the water and act as natural filters for any of the water that runs off.
- Green roofs reduce the amount of stormwater runoff and delay the time at which runoff occurs, resulting in decreased stress on sewer systems at peak flow periods.
- Through the daily dew and evaporation cycle, plants on vertical and horizontal surfaces can cool cities during hot summer months and reduce the Urban Heat Island (UHI) effect. The UHI is also mitigated by the covering some of the hottest surfaces in the urban environment, such as black rooftops.
- Green roofs can also help reduce the amount of dust and particulate matter throughout the city, as well as the production of smog. This plays a role in reducing greenhouse gas emissions and adapting urban areas to a future climate with warmer summers.
- Green roofs help to achieve the principles of smart growth and positively affect the urban environment by increasing amenity and green space.
- The greater insulation offered by green roofs can reduce the amount of energy needed to moderate the temperature of a building, as roofs provide the greatest heat loss in the winter and the hottest temperatures in the summer.
- The presence of a green roof decreases the exposure of waterproofing membranes to large temperature fluctuations, which can cause micro-tearing, and ultraviolet radiation.
- Green roofs have excellent noise attenuation, especially for low frequency sounds. An extensive green roof can reduce sound from outside by 40 decibels, while an intensive one can reduce sound by 46-50 decibels.
- Green roofs can sustain a variety of plants and invertebrates, and provide a habitat for various bird species.

Historically, studies on green roofs have explored their energy performance compared with traditional roofs. Thermal performance indicated a significant reduction (~40%) of a building cooling load during the summer period [Spala *et al.*, 2008]. Similar results were achieved for a nursery school, with reductions ranging from 6% to 49%, and reduction ranging from 12% to 87% on the last floor of the nursery school [Santamouris *et al.*, 2007]. Wong *et al.* [2007] note that green roofs tend to experience lower surface temperatures than the original exposed roof, especially in areas well covered by vegetation. When green roofs are well covered by vegetation, the resulting substrate moisture will tend to keep substrate temperature lower than the original exposed bare roof. These studies determined that over 60% of the heat gain was mitigated by vegetated roof systems. Summertime data have indicated significant lower peak roof surface temperature and higher nighttime surface temperature for green roofs as compared to conventional roofs [Sonne, 2006]. The maximum average daily temperature seen for the conventional roof surface was 54.4°C (129.9°F) in his study, while the maximum average day green roof surface temperature was 32.8°C (~21.7°C lower than the conventional roof). Green roofs offer cooling potential (~3.02 kWh/day) to maintain an average room air temperature of 25.7°C (78.3°F) [Kumar, 2005]. Green roofs help minimize environmental burdens, conserve energy, and extend the life span of the roofing system in overall sustainability [Cavanaugh, 2008]. Up to 30% of total rooftop cooling is due to plant transpiration [Takakura *et al.*, 2000]. Bell and Spolek [2009] compared different types of plants for use in increasing the thermal resistance (*R*-value) of green roofs, and found that ryegrass delivered the highest effective *R*-value compared with bare soil, *Vinca major, Trifolium repens,* and *Sedum hispanicum.* Also, though increasing the depth of bare soil from 5 to 14 cm (2.0 to 5.5 inches) increased the *R*-value, no difference was found for different depths of planted soil. This implies that the bulk of benefit toward *R*-value is from evapotranspiration and leaf shading, rather than the moist soil [Bell and Spolek, 2009].

There are several detailed building simulation programs (BSPs) that take into consideration the complete interaction between all thermal-based elements. The most popular BSPs are A Simplified Energy Analysis Method (ASEAM), Building Design Advisor (BDA), Building Load Analysis and

Systems Thermodynamics (BLAST), Builder Guide, Bus^{++}, Dynamic Energy Response of Buildings (DEROB), DOE-2, Energy-10, Energy Plus, ENERPASS, ENER-Win, ESP, FEDs, Home Energy Saver, Hot 2000, TRNSYS, and VisualDOE [Selim, 2003; Selim, 2008; Mansour, 2001].

UAB has utilized Visual DOE in the past with great success in the analysis of innovative structures designed for energy efficiency. VisualDOE uses the DOE 2 calculating core and provides output in both numerical and graphical forms. This software is a preferred calculation method due to its cost, previous verification/validation success, ease of use, database support and reasonable input/output requirements. We envision that this computer simulation tool will be able to effective capture the differences in roof types being explored in the purposed research.

Results and Discussion

Thermal Performance of Mini-Roof Structures

Mini-Roofs

During this study, 15 mini-roof combinations were observed for trends in internal temperatures. The various 15 mini-roof combinations are summarized in Table 1. Several of the mini-roof structures are depicted in Figure 1. This photo shows the layout of the 15 mini-roofs, and a vegetated roof from which surface temperatures of the mini-roofs were measured periodically using an infrared thermometer (see Figure 1).

(a) (b)

Figure 1. (a) Layout of the 15 mini-roods;
(b) vegetated mini-roof (surface temperatures were measured using an IR thermometer.

The roofing materials used are all standard commercial flat roof materials. Flat roof materials were only looked at during the study, since the primary application for the roofing combinations will be on a commercial flat roof top, and not a slanted roof structure. Each mini-roof is 2.4-m (8.0-ft) long x 1.2-m (4.0-ft) wide x 1.2-m (4.0-ft) deep (see Figure 1). A number of different roofing systems are being examined for their energy performance. All roofs are insulated with 5.1 cm (2.0-in) of extruded polystyrene. Then the particular roofing combination being investigated is applied over the insulation and sealed. The roofs also include a proper drainage spout, to ensure correct water evacuation, such as on a real roof.

The roofing systems being studied using the 15 mini-roofs are listed in Table 1.

Table 1. Mini-roof descriptions.

Mini-Roof No.	Mini-Roof Description
1 (Sensor B)	White TPO/PVC/ Elvaloy fully adhered, FiberTite Membrane.
2 (Sensor C)	Black 60-mil EPDM fully adhered/coated/white urethane, Mule Hide Membrane.
3 (Sensor D)	Black 60-mil EPDM fully adhered, Mule Hide Membrane.
4 (Sensor E)	Beige TPO/PVC/Elvaloy fully adhered, FiberTite Membrane.
5 (Sensor F)	White granular modified, Firestone SBS Modified Membrane.
6 (Sensor G)	Black granular modified, Firestone SBS Modified Membrane.
7 (Sensor H)	Black granular modified coated/white urethane, Firestone SBS Modified Membrane.
8 (Sensor I)	Bituthene IRMA with lightweight "T Clear" pavers.
9 (Sensor J)	Bituthene IRMA with river rock ballast.
10 (Sensor K)	Bituthene IRMA with vegetative green roof, ½-in. drain mat, 350-lbs dry soil.
11 (Sensor L)	Bituthene IRMA with vegetative green roof, 1-in. drain mat, 350-lbs dry soil.
12 (Sensor M)	Black 60-mil EPDM loose, ballasted with river rock, Mule Hide Membrane.
13 (Sensor N)	Black 60-mil EPDM loose, ballasted with #300 marble chips, Mule Hide Membrane.
14 (Sensor O)	White TPO/PVC/Elvaroy loose laid, ballasted with river rock, FiberTite Membrane.
15 (Sensor P)	Bituthene IRMA with vegetative green roof, ½-in. drain mat, 350-lbs dry soil.
Sensor S	Sensor inside mini-roof No.10 (inside the soil of the green roof).
Sensor T	Sensor under the white TPO/PVC/Elvaloy loose laid, ballasted with river rock.

To investigate the thermal properties of the roofing structures an ambient temperature probe was placed inside of each roof (see Figures 2 and 3) recording temperature data every 10 minutes of each day, for more than 3 years. This data was then automatically sent to a data logger and placed into an Excel file for review later. The temperature probe reports the temperature to the nearest hundredth of a degree Centigrade.

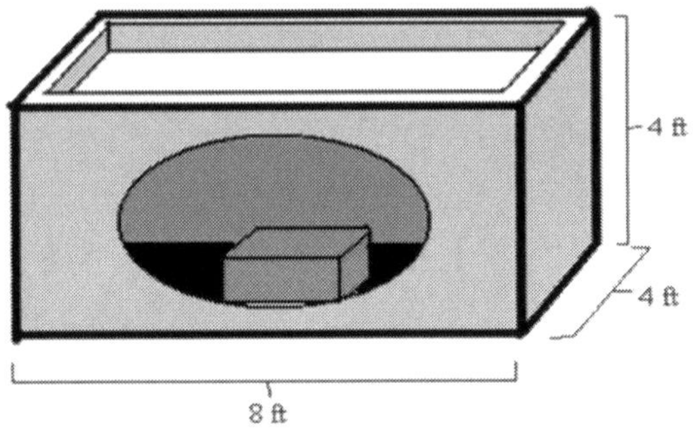

Figure 2. Depiction of a Typical Mini-Roof System.

Figure 3. Mini-roof system showing temperature sensor installed inside a mini-roof.

Figure 4 presents some typical temperature profiles on several different days.

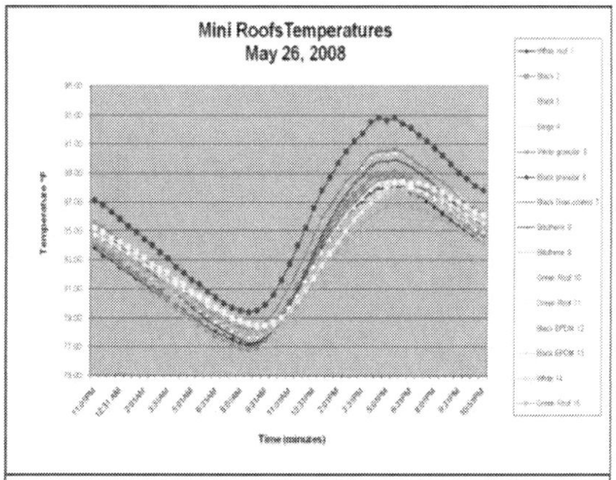

Figure 4a. Temperature profile inside the various mini-roofs on May 26, 2008.

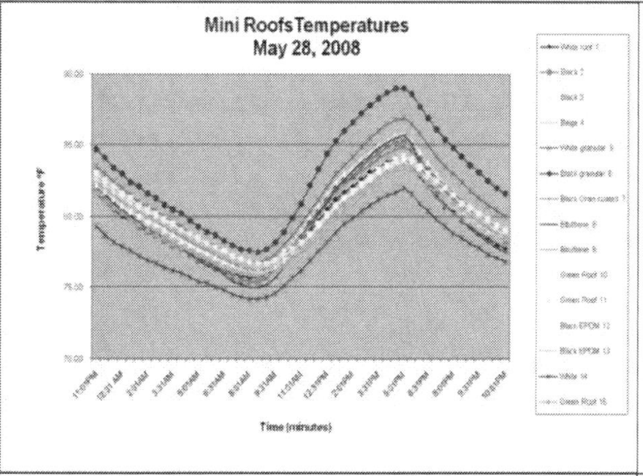

Figure 4b. Temperature profile inside the various mini-roofs on May 28, 2008[Nunez and Peters, 2008].

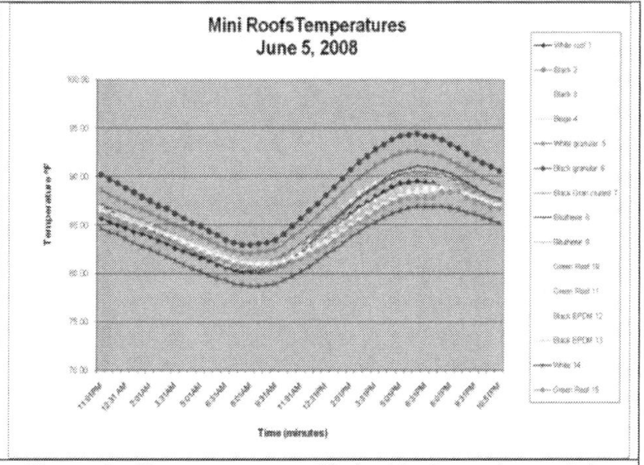

Figure 4c. Temperature profile inside the various mini-roofs on June 5, 2008.

The results from these mini-roof structures have shown the following trends [Nunez and Peters, 2008]:

- Clean white roofs resulted in consistently lower temperatures inside the mini roof than the other roofing materials.
- Black roofs resulted in the highest temperature readings.
- Green roofs resulted in temperatures typically ~1.1-1.7°C (2-3°F) higher than the white roofs; however, they will dampen the drainage of rainfall during a rain storm through the retention of water onto the soil.
- Bituthane (river rock) performs only slightly better than black roofing materials.
- White granular roofing behaved similarly to black granular roofing materials.
- Over time, the reflective (white) roofs become dirty, losing some of their reflectivity, resulting in the roof being less energy efficient.
- White marble chips behaved slightly worse than green roofing materials, but considerably better than black roofing and granular roofing materials.
- While it is too early to come to a definitive conclusion, preliminary evaluations indicate that the "white" and "green" roofs both significantly reduce the roofs surface temperature and therefore, the air temperatures above and around the roof.

Surface temperature measurements during the months of June and July on the various roofing materials used with our mini-roof systems were collected. During this time period, roofing surface temperatures ranging from 20.6°F to 82.2°C were observed. The lower surface temperature values, 20.6°C to 48.3°C, are found in a loose rock or stone roof combination. The higher temperature values occurred on the roofing combinations primarily made of a coating or membrane; these roofs exhibited temperatures roughly between 25.6°C to 82.2°C. The vegetative mini-roofs exhibited surface temperatures ranging between 21.7°C to 52.2°C, while their large counterpart, the pilot roof on top of Hulsey Center, exhibited higher temperatures ranging from 31.7°C to 61.1°C. When looking at the specific roofs contained in each subset of Table 1, it was observed that generally, the lighter colorization of the roof resulted in cooler temperatures. For example, the white Firestone SBS is cooler than the black Firestone SBS on any given day. This thermal property is observed due to the reflectivity of the roof. The darker colored roofs absorb more incoming light radiation than the light colored roofs causing the dark roofs to become hotter. (Since the roofs temperatures observed are taken during the late spring and early summer, it can be inferred that overall roof temperature will increase during late summer).

Photographs of several of these mini-roofs are presented in Figures 5 through 9. Figure 10 presents a typical temperature profile of the 15 mini-roofs during the course of a typical summer week. Series 1 through 5 denoted in the figure refer to the fifteen mini-roofs listed in Table 1. This figure shows a cyclical nature of the temperature readings over each day, generally showing a sinusoidal behavior of temperature; the temperature is cool in the morning, warms up, and is at its hottest during mid-afternoon, and then cools down during evening hours.

Trends seen in statistical comparisons of internal temperatures of similar roofs (using the null hypotheses: $\mu_1 \geq \mu_2$) are summarized below [Peters *et al.*, 2011]:

- Between the two river rock roofs, Roof 9 will most likely always be hotter or equal to roof 12.
- Between the 3 vegetative roofs, all the roofs are statistically equal to each other in thermal properties.
- Both SBS Firestone roofs are statistically the same, but roof 6 is usually hotter.
- The TPO/PVC/Elvaloy Roofs are statistically the same.
- The 60-mil EPDM roofs are also statistically the same.

Figure 5. Black mini-roof.

Figure 6. White (reflective) mini-roof.

Figure 7. Mini-roof equipped with river rocks.

Figure 8. Mini-roof equipped with crushed marble.

Figure 9. Vegetative mini-roof equipped with *Sedum* plants.

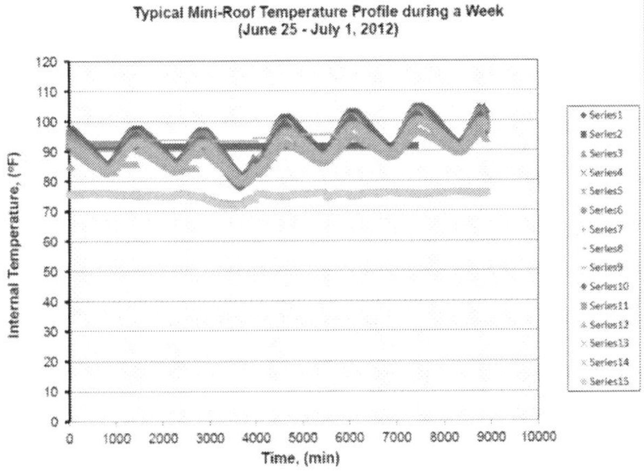

Figure 10. Typical internal mini-roof temperatures during the course of a week.

After acquiring the necessary raw data from the mini-roof sensors, Microsoft Excel (version 2010) was used to plot the data to discover trends in the temperature readings. Subsequent to discovering the trend, a mathematical model was fit to the data. It was noticed that the temperatures cycled in a sinusoidal fashion on yearly and daily time frames, and therefore a general form sine function was utilized as a potential model. Fourier transforms were utilized in order to determine the oscillation frequency of the temperature (ω_0 from the general form $x = A \sin(\omega_0 t + \phi) + C$) by transforming the time domain of the collected raw data into a frequency domain. Two major peaks were discovered from the spectrum: one representing the yearly frequency and the other representing the daily oscillations. After calculating the frequencies, the phase angle (ϕ) and the amplitude (A) of the general form were determined through a regression analysis.

The modeling procedure was applied for all 15 mini-roofs in the study. The developed sine-wave functions indicated that most roofs were statistically different from one another from an amplitude aspect but the phase angles were statistically the same. It was also discovered that almost all roofs had significantly different average mean roof temperatures, but the significance was mostly prevalent in the summer months. During other times of the year, the roofs behaved in a similar fashion (see Tables 2 and 3 for statistics and means of roofs). The fitted sine wave functionalities for the 15 mini-roofs are listed in Table 4.

Table 2.	**Summary of statistics for the phase angles from curve fitting.**	**Table 3.**	**Summary of statistics for the means from curve fitting.**
Statistics of Phase Angles		Statistics of Means	
Mean	4.0303	Mean	66.67
Standard Error	0.00343	Stand Error	0.18
Median	4.0276	Median	66.62
Mode	4.0242	Mode	N/A
Standard Deviation	0.01329	Standard Deviation	0.68
Sample Variance	0.000177	Sample Variance	0.46
Kurtosis	12.0174	Kurtosis	3.11
Skewness	3.3228	Skewness	1.11
Range	0.05512	Range	3.06
Minimum	4.0211	Minimum	65.42
Maximum	4.0762	Maximum	68.47
Sum	60.4541	Sum	1000.12
Number (count)	15	Number (count)	15

Table 4. Fitted sine-wave functionalities describing the internal temperatures in the mini-roofs.

Roof	**Final Equation**	**Amplitude from sinefind.exe**	**Phase Angle**	r^2	**Mean**
1	T = 20.1632*sin ((2*π*3.17894e(-8)*t) +4.021975)+66.68283	20.163	4.022	0.696	66.683
2	T = 20.5499*sin ((2*π*3.18e(-8)*t) +4.02272)+66.63564	20.550	4.023	0.701	66.636
3	T = 20.392*sin ((2*π*3.18e(-8)*t) +4.07624)+66.47947	20.392	4.076	0.712	66.479
4	T = 20.482*sin ((2*π*3.18e(-8)*t) +4.021125)+66.35505	20.482	4.021	0.698	66.355
5	T = 20.4891*sin ((2*π*3.18e(-8)*t) +4.03214)+67.54093	20.349	4.032	0.701	67.541
6	T = 20.7812*sin ((2*π*3.18e(-8)*t) +4.03142)+68.46883	20.781	4.031	0.699	68.469
7	T = 20.2886*sin ((2*π*3.18e(-8)*t) +4.02415)+67.08382	20.289	4.024	0.701	67.084
8	T = 20.7812*sin ((2*π*3.18e(-8)*t) +4.02415)+66.18969	20.781	4.024	0.695	66.190
9	T = 20.3285*sin ((2*π*3.18e(-8)*t) +4.024975)+66.91944	20.329	4.025	0.701	66.919
10	T = 19.8608*sin ((2*π*3.18e(-8)*t) +4.02758)+66.6981	19.861	4.028	0.697	66.698
11	T = 20.0128*sin ((2*π*3.18e(-8)*t) +4.026215)+66.62328	20.013	4.026	0.697	66.623
12	T = 20.0479*sin ((2*π*3.18e(-8)*t) +4.027853)+66.27383	20.048	4.028	0.694	66.274

13	T = 19.9914*sin ((2*π*3.18e(-8)*t) +4.0325)+66.47272	19.991	4.033	0.696	66.473
14	T = 20.0661*sin ((2*π*3.18e(-8)*t) +4.032494)+65.42343	20.066	4.032	0.689	65.423
15	T = 19.4628*sin ((2*π*3.18e(-8)*t) +4.028672)+66.27177	19.463	4.029	0.692	66.272

These results of this study were in agreement with research conducted by Watson [2011] addressing the urban heat island effect for the City of Birmingham by studying the effect from different building facades, building materials and seasonal traits. The data indicates amplitudes (A) ranging from 19.5 to 20.7 degrees and phases angles roughly 4.2 to 4.3 radians. The r^2 values ranged from 0.69 to 0.71showing the data modeling was applicable to understanding the trends. The best model developed was for roof 3 (see Figure 11), having the highest correlation coefficient of all 15 mini-roofs.

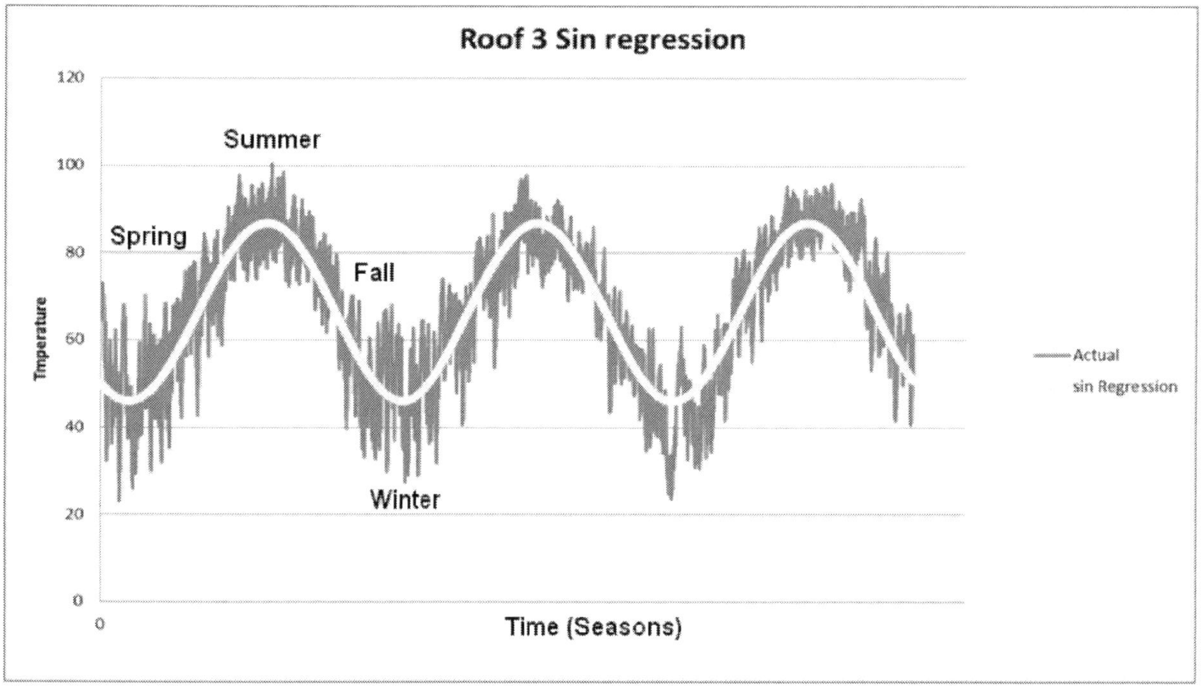

Figure 11. Fitted sine wave regression for a mini-roof system.

Based on the information collected on these mini-roof systems, the UAB Facilities Management Department decided to install an extensive vegetated roof on top of Hulsey Center as a roofing retrofit. UAB wants to obtain more fundamental information and knowledge for establishing green roofs in the southeastern U.S. In the Birmingham, Alabama area, fairly high rainfalls [approximately 132.1 cm/year (52 inches/year) on average] are obtained. However, during the summer months, it is common to have periods of drought with minimal rainfall and very hot and humid days [with temperatures in the 32.2^{+o}C (90+°F range]. Such climatic conditions require plants that can withstand both significant rainfall events and drought conditions. The current pilot green roof on top of Hulsey Center was installed in July 2008. Research conducted at Penn State University shows that green roofs planted with sedum plants reduce the building energy costs [Penn State Center for Green Roof Research, undated (a, b)]. The Facilities Management Department at UAB maintains records of the utilities bills (e.g., natural gas, water, and electricity) for each building on campus. The quantities and costs of utilities prior to (5 years) and after implementation (~3 years) of the pilot green roof have resulted in building energy reductions of 20% to 25%.

Summary and Conclusions

The behavior of different roofing materials affects the heating loads placed upon the buildings and their heating, ventilation, and air conditioning (HVAC) systems. Black roofs resulted in the highest temperature readings. Black roofing materials and bituthane (river rock) perform the poorest of the roofing materials tested, resulting in the highest heating load being placed on the building infrastructure. White granular roofing behaved similarly to black granular roofing materials. Clean white roofs resulted in consistently lower temperatures inside the mini-roof than the other roofing materials; however, over time, the reflective (white) roofs become dirty, losing some of their reflectivity, resulting in the roof being less energy efficient. Green (vegetated) roofs are fairly efficient in terms of energy performance due to evapotranspiration effects. Green roofs resulted in temperatures typically ~1.1-1.7°C (2-3°F) higher than the white roofs; however, they will dampen the drainage of rainfall during a rain storm through the retention of water onto the soil, and thereby lessen the discharge into stormdrains. Green roofs can significantly reduce stormwater runoff, reduce peak flow quantities, and lengthen the time of concentration from roofing structure [Kirby *et al.*, 2011]. "White" and "green" roofs both significantly reduce the roofs surface temperature and therefore, the air temperatures above and around the roof.

Hypothesis testing indicated that, between the 3 vegetative mini-roofs, all the min-roofs are statistically equal to each other in thermal properties. Both SBS Firestone mini-roofs are statistically the same, but mini-roof 6 is usually hotter. The TPO/PVC/Elvaloy mini-roofs are statistically the same. The 60-mil EPDM mini-roofs are also statistically the same.

The temperature varies in a sinusoidal fashion both during the course of the day and on an annual basis. For the various mini-roof structures, the phase angle (ϕ) and the amplitude (A) of the general form were determined through a regression analysis. The developed sine-wave functions indicated that most roofs were statistically different from one another from an amplitude aspect but the phase angles were statistically the same. It was also observed that almost all roofs had significantly different average mean roof temperatures, but the significance was mostly prevalent in the summer months. During other times of the year, the roofs behaved in a similar fashion.

Based on the information collected on these mini-roof systems, the UAB Facilities Management Department installed a vegetated roof on top of Hulsey Center as a roofing retrofit. The vegetated roof is ~1388.0 m^2 (14,940 ft^2) in area, and contains approximately 20,000 sedum plants. Utility bill information both prior to and after implementation of the green roof were gathered for electricity, natural gas, and chilled water were collected. The costs of utilities prior to and after implementation of the pilot green roof indicated utility bill (energy) savings of ~20% to 25% (compared to the case prior to implementation of the vegetated roof system).

Acknowledgments

The financial support and technical support of the UAB's Facilities Management Department is appreciated, including the fabrication and installation of the mini-roofs (on top of the Chemistry Building) and pilot roof on top of Hulsey Center. The input and technical assistance provided by Matt Winslett and Kelly Winnett of the Facilities Management Department in supplying the utility bill information is gratefully acknowledged.

References Cited:

Aitkenhead-Peterson, J.A., B.D. Dvorak, A. Voider, and N.C. Stanley, 2011. "Chemistry of Growth Medium and Leachate from Green Roof Systems in South-Central Texas", ***Urban Ecosystems***, ***14***(1): 17–33.

Bell, H., and G. Spolek, 2009. "Measured Energy Performance of Greenroofs", in ***Proc. Seventh Annual Greening Rooftops for Sustainable Communities Conference***, The Cardinal Group, Toronto, Canada.

Berndtsson, J.C., 2010. "Green Roof Performance towards Management of Runoff Water Quantity and Quality: A Review", ***Ecol. Engrg.***, ***36***(4): 351–360.

Carter, T., and C.R. Jackson, 2007. "Vegetated Roofs for Stormwater Management at Multiple Spatial Scales", ***Landscape and Urban Planning***, ***80***(1–2): 84–94.

Cash, C.G., 1999. "The Relative Durability of Low-Slope Roofing", pp. 119–124 in ***Proceedings of the Fourth International Symposium on Roofing Technology***, Gaithersburg, MD, (September 17–19).

Cavanaugh, L.M., 2008. "Redefining the Green Roof", ***J. Archit. Eng.***, ***14***(1): 4–6.

Clark, C., P. Adriaens, and F.B. Talbot, 2008. "Green Roof Valuation: A Probabilistic Economic Analysis of Environmental Benefits", ***Environ. Sci. Technol.***, ***42***(6): 2155–2161.

Editors of Time Life Books, 1979. ***Roofs and Siding***, Time-Life Books, Alexandria, VA.

Environmental Protection Agency (EPA), 2010. ***Environmental Protection Agency Green Building Design***. www.epa.gov/greenbuilding/.

Environmental Protection Agency (EPA), 2009. ***Buildings and their Impact on the Environment: A Statistical Summary***. www.epa.gov/greenbuilding/pubs/gbstats.pdf.

Fioretti, R., A. Palla, L.G. Lanza, and P. Principi, 2010. "Green Roof Energy and Water Related Performance in the Mediterranean Climate", ***Build. and Environ.***, ***45***(8): 1890–1904.

Givoni, B., 1994. ***Passive & Low Energy Cooling of Buildings***, John Wiley & Sons, Inc., New York, NY.

Green Roofs for Healthy Cities, undated. Website: http://www.greenroofs.org/index.php/about-green-roofs/2577-aboutgrnroofs/, accessed August 3, 2012.

Green Roof Plan, undated. Website: http://www.greenroofplan.com/intensive-vs-extensive-green-roofs/, accessed August 3, 2012.

Hathaway, A.M., W.F. Hunt, and G.D. Jennings, 2008. "A Field Study of Green Roof Hydrologic and Water Quality Performance", ***Transactions Am. Soc. Agric. & Biolog. Engineers***, ***51***(1): 37–44.

Kirby, J.T., R. Peters, S. Watts, and M. Winslett, 2011. "Experimental Investigation of Runoff Mitigation from Vegetated Roofs", Final report submitted to the Alabama Water Resources Research Institute, Auburn, AL.

Kosny, J., D. Yarbrough, P. Childs, and S. Mohiuddin, 2007. "How the Same Wall Can Have Several Different *R*-Values: Relations between Amount of Framing and Overall Thermal Performance in Wood And Steel-Framed Walls", *Oak Ridge National Laboratory*.

Kumar, R., 2005. "Performance Evaluation of Green Roof and Shading for Thermal Protection of Buildings", ***Build. Environ.***, ***40***(11): 1505–1511.

Mansour, H.M.A., 2001. "Sustainable Housing Development with Special Reference to Environmental Aspects in Low Cost Housing in Egypt". M.S. Thesis, Cairo University, Cairo, Egypt. 117 pages.

Niu, H., C. Clark, J. Zhou, and P. Adriaens, 2010. "Scaling of Economic Benefits from Green Roof Implementation in Washington, D.C.", ***Environ. Sci. Technol.***, ***44***(11): 4302–4308.

Nunez, S., and R.W. Peters, 2008. "Effect of Roofing Materials on Building Energy Efficiency", Paper included in the conference proceedings of the 2008 Annual AIChE Meeting, Philadelphia, PA, (November 16–21).

Penn State Center for Green Roof Research, undated (a). Pennsylvania State University website: http://horticulture.psu.edu/cms/greenroofcenter/about_ctr.html; accessed July 5, 2009.

Penn State Center for Green Roof Research, undated (b). "Utility Bills Getting You Down? Plant Stuff Up on the Roof", http://aginfo.psu.edu/news/2000/9/roof.html; accessed July 5, 2009.

Peters, R.W., J.G. Price, R.D. Sherrod, S.A. Watts, J.M. Gohlke, J.T. Kirby, and M. Winslett, 2011. "Roofing Types and Their Effects on the Thermal Properties of Runoff", ***Proc. Alabama Water Resources Conference***, Orange Beach, AL, (September 8–9).

Reid, R.N., 2000. ***Roofing and Cladding Systems: A Guide for Facility Managers***, Fairmont Press, Lilburn, GA.

Roehr, D., and Y. Kong, 2010. "Stormwater Runoff Reduction Achieved by Green Roofs: Comparing SWMM Method to TR-55 Method", pages 1012–1021 in ***Low Impact Development 2010: Redefining Water in the City, Proceedings of the 2010 International Low Impact Development Conference***, San Francisco, CA, (April 11–14).

Rosatto, H.G., D. Laureda, D. Perez, D. Nerrera, M. Meyer, P. Gamboa, G. Villalba, M. Friedrich, M. Bargieta, L.R. Plaza, G. Calvo, M. Miranda, M. Inigo, and E. Quaintenne, 2010. "Water Retention Efficiency of Green Roof Systems", ***Revista de la Facultad de Ciencias Agarias***, ***42*** (1): 213–219.

Santamouris, M., C. Pavlou, P. Doukas, G. Mihalakakou, A. Synneta, A. Hatzibiros, and P. Patargias, 2007. "Investigating and Analysing the Energy and Environmental Performance of an Experimental Green Roof System Installed in a Nursery School Building in Athens, Greece", ***Energy***, ***32*** (9): 178–1788.

Schroll, E., J. Lambrinos, T. Righetti, and D. Sandrock, 2011. "The Role of Vegetation in Regulating Stormwater Runoff from Green Roofs in a Winter Rainfall Climate", ***Ecol. Engrg.***, ***37*** (4): 595–600.

Selim, M.A., 2003. "The Effect of Building Envelope Technology on Energy Conservation in Buildings", M.S. Thesis. Cairo University. Cairo, Egypt.

Selim, R., 2008. "Investigation of Thermal Performance, Indoor Air Quality and Energy Consumption within Tuskegee Healthy House Demonstrative Model", Ph.D. Dissertation,. University of Alabama at Birmingham, Birmingham, AL.

Sonne, J., 2006. "Evaluating Green Roof Energy Performance", ***ASHRAE J.***, ***48*** (2): 59–61.

Spala, A., H.S. Bagiorgas, M.N. Assimakopoulos, J. Kalavrouziotis, D. Matthopoulos, and G. Mihalakakou, 2008. "On the Green Roof System. Selection, State of the Art and Energy Potential Investigation of a System Installed in an Office Building in Athens, Greece", ***Renew. Energy***, ***33*** (1): 173–177.

Takakura, T., S. Kitade, and E. Goto, 2000. "Cooling Effect of Greenery Cover over a Building", ***Energy Build.***, ***31***: 1–6.

United States Green Building Council (USGBC), 2010. United States Green Building Council. www.usgbc.org.

Villarreal, E.L., and A.S.D.L. Bengtsson, 2004. "Inner City Stormwater Control Using a Combination of Best Management Practices", ***Ecol. Engrg.***, ***22*** (4–5): 279–298.

Watson, C.J., 2011. "Sustainable Urban Design Practices that Mitigate Urban Heat Islands and Reduce Energy Consumption", Ph.D. Dissertation, University of Alabama at Birmingham, Birmingham, AL.

Williams N.S.G., J.P. Nicholas, and K.J. Raynor, 2010. "Green Roofs for a Wide Brown Land: Opportunities and barriers for Rooftop Greening in Australia", ***Urban Forestry & Urban Greening***, ***9***(3): 245–251.

Wolf, D., and J.T. Lundholm, 2008. "Water Uptake in Green Roof Microcosms: Effects of Plant Species and water Availability", ***Ecol. Engrg.***, ***33*** (2): 179–186.

Wong, N.H., P.Y. Tan, and C. Yu, 2007. "Study of Thermal Performance of Extensive Rooftop Greenery Systems in the Tropical Climate", ***Build. Environ.***, ***42*** (1): 25–54.

Wong, N.H., S.F. Tay, R. Wong, C.L. Ong, and A. Sia, 2003. "Life Cycle Cost Analysis of Rooftop Gardens in Singapore", ***Building and Environment***, ***38*** (3): 499–509.

(665c) Low-Cost Phase Change Material for Building Envelopes

Ramin Abhari, Syntroleum, Tulsa, OK and Laura Ford, University of Tulsa, Tulsa, OK

Studies have shown that incorporating phase change material (PCM) in building envelopes—attic insulation, wall boards, etc.—can reduce energy consumption for home heating and cooling by more than 30%. However, high manufacturing costs have so far limited PCM use to novelty products. This work has developed a low-cost PCM through a selective process for n-paraffin production and continuous high-throughput equipment for its encapsulation. The n-paraffins are made sustainably via hydrodeoxygenation of bio-oils. The heat of fusion and melting point of our n-paraffins made from bio-oils compare well with those of traditional n-paraffins. Heat cycling tests at 40°C indicate that the n-paraffins encapsulated in high density polyethylene will retain at least 90% of the n-paraffins. The encapsulated n-paraffins have been incorporated into cellulose insulation and exposed to outside conditions for over a month. Results include a more than 10% reduction in heat flux through insulation with our PCM and a 1°C reduction in peak temperature compared with plain insulation. An economic analysis will be presented as well.

Sustainability through Durable Building Materials and Assemblies

Dr. Theresa Weston
DuPont Building Innovations

An opportunity for chemical engineers in transforming the sustainable or green building industry is in developing understanding of building durability. Building durability is a key sustainability attribute. A building cannot be truly energy efficient or sustainable unless it lasts. One of the most important components of building durability is the prevention of moisture damage. Chemical engineering knowledge of combined heat and mass transfer can provide key insights in to moisture damage risk analysis, and can therefore aid in the design on energy efficient and durable buildings.

Sustainable Building Codes & Standards

Despite the overall slow-down of the construction industry, especially the home building industry in North America, one growth area has been green and energy efficient construction. This trend has coincided with a large number of changes in the standards, codes and regulations governing the building construction. More specifically, the industry has introduced new energy efficiency and sustainability codes and standards. While these codes in many ways are an extension of existing building codes, the combination of more stringent, and in some cases new, requirements, lead to the use of new building materials, assemblies and systems. Although a significant amount of research has been conducted moisture durability of traditional wall systems, this research needs to be extended to apply to these new energy efficient and sustainable building systems

Modeling Moisture Damage Risk

Modeling combined heat and mass (water) transport or hygrothermal modeling is a useful tool, which is becoming more prevalent in the construction industry. This type of modeling can highlight the risk of moisture accumulation in building systems. An example (shown below) is the higher risk of accumulation of water from building assemblies when additional layers of low vapor permeability materials such as exterior insulation are added to a wall system.

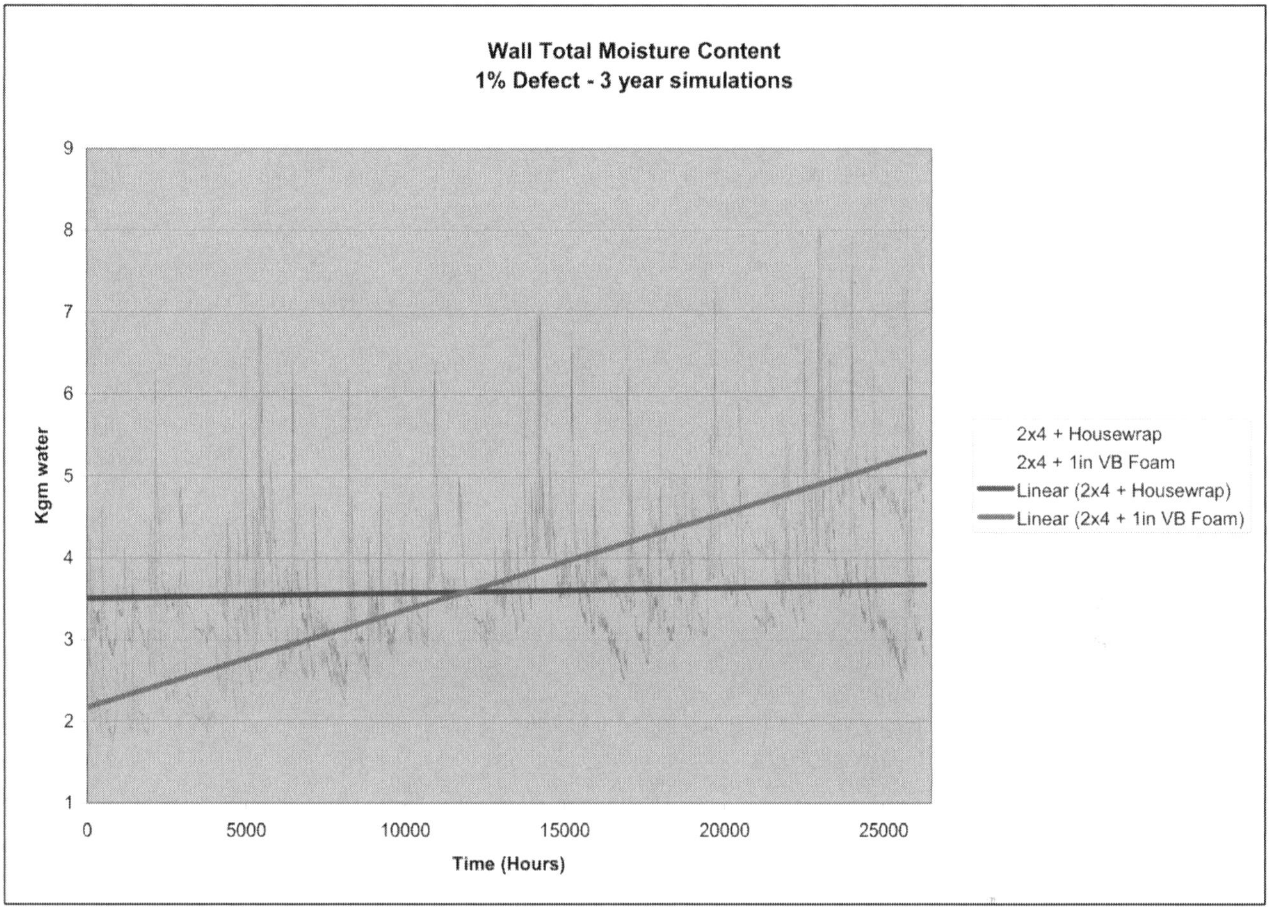

This presentation will provide information on the status of moisture modeling in the construction industry including:

- Standardization of protocols for the use of moisture risk analysis by the development of ASHRAE Standard 160 – *Criteria for Moisture-Control Design Analysis in Buildings*

- Sponsorship of hygrothermal modeling software development by the US Department of Energy, and

- Development of new methodologies to extend the use of hygrothermal simulation to include the management of drying of incidental water leaks.

Examples of specific materials used to help provide durable, energy efficiency to buildings will be discussed.

(667a) A Two-Scale 13C-Based Method for Metabolic Flux Measurement and Prediction in Genome-Scale Models

Hector Garcia-Martin, Fuel Synthesis Division, Joint Bioenergy Institute, Emeryville, CA

Systems biology aims to provide a predictive and quantitative understanding of cell behaviour as the outcome of the interaction of its comprising parts. Metabolic flux profiles (i.e. the number of molecules traversing each biochemical reaction encoded in its genome per unit time) are not only a key phenotypic characteristic but also embody the essence of this complexity since they represent the final functional output of the interactions of all the molecular machinery studied by all the other "omics" fields. Two of the most popular methods for studying metabolic fluxes are Flux Balance Analysis (FBA) and 13C Metabolic Flux Analysis (13C MFA), each of them displaying its own advantages and disadvantages. In this talk I will present a new method: Two scale 13C Metabolic Flux Analysis (2S-13CMFA), which combines the advantages of FBA and 13C MFA. I will showcase its applications and possibilities with data from the KEIO knockout collection.

(667b) Phenotypic and Multi-Omic Approaches to Address Molecular Bottlenecks in the Fermentation of Lignocellulose Into Ethanol by *Saccharomyces Cerevisiae*

Trey Sato[1], Dana Wohlbach[2], Jeffrey Lewis[2], Yaoping Zhang[1], Mingjie Jin[3], Yury Bukhman[1], Wendy Schackwitz[4], Christa Pennacchio[4], David Hodge[3], Venkatesh Balan[3], Bruce E. Dale[3] and Audrey Gasch[2], (1)University of Wisconsin, Great Lakes Bioenergy Research Center, Madison, WI, (2)University of Wisconsin, Madison, WI, (3)Chemical Engineering and Material Science, Michigan State University, East Lansing, MI, (4)The Joint Genome Institute, Walnut Creek, CA

While cellulosic ethanol is being looked to for relief of the global energy demand, a number of molecular bottlenecks currently exist that prevent the efficient bioconversion of lignocellulose into ethanol. For example, it is well known that: 1) native *Saccharomyces cerevisiae* yeast strains cannot sufficiently ferment xylose, and 2) side products generated from pretreatment, including Ammonia Fiber Expansion (AFEX™) and alkaline hydrogen peroxide (AHP), of plant biomass illicit a cellular stress response, which further limits fermentation productivity. At the Great Lakes Bioenergy Research Center, we have taken a multi-comparative approach to facilitate the discovery and understanding of molecular bottlenecks in the fermentation of lignocellulosic hydrolysates by yeast. Through multi-phenotypic and bioinformatic analysis of 111 natural and domesticated isolates, we have identified a wild *S. cerevisiae* strain that maintains rapid growth and cell viability in a variety of distinctly prepared hydrolysates. Following directed engineering of a xylose metabolism pathway, we performed directed evolution that yielded mutants able to ferment 2 to 3-fold more xylose from AFEX™ corn stover hydrolysate (ACSH) than unevolved parents. Furthermore, we employed temporal profiling of gene expression levels during ACSH fermentations, which identified differences in cell physiology between evolved and unevolved strains. Analysis of extracellular metabolite, amino acid and metal concentrations additionally identified limiting nutrients during fermentation. Coupled with comparative genome resequencing of parental and evolved strains, this suite of Omic data is being integrated in metabolic network models to identify and understand genetic differences that impact xylose fermentation in stress-inducing lignocellulosic hydrolysates.

(667c) Towards Industrial Robustness: Understanding of Wild-Type and Populus Hydrolysate- Tolerant Mutant Strains of Clostridium Thermocellum

Jessica Linville[1], Miguel Rodriguez[2], Jonathan Mielenz[3] and Chris D. Cox[1,4], (1)Department of Civil and Environmental Engineering , University of Tennessee, Knoxville, TN, (2)Bioscieces Division, Oak Ridge National Laboratory, Oak Ridge, TN, (3)Biosciences Division, Oak Ridge National Laboratory, Oak Ridge, TN, (4)Center for Environmental Biotechnology, University of Tennessee, Knoxville, TN

An industrially robust microorganism capable of degrading lignocellulosic biomass in the presence of inhibitory compounds produced from the pretreatment process in a CBP scheme is needed to reduce the cost of cellulosic biofuels. *Clostridium thermocellum* has the ability to convert biomasses complex carbohydrates into ethanol by fermentation. However, *C. thermocellum* is inhibited by by-products from the biomass pretreatment process. To increase *C. thermocellum*'s industrial robustness, a 17.5% v/v *Populus* hydrolysate-tolerant mutant (PM) was produced by 117 successive transfers of *C. thermocellum* 27405 strain into increasing concentrations of the hydrolysate. A single colony culture of the PM was then produced and the genome has been sequenced. A study comparing the wild-type (WT) and PM strains of *C. thermocellum* was completed with growth on cellobiose examining sugar utilization and end product formation, and growth on Avicel examining protein concentration. The growth of the PM in 17.5% v/v *Populus* hydrolysate had a 20% higher optical density, a 33% faster sugar utilization rate (g glucan/L/hr), and produced 34% more ethanol (g ethanol/ g glucan) than the WT grown without hydrolysate. The PM retained these characteristics when grown without hydrolysate with a 37% higher optical density, a 40% faster sugar utilization rate and produced 20% more ethanol than the WT grown under the same conditions. The WT was severely inhibited when grown in the presence of 17.5% v/v Populus hydrolysate with a barely detectable optical density, 7% of sugar utilization rate, and produced 7.5% of the expected ethanol versus when it was grown without hydrolysate. A combination of the genomic mutations of the PM and gene expression profiles from triplicate fermentations in various concentrations of the *Populus* hydrolysate will be used to determine which mutations in the genome might be the cause of the increased tolerance to the hydrolysate.

(667d) Development of Advanced Technologies for Feedstock Chracterization and HTP Pretreatment Process Monitoring

Seema Singh, Joint BioEnergy Institute, Sandia National Laboratory, Emeryville, CA

Today, carbon-rich fossil fuels, primarily oil, coal and natural gas, provide 85 percent of the energy consumed in the United States. Fossil fuel use increases CO2 emissions, increasing the concentration of greenhouse gases and raising the risk of global warming. The high energy content of liquid hydrocarbon fuels makes them the preferred energy source for all modes of transportation. In the US alone, transportation consumes around 13.8 million barrels of oil per day and generates over 0.5 gigatons of carbon per year. This has spurred research into alternative, non-fossil energy sources. Among the options (nuclear, concentrated solar thermal, geothermal, hydroelectric, wind, solar and biomass), only biomass has the potential to provide a high-energy-content transportation fuel. Biomass is renewable resource that is carbon-neutral.

Currently, biofuels such as ethanol are produced largely from grains, but there is a large, untapped resource (estimated at more than a billion tons per year) of plant biomass that could be utilized as a renewable, domestic source of liquid fuels. Well-established processes convert the starch content of the grain into sugars that can be fermented to ethanol. Plant-derived biomass contains cellulose, which is more difficult to convert to sugars. The development of cost-effective and energy-efficient processes to transform cellulose in biomass into fuels is hampered by significant roadblocks, including the lack of specifically developed energy crops, the difficulty in separating biomass components, low activity of enzymes used to deconstruct biomass, and the inhibitory effect of fuels and processing byproducts on organisms responsible for producing fuels from biomass monomers.

The Joint BioEnergy Institute (JBEI) is one of three US Department of Energy Bioenergy Research Centers that will address these roadblocks in biofuels production. JBEI draws on the expertise and capabilities of three national laboratories (Lawrence Berkeley National Laboratory (LBNL), Sandia National Laboratories (SNL), and Lawrence Livermore National Laboratory (LLNL)), two leading US universities (University of California campuses at Berkeley (UCB) and Davis (UCD)), and a foundation (Carnegie Institute of Washington at Stanford University) to develop the scientific and technological base needed to convert the energy stored in cellulose into transportation fuels and commodity chemicals. Established scientists from the participating organizations are leading teams of researchers to solve the key scientific problems and develop the tools and infrastructure that will enable other researchers and companies to rapidly develop new biofuels and scale production to meet US transportation needs, and to develop and rapidly transition new technologies to the commercial sector. This talk will present a brief summary of the organization and efforts at JBEI, highlight the efforts on the development and deployment of novel technologies for biomass analysis (compositional and mechanical) and HTP methods for monitoring pretreatment process efficacies.

(667e) Discovery and Functional Determinations of Biomass-Degrading Enzymes by Robotic Cell-Free Translation

Brian G. Fox[1], Taichi E. Takasuka[1], Lai Bergeman[2], Walker Johnnie A.[3], Kai Deng[4] and Trent Northen[4], (1)Great Lakes Bioenergy Center, University of Wisconsin, Madison, WI, (2)Department of Biochemistry, University of Wisconsin-Madison, Madison, WI, (3)Department of Biochemistry, University of Wisconsin-Madison, mad, (4)Joint BioEnergy Institute, Emeryville, CA

Our use of a platform for functional determination of enzymes will be presented. Robotic wheat germ cell-free protein translation system was used for protein synthesis and enzymatic assays were performed using the translation reactions obtained directly from the robot. Fluorescent sugar analogs and purified polysaccharides were used for initial characterizations. Nanostructure initiator mass spectrometry (NIMS)-based assay provided additional functional assignments and also revealed diagnostic differences in time-dependent cascades of products given by different enzymes. The approach was used to evaluate a majority of secreted cellulosomal proteins from *C. thermocellum* in different enzyme formats, their comparative single reactivities, and their synergistic behavior for hydrolysis of various polysaccharide-containing materials. Combinatorial enzyme activity assays revealed synergistic activities between enzymes with different assigned functions

(667f) Characterization of the Interactions of Cellulose, Hemicellulose, and Lignin During Pretreatment Through the Use of Flowthrough Pretreatment

Heather L. McKenzie[1,2], Jaclyn D. DeMartini[1,2], Nancy L. Engle[1,3], Marcus Foston[1,4], Sivakumar Pattathil[1,5], Bruce A. Tomkins[1,6], Michael G. Hahn[1,5], Arthur J. Ragauskas[1,4], Timothy J. Tschaplinski[1,3], Gary A. Van Berkel[1,6] and Charles Wyman[2,7], (1)BioEnergy Science Center, Oak Ridge, TN, (2)Chemical and Environmental Engineering, University of California Riverside, CE-CERT, Riverside, CA, (3)Biosciences Division, Oak Ridge National Laboratory, Oak Ridge, TN, (4)Department of Chemistry and Biochemistry, Georgia Institute of Technology, Atlanta, GA, (5)Complex Carbohydrate Research Center, University of Georgia, Athens, GA, (6)Chemical Sciences Division, Oak Ridge National Laboratory, Oak Ridge, TN, (7)BioEnergy Science Center, Oak Ridge National Laboratory, Oak Ridge, TN

The conversion of lignocellulosic biomass to ethanol requires recovery of the sugars contained in hemicellulose and cellulose at high yields. Hydrothermal pretreatment can be used to hydrolyze and solubilize hemicellulose prior to enzymatic hydrolysis of the cellulose left in the residual solids to glucose. An understanding of the deconstruction of cell wall during pretreatment would support the development of improved energy crops and pretreatment strategies. The results of flowthrough pretreatment of native poplar and model substrates were examined for evidence of the interaction of cellulose, hemicellulose, and lignin during pretreatment. The liquid effluent was analyzed for oligomers and phenols, while the composition of the residual solids was tested using HPLC and NMR. Differences in the release of xylooligomers from native poplar, holocellulose, and isolated xylan demonstrated the importance of the interactions of cellulose-hemicellulose and lignin-hemicellulose on hemicellulose hydrolysis. Pretreatment of the isolated lignin indicates that depolymerization is the primary mechanism for lignin extraction while differences in the release of lignin from native poplar and the isolated lignin indicate that lignin-hemicellulose interactions influence relative reactivity. Recent work has identified the removal of specific cell wall structures, such as methyl glucuronoxylan, as key to improvements in enzymatic digestibility. The removal of these components during pretreatment will be examined closely.

(689a) Development of Intensified Catalytic Reactors for DME and Fischer-Tropsch Biofuels Synthesis in Future Integrated Biorefineries

Martin O'Connell[1], Costa Komodromos[2], Gunther Kolb[3], Ralf Zapf[3], Irene Resa[4], Manickam Jayamurthy[4] and Ashok Bhattacharya[5], (1)1Institut für Mikrotechnik Mainz GmbH, D-55129 Mainz, Germany, (2)Wlfson Centre for Materials processing, Brunel University, Middlesex, United Kingdom, (3)Energy Technology and Catalysis Department, Institut für Mikrotechnik Mainz GmbH , D-55129 Mainz , Germany, (4)Wolfson Centre for Materials Processing, Brunel University, Uxbridge, Middlesex, United Kingdom, (5)Wolfson Centre for Materials Processing, Brunel University, Uxbridge Middlesex , United Kingdom

Intensified catalytic reactors have been developed within a European Union funded project (www.suprabio.eu) to carry out conversion of syngas, derived from thermochemical conversion of biomass, to dimethyl ether (DME) via a direct process (without an intermediate methanol stage) and to middle distillate biofuels via Fischer-Tropsch synthesis. The aim is to provide efficient processes that are scalable to the distributed biomass treatment and conversion processes within an integrated biorefinery system. The approach is to achieve process and engineering integration and to utilise the enhanced heat and mass transfer properties of the reactor designs to overcome the cost penalties associated with downscaling plants.

The catalysts are prepared in a form that allows coating onto arrays of metal plates that are stacked and laser welded to form a mini-reactor. The challenges are to optimise catalyst properties, to achieve stable, adherent and uniform coatings and to design and optimise the reactor geometry to achieve fine control of reaction temperatures.

A range of highly selective, novel nano-catalysts have been prepared and tested for these reactions.

The experiments have been carried out over a wide range of reaction temperatures (210–300 °C), pressure (10–30 bar), $H_2/CO = 2$. The results indicate very promising levels of DME and hydrocarbon yields.

(689b) Drop-in Gasoline From Carbona Wood Gasification, Morphysorb CO2 Removal, and the Tigas Process

Rick Knight, Gas Technology Institute, Des Plaines, IL, Niels R. Udengaard, Syngas Catalyst & Technology, Haldor Topsoe, Inc., Houston, TX, Finn Joensen, Haldor Topsøe A/S, DK-2800 Lyngby, Denmark, Jesper Jensen, Technology Division, Haldor Topsoe A/S, Lyngby, Denmark and Jim Patel, Andritz/Carbona, Vallejo, CA

Drop-In Gasoline from Wood Gasification, CO2 Removal, and the TIGAS Process

ABSTRACT

Biomass conversion is the only near-term option to replace petroleum-based motor fuels with renewable alternatives. One of the pathways under development is thermochemical gasification followed by conversion of the resulting synthesis gas to gasoline. Under U.S. Department of Energy sponsorship, Haldor Topsoe Inc. (HTI), the Gas Technology Institute (GTI), Andritz-Carbona, UPM-Kymmene, and ConocoPhillips have teamed up to build and operate a pilot biorefinery based on this approach. Woody biomass will be converted to gasoline blendstock, offering product compatibility with existing infrastructure. Startup of the 21 bbl/day pilot plant is scheduled for November 2012.

The biorefinery integrates Andritz-Carbona's pressurized, oxygen-blown bubbling-bed gasifier, which features high thermal efficiency and low syngas tar contamination. A cleanup system featuring hot gas filter and catalytic reformer eliminates particulates and heavy organics while also converting residual methane to carbon monoxide and hydrogen. GTI's acid-gas removal process uses Morphysorb® solvent to remove carbon dioxide and hydrogen sulfide with high selectivity that minimizes loss of desired syngas components. The clean, conditioned syngas is then converted to gasoline in Haldor Topsoe's TIGAS process, which employs a unique high-conversion once-through methanol/dimethylether synthesis step. Recycling a portion of the tail gas from the synthesis back to the gasifier results in an essentially power-neutral process, resulting in an overall syngas conversion of well over 90 per cent.

This integrated process is capable of converting any type of biomass, but the commercial case under consideration is to supply the plant with woody biomass such as mill wastes, non-merchantable roundwood, whole tree chips, or other available forestry byproducts. This strategy will leverage the resources, facilities, and expertise of the pulp and paper industry to provide value-added revenue streams while helping meet U.S. transportation fuel needs with home-grown sustainability.

In this presentation, we will discuss the proposed process layout of the wood-to-gasoline process, summarize the engineering and construction status, and present an updated prediction of operating cost for the early commercial plants.

(689c) The Effects of Heating Rate and Reaction Time On Hydrothermal Liquefaction of Microalgae

Julia L. Faeth and Phillip E. Savage, Chemical Engineering, University of Michigan, Ann Arbor, MI

Hydrothermal liquefaction (HTL) is one of many methods for biomass conversion to biofuels or biofuel precursors. HTL avoids energy-intensive drying steps, and is therefore more energy efficient for biomass with very high moisture content, like microalgae. The effects of many different parameters (including biomass concentration, nominal reaction temperature, catalysis, and reaction time) on product yields for HTL of microalgae have been reported in the literature. However, a majority of the reaction times reported in HTL literature have been defined to ignore the time necessary to reach the nominal reaction temperature. Furthermore, the heating times reported vary greatly, from several minutes to a few hours.

The time necessary to heat an HTL reaction mixture to the nominal reaction temperature is dependent on the heating rate. The heating rate of a reaction mixture has been observed to influence the product yields for HTL of woody biomass. Product yields for supercritical water gasification of glucose and cabbage are also influenced by heating rate. This presentation explores the influence of heating rate on the product yields for HTL of microalga *Nannochloropsis* sp. The influence of reaction time, including time necessary for the reaction mixture to reach the nominal reaction temperature, will also be discussed. These effects need to be considered during the development of an efficient HTL flow reaction process.

(689d) Liquefaction of Low-Lipid Macroalgae for Production of Liquid Fuel

Shicheng Zhang, Department of Environmental Science and Engineering, Fudan University, Shanghai, China

Liquefaction of Low-Lipid Macroalgae for Production of Liquid Fuel

Shicheng Zhang*, Dong Zhou, Changjun Li, Xiao Yang, Jianmin Chen

Department of Environmental Science and Engineering, Fudan University, Shanghai 200433, China

*Corresponding author phone: +86-21-65642297; fax: +86-21-65642080; e-mail: zhangsc@fudan.edu.cn

Low-lipid marine macroalgae Enteromorpha prolifera, one of the main algae genera for green tide, was converted to bio-oil and ester compounds in sub-/supercritical water or alcohols in a batch reactor. Effects of the temperature, reaction time, Na_2CO_3 catalyst, and algae/solvent ratio on the yields of liquefaction products were investigated. The raw algae and liquefaction products were analyzed using elemental analysis, Fourier transform infrared (FTIR) spectroscopy, gas chromatography-mass spectrometry (GC-MS), and 1Hnuclear magnetic resonance (NMR). The bio-oil obtained in water was a complex mixture of ketones, aldehydes, phenols, alkenes, fatty acids, esters, aromatics, and nitrogen containing heterocyclic compounds. And acetic acid was the main component of the water-soluble products. Otherwise, the bio-oils obtained in alcohols were mainly composed of ester compounds. A variety of fatty acid (C3—C22) esters (methyl or ethyl) in the bio-oils respectively obtained in methanol and ethanol were qualified by GC-MS, and their relative contents are above 60% of the total area for each bio-oil. In addition, some N-containing compounds, sugars, fatty alcohols/ketones and very few hydrocarbons were also qualified. Overall, bio-oils obtained in two alcohols are much similar to biodiesel on the composition. Moreover, the bio-oils are found to contain a considerable fraction of light components using thermogravimetric analysis (TGA), and the contents of low-boiling (bp<350 °C) compounds are up to 70% of the weight for both bio-oils, therefore, it might help to the further separation and refining of bio-oils to produce fuels and chemicals.

References

1. Dong Zhou, Liang Zhang, Shicheng Zhang, Hongbo Fu, Jianmin Chen. Hydrothermal Liquefaction of Macroalgae Enteromorpha prolifera to Bio-oil. Energy & Fuels. 2010, 24, 4054–4061.
2. Dong Zhou, Shicheng Zhang, Hongbo Fu, Jianmin Chen. Liquefaction of Macroalgae Enteromorpha prolifera in Sub-/Supercritical Alcohols: Direct Production of Ester Compounds. Energy & Fuels. 2012. 26 (4), 2342-2351.

(689e) Renewable Liquid Alkanes Production in Continuous Flow Microchannel Reactor

Lalita Attanatho, Department of Chemical, Biological and Environmental Engineering, Oregon State University, Corvallis, OR

The study focused on the development of microtechnology-based chemical reaction process for liquid biofuel production from oil-based biofuel feedstock. A hydrotreating reaction of 5 wt% of oleic acid in dodecane was performed in continuous flow microchannel reactor. The microchannel reactor was fabricated from SS-316 and the wall of the microreactor was coated with a thin layer of $NiMoP/Al_2O_3$ catalyst. The reactions were carried out in the temperature range of 275-325 °C, residence time in the range of 36-40 s and at constant system pressure of 500 psi.

The results showed that microchannel reactor was well suitable for the hydrotreating process. The almost complete conversion of fatty acid hydrotreating reaction was achieved at a reaction temperature of 300 °C and above. The conversion of fatty acid increased with increasing temperatures and decreasing initial feed concentrations. Hydrotreating of fatty acid occurred primarily via hydrogenation and dehydration and the main liquid products were octadecane and heptadecane. Fatty alcohol, fatty acid and long chain ester were formed as intermediates.

(689f) Novel Reactors with Intensified Catalytic Properties for Renewable Diesel Production

Martin O'Connell[1], Costa Komodromos[2], **Jens Ole Wiborg**[3], Manickam Jayamurthy[4], Hans-Joachim , Kost[1] and Ashok Bhattacharya[5], (1)1Institut für Mikrotechnik Mainz GmbH, D-55129 Mainz, Germany, (2)Wlfson Centre for Materials processing, Brunel University, Middlesex, United Kingdom, (3)Energy Technology and Catalysis Department, Institut für Mikrotechnik Mainz GmbH , D-55129 Mainz , Germany, (4)Wolfson Centre for Materials Processing, Brunel University, Uxbridge, Middlesex, United Kingdom, (5)Wolfson Centre for Materials Processing, Brunel University, Uxbridge Middlesex , United Kingdom

Before feedstocks derived from renewable organic material can be used in conventional automobile engines and distributed using existing fuel infrastructure, it is desirable to convert the material into hydrocarbons whose composition resembles that of petroleum-derived fuels. A well-established method is the conversion of vegetable seed oils into green diesel by hydrotreating, an exothermic process involving catalytic hydrogenation of the double bonds present in fatty acid chains of triglycerides in the seed oils.

This paper describes the development and use of improved catalysts and reactor design concepts to alleviate the mass transfer limitations in conventional fixed bed reactors, enable very careful control of reaction temperature, and enhanced yields of green diesel. The focus is on minimising side reactions, and improving selectivity towards C_{15}-C_{18} alkanes.

Two reactor concepts have been developed and fabricated: a tubular foam reactor designed to allow staged hydrogen addition, and a membrane reactor.

Experimental work was carried out in the temperature range of 280-325°C, pressures of 30-50 bar using rapeseed and Jatropha oils. The results show that the reactor designs enable better control of temperature, with improved mixing, mass and heat transfer, leading to an increased selectivity of the hydrogenation process compared to published data.

(704a) Key Ingredients and Co-Products in Biofuel Life Cycle Analysis

Jennifer B. Dunn[1], Steffen Mueller[2], Zhichao Wang[1], Jeongwoo Han[1] and Michael Q. Wang[1], (1)Energy Systems Division, Argonne National Laboratory, Argonne, IL, (2)Universityof Illinois at Chicago, Chicago, IL

In many cases, life cycle analyses (LCA) of biofuels have treated key ingredients and co-products less rigorously than feedstocks and fuel products. Key ingredients and co-products, however, can significantly impact LCA results and merit attention. For example, we have recently examined the contribution of enzymes to the life cycle energy consumption and greenhouse gas (GHG) emissions of corn and cellulosic ethanol with Argonne National Laboratory's Greenhouse gases, Regulated Emissions, and Energy use in Transportation (GREET) model. We estimate that the contribution to farm-to-pump GHG emissions in the case of corn ethanol is approximately 1%. In the case of cellulosic ethanol, this contribution is much higher (up to 20%) because enzymes are consumed at a much higher rate. We have also added new data to GREET for the production of yeast and determined that it can contribute up to 11% of farm-to-pump emissions for cellulosic ethanol.

Additionally, we examine key co-products from biofuel production with a focus on biochar, which has not received rigorous treatment in previous biofuel LCAs. We discuss the properties of biochar including GHG mitigation, yield increase, and fertilizer offsets and how these properties can be incorporated into LCAs of biofuels that produce biochar as a co-product.

(704b) Life Cycle Assessments of Pyrolysis-Based Biofuels From Diverse Biomass Feedstocks

Matt Mihalek[1], Jiqing Fan[2], Gaurav Bhardwaj[1], Robert Handler[3], Tom N. Kalnes[4], David R. Shonnard[2] and Edwin.A Maleche[2], (1)Michigan Technological University, Houghton, MI, (2)Chemical Engineering, Michigan Technological University, Houghton, MI, (3)Sustainable Futures Institute, Michigan Technological University, Houghton, MI, (4)Research and Development, UOP LLC, Des Plaines, IL

Life Cycle Assessments of Pyrolysis-Based Biofuels from Diverse Biomass Feedstocks

Matt Mihalek[1], Jiqing Fan[1], Gaurav Bhardwaj[1], Kyle Andrews[1], Robert Handler[2], Tom Kalnes[3], David Shonnard[1,2]

1 Department of Chemical Engineering, Michigan Technological University, Houghton, MI

2 Sustainable Futures Institute, Michigan Technological University, Houghton, MI

3 UOP LLC (Honeywell), Des Plaines, IL

Abstract

Renewable hydrocarbon biofuels are being investigated as possible alternatives to conventional fossil fuels like gasoline, kerosene and diesel. Feedstocks such as corn stover, sugarcane bagasse, switchgrass, waste wood, and algae, are being evaluated as candidates for pyrolysis and upgrading to drop-in hydrocarbon fuels. This research is developing preliminary life cycle assessments (LCA) for each feedstock and comparing the greenhouse gas (GHG) emissions of the hydrocarbon biofuels to current fossil fuels. As a comprehensive study, this analysis attempts to account for all of the emissions associated with each feedstock through the entire life cycle. Emissions from all stages including feedstock production, land use change, pyrolysis, stabilizing the pyrolysis oil for transport and storage, and upgrading the pyrolysis oil to a hydrocarbon fuel are included. In addition to GHG emissions, the energy requirements and water use have been evaluated over the entire life cycle. The goal of this research is to help rank the feedstocks and the resultant hydrocarbon biofuels based on three environmental indicators; GHG emissions, energy demand, and water utilization. Preliminary results indicate that liquid hydrocarbon biofuels produced through this pyrolysis-based pathway can achieve greenhouse gas emission savings of greater than 50% compared to petroleum fuels, thus qualifying these biofuels under the US EPA RFS2 program.

(704c) Sustainability Analyses of Production Scale Processes for Forming Biofuels From Algal Biomass Based On Experimental Data

Paul Blowers, Chemical and Environmental Engineering, The University of Arizona, Tucson, AZ, Christina Canter, Dept. of Chem. and Env.Engr., The University of Arizona, Tucson, AZ, Robert Handler, Sustainable Futures Institute, Michigan Technological University, Houghton, MI, David R. Shonnard, Chemical Engineering, Michigan Technological University, Houghton, MI, Micaela Taborga, Engr. Management, U of Arizona, Tucson, AZ and Michael Galka, U of Arizona, Tucson

Algal biomass is being investigated widely as a route to generating liquid transportation fuels using a variety of technologies. The sustainability analyses and life cycle assessments that have been completed in the literature are often based on a range of assumptions regarding the processes and technologies that cultivate, harvest, separate, and transform the cells into fuels and high value products. In this work, we use both bench and pilot scale data of several algal species to rigorously analyze novel harvesting, dewatering, oil extraction, and conversion technologies regarding the material and energy balance requirements. This life cycle inventory for bench scale and pilot plant data is transformed into an environmental footprint using life cycle assessment principles so opportunities and challenges regarding sustainability can be characterized.

Results show that energy requirements for harvesting range over three orders of magnitude on a kWh/gram of algae basis for some emerging technologies, including membrane technologies, electrochemical flocculation, and flocculation with chitosan, which has lower energies than some existing technologies. However, material needs for these processes incur other embodied energy, water, and greenhouse gas emissions that increase the environmental footprint by several factors for some of the technologies. Environmental impacts for extraction methods are generally larger than for algae harvesting, due to lower efficiencies and more energy intensive stages.

Scenarios are investigated that combine different commercially available technologies that have been previously modeled environmental footprints to the experimentally based emerging technologies. Modified processing trains have the potential of enhancing performance and increasing environmental sustainability so sustainable production of algal based biofuels is possible.

(704d) Evaluating the Sustainability and Resource Consumption of Biomaterials Using Eco-LCA and Spatial Analysis

Erin F. Landers and Bhavik R. Bakshi, William G. Lowrie Department of Chemical and Biomolecular Engineering, The Ohio State University, Columbus, OH

Ecosystem goods and services are essential to all human activity. In spite of this importance, the Millennium Ecosystem Assessment has found that many ecosystem services are highly degraded. In many engineering analyses, these necessary goods and services are undervalued, or even ignored completely. It is therefore crucial to develop and modify existing methods to allow these services to be valued and accounted for in decision making. Often, conventional life cycle assessment (LCA) tends to only look at energy consumption and emissions. However, Ecologically Based LCA (Eco-LCA) is a thermodynamically based LCA that allows for many ecosystem goods and services used throughout our economy to be quantified, in addition to the traditional LCA results of energy consumption and emissions. Eco-LCA looks at resource use throughout the entire economy to determine how much various industries and commodities consume. The commodities that we directly get from ecosystems are called provisioning services and include valuables such as food, water, energy, and other materials. There have been recent additions to the Eco-LCA model and it now accounts for even more ecosystem services (particularly provisioning services) than it used to, including many valuable intermediate flows throughout the economy, such as grains, livestock, and other food sources. This enables us to present our results on sustainability from an engineering perspective to others in the sustainability field with different backgrounds, such as ecology, business, or public policy.

In addition to the insight we can get from the economy-scale Eco-LCA, we have also begun incorporating spatial analysis into LCA. By using specific, spatial data, we can achieve levels of precision that would otherwise be difficult to realize. We do not need to make assumptions about the general conditions of different land uses, we can choose a specific plot of land and know the exact conditions on the piece of land and how that affects the sustainability of the resources coming from that land. We can also introduce time through multiple sets of spatial data, allowing us to see the effects of our decisions on ecosystems. The ecological models that we are using come from the InVEST (Integrated Valuation of Ecosystem Services and Tradeoffs) software from the Natural Capital group. InVEST uses spatial data such as Digital Elevation Models (DEM) and Land Use/Land Change (LULC) maps to feed into the ecological models, which include marine, freshwater, and terrestrial models. From InVEST, we can get new maps of different possible future scenarios as well as aggregated data on total ecosystem service use for the parcel of land.

A case study will be presented using both of these techniques to achieve a very comprehensive LCA. First, we will use spatial analysis to look at the farmland that will be growing our feedstock. It will allow us to better quantify the ecosystem services required for our product, such as pollination, water use, carbon sequestration, and sediment runoff. We will then use the results from that to feed into Eco-LCA, which will allow us to see what resources and valuable intermediates will be necessary for both the inputs to the farming as well as the industrial processing of our feedstock to various biomaterials. From these two methods, we can get information to calculate metrics such as Carbon, Nitrogen, and Water Footprints; Return on Investment; and Renewability. In using both of these methods, we will get a more complete picture of ecosystem service requirements for biomaterials and their relative sustainability.

(704e) Economic and Environmental Assessment of Bio-Based Chemicals Production in Singapore's Petrochemical Cluster

Ang Li[1], Arief Adhitya[2], Iskandar Halim[2], Balaji Balagurunathan[2] and Rajagopalan Srinivasan[1], (1)Department of Chemical and Biomolecular Engineering, National University of Singapore, Singapore, Singapore, (2)Process Science and Modeling, Institute of Chemical and Engineering Sciences, Singapore, Singapore

Singapore is one of the major centers for oil, gas, and petrochemicals in the world. The centerpiece of Singapore's petrochemical industry is the Jurong Island complex, which houses many of the world's leading petroleum and petrochemical companies. Currently, the total refining capacity of Singapore is 1.3 million bpd [1] and the ethylene capacity is 2.1 million tons per year, projected to reach 4 million tons per year by 2014 [2]. The petrochemical industry in Singapore is heavily dependent on fossil fuel-based feedstock, mostly crude oil and natural gas. Use of fossil fuels has negative environmental impact due to the greenhouse gas emission. Another crisis arising from overdependence on fossil fuels is their rapid depletion. Fossil fuels are non-renewable, which means they are finite and might well run out in the near future – a study found that fossil fuels reserve depletion times for oil, coal, and gas are approximately 35, 107, and 37 years, respectively [3]. This rapid depletion might render petrochemical production increasingly unsustainable as fossil fuels prices continue to rise. Thus, alternative sources have to be found for continued production and growth of the petrochemical industry.

Biomass is one possible source of renewable feedstock. Biomass is abundantly available on earth in the form of plants and organisms and considered to be carbon neutral [4]. Due to the complex carbon backbone and existing chemical functionality, the processing of biomass along the chemical synthesis chain can significantly reduce the energy requirement as compared with fossil fuels-based feedstock [5]. In this paper, we investigate the feasibility of incorporating bio-based raw materials into the slate of feedstocks available for chemicals production in Jurong Island. We survey the chemicals currently produced in the cluster and identify potential bio-based chemicals that can be produced. A techno-economic assessment based on current technology is then conducted taking into account production quantities, costs, and conversion yields. After the necessary supply chain is established, cost and carbon footprint assessment of the supply chain using various bio-renewable feedstocks sourced from the region is presented.

References:

1. http://www.edb.gov.sg/edb/sg/en_uk/index/industry_sectors/energy/industry_background.html, accessed on 1 May 2012.
2. http://www.siew.sg/energy-perspectives/energy-singapore/singapores-role-global-ethylene-production, accessed on 1 May 2012.
3. Shafiee, S. and Topal, E. (2009). Energ Pol, 37(1), 181-189.
4. Boom, R., Brehmer, B., and Sanders, J. (2009). Maximum fossil fuel feedstock replacement potential of petrochemicals via biorefineries. The Institution of Chemical Engineers. Elsevier B. V.
5. Sanders, J., Scott, E., Weusthuis, R., and Mooibroek, H. (2007). Macromol Biosci, 7(2), 105-117.

(704f) Analysis of the Life Cycle of 2-Methyltetrahydrofuran Produced From Agricultural Waste

Mariano J. Savelski[1], C. Stewart Slater[1], Dave Aycock[2], Steve Prescott[2] and Jeff Shifflete[2], (1)Chemical Engineering, Rowan University, Glassboro, NJ, (2)PENNAKEM, Memphis, TN

The life cycle emissions to manufacture 2-methyltetrahydrofuran from agricultural by-products have been evaluated. This solvent has the potential as a substitute for other ether and chlorinated solvents commonly used in pharmaceutical and fine chemical syntheses. The starting material, 2-furfuraldehyde, is produced from corn cob waste by acid hydrolysis. The process to convert 2-furfuraldehyde to 2-methyltetrahydrofuran occurs in two hydrogenation steps. Cradle to gate life cycle emissions for 2-methyltetrahydrofuran are not available in commercially used environmental databases, so a new inventory was created based on manufacturing information. The raw materials, energy and wastes have been analyzed for each production stage and an evaluation of the life cycle emissions performed. Our study shows that 2-methyltetrahydrofuran exhibits lower impact scores in the emissions categories when compared to solvents such as tetrahydrofuran, made through a conventional chemical route. A further assessment has been performed on the methods to manufacture tetrahydrofuran. A production pathway starting with the bio-based 2-furfuraldehyde process is greener than a 1,4-butanediol route.

(724a) Simultaneous Production of Liquid Fuels and Hydrogen From Hybrid Feedstock: Switchgrass and Shale Gas

Mariano Martín, Chemical Engineering, University of Salamanca, Salamanca, Spain and Ignacio E. Grossmann, Chemical Engineering Department, Carnegie Mellon University, Pittsburgh, PA

Over the last decades there has been an enormous effort to develop alternative fuels for transportation. Bioethanol and biodiesel are the most promising ones due to their compatibility with the automobiles and the crude supply chain. However, there are a number of challenges that biofuels must overcome mainly due to the large demand of liquid fuels and the limited production capacity of biofuels from biomass resulting in the need for large harvesting areas and for raw materials that do not compete with the food market. In the mean time hybrid first and second generation biofuels, using corn grain and stover for the production of bioethanol (Cucek et al 2011) or hybrid biomass – fossil fuel feedstock (Baliban et al 2012) can be used to reduce the dependency on fossil fuels and to serve as a bridge between technologies. Recently a new and plentiful source has been found, shale gas. Shale gas is natural gas that is trapped within shale formations. Of the natural gas consumed in the United States in 2009, 87% was produced domestically; thus, the supply of natural gas is not as dependent on foreign producers as is the supply of crude oil, and the delivery system is less subject to interruption and can be used as raw material for liquid fuels.

In this paper, we focus on the conceptual optimal design of multiproduct facilities for the production of green diesel from lignocellulosic biomass and shale gas feedstocks in an attempt to increase the yield to hybrid fuels while reducing the dependency on foreign supply in order to reduce its production cost to make it competitive with bioethanol from switchgrass (Martin & Grossmann 2011). We consider swichgrass as lignocellulosic feedstock since it is the raw material of choice by the DOE due to the harvesting possibilities in the US and its high yield to biofuels. We propose a limited superstructure optimization approach where we first construct a flowsheet embedding alternative routes for the production of liquid fuels. The process is based on FT technology in which the shale gas is reformed with steam to generate raw syngas. In parallel the switchgrass is gasified, reformed (we consider steam reforming or partial oxidation) to produce raw syngas. The total raw syngas is cleaned up from solids and sour gases. Next its composition may need to be adjusted for the optimal operation at the FT- reactor (using either WGSR, or PSA). The syngas is transformed into liquid fuels in an FT - reactor. The heavy liquids are upgraded using hydrocraking to increase the yield towards FT-diesel. The goal is to optimize the flowsheet, the product distribution (hydrogen, FT –diesel and gasoline production) and the operating conditions to maximize FT-diesel and hydrogen production determining the optimal consumption of raw materials (biomass and shale gas) as function of their price while minimizing the energy input for a number of scenarios where we fix the production level of liquid fuels and/or the maximum availability of biomass and shale gas devoted to fuels production to evaluate the profitability and competitiveness of this kind of hybrid facilities.

It turns out that the selected technology for reforming the raw syngas from the gasifier is steam reforming due to the higher yield to hydrogen. The process operates in such a way that the shale gas that is not required to meet the demand established for liquid fuels is used for the production of hydrogen that is sold to help in the economy of the plant as long as its price is low, otherwise it is better not to produce it. Thus, for this process to produce liquid fuels at around $1/gal, the biomass price must be kept below $100/t and the shale gas should not go beyond $11.5/MMBTU.

References:

Baliban, R.C.; Elia, J.A., Floudas, C. A. (2012) Simultaneous process synthesis, heat, power, and water integration of thermochemical hybrid biomass, coal, and natural gas facilities. Comp. Chem. Eng. 37, 10, 297–327

Cucek, L., Martín, M., Kravanja, Z., Grossmann, I.E., (2011) Integration of Process Technologies for the Simultaneous Production of fuel Ethanol and food from Corn grain and stover. Computers and chemical engineering 35,8, 1547-

1557

Martín, M., Grossmann, I.E. (2011) Energy optimization of lignocellulosic bioethanol production via gasification. AIChE J. 57, 12, 3408, 3428

(724b) Streamlining Lifecycle Greenhouse Gas Emissions of Coal, Conventional and Unconventional Natural Gas for Electricity Generation

Ozge Kaplan, Office of Research and Development, U.S. Environmental Protection Agency, Research Triangle Park, NC

An analysis of the lifecycle greenhouse gas (GHG) emissions associated with natural gas use recently published by Howarth et al. (2011) stated that use of natural gas produced from shale formations via hydraulic fracturing would generate greater lifecycle GHG emissions than petroleum per unit of energy input. The analysis also suggested that the lifecycle GHG emissions from natural gas (at the point of input into a power plant) could exceed (or be "quite similar to") those from coal. We re-evaluated the estimated lifecycle GHG emissions from Howarth's study, reviewed other published literature on lifecycle analysis of natural gas and coal, and calculated normalized full lifecycle emission factors per MWh of base load electricity generated. The CO_2equivalents were presented on both 20- and 100-yr time horizons to evaluate both near- and long-term impacts. Howarth et al. (2011) raise valid points regarding the need for considering emissions over the full lifecycle, including fugitive CH_4 emissions from hydraulically-fractured natural gas wells and distribution systems, as well as the need to recognize the time basis (e.g., 20-yr, 100-yr, 500-yr) for comparing the climate impacts of different energy sources. However, because their analysis did not account for the efficient use of natural gas in advanced combustion technologies, their estimates appear to be much more representative of worst-case conditions than average industry emissions.

(724c) Biomass and Natural Gas to Liquid Transportation Fuels (BGTL): Process Synthesis, Global Optimization, and Topology Analysis

Richard C. Baliban, Josephine A. Elia and Christodoulos A. Floudas, Chemical and Biological Engineering, Princeton University, Princeton, NJ

Heavy dependence on petroleum and high greenhouse gas (GHG) emissions from the production, distribution, and consumption of hydrocarbon fuels pose serious challenges for the United States transportation sector [1]. Depletion of domestic petroleum sources combined with a volatile global oil market prompt the need to discover alternative fuel-producing technologies that utilize domestically abundant sources. A recent review highlights the single and hybrid feedstock design alternatives that can produce gasoline, diesel, and kerosene from coal, biomass, or natural gas [2]. Biomass and natural gas are two domestic energy sources that have received considerable attention as feedstock candidates for processes that can generate synthetic liquid fuels. The combination of these two feedstocks offers an important advantage over single feedstock systems because the benefits of both natural gas (e.g., cheaper, large domestic reserves) and biomass (e.g., sustainable, reduces GHG emissions) can be harnessed in a hybrid refinery.

We present an optimization based framework to perform a comprehensive technoeconomic and environmental assessment of an alternative energy refinery that will convert biomass and natural gas to liquid fuels (BGTL). The framework will incorporate process design, global optimization, and process synthesis strategies [3-10] to determine the optimal plant design of a BGTL refinery under different scenarios (e.g., feedstock composition, plant capacity, liquid fuel type/composition, etc.). Using this framework, thousands of distinct process designs will be analyzed at once to determine the design that produces the liquid fuels in the most optimal way (i.e., lowest cost or highest profit). This superstructure of postulated topologies will include multiple components for biomass conversion, natural gas conversion, hydrocarbon production, and liquid fuels upgrading and will identify the economical and environmental tradeoffs between each component. A simultaneous heat, power, and water integration will be included in the optimization model to ensure that the cost of utility generation and wastewater treatment is adequately compared to the cost of producing the liquid fuels. Nine case studies are presented to investigate the effect of biomass type and plant capacity on the optimal process topology, total plant cost, and break-even oil price. The key topological decisions from the mathematical model along with the major components of the plant capital and operating costs will be discussed.

[1] National Academy of Sciences, National Academy of Engineering, and National Research

Council. Liquid Transportation Fuels from Coal and Biomass: Technological Status, Costs, and Environmental Issues. Washington, D. C., EPA, 2009.

[2] C.A. Floudas, J.A. Elia, R.C Baliban. Hybrid and Single Feedstock Energy Processes for Liquid Transportation Fuels: A Critical Review. *Comp. Chem. Eng.*, 2012: 41(11), 24-51.

[3] R.C. Baliban, J.A. Elia, C.A. Floudas. Toward Novel Hybrid Biomass, Coal, and Natural Gas Processes for Satisfying Current Transportation Fuel Demands, 1: Process Alternatives, Gasification Modeling, Process Simulation, and Economic Analysis. *Ind. Eng. Chem. Res.*, 2010:49(16), 7343-7370.

[4] J.A. Elia, R.C. Baliban, C.A. Floudas. Toward Novel Hybrid Biomass, Coal, and Natural Gas Processes for Satisfying Current Transportation Fuel Demands, 2: Simultaneous Heat and Power Integration. *Ind. Eng. Chem. Res.*, 2010:49(16), 7371-7388.

[5] R.C. Baliban, J.A. Elia, C.A. Floudas. Optimization Framework for the Simultaneous Process Synthesis, Heat and Power Integration of a Thermochemical Hybrid Biomass, Coal, and Natural Gas Facility. *Comp. Chem. Eng.*, 2011:35(9), 1647-1690.

[6] J. A. Elia, R.C. Baliban, X. Xiao, C.A. Floudas. Optimal Energy Supply Network Determination and Life Cycle Analysis for Hybrid Coal, Biomass, and Natural Gas to Liquid (CBGTL) Plants Using Carbon-based Hydrogen Production. *Comp. Chem. Eng.*, 2011:35(8), 1399-1430.

[7] R.C. Baliban, J.A. Elia, C.A. Floudas. Simultaneous Process Synthesis, Heat, Power, and Water Integration of Thermochemical Hybrid Biomass, Coal, and Natural Gas Facilities. *Comp. Chem. Eng.*, 2012:37(10), 297.

[8] R.C. Baliban, J.A. Elia, R. Misener, C.A. Floudas. Global Optimization of a Thermochemical Based Hybrid Coal, Biomass, and Natural Gas to Liquids Facility. *Comp. Chem. Eng.*, 2012, http://dx.doi.org/10.1016/j.compchemeng.2012.03.008.

[9] R.C. Baliban, J.A. Elia, V.W. Weekman, C.A. Floudas. Process synthesis of hybrid coal, biomass, and natural gas to liquids via Fischer-Tropsch synthesis, ZSM-5 catalytic conversion, methanol synthesis, methanol-to-gasoline, and methanol-to-olefins/distillate technologies. *Comp. Chem. Eng.*, submitted.

[10] J.A. Elia, R.C. Baliban, C.A. Floudas. Nationwide Energy Supply Chain Analysis for Hybrid Feedstock Processes with Significant CO_2 Emissions Reduction. *AIChE J.*, submitted.

Investment Opportunities in the Liquefied Natural Gas (LNG) Projects and Development: The Emerging Cleaner Fossil Fuel of Higher Energy Efficiency

M. N. Idris

Process Refining Technology, Department of Chemical Engineering, University of Maiduguri, Nigeria

ABSTRACT

The LNG remains one of the cleanest, cheapest (except electricity) and higher energy efficiency products toward the actualisation of global energy and utilisations. In these studies, concise effort was used to review the LNG process description and most importantly, the liquefaction processes in a modern scope. The focus on this paper is to highlight the various LNG investment opportunities globally. Apart from the usefulness of LNG production for economic gains, it help in reducing the flaring of global natural hydrocarbon which are further transform into economy values and environmental friendliness. In term of financial gain, Nigeria alone generates an approximate sum of revenue above $200 million US in a month, while the global world through LNG production has an approximate benefit to the tune of $5 billion US monthly. The interest on this study is to highlights the investment opportunities and possible gains to the investors in the oil and gas projects, because of its viability and high economic advantages in various ways when LNG is compared to other fossil fuel.

Keywords: LNG, NLG, MCHE, PMR, Investment Opportunities

Correspondence Author: *Dr. (Engr.) M. N. Idris; MAIChE, AMIChemE, MNSE (muhinu@hotmail.co.uk or idrismn@hotmail.com), Department of Chemical Engineering, University of Maiduguri, Nigeria and/or IPSE-ERRI, SPEME, University of Leeds, United Kingdom, LS2 9JT, Tel: +234(0)705 8771 096

1.0 INTRODUCTION

Natural gases (NG) can be classified as associated or non-associated. The associated natural gases are those that are present with liquid fossil fuel, while non-associated gases are those that occur naturally without any impurities. The various importance uses of natural gas is in lighting cooker/stove and heating houses domestically and industrially use in the high powered engine etc. Interestingly, the exportation of natural gas is increasingly growing, as domestic production has lagged behind this growing demand for this clean fuel. Liquefied natural gas (LNG) processing is typically created using three-step processes Idris et al (2007). Firstly, the subterranean gaseous-form natural gas is 'frozen' into a liquid state through a complex cryogenic process system (called '*liquefaction*'), this involve temperatures as low as -161 ^{0}C. The essence of the liquefaction process of LNG is to make transportation of natural gas easier, cheaper and convenient (Idris and Abowei 2004). Secondly, after the gas is liquefied, it can be stored in cryogenic holding tanks or pumped directly from the cooling vestibule into special insulated transportation vessels, such as railcars, trucks, or ships. Finally, upon delivery of the vessel to its final destination, LNG is pumped from the vessel into either another cryogenic storage tank for later delivery or directly into a re-gasification unit that uses sea water or air to reheat the LNG in order to convert it back into gaseous natural gas (Idris 2003).

The LNG processing sounds fluid, but usually is not quite as simple. Raw natural gas can be 'heavy' (also called 'wet' or 'hot' gas) with various higher carbon molecules. These heavier particles create a higher heat content (ranging from 11.28 – 13.66 kWh/m^3) than permissible by many pipeline distribution standards (PDSs). Although this heavier form of gas can be frozen, stored and shipped, the extra heat content may have to be 'stripped off' prior to entering another distribution system to conform to the gas heat standards of the pipeline network Idris (2003).

In the United States, state public utility commissions (PUCs) oversee the gas pipeline heat content standards. It is estimated that the average heat content of current distributable gas in the United States is approximately 10.66 – 10.97 kWh/m^3, HEA (2011). Any gas disseminated into a system that does not conform to these independent standards will not be permitted into the inherent system without the heavier particles being stripped (through a process called '*fractionation*') or diluted with inert gas. This fractionation process is an additional cost to the overall LNG project. Fractionation can occur either upstream or downstream, but is usually contingent upon which location is most cost effective, politically sound, environmental-friendly or where there may be access to a viable purchasing market for the fractionated heavier gas, although guaranteed the economies and long-plant life. Fractionated gas products can either be marketed or injected back into the ground. Any marketing and sales of this heavier gas can serve as a revenue enhancement of or a contra-expense item to the total LNG projects.

In a similar vein, such adopted standards on gas pipeline heat content measurements can be initiated in the Nigeria LNG project through the appropriate Federal Ministries in order to establish resource data for future research and development on the LNG project and all other energy efficiency projects. The primary aim of this study is to discuss the global investment opportunities in LNG projects and further buttress the cleaner and the environmental friendliness of natural gas over the conventional hydrocarbon products. Once this LNG turns into gaseous gas, it is pumped through a pipeline system that leads to an ultimate market. Natural gases have various important uses which are described as follows:

a) **Domestically:** They are use in heating building, heating water, cooking, drying clothes, lightning etc. Household appliances that use natural gas include: heater (furnace), pool and spa heaters, clothes dryers, outdoor lights, barbecues, water heaters, stoves etc.
b) **Industrially:** They are use in high powered engine, HPE; transport & locomotive engines etc.

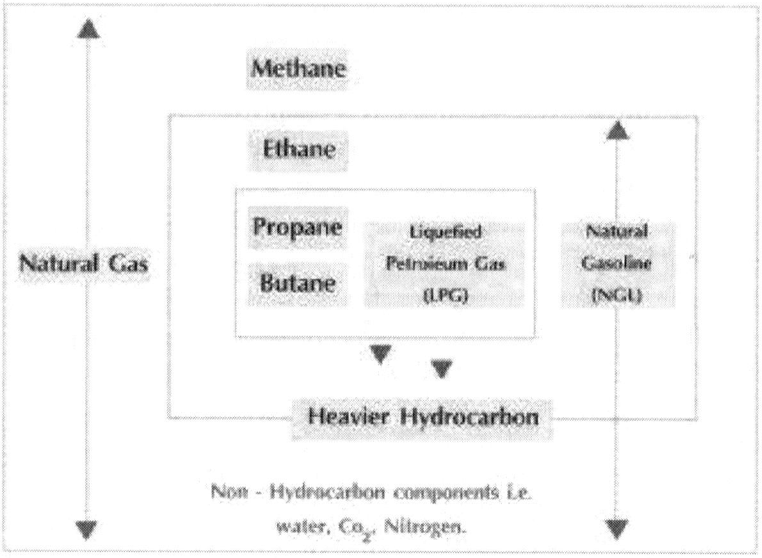

Figure 1: Composition of Natural Gas (Natural Gas Information, 2010)

Table 1 represent a detailed comparison of NG with other conventional fossil fuel. From this table it is clearly observed that NG is easy on engine with non-toxic explosive limit. NG gives a longer service life with lower maintenance costs involved. In addition, NG is the least expensive alternative fuel (except electricity) and it has a higher octane rating (Idris et al., 2006).

Table 1: Comparison of natural gas with gasoline, diesel oil and liquefied petroleum gas, LPG

Properties	Physical state	Boiling Range (°C @ 1 atm)	Density kg/m³	Net Content Energy *10⁶ kJ/m³	Auto Ignition Temperature (°C)	Flash Point (°C)	Octane Number Range N_R=(R+M)2	Flammability Limits	Volume % in Air	Explosive limit for Fuel (PPM)
Gasoline	Liquid	26.67 - 215.56	688.79 - 784.90	5.212 - 5.323	232.2 - 482.2	-42.7	87 - 93	L = 1.4	H = 7.6	500
Diesel No. 2	Liquid	160 - 382.2	784.90 - 881.01	5.268	204.4 - 260	51.67	n/a	L = 0.7	H = 5.0	n/a
LPG (HD-5)	Gas	-42 - 0.56	496.57[b]	5.519	494.3 - 548.89	73.3 - -65.5	104[c]	L = 2.4	H = 9.6	n/a
NG	Gas	-161.67	128.15	5.937	732.2	-148.89	120[c]	L = 5.3	H = 14	nontoxic

Source: NRIS (2005)

Table 2 expresses the thermo-physical properties of methane, ethane and ethylene. This comparison is to further represent qualitatively the high grade and environmental acceptability of methane with other conventional fuel products.

Table 2: Thermophysical and chemical properties of methane, ethane and ethylene

Property	**Methane**	**Ethane**	**Ethylene**
Chemical Symbol	CH_6	C_2H_6	C_2H_4
Molecular weight	16	30	28
Normal boiling point (K)	111.7	184.6	169.3
Freezing temperature (K)	89.9	90.6	104.2
Critical temperature (K)	191	305	283
Critical pressure (bar absolute)	47	49.7	52.5
Expansion ratio-increase in volume as liquid as 1 bar abs boil to gas as 1 bar, 15°C	626	437	489
Density of standard liquid at 1 bar abs (kg/m³)	424	546	565
Latent heat of vaporisation (cooling potential of phase change) (kJ/kg)	512.4	488.3	483.4
Relative gas density (referred to dry air at 1 bar abs,) 15°C at density 1.21kg/m³	0.56	1.05	0.97
Air liquefaction hazard	No	No	No
Spontaneous ignition temperature in air (K)	813	783	753
Flammability limits in air (K)	5 - 15	3 - 12.4	2.7 - 36
Minimum ignition energy (MJ)	0.28	0.24	0.085
Flame temperature (K)	2153	2168	2248
Limiting oxygen index (vol %)	11.5	11.0	11.5

Source: Tiratsoo (1979) and BCC (2004)

2.0 THE LNG PROCESS AND OPERATIONS

2.1 The Main Cryogenic Heat Exchanger (MCHE) and LNG Production

The MCHE is the liquefaction unit in an LNG process that cryogenise the treated natural gas from -32 ^{0}C to the desirable temperature -161 ^{0}C by a physical process using the principles of appropriate simulation model. The MCHE is the most important and delicate unit of LNG process plant, because of it significant values in the LNG processing. The treated and dried natural gas (NG) from the mercury removal unit is further processed and liquefied in the liquefaction unit. The processing includes removal of heavy ends in the natural gas section, to help preserve the life of the process equipment and the quality of the LNG. The essence of the MCHE unit helps in liquefaction of the LNG by one-six hundredth of the ideal volume of treated natural gas for economic and easy transportation overseas (Idris et al., 2006).

However, the LNG production from every process train is strongly influenced by the ambient air and the cooling water temperature. At high site temperatures, the power required to produce a quantity of LNG is greater than that required to produce the same quantity at lower site temperatures. Also, at a high site temperature, the power available from the gas turbines to drive the refrigerant compressor is less than that available at lower site temperatures. These factors result in variations in the maximum LNG production rate from day to night and from season to season Idris and Abowei (2004). Figure 1 represents the flow sheet of the Main Cryogenic Heat Exchanger MCHE) or liquefaction unit of LNG Train, while Figure 2 depicts the existing train of the Nigeria LNG Production unit.

2.2 The LNG Energy Efficiency and Opportunity

The LNG energy efficiency can be easily being compared with the conventional gasoline and diesel oil as represented in Table 3. The natural gas which is being liquefied in the LNG plant is an emergent fuel of choice, and considered environmentally friendly, higher economic value and reduced hazard effects.

Table 3: Fossil fuel consumption per 1.05587×10^{-9} kg of gas emission

Gases	Natural Gas	Gasoline Oil	Diesel Oil
Carbon Dioxide	**52.153**	**74.828**	**95.235**
Carbon Monoxide	**7.710**	**14.966**	**9.070**
Hydrocarbons	**0.635**	**2.2675**	**10.431**
Nitrogen Oxide	**45.350**	**149.655**	**378.219**
Sulphur Dioxide	**0.272**	**453.500**	**770.950**
Particulates	**2.268**	**37.641**	**1405.850**

Sources: Modified approach to GEH (1965) and HEA (2011)

From Table 3, it is clearly observed that the natural gas production has various advantages over the conventional gasoline and diesel oil. Therefore, for economics, environmental and health advantages, increase in global production of natural gas could give the energy consumption markets greater opportunities.

3.0 OPPORTUNITIES AND CAPITAL INVESTMENT IN LNG PRODUCTION

3.1 The LNG and the Global Investment

In the year 2012 and beyond, it has been reported that the new LNG-related capital investment would be above $62 billion US as LNG becomes a significant component of the energy industry's marketing strategy HEA (2011). Historically, the price of LNG has been prohibitive when compared to United States (US) gas prices reference. The costs of delivery ranged are $0.726 to $0.871/mkWh range (not including the netback price to the owner of the stranded gas reserves from which the gas was initially purged). Assuming a $0.363mkWh netback to such owner, a total deliverable gas price of around $0.726 to $0.871/mkWh could possibly be attained on a cost basis HEA (2011).

In the 1970's and 1980's, physical gas prices had spiked in the US and Europe, feeding fears of gas shortage. This phenomenon inspired the construction of four major LNG import re-gasification terminals, all on the United States East Coast (Everett, Massachusetts MA, Elba Island, Georgia GA, Cove Point, Maryland MD, and Lake Charles, Los Angeles LA, in total constituting approximately 1.05×10^2 billion m^3/day of installed capacity).

However, deregulation led the producers of natural gas to become motivated to explore and develop newer, easy access gas reserves in the mountain states and the Gulf Coast, flooding the market with excess gas inventory. This was exacerbated by Canadian policy changes that allowed increased gas exports, in effect emigrating additional low-cost gas from the Western Canadian Sedimentary Basin (WCSB) to compete for demand in the United States. The result was a gas supply 'bubble' which kept prices very low for quite some time. Imagine how times have changed. Based on the Department of Energy's Energy Information Administration (EIA), it was anticipated in 2011 medium-term Henry Hub gas prices to range from $0.769/mkWh to $0.726/mkWh, and long-term gas prices to be close to $0.871/mkWh.

In analytical forecasting of natural gas prices, one can label the expected gas price a 'dependent' variable and the underlying fundamentals as 'independent' variables that influence the dependent variable. Using quantitative techniques, Hybrid Energy Advisors, Inc. (HEAI) predicted in the spring of 2002 that Henry Hub financial prices, the dependent variable, would range between $3.35 and $4.10 through the winter months of 2002/2003. They also calculated ten-year price forecasts of $0.5445 - $0.6776/mkWh during the summer months and $0.8954 - $1.0285/mkWh during the winter months. HEAI believes that ten-year city gate prices of gas on the West Coast and Midwest United States will trade approximately equivalent ('flat') to Henry Hub prices, and upper East Coast prices to trade at approximately 13.06 – 18.87 basis points (cents/mkWh) above Henry Hub prices HEA (2011). These forecasts include a greater gas import as a percentage of total domestic gas. HEAI forecasted that by the year 2011, approximately 6 - 10% of the total US gas supply will come from LNG produced from foreign stranded gas reserves. In a similar vein, a forecast in the European and Asian gas consumption of similar trend has been predicted.

Today the total cost of LNG production has been quite streamlined and reduced to approximately $0.484/mkWh due to competition and improvements in technology. Much of the technology pertaining to lower-cost land-based terminals and offshore and ship-board re-gasification units are still relatively unproven. However, major energy industry players and engineering firms do not foresee problems with the design and implementation of these methods HEA (2011).

Therefore, assuming a netback of $3.44/mkWh, the total delivered price of approximately $10.45/mkWh can now be theoretically achieved. This LNG price is almost $3.44/mkWh less than a decade ago. When comparing this new price to industry average price forecasts for Henry Hub and East and West coast city gates, LNG can certainly be considered an economical source of natural gas supply.

3.2 The Nigeria LNG Gas Reserve and Reduce Flaring

Nigeria is highly blessed with massive reserves of associated and non-associated gas, estimated in excess of 160 trillion cubic feet (tcf). It is ranked amongst the 10th largest in terms of proven natural gas reserves in the world, and its natural gas reserves / production is estimated at 109 years. The expert geologists has forecast that there is a lot more gas still to be found, if companies deliberately explore for gas, as opposed to finding it by chance whilst in search of oil NLNG (2012). Unfortunately, Nigeria is one of the biggest gas flares in the world, in spite of the associated environmental hazards of gas flaring. The government is, however, working to achieve a target of zero flares in the near future. It is expected that all the NLNG trains, including the proposed Train 7, operating at full capacity will play a significant part to help put out these flares NLNG (2012).

3.3 The Nigeria LNG and the Progress So Far

The Nigeria LNG Ltd remains the largest consumer of the associated gas in the Nigeria gas project through the stakeholders companies (NNPC, SPDC Shell Gas B.V., Total LNG, and ENI International). A brief of the stakeholders are described as follows:

The Nigerian National Petroleum Corporation (NNPC) was established in 1977 under the constitution of the Federal Republic of Nigeria. It is the corporate entity through which the Nigerian government participates in the oil and gas industry. NNPC and its subsidiaries dominate all sectors of the industry - exploration, production, refining, pipelines, marketing, crude /product exports, and petrochemicals. NNPC owns 49% of the shares in Nigeria LNG Limited.

The Shell Gas B.V (SGBV) is a company incorporated under the laws of the Netherlands. Besides its interest in Nigeria LNG, Shell Gas and Power has interests in many of the world's major LNG production companies including those in Brunei, Malaysia, Australia and Oman and is one of the Royal Dutch Shell Group of Companies which operate throughout the world in all aspects of the petroleum industry. It sells over 80 billion cubic metres of gas yearly. SGBV owns 25.6% of the shares in Nigeria LNG Limited.

Total is a major integrated oil and gas company active in all aspects of the petroleum industry. Total explores in 25 countries, and has reported proven reserves of 3,639 million barrels of oil equivalent (492 million tons). Of these, crude oil accounts for 71% and natural gas for 29%. Total owns 15% of the shares in Nigeria LNG Limited.

Eni is a fully integrated oil and gas company engaged in all aspects of the petroleum business. Eni is involved in exploration, development and production of oil and natural gas. Eni owns 10.4% of the shares in Nigeria LNG Limited.

Currently, the Nigeria LNG Ltd is made up of 6-functional cryogenic plant in Bonny Island, Rivers State. Other proposed LNG projects include Olokola LNG and Brass LNG among others. The Nigerian Gas Company currently caters for Independent Power Producers, and local industries in Lagos and other parts of the country. The proposed West African Gas Pipeline Project (WAGPP) will make gas available to other West African countries for domestic and industrial uses.

4.0 MAJOR SUPPLY AND DEMAND ZONES OF LNG

New technologies in LNG and the upward gas price trends are creating the opportunities for re-gasification terminals to spring up in a variety of areas. These areas include the United States East and West Coasts, Northwest Mexico, Italy, Germany, England, Spain and even emerging markets such as China and India. Asian counterparties such as Japan, Korea and Taiwan, who have historically constituted approximately 70% of demand in the LNG market, are also building new terminals. The gas supply to meet expected growth in LNG demand is from stranded gas reserves associated with crude oil production that historically did not have much hope of coming to market. These reserves are typically owned and operated by many major worldwide energy companies. Longstanding production zones include Algeria, Libya, Nigeria, Qatar, Malaysia, Australia, Brunei, Indonesia and the United States (Alaska, Cook Inlet). Many proposed new areas of LNG liquefactions include Norway, Trinidad & Tobago, Venezuela, Bolivia and various countries of the Middle East.

The ideal economic scenario when evaluating the LNG terminal projects make sense with contingent upon studies related to where and how much gas demand is going to be, where the supply will come from, and how much gas to produce.

5.0 THE FUTURE OF GLOBAL LNG PRODUCTION AND CONSUMPTION

5.1 The LNG Production and Consumption

While natural gas consumption growth rates are expected to increase over the next twenty years by 50% as compared to a current level of 1.84 billion m^3/day, the solution to providing the supply to satiate such growth is not quite as transparent as just evaluating the face-value economics of alternative sources.

In any energy infrastructure project, due diligence on regulatory, permitting, and environmental factors is quite important. A hiccup in procuring the necessary political approvals, permits and rights-of-way can be devastating to the economics and viability of a project. Projects have been known to come to a dead halt, much to the chagrin and expense of the lead project developer and production consortium, over matters such as residential noise or particulate pollution, wildlife endangerment, political vacillation, or just plain general nuisance.

LNG re-gasification terminals are viewed as large, obstructive and generally displeasing to the eye for local residents and businesses. Although designs to bury such large facilities underground have been forthcoming, general apprehension related to the "don't build in my backyard" philosophy still emphatically exists. One mitigant to that problem is sighting the terminal off-shore or aboard the LNG transport ship itself, far from residential or commercial areas. Although such new technologies regarding offshore and ship-board re-gasification is being studied and developed by a variety of industry players, the educating process required with political parties is an on-going and sometimes frustrating process.

The guidelines regarding the oversight, regulatory standards and environmental compliance of on-shore or off-shore terminals are still in debate. In the United States, three of the four import terminals are overseen by the federal energy regulatory commission, FERC. Regulatory oversight can be an issue to contend with as most merchant energy firms and pipeline companies vie for ownership, control, and equity interest. For such market participants, FERC jurisdiction impedes profitability and ease of management control of the owners and its merchant gas off-take counterparties. Many of the industry participants building and sighting these terminals contend to the FERC that if they have the burden of finding, constructing, funding, and permitting a terminal, they should not be subject to losing the economic potential to sell such re-gasified gas into open market by subjecting the capacity to a FERC-controlled 'open-season'. After all, if one firm does all of the work, why would that firm surrender the market opportunity to profit from the off-take gas sales to another firm who has not incurred the same 'finders' costs? For many of these project leaders, equity ownership of a FERC-jurisdiction terminal at a regulated rate of return is not enough. They want the most facile way to attain the real 'juice', that is, they prefer the sole ability to sell their upstream LNG into the interstate domestic markets as gaseous gas and take merchant positions in key locations.

5.2 The LNG non-Operational Risks

The two LNG-related non-operational risks that are worth mentioning at this time are terrorism and/or accident risk and sovereign government risk. Terrorism has been shown to be a more significant influence on energy endeavours than ever before. With the threat of oil tanker, nuclear reactor and LNG ship sabotage, political risk to sighting an on- or off-shore re-gasification unit is high. Man-made or accidental explosion of such a tanker can be catastrophic in the mind of the common person, and although studies have shown that LNG volatility is much lower in its liquid cooled form, the image of a giant vapour cloud can have devastating effects on the human psyche and at the very least create an apocalyptic image that few are willing to internalize. However, the Nigeria LNG Ltd has been having smooth operations of the production trains since the official commissioning in October 2000. This success is due to a trusted and fulfil memorandum of understanding, MOU with the hosting communities.

Secondly, since much of the gas that is being liquefied in the western countries (US and Europe) comes from foreign and third-world countries, sovereign risk can be quite disenchanting as compared to reliable, high-credit, highly-liquid United States gas. Changing governments, political coups, local customs, and draconian regimes can portend inconsistent supply. Due to this factor, price concessions are usually made in relation to United States-priced gas supply, however, the sovereign risk inherent in such supply must to be evaluated concomitantly with the discounted pricing.

6.0 CONCLUSIONS

The global world is in need of additional natural gas sources based on forward fundamental supply/demand and price expectations. More so LNG appears to be one of the viable and constantly improving choices of energy source. While the economic viability of LNG as a necessary gas supply-stack contributor as compared other hydrocarbon in many ways. However, the issues on politics, sovereignty and regulatory risk can impede the progress of such beneficial projects and turn what was once a supply saviour into a bureaucratic roundabout of the hosting communities. To the global world, the LNG and the reduced gas flaring has an approximate benefit to the tune of $5 billion US in a month. Nigeria alone, generate a sum of $200 - $300 million US monthly. This study is to serve as an eye-opener to investors in the oil and gas projects because of its viability and high economic advantages in various ways over the conventional hydrocarbon products (gasoline and diesel oil etc.).

Acknowledgement

The authors wish to acknowledge the support of the Hybrid Energy Advisors HEA, Inc. based in Houston, Texas USA for providing some of the information used in this report. The support of the petroleum technology development fund (PTDF) toward the research work carried at the University of Leeds is highly appreciated.

Abbreviations

EIA:	-	Energy Information Administration
FERC:	-	Federal Energy Regulatory Commission
HEAI	-	Hybrid Energy Advisors Inc.
LNG:	-	Liquefied Natural Gas
NLG:	-	Natural Gas Liquid
MCHE:	-	Main Cryogenic Heat Exchanger
PMR:	-	Propane Mixed-Refrigerant
PUCs:	-	Public Utility Commissions

REFERENCES

1. BCC British Cryogenic Council (2004), 'Cryogenic Safety Manual: A guide to good practise', 3rd Ed., pp52-76
2. Gas Engineers Handbook (1965), American Gas Association, New York, USA
3. M. N. Idris, Z. Abdu and A. S. Ahmed. (2007), 'Compressed Natural Gas (CNG) development in Nigeria's growing economy: A review', International Petroleum Technology Training Journal, Jan 2007, vol. 4, No.1
4. M. N. Idris and M. F. N. Abowei (2004), 'Performance optimization of a liquefied natural gas (LNG) process train', Journal series of Engineering, University of Maiduguri, Oct 2004, vol. 1, No. 3
5. M. N. Idris, A. S. Ahmed and S. S. Adefila (2006), 'Advances in Compressed Natural Gas (CNG) development in Nigeria's growing economy', Journal of Engineering series of university of Maiduguri, Oct 2006, vol. 4 No.1
6. Natural Gas Information
7. NRIS (2005), Natural Resources Information Services 2005, P.C. McKenzie Company, Pittsburgh, USA PA 15241
8. NLNG (2012), the copyright reserve of the Nigeria LNG Limited.
9. Idris M. N. (2003), 'Modelling and Simulation of an Main Cryogenic Heat Exchanger (MCHE) of an existing Liquefied Natural Gas (LNG) Process Unit', An M.Tech Thesis submitted to the Rivers State University of Science and Technology (RSUST) Port Harcourt, Rivers State, Nigeria.

10. HEA (2011) Hybrid Energy Advisors, Inc. based in Houston, Texas USA (2011) www.hybrid-advisors.com and directly by e-mail at information@hybrid-advisors.com

11. Tiratsoo E. N. Natural Gas (1979), Beaconsfield Science Press Int. London 3rd Ed pp15-89

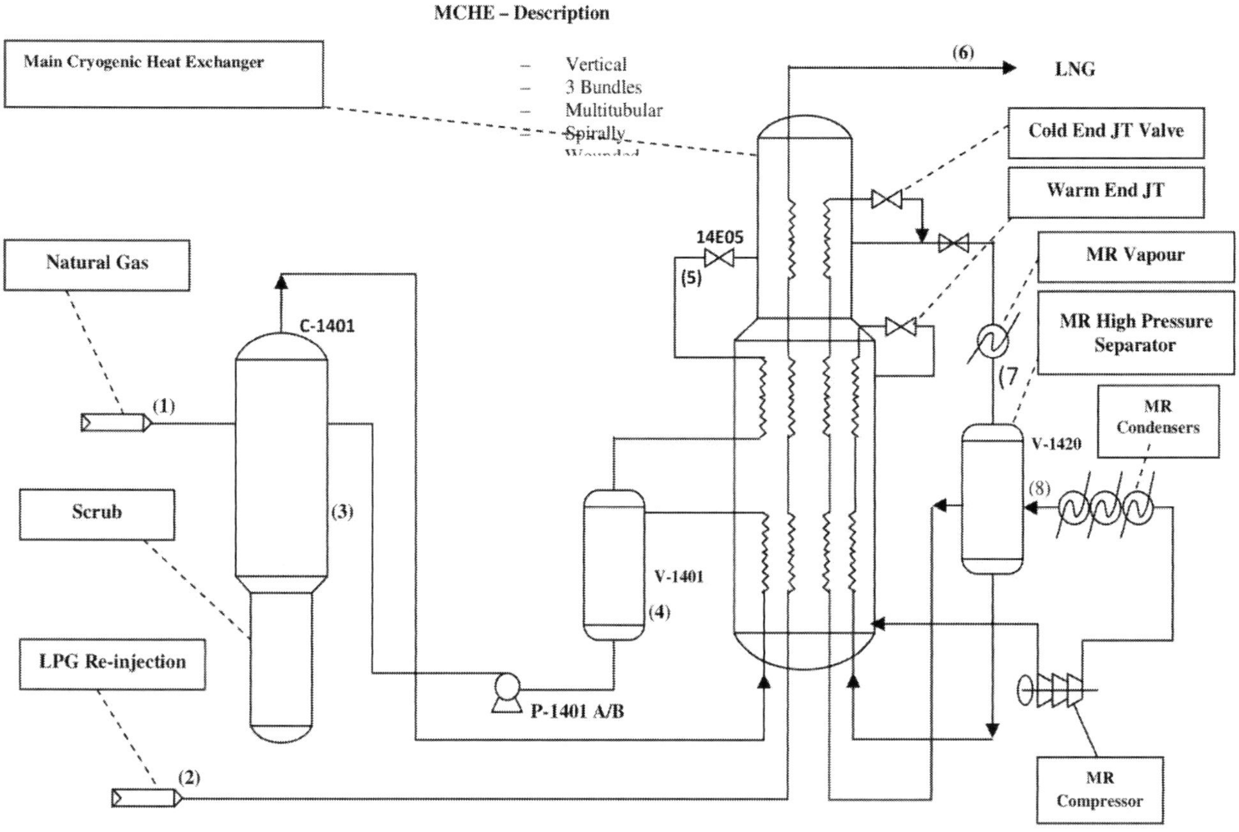

Figure 2: Flow sheet of the Main Cryogenic Heat Exchanger MCHE) or liquefaction unit of LNG Train

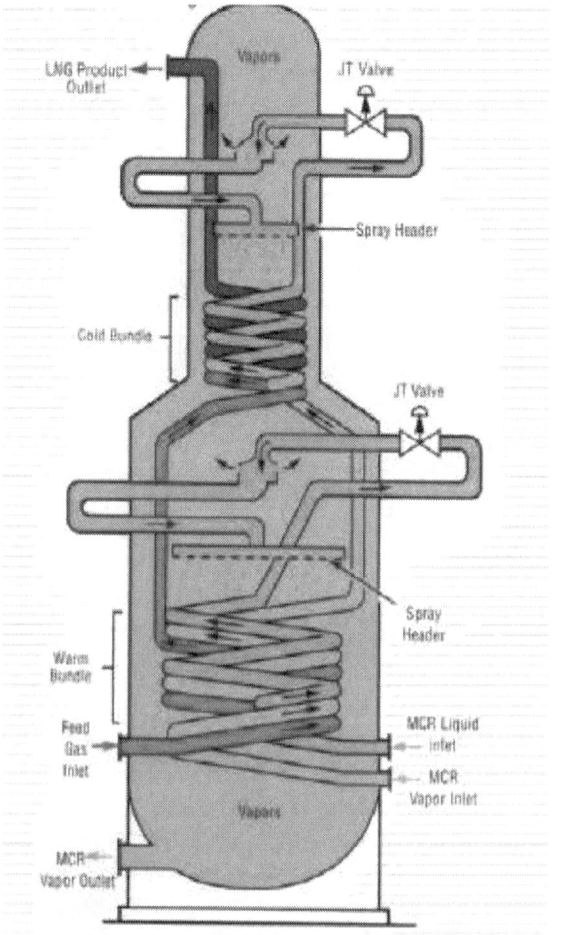

Figure 3: Design flow of MCHE (LNG and World of Energy, 2000)

Figure 4: LNG Production train, the Nigeria LNG Ltd. Source: http://www.nlng.com/

(724e) Co-Fire Pellets Consisting of Coal and Pyrolysis-Derived Binder

Andrew Friend[1], Robert Brown[1] and R. Christoper Williams[2], (1)Center for Sustainable Environmental Technologies, Iowa State University, Ames, IA, (2)Department of Civil, Construction, and Environmental Engineering , Iowa State University, Ames, IA

The goal of this research is to allow utilities with coal-fired plants to meet renewable energy mandates and reduce greenhouse gas emissions through the utilization of a co-fire pellet consisting of coal and a fast pyrolysis-derived binder. In the past decade, carbon dioxide and other greenhouse gases (GHGs) have been subjected to intense study for their contribution to climate change. Electrical utilities are facing federal, state, and local mandates for renewable power production. At the same time, utilities are also facing existing and upcoming regulations that demand emissions reduction from their current sources of power generation. In order to meet these mandates and regulations in a timely and inexpensive manner, electrical utilities are adding more "green" options to their power portfolios. Biomass is a near term option due to its low cost and easy integration into the existing coal-fired infrastructure.

Biomass co-firing is heralded as a simple and inexpensive way to reduce CO_2 and other greenhouse gas emissions from power plants. However, there are many practical difficulties in storing and co-firing two fuels with large differences in heating value, bulk density, moisture content, and ash composition as often occurs for coal and biomass. This can lead to reductions in boiler capacity and combustion efficiency, increases in emissions of regulated pollutants, and loss of markets for coal ash.

Our approach to solving several of the problems of conventional co-firing is to create a pellet from coal and a co-product of fast pyrolysis. These pellets have many advantages over both traditional coal combustion and co-firing with biomass. Co-fire pellets have approximately the same or greater heating value than the coal used in the pellets. They can be handled with existing coal infrastructure, injected into stoker-grate boilers or pulverized for use in PC boilers with little loss in boiler capacity or ash properties. They also contain substantially lower alkali levels than biomass, allowing for a wide utilization of annual or perennial biomasses without fear of boiler corrosion or fouling. Greenhouse gas emissions are much lower for the co-fire pellets than for coal because the biomass contributes relatively little net carbon dioxide emissions. Co-fire pellets allow electrical utilities to harness all the advantages of biomass co-firing without experiencing the downsides. They have the potential for utilities to meet both their renewable energy mandates and their emission reduction regulations without expensive facility retrofits.

(724f) Hybrid Energy Systems for Photosynthetic Capture and Re-Utilization of Fossil-Fuel Derived CO2

Erik Shuster, National Energy Technology Laboratory, U.S. Department of Energy, Pittsburgh, PA and **John J. Marano**, JM Energy Consulting, Inc., Gibsonia, PA

A broad examination of fossil energy production and use, applying established principles of Industrial Ecology would indicate that society is currently inefficiently using fossil carbon resources. Carbon efficiency could be dramatically improved, if it were possible to re-use the carbon in CO_2 versus the current practice of disposing of it directly to the atmosphere. Large-scale re-use could significantly reduce anthropogenic CO_2 emissions, as one aspect of any comprehensive strategy to reduce impacts of climate change over the next fifty to one hundred years. In a separate paper at this conference, arguments were presented for utilizing CO_2 versus disposing of it via sequestration. There are three major requirements for CO_2 re-use: 1) a source of non-fossil energy, identified as direct solar or nuclear energy, or electricity generated from these primary energy resources; 2) a source of fossil fuel-derived CO_2, identified primarily, but not exclusively as large, stationary, coal or natural gas-fired power plants; and 3) a useful product from the conversion of CO_2, identified as a substitute for petroleum, a "synthetic" oil that may be refined and further processed into a wide variety of finished products.

This paper examines the hybridization of primary energy systems to capture CO_2 from existing power plants and convert it into supplemental feedstocks for existing petroleum refineries and petrochemical facilities. It is argued that policy choices along with successful, targeted R&D could make this approach cost effective versus other options. Both biological (algae-based) and non-biological (syngas-based) conversion schemes employing current and near term technologies are described, analyzed, and compared.

The potential of Synthetic Biology in the longer term to go beyond algae to produce what we refer to as synthetic biofuels produced from sunlight and waste CO_2 is then explored. Desirable attributes of this merging of the biological and the chemical into a truly hybrid technology are described. Systems envisioned would employ process intensification to convert sunlight and CO_2 directly to liquid products and to separate these products, within a single solar-driven reaction chamber. Recent advances within the new field of Synthetic Biology will be reviewed that provide credence to these possibilities. Finally, results of a sensitivity analysis are presented, in order to quantify critical process performance targets, such as process yield, energy requirements, and production costs, which will need to be met for these fuels to become a commercial reality.

(724g) Polygeneration of Fischer-Tropsch Fuels and Electricity by Hybridised Solar Gasification of Coal – a Pseudo-Dynamic Process Model

Ashok A Kaniyal[1], Philip van Eyk[2], Graham J Nathan[2], Peter J Ashman[2] and Jonathan J Pincus[2], (1)School of Mechanical Engineering, University of Adelaide, Adelaide, Australia, (2)University of Adelaide, Adelaide, Australia

Polygeneration of Fischer-Tropsch fuels and electricity by hybridised solar gasification of coal – a pseudo-dynamic process model.

Ashok A Kaniyal[1,2*], Philip J van Eyk[1,3], Graham J NATHAN[1,2], Peter J Ashman[1,3], Jonathan J Pincus[1,4]

*Email: ashok.kaniyal@adelaide.edu.au; Phone: +61403691321.

[1] Centre for Energy Technology, The University of Adelaide, South Australia, AUSTRALIA, 5005

[2] School of Mechanical Engineering, The University of Adelaide, South Australia, AUSTRALIA 5005

[3] School of Chemical Engineering, The University of Adelaide, South Australia, AUSTRALIA, 5005

[4] School of Economics, The University of Adelaide, South Australia, Australia 5005

<>Abstract

Presented is a comparative energetic and environmental performance analysis of a solar hybridised gasification, coal to liquids polygeneration system and a non-solar reference. Using AspenPlus and HYSYS software, the reference system was configured, assuming the integration of a pressurised, Shell entrained flow gasifier with a Fischer Tropsch liquids (FTL) polygeneration facility. The hybrid plant assumes the feasibility of integrating an atmospheric, continuously operational, directly irradiated, oxygen blown hybrid solar reactor with this polygeneration facility. To mitigate the diurnal and stochastic impacts of the solar-boosted production of syngas, the hybrid polygeneration model was also assumed to be configured with a pressurised syngas storage plant. Here, the dynamic operation of the polygeneration system was modeled using a pseudo-steady state approximation for two, six day time-series of validated solar insolation model data. The two time-series were selected to represent a period in 'summer' characterised by a steady, low cloud period of solar insolation, and a period in 'winter' marked by intermittent solar insolation and high cloud. Using these data, a MATLAB model was used to predict the maximum steady rate of liquids that could be produced for the given solar insolation scenario.

For the summer time series, the hybrid solar gasification system was shown to improve the steady rate of liquids production by 32% and decrease the source-to-wheel (STW) GHG emissions per GJ FTL by 26%, relative to the reference. For the winter time-series, only a marginal improvement in liquids production was predicted .This result followed from the assumed difference in gasification temperatures between the hybrid system and the reference. On a total energetic output basis, GHG emissions were shown to decrease relative to the reference by 17-21 percentage points for the summer-time series, but increase by 18 percentage points for the winter time-series. This increase in GHG emissions follows the large parasitic impact of SG compression on the hybrid system's net electrical output. Additional sensitivity analyses identified the potentially significant energetic and environmental value of either operating the hybrid gasifier at just 2 bar-g instead of 1 bar-g or incorporating an O_2 storage system to reduce the parasitic load of the hybrid system's ASU.

Importantly, the present analysis found the difference in the steady output from the FT reactor and electricity generating plant between the summer and winter time-series to be within the feasible operational limits of the respective plant components. This provides assurance as to the feasibility of developing a solar hybridised C2L polygeneration facility using commercially available plant components.

Keywords

Coal gasification, concentrated solar energy, hybrid solar gasification, Fischer-Tropsch liquids production, pseudo-dynamic process modeling, lifecycle GHG assessment.

(737a) Catalytic Performance of Char for Reduction of Tars in Biomass Gasification

Naomi Klinghoffer, Earth and Environmental Engineering, Columbia University, New York, NY, **Marco J. Castaldi**, Department of Earth and Environmental Engineering (HKSM), Columbia University in the City of New York, New York City, NY and Ange Nzihou, LGPSD UMR CNRS 2392, Ecole des Mines d'Albi-Carmaux, ALBI, France

Biomass has the potential for meeting a significant portion of the demand for energy and transportation fuels in the near future. Gasification is a thermochemical conversion method that creates synthesis gas from biomass, which enables the production of a wide variety of products such as fuels, electricity, or chemicals. Two of the main by-products of gasification are tar anc ash or char. Tar can be converted to synthesis gas via catalytic steam reforming, but the tars cause rapid catalyst deactivation. Char has been shown to have catalytic activity for tar decomposition. However, char properties vary based on gasification conditions, which will impact its catalytic performance. In our experiments, poplar wood was gasified in a fluidized bed reactor under 10% CO_2/90% N_2 and 90% H_2O/10% N_2 at temperatures ranging from 550-920^oC. Data is presented that demonstrates the char's catalytic performance for decomposition of hydrocarbons such as methane and propane. The surface area of the char ranged from 400-680 m^2g^{-1}, which is significantly higher than typical catalyst carriers. For example, the surface area of gamma alumina is around $150m^2g^{-1}$. The catalytic performance of the char was related to the surface area, with higher surface area giving higher catalytic performance. The presence of iron was identified in the char and carbon deposition was observed on the iron, which suggests it may be a catalytic site. However, the performance of ash is lower than that of char and therefore it is likely that the metal-carbon interactions are important to the catalytic performance of the char. This presentation will discuss the catalytic performance of the char as well as the relevant properties of the char and how they impact its performance.

(737b) Coupling Biomass Conversion with CO2 Reduction in Hot Temperature Water

Hongfei Lin, Chemical and Materials Engineering, University of Nevada, Reno, Reno, NV

The global carbon balance is broken by anthropogenic activities as humans consume fossil fuels and release a significant amount of CO_2 within several hundred years, while the fossil fuel formation takes hundreds of millions of years. To mitigate the imbalance of global carbon, it is appealing to develop technologies that can efficiently and rapidly convert CO_2 into fuels and chemicals without a high excess energy input. Herein our approach is to couple biomass conversion with CO_2 reduction over heterogeneous catalysts under hydrothermal environment. A variety of biomass feedstocks and metal and metal oxide catalysts were evaluated for the simultaneous biomass partial oxidation and CO_2 reduction. In presence of CO_2, increase of yields of the carboxylic acids was observed and a high formic acid yield was attributed to CO_2 reduction. The results suggest that CO_2 reduction coupling with biomass partial oxidation is an attractive alternative to CO_2 capture and sequestration processes.

(737c) Alkene Metathesis of Camelina FAME with Supercritical Ethylene

Randy Maglinao[1], Akash Narani[2] and Nestor Soriano Jr.[1], (1)Bio-Energy Center, Montana State University-Northern, Havre, MT, (2)State University of New York at Buffalo and Bio-Energy Center, Montana State University - Northern, Amherst/Havre, NY/MT

Akene metathesis of unsaturated FAME with ethylene is a powerful synthesis reactor to produce terminal alkenes which can be used for synthesis of polymers and other valuable compounds. Although commercial metathesis catalysts have been proven to be efficient in converting pure FAME, this process has not been used effectively in a mixture of methyle esters like biodiesel. Moreover, there has been no study in the literature that has used supercritical ehtylene in alkene metathesis. We anticipate to minimize mass transfer limitations with the use of supercritical ethylene during metathesis with FAME.

In this study FAME derived from camelina oil, which contains as much as 90S% unsaturated components, was allowed to react with supercritical ethylene at room temperature in the presence of Grubb's catalyst. A high pressure syringe pump is used to deliver 750 psi of ethylene and the reaction is performed in a 500 mL high pressure and high temperature reactor fro 4 hours. Different commercial first and second generation Grubbs catalysts were tested. Depending on the nature of the catalyst, self metathesis of FAME, isomerization and cyclization were observed as well at varying degrees as evidenced by GC/MS analyses. Percent conversion of unsaturated FAME as high as 87% was observed.

(737d) Kinetics of Mixed Alcohol Ethanol/Butanol Esterification of Butyric Acid

Arati Santhanakrishnan[1], Dennis Miller[1], Lars Peereboom[1], Carl T. Lira[1] and Abigail Shannon[2], (1)Chemical Engineering and Materials Science, Michigan State University, East Lansing, MI, (2)MichiganState University, East Lansing, MI

The growing need to reduce our dependence on fossil sources for fuels and chemicals has led to the exploration of alternate pathways for the use of greener chemicals and their manufacture from bio based sources. Esters of higher alcohols (alcohols with more than two carbons) are an industrially important class of compounds and economically viable processes for making them need to be designed. Blends of esters are being studied as fuel additives to biofuels for their high energy density and favorable fuel properties(1). One route for producing esters that is currently being extensively studied is the esterification of carboxylic acids with alcohols using acid catalysts(2). The catalyst used may be homogeneous such as sulfuric acid, or para-toluene sulfonic acid (pTSA). Heterogeneous catalysts used include cationic exchange resins.

Mixed alcohols are obtained from the condensation of alcohols to higher alcohols(3), the Fischer-Tropsch synthesis of alcohols from syn gas(4) and fusel alcohols produced from ethanol fermentation. The simultaneous esterification of these alcohol mixtures would be a good use of these mixed alcohol streams and improve the profitability of these processes. Since these reactions are equilibrium limited, one option for driving them forward is reactive distillation where the products are separated from the reaction mixture as and when formed. The design of a reactive distillation process requires a good understanding of the kinetics of the reaction system.

It is not known whether the presence of one alcohol would accelerate or inhibit the rate of reaction of another, or if the formation of these mixed esters over complicates the recovery process. In this study, the kinetic behavior of butyric acid undergoing esterification by ethanol and butanol has been investigated under homogeneous (pTSA) and heterogeneous catalysis (Amberlyst 70). The fermentation of butyric acid from biomass has sparked interest in the production of many chemicals using butyric acid as a building block. Apart from their potential as biofuel components, ethyl butyrate and butyl butyrate serve as food flavoring agents and green solvents(5).

The kinetics were first studied separately, and then in alcohol mixtures of varying relative compositions of ethanol and n-butanol. Reaction equilibrium constants were first determined experimentally from samples taken after reactions had reached equilibrium. UNIFAC (UNIversal Functional Activity Coefficient) was used to calculate the activity coefficient values. Concentration profiles of activity-based fits and experimental data of mixed alcohol esterification runs show that the models fit experimental data reasonably well. Activity and mole fraction based fits were found to be indistinguishable. An additive combination of individual models was found to predict the mixed alcohol esterification system reasonably well. The kinetic model fitted to experimental data is useful for simulating reactive distillation columns to drive the equilibrium limited simultaneous esterification reactions forward.

References

1. I. K. Mbaraka, D. R. Radu, V. S. Y. Lin, B. H. Shanks, Journal of Catalysis 219, 329 (2003).

2. A. Chakrabarti, M. M. Sharma, Reactive Polymers 20, 1 (1993).

3. W. Ueda, T. Ohshida, T. Kuwabara, Y. Morikawa, Catalysis Letters 12, 97 (1992).

4. T. J. Mazanec, Journal of Catalysis 98, 115 (1986).

5. D. W. Armstrong, H. Yamazaki, Trends in Biotechnology 4, 264 (1986).

(737e) In-Situ Catalytic Upgrading of Pinewood Pyrolysis Vapors: Effect of Process Variables

Vamshi Krishna Guda[1], Hossein Toghiani[1], Rebecca K. Toghiani[1] and Charles U. Pittman Jr.[2], (1)Dave C. Swalm School of Chemical Engineering, Mississippi State University, Mississippi State, MS, (2)Department of Chemistry, Mississippi State University, Mississippi State, MS

The main objective of this research effort was to produce high quality (i.e., deoxygenated, less acidic) bio-oil through in-situ catalytic modification (upgrading) of pine wood pyrolysis vapor. In-situ catalytic upgrading of pine wood pyrolysis vapors was performed using an integrated reactor set-up where fast pyrolysis of pine wood chips was carried out in an auger reactor followed by catalytic upgrading in a packed bed reactor mounted on the top of auger reactor. The pyrolysis process was carried out at 450 °C under a nitrogen atmosphere. The effects of process variables such as catalyst composition (γ-Al_2O_3, Si/Al, HY, HZSM-5), temperature and Weight Hourly Space Velocity (WHSV) were studied. The WHSV was varied by changing the biomass feed rate while holding the amount of catalyst constant. Compared to non-catalytic experiments, deoxygenated products increased over all the tested acid catalysts. While Si/Al and γ-Al_2O_3 yielded higher amounts of liquid product, the zeolites (HZSM-5, HY) favored the formation of aromatic hydrocarbons and higher gaseous yield. The GC-MS results, elemental analysis, acid value and Karl-Fisher titration analysis showed that the liquid fuel was deoxygenated to the largest degree. The largest concentrations of gasoline-range hydrocarbons were obtained using the HZSM5 catalyst at low WHSV (6 h^{-1}) and moderate reaction temperatures (370-400 °C). The occurrence of deoxygenation could be noticed through the reduced oxygen content in the form of H_2O (significant increase in water yield), CO, CO_2.

(737f) Catalytic Hydrodeoxygenation of Guaiacol

Danni Gao, Hyun Tae Hwang and Arvind Varma, School of Chemical Engineering, Purdue University, West Lafayette, IN

Pyrolysis of biomass is a promising technology to convert solid biomass into liquid bio-oil. However, bio-oils have high water and oxygen content which subsequently lowers their energy density relative to conventional hydrocarbons. For this reason, an upgrading process is required. Catalytic hydrodeoxygenation (HDO) is a rapidly developing technology for achieving oxygen removal from bio-oils. While various catalysts have been studied for this purpose, due to the presence of more than 400 oxygenated organic compounds, the catalytic activity and reaction pathways in the HDO process are not well understood. In this context, it is important to select model compounds which represent the raw bio-oil. In this study, Guaiacol, a phenol derived compound produced by the thermal degradation of lignin, is selected as a model compound, because it is one of the major components of bio-oil and, in addition, is thermally unstable and leads to coke formation. Conducted in a fixed-bed reactor, this work identifies the optimum noble metal catalyst(s) along with the optimum operating conditions which minimize hydrogen consumption while maximizing oxygen removal. Further, the HDO reaction kinetics are measured and, based on analysis of both the liquid and gas the gas phase products, the reaction pathways are proposed.

(745a) Plug-Flow Reactor Design for Hydrothermal Liquefaction of Biomass

Mitchell James Minarick, Agricultural and Biological Engineering, University of Illinois at Urbana-Champaign, Urbana, IL

Direct liquefaction of biomass is a technology that has shown good potential for producing oil from biomass, especially for wet biomass sources like algae and livestock waste. Previous research at the University of Illinois at Urbana-Champaign proved that up to 70% of swine manure volatile solids could be converted into oil while reducing the manure's chemical oxygen demand by up to 75%. A continuous stirred-tank reactor design was implemented and produced similar conversion rates to the batch process. Algae conversion rates are similarly high for batch processes, and integrated systems that utilize a manure source to provide nutrients for enhanced algae growth could simultaneously clean the waste stream while producing algae for oil production.

However, the high pressures required for the direct liquefaction process preclude the scale-up of continuous stirred tank reactor systems. A plug-flow reactor was designed and tested to explore the feasibility of a scaled-up continuous reactor for the liquefaction of the biomass from this integrated bioenergy system.

(745b) Conceptual Design of a Tubular Coil Reactor for Biodiesel Production

Rogelio Sotelo-Boyás, Alberto Castillo-Bertran, Rubi Morales-Avila and Lilia Arteaga-Espinoza, Instituto Politécnico Nacional (National Polytechnic Institute of Mexico), Mexico City, Mexico

The continuous production of biodiesel is aimed to be performed in a tubular reactor. The transesterification reaction of vegetable oil was carried out with methanol and potassium hydroxide as catalyst. Tubular reactors are commonly used for relatively fast reactions, however the alkaline transesterification of vegetable oils requires at least 30 minutes to reach a high oil conversion, and it is rather performed in tank reactors, either in batch or continuous modes. To reduce the space needed for the reactor, a coil tubular reactor was conceptualized. Simulation tools used in this work are essential for chemical engineers to perform innovative designs for the chemical industry. In this context, a biodiesel reactor was simulated and optimized. The first stage in this work was an experimental design to determine the best reaction conditions at which the vegetable oil could be effectively transformed into methyl esters. Optimization was performed by Taguchi method and using the response surface methodology. The results taken from here were the methanol to oil molar ratio and the amount of catalyst that maximized the yield of methyl esters. In a second stage the kinetic parameters of the transesterification reactions were estimated with Marquardt algorithm, i.e. the activation energy and the pre-exponential factors for each reaction considered. Thirdly, a simulation of the tubular reactor was performed in ASPEN Plus as a plug flow reactor. From here, the size of the reactor was calculated. The transport phenomena were then evaluated by a multiphysics simulation. A prototype of the reactor was finally build. The reactor shows a very good performance for a production of 60 L of biodiesel in one hour. More innovative solutions could be given at industrial scale by following this methodology.

PRODUCTION OF BIODIESEL FUEL FROM MICROALGAE OIL VIA TRANSESTERIFICATION REACTIONS IN SUPERCRITICAL METHANOL

Jiuxu Liu, Ronghong Lin and Lawrence L. Tavlarides, Syracuse University, Syracuse, NY, USA

Introduction

Biodiesel production is mostly produced from refined vegetable oils which are expensive, and which makes the cost of commercial biodiesel prohibitively high without government subsidies. Furthermore, biodiesel plants for feedstock oil must also compete for food, cosmetics, industrial, and livestock feed demands (McNeff et al., 2008). Although countries that use biodiesel are not expected to take full responsibility, international concerns about the upward pressure on global food prices and intensified competition for cropland currently make the use of crop-based biodiesel a politically stressful situation (Xu & Mi, 2011). To decrease competition with food stock and production costs to make biodiesel profitable in comparison with petroleum fuels, it is inevitable for production processes to use inexpensive triglyceride sources, for example microalgae oil (Chisti, 2007) which potentially offers greatest opportunities in the longer term (Schenk et al., 2008). Also biodiesel synthesis under supercritical alcohols conditions provides both energy and economic benefits over the conventional base catalyzed biodiesel production process (Deshpande, Anitescu, Rice, & Tavlarides, 2010).

Experimental

Materials

Microalgae oil was provide by Soley Institute in Istanbul Turkey. N-Methyl-N-(trimethylsilyl) trifluoroacetamide (MSTFA), reference FAME standards (methyl linoleate 20 wt%, methyl linolenate 20 wt%, methyl oleate 20 wt%, methyl palmitate 20 wt%, and methyl stearate 20wt%) for gas chromatography, a kit with gas chromatography (GC) standard solutions and internal standards to analyze total and free glycerol (according to the ASTM method D-6584) were purchased from Sigma Aldrich. *N*-heptane and HPLC-grade methanol were purchased from Thermo Fisher Scientific and used without further purification.

Experimental setup

A schematic diagram of the experimental setup is illustrated in **Figure 1**. The experimental setup consists of two high pressure syringe pumps (Teledyne ISCO) for triolein and methanol delivery, two preheaters (red dot rectangle), a coiled tubular reactor (316 SS, ID1.753mm×6m) placed in a furnace (Briskheat)(red rectangle), a micrometering valve (Autoclave Engineers) for pressure control, a water bath to cool the products, several thermocouples (T1-T5), a pressure transducer, and a data acquisition system (National Instruments) for recording and monitoring temperature and pressure.

Experimental Conditions

In this study, supercritical transesterification of microalgae oil with methanol was investigated at various temperatures (350, 385, and 400 °C), pressures (150, 200, and 300 bar), methanol-to-microalgae oil molar ratios (from 6:1 to 12:1), and residence times (3-12 min) conditions.

Analytical Methods

Components of microalgae oil, free and total glycerol were analyzed by GC-FID (HP 5890 series II) equipped with a Restek Rtx-Biodiesel triglyceride column (10 m × 0.32 mm I.D. × 0.1 μm film thickness). The ASTM standard method D6584 was followed to prepare calibration curves and perform sample analysis.

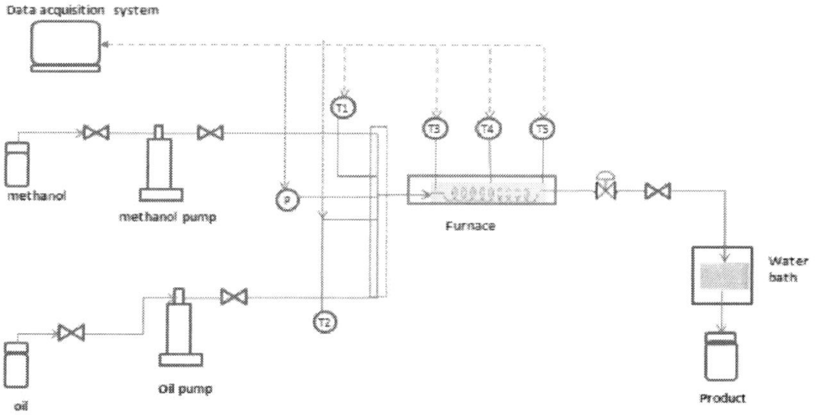

Figure 1. Schematic diagram of experimental setup for supercritical transesterification reactions.

Results and Discussion

Component profiles

Table 1 summarizes experimental reaction conditions and content profiles of different components in biodiesel product samples. Runs 1-9 were performed at 385°C, 200 bar, molar ratio of 9:1, and residence time from 3 to 12 min. Runs 10-18 were performed at 400 °C, 200 bar, molar ratio of 9:1, and residence time from 3 to 12 min. Runs 19-26 were performed at 400°C, 200 bar, molar ratio of 12:1, and residence time from 3 to 12 min. The transesterification conversion was defined as the ratio of mass fraction of the reacted triglyceride to the mass fraction of initial triglyceride (assumed to be 99.9% purity):

$$conversion(\%) = 100\left(1 - \frac{C_{TO}+C_{DO}+C_{MO}}{C_0}\right) \quad (1)$$

Here C_{TO}, C_{DO}, C_{MO} are concentrations of triglyceride, diglyceride and monoglyceride after reaction, respectively, in weight %. C_0 is the initial concentration of triglyceride before the reaction (99.9 wt%).

As expected, reaction conversion increases with temperature and residence time, while glycerol content decreases with temperature and residence time. The reaction conversion was almost complete at 400 °C and residence time longer than five minutes. For 400 °C, 200 bar, molar ratio of 9, residence time longer than 6 min, both free glycerol concentration and total glycerol concentration satisfies the limitation of ASTM standards which is 0.024% and 0.24%, respectively.

Table 1. Content of each compound in final biodiesel product.

Run #	T (°C)	P(bar)	Molar ratio	τ[a] (min)	GL[b] (wt%)	MG[c] (wt%)	DG[d] (wt%)	TG[e] (wt%)	conversion[f] (%)	BG[g] (wt%)	TG[h] (wt%)
1	385	200	9:1	3	0.96	4.26	5.76	3.09	86.87	2.28	3.24
2	385	200	9:1	4	0.79	3.08	1.61	0.23	95.07	1.06	1.85
3	385	200	9:1	5	0.56	2.89	1.16	0.13	95.82	0.93	1.5
4	385	200	9:1	6	0.37	1.8	0.42	0.06	97.72	0.54	0.91
5	385	200	9:1	7	0.25	1.54	0.25	0.03	98.18	0.44	0.69
6	385	200	9:1	8	0.23	1.3	0.19	0.02	98.49	0.37	0.6
7	385	200	9:1	9	0.22	1.23	0.16	0.01	98.59	0.35	0.56
8	385	200	9:1	10	0.22	1.32	0.16	0.01	98.5	0.37	0.58
9	385	200	9:1	12	0.1	0.66	0.03	0.01	99.29	0.18	0.28
10	400	200	9:1	3	0.24	2.06	0.58	0.06	97.3	0.63	0.86
11	400	200	9:1	4	0.12	1.41	0.32	0.04	98.23	0.42	0.54
12	400	200	9:1	5	0.05	0.65	0.05	0.03	99.27	0.18	0.23
13	400	200	9:1	6	0.04	0.59	0.05	0.02	99.34	0.16	0.21
14	400	200	9:1	7	0.02	0.51	0.06	0.02	99.41	0.14	0.17
15	400	200	9:1	8	0.02	0.37	0.04	0.01	99.58	0.1	0.12
16	400	200	9:1	9	0.02	0.31	0.01	0.01	99.67	0.08	0.1
17	400	200	9:1	10	0.02	0.32	0.02	0.01	99.65	0.09	0.1
18	400	200	9:1	12	0.01	0.2	0.02	0.01	99.77	0.06	0.07
19	400	200	12:1	3	0.57	2.48	0.76	0.51	96.25	0.81	1.38
20	400	200	12:1	4	0.3	1.54	0.51	0.36	97.59	0.51	0.82
21	400	200	12:1	5	0.18	1.07	0.12	0.06	98.74	0.3	0.48
22	400	200	12:1	6	0.12	0.79	0.11	0.06	99.04	0.23	0.35
23	400	200	12:1	7	0.11	0.69	0.12	0.04	99.15	0.2	0.31
24	400	200	12:1	9	0.09	0.57	0.08	0.04	99.31	0.16	0.26
25	400	200	12:1	10	0.12	0.52	0.09	0.04	99.35	0.15	0.27
26	400	200	12:1	12	0.06	0.38	0.04	0.02	99.56	0.11	0.16

a- the residence time; b- free glycerol; c- monoglyceride; d- diglyceride; e- triglyceride; f- reaction conversion; g- bound glycerol; and h- total glycerol (GL+BG).

Temperature effect

Temperature is an important factor which affects the biodiesel synthesis process. The reaction conversion and content variation of triglyceride and the consequent reaction products (diglyceride, monoglyceride and glycerol) with temperature at 200 bar, molar ratio of 9, residence time 8 min are shown in **Figure 2** indicating, as expected, higher conversion with temperature.

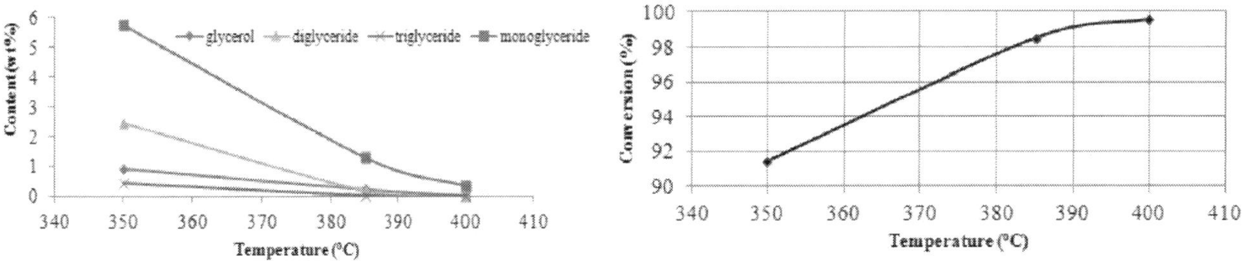

Figure 2. Temperature effect (350, 385, 400 °C) at 200 bar, molar ratio of 9:1, residence time of 8 min on conversion and content of each component.

Figure 3 clearly shows that with increasing temperature, the content of glycerol and unreacted glycerides decreased, and FAME partially decomposed into smaller molecules, which is thought to improve biodiesel fuel quality (Marulanda, Anitescu, & Tavlarides, 2010).

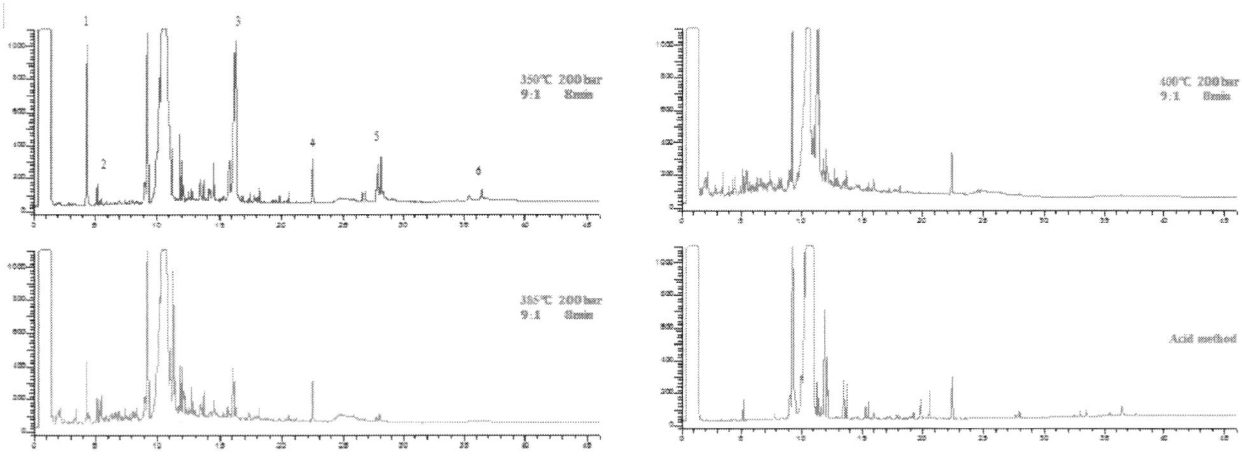

Figure 3. GC-FID chromatography of biodiesel product made at different temperatures. Peak identification : (1) Glycerol, (2) Internal Standard, Buanetriol, (3) Monoglyceride, (4) Internal Standard, Tricaprin, (5) Diglyceride, (6) Triglyceride.

Residence time effect

Figure 4 illustrates how residence time influences conversion and content of components in biodiesel samples made at 400 °C, 200bar, and molar ratio of 9:1. Free glycerol content decreased with residence time, and when residence time is longer than 6 min, free glycerol concentration is lower than 0.024 wt%, which satisfies ASTM standards limit. The reaction conversion increased with residence time, and was near complete (> 99%) after 6 min.

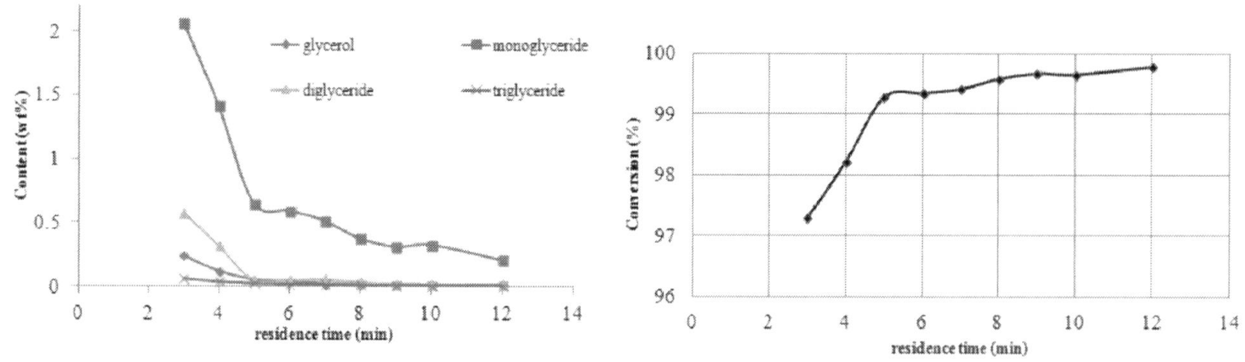

Figure 4. Residence time effect (3-12min) at 400 °C, 200bar, molar ratio of 9:1.

Molar ratio effect

From the observation of product samples, transesterification runs performed at a molar ratio of 9:1 produced an upper biodiesel phase, and a lower second film phase which was considered to be water, glycerol, and some unreacted methanol. With a higher molar ratio of 12:1, the biodiesel phase is present as a lower phase. Without analysis, the upper phase was thought to be an aqueous solution of some of the glycerol decomposition products and unreacted methanol. Sample products made at molar ratio 9:1 have less bound glycerol. There is no obvious difference in the free glycerol content of the sample products with molar ratios of 9:1 and 12:1.

Conclusions

The transesterification of microalgae oil in supercritical methanol was conducted at temperature of 350, 385, and 400 °C, pressure of 150, 200, 300 bar, methanol to oil molar ratio of 6:1, 9:1, and 12:1, and residence times of 3 – 12 min. According to experimental data, a good quality biodiesel was produced at temperature of 400 °C, pressure of 200 bar, a molar ratio of 9:1, and residence time of 7 – 12 min. Unreacted glycerides content decrease with temperature and residence time. Free glycerol decomposed into smaller molecules with temperature and residence time. Free glycerol content satisfies ASTM standards limitation for reactions with residence time of 7-12 min, temperature of 400 °C, and 200 bar. Because of the high temperature, biodiesel molecules will be decomposed into smaller molecules, which are thought to improve viscosity and cold flow properties of the biodiesel product.

References

1. McNeff, C. V., McNeff, L. C., Yan, B., Nowlan, D. T., Rasmussen, M., Gyberg, A. E., et al. (2008). A continuous catalytic system for biodiesel production. *Applied Catalysis A-General, 343*(1-2), 39-48.
2. Xu, R., & Mi, Y. (2011). Simplifying the process of microalgal biodiesel production through in situ transesterification technology. *Journal of the American Oil Chemists Society, 88*(1), 91-99.
3. Chisti, Y. (2007). Biodiesel from microalgae. *Biotechnology Advances, 25*(3), 294-306.

4. Schenk, P. M., Thomas-Hall, S. R., Stephens, E., Marx, U. C., Mussgnug, J. H., Posten, C., et al. (2008). Second generation biofuels: High-efficiency microalgae for biodiesel production. *Bioenergy Research, 1*(1), 20-43.
5. Deshpande, A., Anitescu, G., Rice, P. A., & Tavlarides, L. L. (2010). Supercritical biodiesel production and power cogeneration: Technical and economic feasibilities. *Bioresource Technology, 101*(6), 1834-1843.
6. Marulanda, V. F., Anitescu, G., & Tavlarides, L. L. (2010). Investigations on supercritical transesterification of chicken fat for biodiesel production from low-cost lipid feedstocks. *Journal of Supercritical Fluids, 54*(1), 53-60.

(745d) Simulation and Heat-Integration of Glycerol-Free Biodiesel Plant From Canola Oil with Dimethyl Carbonate

Mohammad Rafiqul Islam, Yogesh Kurle and Tracy J. Benson, Dan F. Smith Department of Chemical Engineering, Lamar University, Beaumont, TX

Biodiesel, a mixture of fatty acid methyl esters (FAMEs), has grown in the marketplace over the last couple of decades. However, the production process produces 1 lb of glycerol byproduct for every 10 lb of biofuel produced. Hence, the glycerol market has become saturated, with little market value for this byproduct. Replacement of methanol with dimethyl carbonate (DMC) eliminates glycerol production, along with process wastewater during manufacturing. The DMC reaction produces 2 FAMEs and 1 fatty acid glycerol carbonate (FAGC) for every triglyceride molecule converted. In early attempts, sodium methoxide, the usual catalyst for base-catalyzed biodiesel production, was found to have poor conversions. However, triazabicyclodecene (TBD) was found to easily ionize the dimethyl carbonate and ease of separation post reaction.

This paper reports on the development of specific processing steps to convert canola oil to FAMEs and FAGCs, meeting ASTM D6751 standard. A process flow diagram was developed using Aspen Plus simulation software and verified through laboratory testing of reactions and downstream unit operations. Heat integration techniques have been applied via pinch analysis to identify the optimal heat load to be removed and added from the hot and cold streams, respectively. It was found that about 31.6% energy savings was possible through heat integration techniques. A comparison with similar process plants indicates that this process is simple, environmently friendly, and energy efficient. Complete reaction details and Aspen modeling results will be reported, illustrating a commercially viable biofuel process.

(745e) Znla Catalyst Durability and Regeneration Method in the Process of Converting DDGS Oil and Yellow Grease Into Biodiesel

Shuli Yan[1], Like Yang[2], Huali Wang[3], Manhoe Kim[4], Steven O. Salley[3] and Simon K. Y. Ng[3], (1)NextCAT inc., Detroit, MI, (2)NextCat INC, Detroit, MI, (3)Chemical Engineering and Materials Science, Wayne State University, Detroit, MI, (4)NextCat Inc, Detroit, MI

This research focuses on using heterogeneous catalyst for biodiesel production. Currently biodiesel producer suffer from high oil feedstock cost. Development of catalytic technologies using some crude and inexpensive oil as feedstocks becomes attractive to all biodiesel producers. However, these cheap oil feedstocks generally contains a high content of impurities, such as free fatty acids, moisture, phosphor lipid, steroidal, some lipid soluble protein and polysaccharide. These impurities not only poison the catalysts but also complicate the whole biodiesel production process. Therefore, investigate the catalyst tolerance to poisons, the regenerate method for the foiled catalysts and the structure of the regenerated catalyst become important. In last decades, there are multiple studies on the effect of FFA and water to homogeneous biodiesel catalysts. However, few focus on heterogeneous catalysts. In our previous studies, a series of Zn based catalysts were developed with high activity and long life in refined soybean oil system. In this study, crude extracted corn oil from DDGS and yellow grease from several restaurants were used as raw materials. The compositions of oil feedstocks were tested by HPLC, GC-MS, ICP, UV-VIS etc. Surface coverage of catalyst active site by long carbon chain species are considered as the reason for catalyst deactivation. A deactivation mechanism is proposed here. The foiled catalyst was regenerated by simple calcination method and hydrothermal calcination method. The regenerated catalysts were characterized by XRD, XPS, EDS, HRTEM etc. and evaluated in batch and continuous reactors. It was found that the regenerated catalyst particles have the same structure and activity as fresh catalyst.

(745f) Theoretical Yields and Properties of Hydrotreated Renewable Diesel From Common Lipid Feedstocks

Andrew M. Duncan, Chemical and Petroleum Engineering, University of Kansas, Lawrence, KS, David A. Slade, Renewable Energy Group, Ames, IA and Susan M. Stagg-Williams, Chemical & Petroleum Engineering, University of Kansas, Lawrence, KS

Rising petroleum prices and growing concerns about greenhouse gas emissions have led to increased biofuel usage globally. In the United States, the Environmental Protection Agency's Renewable Fuel Standard specifies annual target quantities of various bio-derived fuels for inclusion in the U.S. fuel supply, including two types of biomass-based diesel: biodiesel and hydrotreated renewable diesel (HRD). Although biodiesel has been studied extensively and has been in commercial production and use for a relatively long time, HRD fuels have only recently been produced in the United States in commercial quantities and little is known about the physical properties, feedstock requirements, and theoretical yields of HRD.

This work presents a model that uses pure triglycerides and hydrogen as its feedstocks for HRD production from three common lipid feedstocks. The HRD model is derived from mass balances and produces theoretical feedstock requirements, fundamental production constraints, and yield estimates for HRD and its co-products. The corresponding theoretical limitations for biodiesel are also included.

Key fuel properties, including density, viscosity, flash point, and cloud point, are predicted using literature methods for the theoretical HRD fuels determined by the model. The analysis also includes estimates of the net energy content and renewable energy content of each of the fuels considered.

Energy Conservation in Higher Education Facilities

Robert W. Peters, Ph.D., P.E.
Department of Civil, Construction, and Environmental Engineering
University of Alabama at Birmingham
1075 13th Street South
Birmingham, Alabama 35294-4440

Atul Kajale and Zhuo Li
Department of Civil, Construction, and Environmental Engineering
University of Alabama at Birmingham
1075 13th Street South
Birmingham, Alabama 35294-4440

Matthew Winslett
Engineering Manager, Energy Management
Facilities Management Department
University of Alabama at Birmingham
1530 3rd Avenue South
Birmingham, Alabama 35294-4564

Abstract

Sustainability has become an important aspect in planning and design of many new construction projects. Initially, commercial and industrial buildings became the main focus of the sustainability and green house reduction programs as these buildings had great potential for improvements. Leadership in Energy and Environment Design (LEED) certification helps in incorporating energy and water conservation techniques in the new and already existing structures. Higher education buildings such as universities and research institutes that are built prior to 10 years ago have tremendous potential for energy and water conservation. Higher education facilities operate buildings which can be categorized as administration buildings, education buildings (classrooms), research buildings (laboratories both medical/clinical and engineering laboratories), libraries, dormitories, and cafeterias. Resource consumption and occupancy of these buildings vary from each other as each building type serves different purposes.

Introduction

The purpose of this paper is to highlight the importance of sustainability in higher education institution buildings and also demonstrate green initiatives of the University of Alabama at Birmingham (UAB). UAB is one of the leading universities in the medical research field throughout the United States and it operates more than 155 buildings in its Birmingham, Alabama campus. The energy Management Division (EMD) of the Facilities Management Department at UAB is very proactive and is constantly incorporating new technologies in operating the campus buildings efficiently to create more sustainable environment for education, and research. Facilities Management and the various departments spent billions of dollars on utilities every year to ensure the smooth operation of the campus buildings. EMD realized that in order to minimize the money spent on the utilities, it is necessary to reduce the utility demand of the buildings without compromising the normal operations of the buildings. EMD started to analyze the performance of the campus buildings using historical utility bills and utility consumption data for every season.

Many universities are currently employing the conservation practices in operating their campus buildings. However, the green initiatives of the University of Wyoming are discussed briefly as background studies. The University of Wyoming (UW) has performed similar projects in its campus buildings to reduce greenhouse gas emissions in the energy, water, transportation, purchasing and procurement, recycling, and waste management, and behavioral change fields [Sharma and Peters, 2011]. In a three phase energy efficiency program, the UW has completed first phase with installation of condensate, water, and electric meters for core campus buildings. It involves replacement of T-12 bulbs with T-8 bulbs, lighting control upgrades, installation of vending misers, and steam trap upgrade [Peters *et al.*, 2011]. All new buildings will be constructed with the LEED standards which will further reduce the UW's greenhouse gas emissions. Two buildings are already gold and silver certified. Other projects include campus wide tree plantation, use of environmentally friendly fertilizers and pesticides, community awareness, upgrading plumbing fixtures, purchase of new electronic fleet, campus shuttle system, and installation of solar shed [Peters *et al.*, 2011].

Significance of the study: UAB expanded dramatically since the Fiscal Year 2001–2002, which can be explained using the increase in the gross square footage of the campus buildings. Figure 1 shows the total gross square footage of the campus buildings from fiscal year 2001–2002 through 2010–2011. There has been a 36.6% increase in the gross square footage from the fiscal year 2001–2002 through 2010–2011.

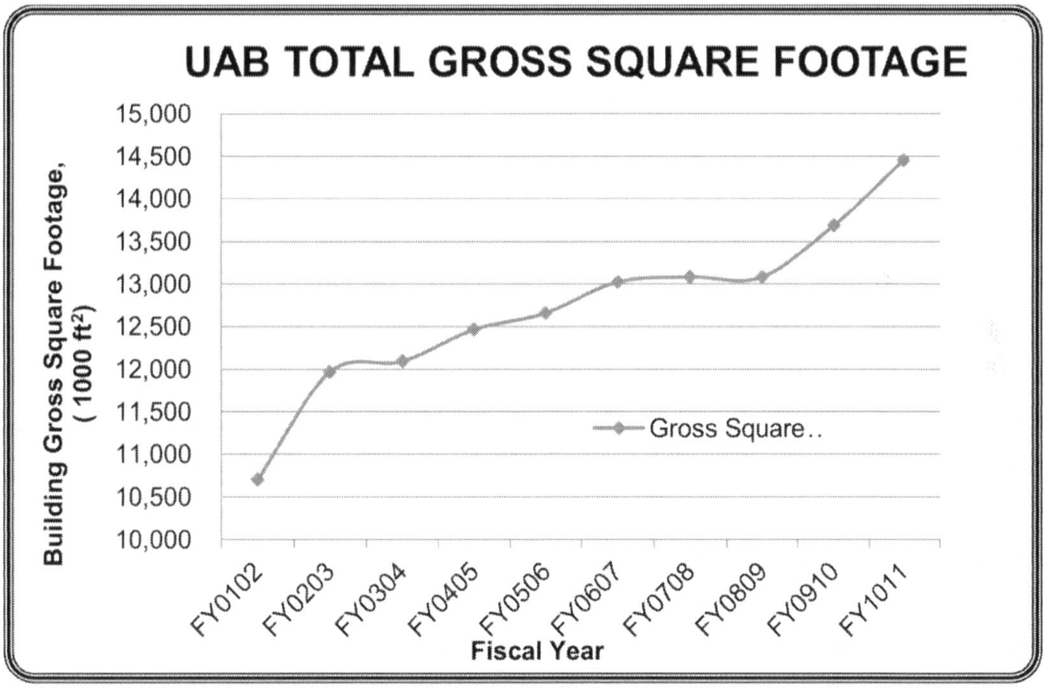

Figure 1. UAB total gross square footage of campus buildings.

As a result, the energy consumption and utility cost have also significantly increased. These utilities include electricity, water and sewer, natural gas, chilled water. Detailed consumption and utility costs of each utility are shown in Figures 2 to 6. In addition, the combined total utilities cost including each component is also shown in Figure 7. It can be seen that with the increase in the gross square footage of the campus buildings, the utility consumption and the utility costs have also increased from fiscal year 2001–2002 to fiscal year 2010–2011, except for steam. Steam consumption decreased because of a change in boiler operation at the hospital, while natural gas consumption increased accordingly. In addition, these trend lines generally are in accord with the total gross square footage as well. The consumption of steam has a different trend with other utilities that declined in the FY 2007–2008 and kept decreasing since then. If the utility cost and consumption value of FY 2001–2002 is assumed as the baseline utility cost and consumption for the study period, an average 80% increase in the utility cost and consumption has been observed at the end of fiscal year 2010–2011. Among all the utilities, consumption and cost of natural gas is the largest and that of steam is lowest. Electricity costs decline after the fiscal year 2009–2010, since EMD implemented energy conservation projects during this period. It is interesting to note that the difference between the electricity cost during fiscal year 2008–2009 and 2009–2010 equals the cost of electricity saved during this period. Thus, if the consumption of all the utilities is reduced using energy efficient coupled with control equipment, the utility cost will reduce significantly. Therefore, incorporation and implementation of energy and water conservation projects is very important in higher educational institutional buildings.

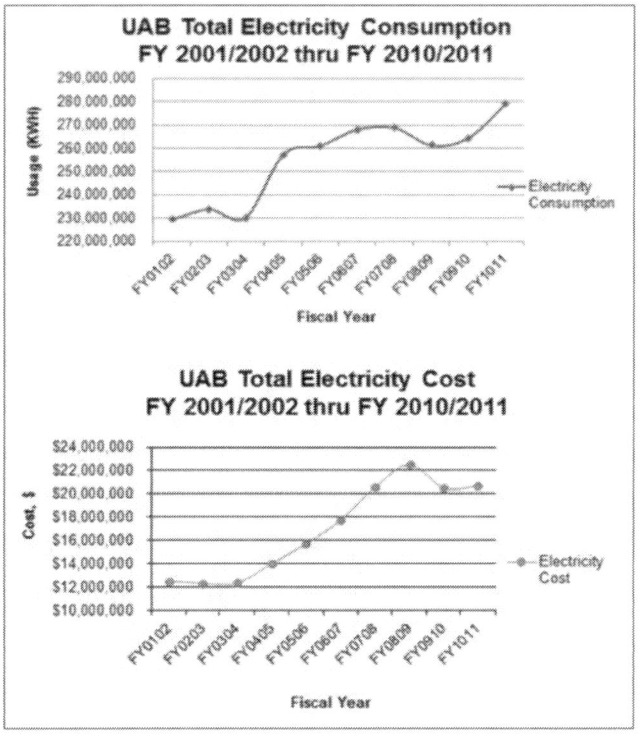

Figure 2. UAB total electricity consumption and cost through FY 2001-2002 and FY 2010-2011.

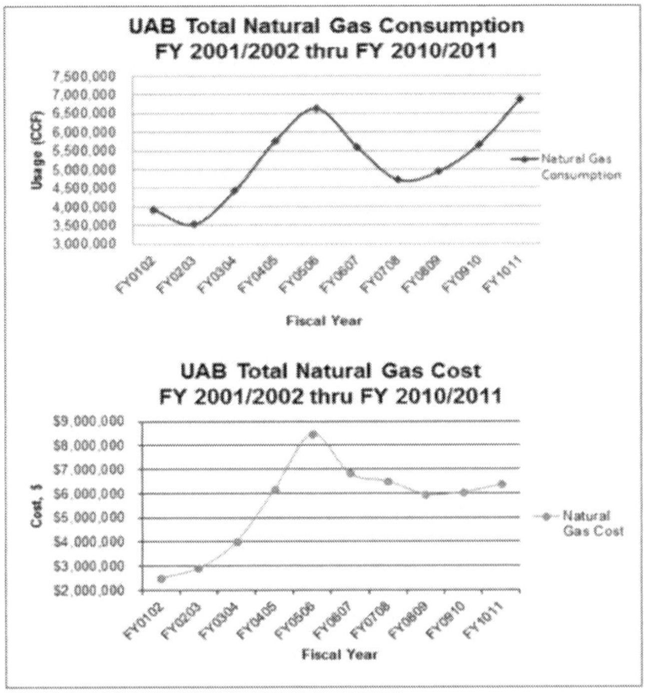

Figure 3. UAB total natural gas consumption and cost through FY 2001/2002 and FY 2010/2011.

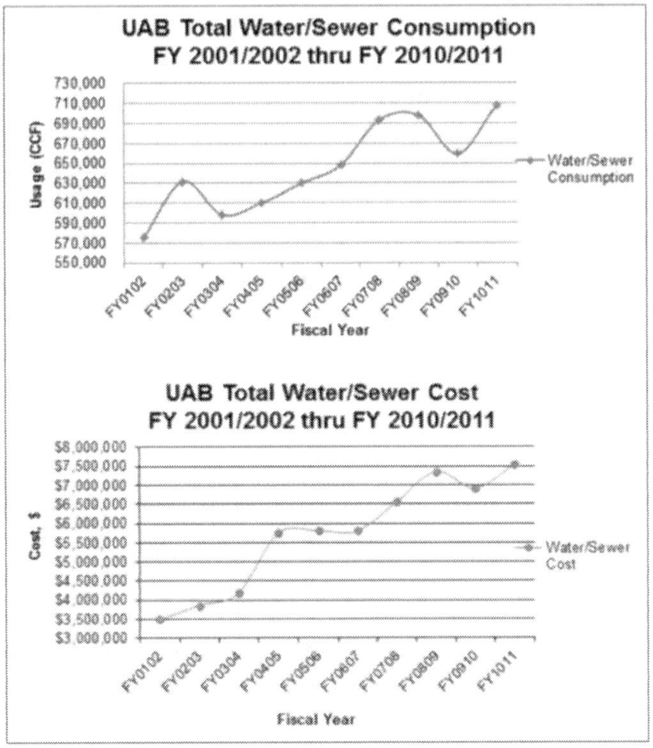

Figure 4. UAB total water/sewer consumption and cost through FY 2001–2002 and FY 2010–2011.

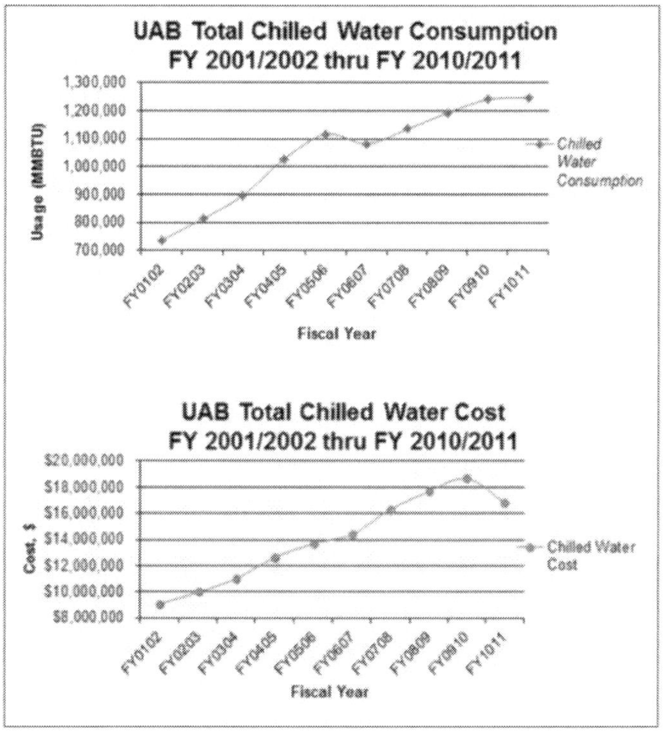

Figure 5. UAB total chilled water consumption and cost through FY 2001/2002 and FY 2010/2011.

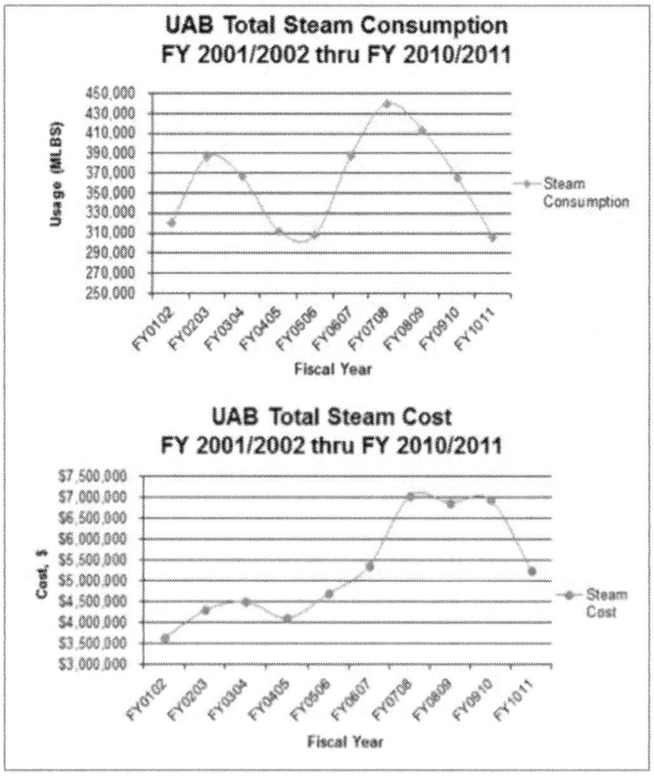

Figure 6. UAB total steam consumption and cost through FY 2001/2002 and FY 2010/2011.

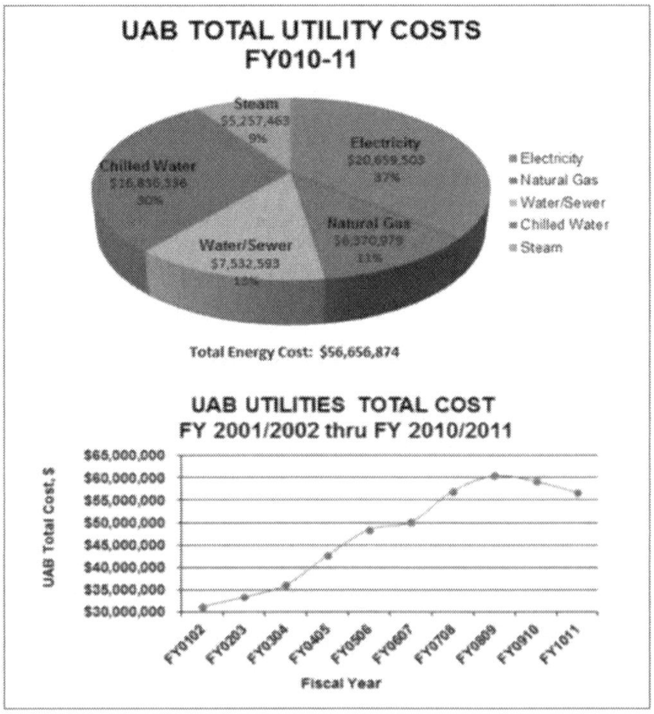

Figure 7. UAB total utilities cost including each utility.

Based on the preliminary analysis of the utility consumption, new conservation approaches have been researched and energy and water conservation strategies were evaluated which are summarized below and are explained briefly in the next paragraph.

- Programmed equipment setbacks during unoccupied times;
- Installation of efficient lighting upgrades with occupancy sensed lighting controls;
- Installation of variable frequency drives (VFD's precisely control electric motor speeds based on needed load);
- Installation of more accurate and efficient control valves for hot and chilled water distribution;
- Installation of more efficient vacuum and compressed air pumping systems that drastically reduce water usage;
- Recovery of condensate water (from conditioned air) and ground water for use in evaporative cooling towers;
- Grid-free solar-charged electric cars [Winslett, 2011]; and
- Pilot vegetated roof on Hulsey Center.

1. **Programmed Equipment Setback during Unoccupied Time:** Air conditioning equipment setbacks during unoccupied times and reduction of air exchanges during unoccupied times have achieved a great payback after being carried out. Since the year of 2007, numerous buildings at UAB were surveyed by graduate students from the Civil, Construction, and Environmental Engineering Department at UAB at intervals around the clock to determine occupancy frequency habits for both students and employees. After gathering the occupancy data, the graduate students analyzed and determined the buildings' least-occupied times are from 10 p.m. to 6 a.m., and holidays and weekends. Every surveyed building was setup with a programmed automation system to shutoff air handling motors and decentralized chillers to conserve unnecessary energy consumption. The installed sensors automatically detect the motion, and turn off lightings if the room or corridor is not occupied for 30 minutes. In addition, a timer is also used to control lighting fixtures. There is individual system to

insure safe operation when internal temperature exceeds 85°F in the summer or below 55°F in the winter. Currently, there are 43 buildings at UAB employed energy during the unoccupied time leading to a cost saving over $1,500,000 per year.

2. **Installation of efficient lighting upgrades with Sensor Lighting Control:** LED exit signs (Light-Emitting Diode) use 1/10 the energy for the same lumen output, utilize 2 to 5 watts of power, and have an expected service life of 20 to 25 years, which greatly decreases required maintenance. In the last two years, over 60,000 fluorescent lamps has been upgraded from T-12* lamps to a more energy efficient T-8* lamps saving over 420,000 kWh's of electricity and $368,000 annually. With energy savings of $20/fixture, the labor/time savings for maintenance can be greater. The UAB Sterne Library pedestrian ramp LED Lighting has been upgraded by installing 13, 46/50 watt 4,000/4,500 lumen LED lights to replace the 250 2' x 2' dual tube fluorescents overhead. The UAB Campus Recreation Center (REC) Pool Lighting has been upgraded for replacing 36 - 1000 watt metal-halide lamps with 200 watt 'compact' fluorescents reducing the wattage by 80% resulting in savings of almost $20,000 per year in electricity costs. At $65 per lamp, the lighting upgrade of the REC center will result in less than a 2 month payback period.

3. **Installation of Variable Frequency Drives (VDF) to Precisely Control Electric Motors:** Installation of VDFs aims to achieve a precise control for electrical motor speeds based on the needed load. VFDs are used to vary the speed of an electric motor by changing the frequency of the electric power going to the motor. A typical application includes air handling units (AHUs) and water pumping systems. UAB has replaced more than 40 motors with variable frequency drives in the last four years. The drive provides a three phase voltage with output varying between 3 and 66 Hz. The voltage/hertz pattern is selectable to that best suited to the motor. The drives automatically adjust this relationship during operation for the maximum operating efficiency according to speed and load [UAB DFP, 1998]. Nowadays, VDFs are very economical and save energy by allowing the motor horsepower to be used only as needed.

4. **Installation of Accurate and Efficient Valves for Chilled and Hot Water Distribution:** UAB has conducted another project to utilize accurate and efficient valving for hot and chilled water distribution. In the chilled water distribution system, butterfly valves used above manhole or other limited access space were high-performance type (Section 231400, UAB Department of Facility Management). UAB has upgraded to ball valves from butterfly valves to provide water. Butterfly and globe valves can precisely control water flow. Depending on different weather conditions, UAB Central Utilities Plants (CUP) varies chilled water temperature set point. The chilled water supply temperatures at the CUP generally are 43°F in summer and 45°F in winter (Section 231400, UAB Department of Facility Management). The Direct Digital Controls have been installed to automatically adjust the heating hot water temperature based on outside air temperature. The heating hot water temperature is adjusted to a lower temperature during the warmer months thereby reducing heat loss from the piping and lowering the cooling load that would be incurred to that heat loss.

5. **Installation of more efficient vacuum and compressed air pumping systems that drastically reduce water usage:** Domestic water conservation and recovery projects have also been very beneficial in reducing costs at UAB. Compressed air and vacuum pumping systems are found throughout the UAB campus laboratories and hospital settings. These traditionally inefficient pumps require water for compression as well as a cooling source. With escalating water and sewer costs, this method of cooling was deemed unfeasible. These projects replace liquid-ring pumps with pumps that do not require water for cooling including claw-type vacuum systems and scroll-type compressed air systems. In most applications less electricity is required to do the same amount of work due to increased pump efficiencies. Depending on building type which determines electrical rates ($0.08–$0.10kW-hr), for every horsepower that is saved by a more efficient pump, electrical costs are reduced approximately $400 - 500 per year for continuous duty operations. A constant single-pass through water flow pump for cooling is especially expensive if sewer costs are also required. The

local cost for each constant year-round one gallon per minute cooling water source is approximately $7,500 per year including sewer charges. Therefore, system paybacks are very short – usually within one to two years. Table 1 shows the annual savings of water and energy due to this upgrade.

Table 1. Savings associated with water and energy using efficient vacuum, and compressed air pumping.

Building Location	Pump System Type	Required Cooling Water (gpm)	Annual Savings		Horse Power Reduction	Annual Electrical Savings ($)
			Volume (gallons)	Water-Sewer, ($)		
Tinsley-Harrison Research	Vacuum	4.5	2,366,820	33,750	0	0
School of Dentistry	Medical Air	4.5	2,366,820	33,750	50	20,000
Bevill Biomedical Sciences	Lab Air	3.5	1,840,860	26,250	5	2,000
McCallum Health Science	Vacuum	2.5	1,314,900	18,750	0	0
Spain-Wallace Hospital	Medical Air	3.5	1,840,860	26,250	15	6,000
Kaul Human Genetics	Vacuum	2.5	1,314,900	18,750	15	6,000
Biomedical Research II	Vacuum	2.5	1,314,900	18,750	15	6,000
TOTAL		23.5	12,360,060	176,250	100	40,000

6. **Recovery of Fin water condensate (from conditioned air) and ground water for use in evaporative cooling towers:** EMD has successfully installed pilot condensate recovery systems in five buildings on the UAB campus: McCallum Building (MCM), Kaul Building (KB), Zeigler Building (ZB), Tinsley Harrison Tower (THT), and Bevill Building (BB). Condensate is formed on the outside surfaces of the pipes and other surfaces in the close proximity of the cooling coils (heat exchanger) in the Air Handling Units (AHU). The AHUs at the medical campus buildings use 100% outside air with no recirculation. Condensate from each AHU is then taken to small tanks from which it is pumped to a 500-gallon storage tank. The recovered condensate water is supplied to the chilled water plant to reduce the make-up water requirement. The temperature of the condensate water is nearly same as that of chilled water and hence less energy is required to bring this condensate water to the desired temperature. Table 2 summarizes building information, the condensate collection system, and initial cost of installation. The installation of condensate collection system (CCS) in the MCM, KB, ZB, and THT was completed before the cooling season of 2011. Therefore, there was an opportunity to collect data during this period except for the BB where system installation work was finished in August of 2011.

Table 2. Summary of the condensate recovery system at UAB.

Building Name	System	Installation Cost, ($)	Remarks
McCallum Building (MCM)	2 AHUs drains into 2 storage tanks	25,000	McCallum and Kaul share one 500-gallon tank; started in May 2011
Kaul Building (KB)	5 AHUs drains into one 500-gallons tank	16,220	
Zeigler Building (ZB)	3 AHUs drains into two storage tanks	5,500	Zeigler and Tinsley Harrison Tower share a one 500-gallon tank; started in May 2011
Tinsley Harrison Tower (THT)	10 AHUs drain into a 500-gallon tank	2,475	
Bevill Building (BB)	6 AHUs drains into two 500-gallon tanks	29,965	Started in August 2011
Total		79,160	

[Kajale *et al.*, 2012]

During the pilot study of the project, monthly totals of the condensate water were collected from the water meter installed across each storage tank. The monthly totals of the water recovered in each building are summarized in Table 3.

Table 3. Condensate recovery at the pilot CRS plant during the cooling season of 2011.

Month	Buildings					
	MCM & KB		ZB & THT		BB	
	Vol. of water, (gallons)	Savings, (dollars)	Vol. of water, (gallons)	Savings, (dollars)	Vol. of water, (gallons)	Savings, (dollars)
May	166,200	$ 688.70	183,700	$ 773.50		
June	430,300	$ 1,811.85	285,100	$ 1,200.46		
July	615,604	$ 2,592.10	410,497	$ 1,728.47		
August	456,040	$ 1,920.23	311,379	$ 1,311.12	125,434	$ 528.16
September	221,457	$ 932.50	122,730	$ 516.77	129,175	$ 543.91
Total	1,889,601	$ 7,945.38	1,313,406	$ 5,530.32	254,609	$ 1,072.07

[Kajale *et al.*, 2012]

Table 4 shows the amount of condensate recovered during 2012. The first three columns 2,3, and 4 represents the condensate data from one air handling unit whereas the column 5 and 6 represent condensate data for the group of air handling units. It can be seen that more than 1.5 million gallons of condensate water has been recovered from June 14th, 2012 through August 30th, 2012 which is equivalent to a saving of more than $5,900.

Table 4. Condensate recovery at CRS plant during cooling season 2012.

Month	Kaul #4	MCM #1	MCM #2	MCM Total		THT Total	
	Total, (gallons)	Total, (gallons)	Total, (gallons)	Total. (gallons)	Savings, ($)	Total, (gallons)	Savings ($)
June	17,776	27,540	33,422	101,888	432	93,703	397
July	53,309	82,795	85,982	343,226	1,455	232,297	984
August	46,817	78,276	88,270	385,282	1,633	235,157	997
Total	117,902	188,611	207,674	830,396	3,520	561,157	2,378

[Kajale *et al.*, 2012]

7. **UAB's Solar Powered Vehicles:** The University of Alabama at Birmingham Facilities Management Department wanted to incorporate a solar collector project as a part of its sustainability efforts but the local cost of electricity, combined with the initial cost of solar panels and ancillary equipment, made project payback times undesirable when only considering energy costs. One way achieve an acceptable return on investment was to utilize solar electricity to recharge batteries in electrical vehicles. Most electric vehicles only transfer their carbon footprint from petroleum fuel to the fossil fuel required to generate the electricity required during a battery recharge. By using solar derived electricity, the carbon is totally eliminated, yielding a mode of grid-free transportation around campus. The solar charging system was installed in November 2008. It consists of 12 Kyocera KDKD205GX-LPPhotovotaic Modules where each module can produce 205 watts totaling 2,460 watts. The panels are cover 192,075 square feet of roof area. WattPlot™ software is used for performance monitoring of renewable energy components from Outback Power Systems. Cars are driven during the day while the solar panels charge the 8-battery bank located in the Facilities Administration Building (FAB) Parking Garage. Car batteries are recharged during the night from the solar charged battery bank.

8. **Pilot vegetated roof on Hulsey Center:** A pilot green roof system has been installed on the top of the Hulsey Center in July 2008. The roof occupies approximately 14,940 ft^2 (i.e., occupying ~81.2% of the roofing area, or approximately 0.34 acres. Prior to installing the pilot green roof, the civil engineering senior design class investigated the loading of a wet vegetative roof on top of Hulsey Center, and found that the current building infrastructure could withstand the loading associated with the vegetative roof. Hulsey Center was originally designed to have two additional floors in the building. This pilot green roof contains more than 20,000 sedum plants. In the construction of this vegetated roof, the existing roof was removed down to the structural concrete deck. A waterproof roofing membrane was installed directly to the concrete deck. A layer of 3-inches of extruded polystyrene roofing insulation was then applied, to which an 8-oz non-woven geo-textile scrim sheet was applied. Figure 8 shows the green roof during its construction phase, and Figure 9 shows the pilot vegetated roof system on top of Hulsey Center, while Figures 10 and 11 show close-up photos of the green roof.

Figure 8. Photograph of roof system on top of Hulsey Center on the UAB campus prior to installation of the green roof.

Figure 9. Close-up of general layout of pilot vegetated roof system on top of Hulsey Center on the UAB campus.

Figure 10. Pilot vegetated roof system on top of Hulsey Center.

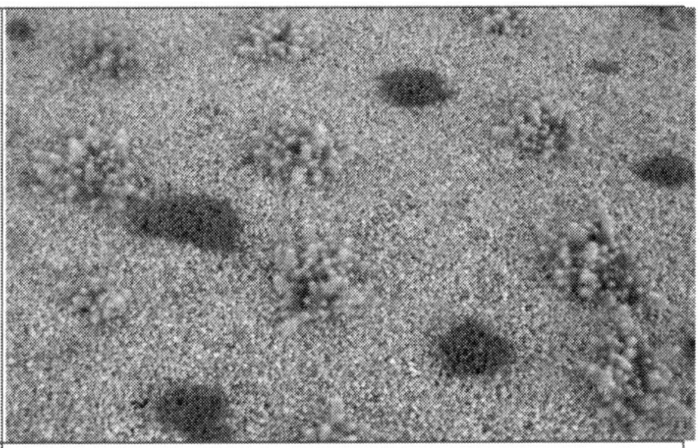

Figure 11. Close-up of pilot vegetated roof system on top of Hulsey Center.

Then, after application of the roofing membrane, 2-ft x 2-ft x 2-in prestressed pavers or brick pavers were installed for design, decoration, and access to the vegetative roof. Then, 3½-inches of light weight engineered soil were applied. Sedum plants were planted at the rate of 150 plants per 100-ft^2. The cost of retrofitting the roof and installing a pilot green roof on top of Hulsey Center was ~$150,000. The usage of natural gas, water, and electricity are significantly lower than that compared to the utility usage prior to installation of the green roof on top of Hulsey Center. Implementation of the vegetated roof system has reduced the utility bills of Hulsey Center by 20–25% compared with the utility bills prior to implementation of the vegetated roof system [Peters *et al.*, 2012; Kirby *et al.*, 2012; Nunez *et al.*, 2009].

Discussion

The University of Alabama at Birmingham has been actively incorporating the new sustainability and conservation techniques in the campus buildings to improve the building performance and reduce its carbon footprints. Since 2008, several projects were successfully implemented to reduce the utility consumption even when new buildings were added to campus. Starting with the energy conservation, EMD has broadened their conservation scope and implemented projects related to water conservation and reduction in utility consumption. New lighting bulbs and fixtures along with the occupancy sensors have reduced the electricity consumption significantly after year fiscal year 2008-2009. Air handler condensate recovery and recycling is another example of reduction in water consumption which also helps in saving energy cost and sewer bills. All air conditioning equipment and pumps has been upgraded and repaired which will further help in lowering the utility consumption in campus buildings.

As a result, higher education buildings have great potential to reduce their energy and water demand as these buildings are occupied for certain hours. Also, given the nature of university campuses these savings are multiplied by the number of buildings.

This paper has shown that the University of Alabama at Birmingham (UAB) is very enthusiastic in implementing new ideas to lower the carbon footprints and utility consumption to create environment friendly campus for education and research.

Acknowledgement: The authors would like to acknowledge the Facilities Management Department of UAB and the Energy Management Division for providing the technical information in preparing this paper.

References Cited

Kajale, A., R.W. Peters, and M. Winslett, 2012. "Estimation of Air Handling Condensate Recovery in the Medical Campus Buildings at University of Alabama at Birmingham", ***Proc. 2012 Annual AIChE Meeting***, Pittsburgh, PA, (October 28–November 2).

Kirby, J.T., R.D. Sherrod, M. Winslett, and R.W. Peters, 2012. "Variations in Temperature in Mini-Roof Structures Employing Different Roofing Materials", ***Proc. 2012 Annual AIChE Meeting***, Pittsburgh, PA, (October 28–November 2).

Nunez, S., R.W. Peters, and M. Winslett, 2009. "Effect of Green Roofs on Building Energy Costs", ***Proc. 2009 Annual AIChE Meeting***, Nashville, TN, (November 9–14).

Peters, R.W., R.D. Sherrod, and M. Winslett, 2012. "Thermal Performance and Energy Conservation Associated with Implementation of Vegetated Roofs in the Soutgheastern United States", Book chapter in ***New Developments in Renewable Energy***, InTech Publishers, Rejeka, Croatia, [*in press*].

Peters, R.W., M. Winslett, and M.P. Sharma, 2011. "Sustainability Initiatives at Several U.S. Institutions of Higher Education", ***Proc. 2011 AIChE Annual Meeting***, Minneapolis, MN, (October 16–21).

UAB Deparment of Facilities Management, undated. "*Design and Construction Standards: Section 231400 Chilled Water*", University of Alabama at Birmingham, Birmingham, AL.

UAB Department of Facilities Planning, 1998. *Facility standard: Variable Frequence Drives.* University of Alabama at Birmingham, Birmingham, AL.

Winslett, M., 2011. Personal communication with Mr. Matt Winslett, Energy Management Division of UAB Facilities Management. Birmingham, AL, (September 30).

ENVIRONMENTAL PERFORMANCE FOR THE ELABORATION PROCESSES FOR CATALYSTS OF Á - AL2O3 AND SILICA

Johana Garcia[1], Gabriel Camargo[2], Luisa Martin[1], Lady Gaitan[1], Daniel Portela[3], Nikolay Agudelo Valencia[1] and Juan Carlos Moreno[4], (1)Environmental Engineering, Universidad Libre, Bogotá, Colombia, (2)Mechanical Engineering, Universidad Libre, Bogotá, Colombia, (3)Mechanical Engineering, Universidad Libre, Bogotá , Colombia, (4)Chemistry, Universidad de los Andes, Bogota

ABSTRACT

The processes for catalyst's preparation can be used as a tool for green engineering concepts in several courses in the engineering curriculum. In this job two catalysts were prepared and the raw materials, utilities (water, electrical power) and waste were measured. The water used in entire processes was registered for evaluated the process impact in the waste generation. The materials balance was realized and the Green Chemistry Expert System was used for the environmental impact.

The catalysts manufactured were α - Al_2O_3 and Silica. In the case of alumina the raw material was aluminum shaving (recycled materials). The quantities of catalyst prepared were 1 g. The aluminum was treated with nitric acid and ammonia. For the silica, the starting material was sodium silicate. The raw materials were characterized by atomic absorption. The final catalysts were analyzed by area BET, TGA, XRD, and infrared.

The processes waste waters was treated for minimizing its environmental impact

INTRODUCTION

The chemical processes impacts on the environment and the different ways to evaluate and avoid can be used to teach and learn about green engineering concepts, and use every tools for measure the different variables and formulate new process benign with the environment. The idea for the present work was originated when one of the authors was teaching a catalysts heterogeneous course for chemical engineering students and simultaneously material balance course for environmental engineering program. The project is an interdisciplinary work between different universities in Bogota Colombia.

The elaboration of catalytic supports can be used as a tool to measure the water footprint, and at the same time to evaluate important items such as waste disposal strategies, material balance, metals recycling in environmental engineering program. On the other hand for chemical engineering courses other aspects can be illustrated with these processes: the reactions efficiency, atoms economy reaction, materials characterization, and the effect of the variables process in the properties of the products.

The main purpose in this project is evaluating the environmental impact of the preparation catalytic supports from raw materials to end products. To reach this goal, two different catalytic supports were chosen α-Al_2O_3 and silica gel. These materials can be obtained from easily raw materials, aluminum shaving (from solid wastes) and sodium silicate. The experimental procedure is available from different sources.

EXPERIMENTAL PROCEDURE

For the α- Al_2O_3, there are several sources for raw material, soft drinks can, mechanical and industrial wastes, etc. the first step was a cleaning process with toluene for degreasing process after that several acid treatments steps are followed for obtain de catalytic support. The block diagram for this process is shown in fig No 1.

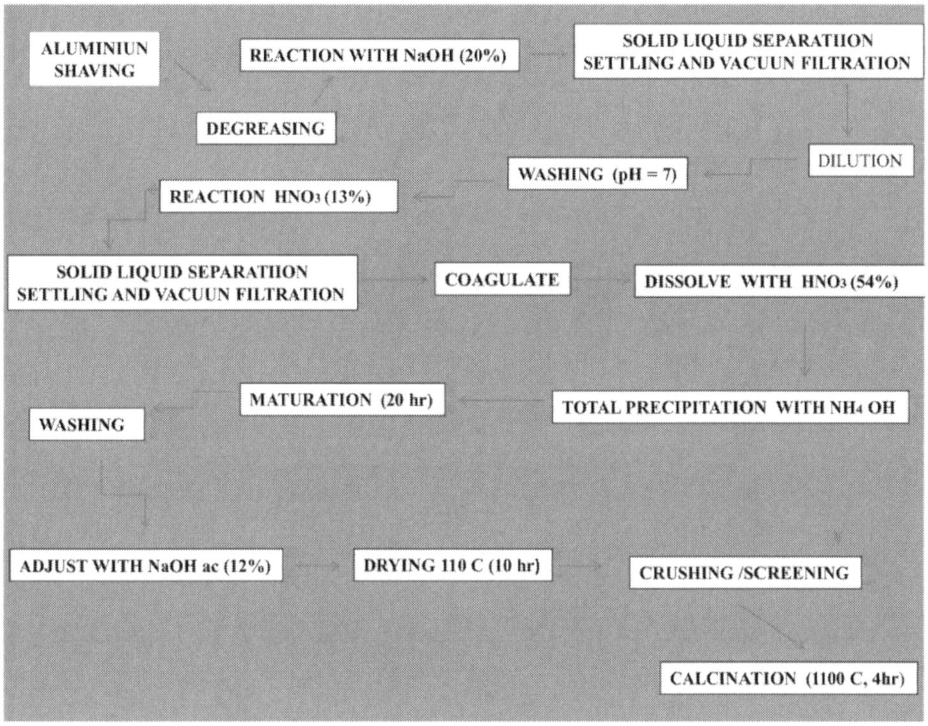

Block Flow Diagram for α- Al_2O_3 Process

For the silica gel support, the raw material was sodium silicate, and the first step in the process was a bleach using diatomaceous earth as adsorbent with constant agitation by 30 minutes. After that, the sodium silicate reacted with HCl (6N) adding the silicate to the acid until the pH was 8, at temperature of 20 c with permanent agitation. The block flow diagram for the entire process is shown in the figure No 2.

RESULTS AND DISCUTION

In the all processes the water employed in every step was measure and both catalytic support properties were determinate by XRD, area BET and TGA analysis. The water quantities were 3.5 Liters for 0.22 grams of α- Al_2O_3 from 3 grams of aluminum shaving. And for the silica gel the final support obtained was 0.6 gram and the water employed was 4.8 L.

The XRD result for the α- Al_2O_3 s material is shown in the figure 3. In this figure the characteristic peaks for the alumina can be observed. The area bet was evaluated and the TGA analysis was obtained

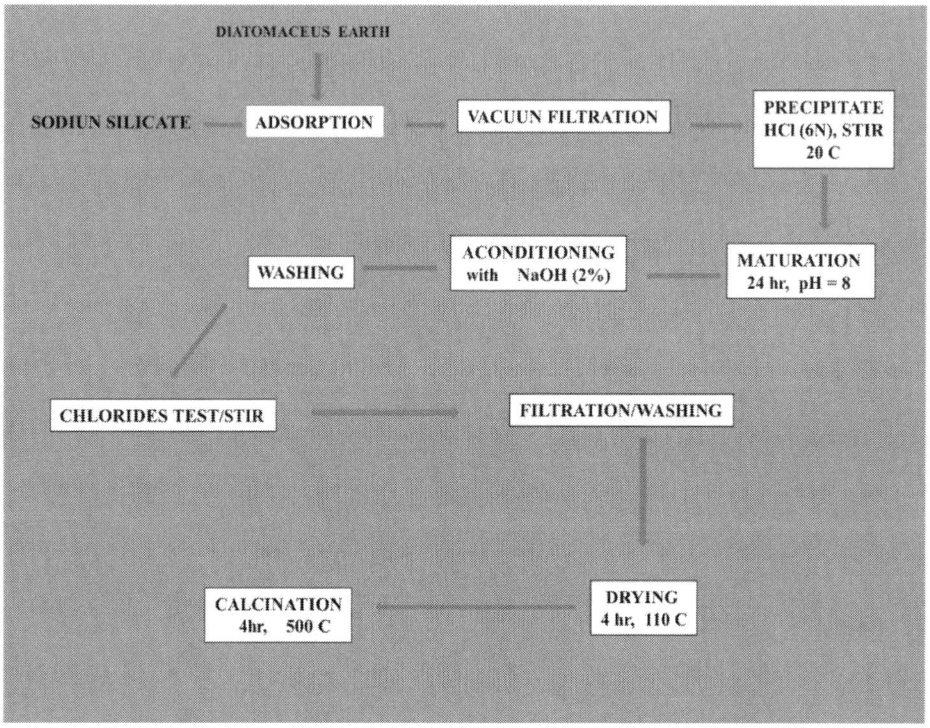

Figure No 2. Block Flow Diagram for Silica Gel process.

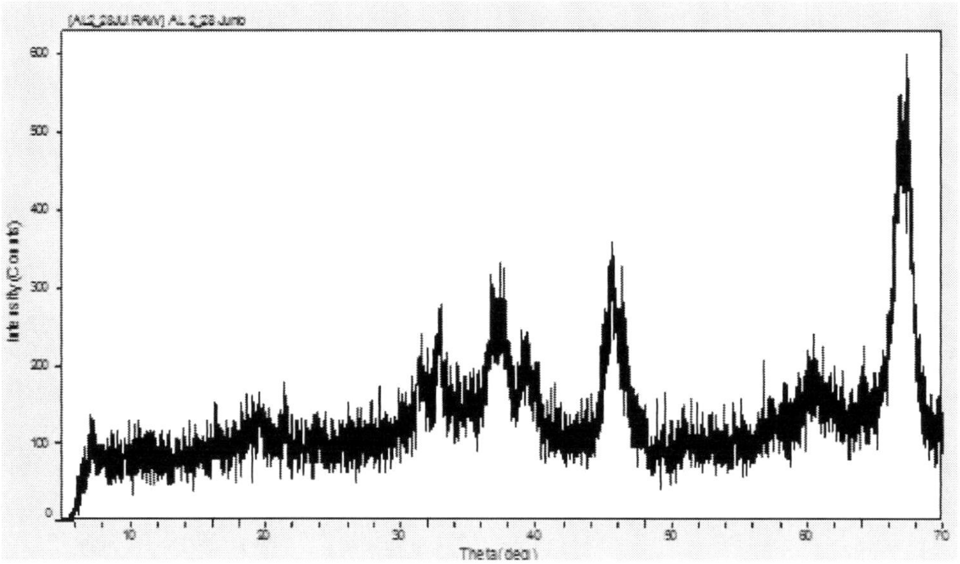

Figure No 3 XRD for α-Al_2O_3

The higher water quantities, in both processes were in the washing steps.

(758a) Thermochemical Conversion to Biocrude From Pilot Scale Grown Wastewater Fed Algae

Griffin W. Roberts[1], Marie-Odile Fortier[2], Belinda S.M. Sturm[2] and Susan M. Stagg-Williams[1], (1)Chemical & Petroleum Engineering, University of Kansas, Lawrence, KS, (2)Civil, Environmental & Architectural Engineering, University of Kansas, Lawrence, KS

Advantages of algae as a biofuel feedstock are well known; high growth rate, high lipid content, and CO_2 neutral. Thermochemical conversion or hydrothermal liquefaction (HTL) of algae to a hydrocarbon rich biocrude is growing in popularity primarily because it uses subcritical water as the reaction media eliminating the high energy intensive drying step associated with lipid extraction. HTL efficiently extracts existing lipids in addition to converting other macromolecules such as carbohydrates and proteins to smaller molecules that can be readily upgraded to fuels and other chemicals. Such a process also eliminates the need for a high lipid producing algae and can solely focus on high growth rate algae. Recent studies have shown that biofuels from algal feedstocks are most economically favorable if the algae is grown in conjunction with wastewater treatment plants; using free nutrients such as nitrogen and phosphorous found in the wastewater effluent rather than purchasing fertilizers for algal growth. There is an additional environmental benefit of removing such nutrients before they enter rivers, lakes and oceans by minimizing unnatural algal blooms which can cause anoxic zones.

This study involves pilot scale algae growth tanks utilizing effluent from the second clarifier at the Lawrence, KS wastewater treatment plant. Algae was harvested and hydrothermal treated toward biocrude. Both a mixed-culture microalgae and macroalgae identified as *Cladophora sp.* were studied. Hydrothermal treatment was performed with 3g of freeze dried algae and 50mL of water at 350°C for 1 hour. The starting algae and product yields were evaluated on an ash free dry weight basis. High oil yields were obtained for both the micro- and macroalgae (54% and 57%). Residual solids were determined to have higher heating values of 12.4 and 10.8 MJ/Kg for micro- and macroalgae, respectively. Aqueous co-products had carbon, nitrogen, and phosphorus concentrations of 3100-4000 ppm, 1200-1600 ppm, and 3.5 ppm respectively. The biocrude produced was very similar to that of petroleum crude in terms of energy density (39-42 MJ/Kg) and elemental percentages of carbon, hydrogen, and oxygen.

(758b) Hybrid Thermochemical/Biochemical Gas Fermentation Process for Production of Fuels and Chemicals

Derek W. Griffin and Michael A. Schultz, LanzaTech Limited, Roselle, IL

The ever increasing demand for renewable feedstock-based fuels and chemicals is driving the interest and rapid development of 2nd generation biofuel processes worldwide. There is significant research into improving the overall carbon and energy efficiency of biorefineries in order to compete with petroleum derived products in terms of economics and sustainability. A range of 1st generation renewable thermochemical or biochemical processes face various challenges including high energy demands, high capital costs, poor carbon utilization, or a combination of these factors. Integrated thermochemical/biochemical processes have the potential advantage to overcome these challenges to produce sustainable fuels and chemicals from renewable feedstocks.

The hybrid thermochemical/biochemical LanzaTech process captures various gas streams with renewable resources and converts them into low-carbon fuels and chemicals. The gas feedstock can come from a variety of sources including gaseous waste carbon from industrial processes or from thermochemical processes such as gasification of biomass/MSW or reformed natural gas. The CO, H_2, and/or CO_2 contained in the feed gas are fermented via a biocatalyst to chemicals/fuels such as ethanol, acetic acid, or 2,3-butanediol which also serve as building blocks to longer chained hydrocarbon drop-in fuels. The LanzaTech process offers superior carbon conversion, energy efficiency, and greenhouse gas emission performance compared to conventional and emerging routes to the same products.

The LanzaTech's integrated thermochemical/biochemical gas fermentation process is described including a comparison to current stand-alone biochemical and thermochemical technologies for the case of producing ethanol from renewable feedstocks, such as biomass. The different routes are compared in terms of product yields, overall carbon capture, thermal efficiencies, and green house gas (GHG) emissions based on overall mass and energy balances of the various processes. A previous analysis comparing gas fermentation to the thermochemical route to ethanol[1] is expanded here to include an assessment of ethanol production via biochemical routes. It is demonstrated that engineering combined thermochemical & biochemical processes can result in benefits that neither process can achieve independently.

[1] Griffin, D.W. and Schultz, M.A. (2012). Fuel and Chemical Products from Biomass Syngas: A Comparison of Gas Fermentation to Thermochemical Conversion Routes. Environmental Progress & Sustainable Energy. *31*, 219-224.

(758c) Effects of Hydrolysis and Torrefaction On Pyrolysis Product Distribution of Spent Mushroom Compost (SMC)

Rene Garrido, Chemical Engineering Department, Villanova University, Villanova, PA, Maria Nydia Ruiz-Felix, Chemical Engineering, Villanova University, Villanova, PA and Justinus A. Satrio, Department of Chemical Engineering, Villanova University, Villanova, PA

Energy based in biomass is a source of renewable energy from organic and inorganic matter formed in a biological or mechanical process. Although fossil fuels, such as petroleum, oil and coal, have their foundation in ancient biomass, they have been "out" of the carbon cycle for a long time. Their combustion consequently disturbs the carbon dioxide content in the atmosphere.

Lignocellulosic biomass can be classified as natural (without human intervention), residual (by-product of waste generated by agricultural activities, forestry, and industry wood processing), and energy-based crops (for the production of biofuels); it is the only carbon-containing renewable energy source. The complex structure of lignocellulosic biomass has been extensively studied and poses highly valuable carbohydrates (cellulose and hemicelluloses) and lignin, whereas there is a substantial challenge to the large-scale biomass utilization. The carbohydrate polymers are strongly bound to the lignin, which makes the separation process technically difficult.

While it is scientifically viable to exploit lignocellulosic materials and organic wastes into energy, chemicals and fuels, the commercial production is far from an economical process and the cost needs to be lowered. It also must be demonstrated that a commercial scale biomass utilization process is environmentally sound, and there is a need to find an inexpensive and widely available lignocellulosic source of biomass

This study investigates the potential use of a unique biomass material, produced from mushroom farms, known as spent mushroom compost (SMC) as a biomass-based energy feedstock. It was found that, even though SMC has a highly probability for bio-oil conversion its high moisture content and high ash content in feedstock results in low yield and quality of bio-oil. The removal of these recalcitrants is a key element to make SMC a desirable feedstock for biomass conversion. This investigation has revealed that using torrefaction and hydrolysis for SMC treatment after pyrolysis shows a higher yield with a wider product distribution after the combined pre-treatments proposed.

Analytical pyrolysis combined with gas chromatography and mass spectrometry (Py-GCMS) is used to study the chemical composition of volatile products from previously hydrolyzed and torrefied biomass. Effects of the severity level of hydrolysis pre-treatment and torrefaction temperatures on the chemical composition and yield distribution of the fast pyrolysis products will be evaluated. The chemical composition study of pyrolyzed biomass would then aid in the understanding of the chemistry behind the different compounds to further enhance the process of upgrading bio-crude oil. By this, ground work is laid as a preliminary attempt to understand SMC as a potential energy feedstock.

(758d) Mixed Catalysts for the Conversion of Hemicellulose and Cellulose

Nathan S. Mosier[1], Eurick Kim[2], Shuo Liu[2] and Mahdi Abu-Omar[3], (1)Agricultural and Biological Engineering & Laboratory of Renewable Resources Engineering (LORRE), Purdue University, West Lafayette, IN, (2)Chemistry, Purdue University, West Lafayette, IN, (3)Chemistry Department, Purdue University, West Lafayette, IN

Direct catalytic conversion of lignocellulosic biomass to biofuels could improve the carbon efficiency of biofuel production. We report the use of maleic acid, a dicarboxylic acid, mixed with Lewis acids such as lithium chloride and aluminum chloride to catalyze the fractionation of biomass into an aqueous solution of pentose (primarily xylose) and insoluble cellulose and lignin, followed by the conversion of the xylose to furfural under higher temperature and pressure. This method achieved 80-90% yield of xylose through hydrolysis of the hemicellulose from various biomass sources (switchgrass, poplar, pine) and achieved 54-61% yield of furfural (based on original biomass). The residual solids contain cellulose that has increased reactivity for hydrolysis by cellulolytic enzymes. SEM imaging suggests that maleic acid pretreatment is effective in fibrillating biomass, thus significantly increasing the accessible surface area of cellulose for enzymatic or catalytic conversion. This hybrid approach combines pretreatment for biochemical conversion with thermochemcial processing. We present a kinetic analysis of biomass hydrolysis and formation of aldehydes and discuss application of results from pure sugars to results from biomass conversion.

(758e) Aspen PLUS Modeling of the Utilization of Equine Waste for Localized Heating Via Fast Pyrolysis

Nicole L. Hammer[1], Akwasi A. Boateng[2], Charles A. Mullen[1], M. Clayton Wheeler[3] and Jose L. Carrasco[4], (1)Eastern Regional Research Center, Agricultural Research Service, USDA, Wyndmoor, PA, (2)U.S. Department of Agriculture, Crop Conversion Science & Engineering, Agricultural Research Service, Wyndmoor, PA, (3)Chemical and Biological Engineering and Forest Bioproducts Institute, University of maine, Orono, ME, (4)Chemical and Biological Engineering, University of Maine, Orono, ME

ARS/USDA is investigating the local utilization of equine waste as an alternative and replacement fuel source for heat generation using bio-oil produced via fast pyrolysis at the equine rehabilitation facility at Morrisville State College (MSC) with potential expansion to local farms. The Aspen Plus based simulation models the production of pyrolysis oil and its combustion in an existing boiler for the generation of hot water. Ultimate and proximate analysis on the manure and its pyrolysis products enabled the defining of the manure as a non-conventional compound and the pyrolysis products as conventional compounds based on stoichiometric formula. This results in quantifying the production of pyrolysis oil and bio-char which are valuable and sustainable energy sources that can be utilized at the local facility. Pyrolysis oil can be used as fuel in an existing hot water boiler system at MSC that produces hot water for the facility via diesel. The results show potential for displacing the diesel fuel by the pyrolysis oil produced from the locally generated equine waste, while also alleviating a waste disposal problem. Based on mass and energy balances of the system, heat integration is possible. Using a portion of the co-product bio-char for combustion provides all the heat required to run the pyrolyser. The fluidized bed pyrolyser will circulate sand from the pyrolysis unit through a combustion unit for reheating and then return it hot to the pyrolysis unit. Excess bio-char can be utilized in surrounding soils acting as a soil amendment. By incorporating the heat sources and heat integration in the model, energy requirements to run the system are significantly reduced. The mass and energy balances generated by Aspen Plus will be the basis for a Techno Economic Analysis for locating such a system at MSC; this will be presented and discussed further.

Critical and Adequate pretreatment for Biomass Sugar Release – Selective Lignin Unlocking Process Inspired from Wood-Feeding Termite

Jing Ke, Shulin Chen

Department of Biological Systems Engineering, Washington State University

Currently, the limiting factor for lignocellulosic biofuels is the lack of efficient pretreatment technology within an industrially acceptable economics. The heterogeneous and irregular arrangement of lignin construction results in resistance to saccharification, thus severely limiting biochemical conversion of lignocellulosic biomass to fermentable sugars. Lignin framework also inhibits cellulase function by competitive adsorption, chemical inhibition, and steric hindrance. Since lignocellulosic cellulose can only be digested after the lignocellulose is exposed out of lignin framework. It is therefore important to understand to which extent of lignin content and composition can be modified in useful ways without deleterious consequences to plant growth and development. Existing biomass conversion schemes typically rely on a combination of thermochemical treatments that are not sustainable and environmentally friendly as the processes still consume a great deal of energy and require input of chemicals because of the needs for high temperature and pressure. Nature, on the other hand, has developed through evolution efficient biological systems for extracting energy from plant materials to power the lives by effectively deconstructing lignocellulose under natural conditions. Biological systems are the proof-of-concept of what can be physically achieved as they are built of the same fundamental elements of matter and follow the same laws of physics as inorganic and human-engineered systems. With adequate understanding, these biological systems can be harnessed for the production of bioenergy sustainably from renewable lignocellulosic biomass. The high efficiency of cellulose utilization by wood-feeding termite in terms of rate and extent (95% of cellulose in wood within 24 hrs) makes it a promising model system for biofuel production by understanding and mimicking its unique pretreatment system. This work explored a unique lignin unlocking process in wood-feeding lower termite workers *Coptotermes formosanus* (Shiraki) for efficient woody carbohydrate utilization and how the termite workers manage the selective lignin modification. The rationale is based on its superior performance on cellulose utilization and its stepwise introduction of specific reactions on lignin, as well as the varied micro-parameters for lignin modification in each gut section. Obtained results provide that termite-induced lignin pretreatment is not achieved by savage degradation, but selectively modification of functional group, cleavage of ether linkage within lignin and ester/ether linkage between lignin-hemicellulose and lignin-cellulose. The demonstrated structural modification results in alleviation of both physical and chemical inhibition effect of lignin on cellulases, together with alteration of the three dimensional structure of lignin, which unlocks cellulose fiber from protection. The results also demonstrate that during the lignin structural modification by termite, a non-enzymatic system of $\cdot O_2^-$ and catalysts, generating from the initial enzymatic lignin aliphatic dehydroxylation and carbonylation, penetrate inside the plant cell wall to induce reactions profiting from their small molecular sizes. This non-enzymatic system is proved to take a lead for the selectivity of lignin modification as what termites do, including modification on the specific functional group and ether/ester linkages within lignin framework or between lignin-hemicellulose and lignin-cellulose, which is partially reverse reaction of lignin synthesis. The *in-vitro* mimicking of this non-enzymatic system with umol level concentrations of the chemicals results in alleviation of the lignin inhibition effect on enzymatic cellulose hydrolysis.

Notably, the lignin-modifying system mimicking from termite is advanced in its low chemical concentration, rate, selectivity, and neutral working condition. It leaves lignin backbone alone, thus consuming less energy and producing negligible lignin-derived inhibitors for subsequent cellulose hydrolysis and sugar fermentation. Meanwhile, the residual lignin backbone could be applied for other purposes such as value-added lignolytic co-products. On the basis of this, this new pretreatment process inspired by termites is from but beyond the nature, which performs lignin pretreatment as efficiently, robust and environmentally-friendly as termites. This effort also made major contributions to advance the fundamental science by supporting the new concept of critical and selective lignin modification for farthest alleviation of the inhibition effect, which is critical for filling the existing knowledge and technology gaps in sustainable and economical biofuel and value-added chemicals production from lignocellulosic biomass.

(774a) Development of Proton Selective Membranes for Efficient and Economical Vanadium Redox Flow Batteries

Justin Walls[1], Pedram Jahanian[2], Guangzhao Mao[2], Dr. Steven Salley[3] and Dr.Simon Ng[3], (1)Chemical Engineering, Wayne State University, Detroit, MI, (2)Chemical Engineering and Materials Science, Wayne State University, Detroit, MI, (3)Chemical Engg., Wayne State University, Detroit, MI

Development of Proton Selective Membranes for Efficient and Economical Vanadium Redox Flow Batteries

Justin Walls, Padram Jahanian, Guangzhao Mao, Steven Salley, K. Y. Simon Ng

Department of chemical Engineering and Materials Science, College of Engineering,

Wayne State University, Detroit MI 48202

Flow Redox Batteries (FRBs), especially the Vanadium Redox Flow Battery (VRFB), provide a promising chemistry for storing intermittent renewable energy on a very large scale. While these batteries have already been developed on the megawatt hour level; none of the FRB technologies has seen significant market penetration. This is mainly because the technology is still very expensive, due to high material costs, and less than satisfactory performance parameters and reliability.

The key component of the VRFB, which is one of the leading causes of the elevated cost and unreliability, is the ion exchange membrane (IEM). The membranes need to possess good ion conductivity while maintaining high mechanical and chemical stability. Perfluorosulfonic polymers such as Nafion are the most common membranes for proton exchange in the VRFB due to their good chemical stability and high proton conductivity. However, due to the tendency of vanadium to pass through the membrane, Nafion membranes suffer from decreased energy efficiency. This project is focused on decreasing the vanadium ion diffusion through the membrane by modifying the nanostructure of Nafion. There has been extensive research activity in the modification of Nafion membranes, for example, modification by layer-by-layer polyelectrolyte multilayer films and in situ sol-gel reaction to close the pores by silica nano particles. Our project combines these two methods using hyper-branched polyelectrolyte multilayer films to further lower the vanadium ion diffusion rate. Nafion modification by the sol-gel chemistry was conducted by treating the Nafion in the tetraethylorthosilicate (TEOS) methanol solution. Poly(diallyldimethylammonium chloride) (PDDA) and poly(sodium styrene sulfonate) was used in the layer-by-layer deposition. Diffusivity and ion crossover of these membranes was tested and compared to that of the unmodified membranes as well as other modification techniques that have been reported. A small scale VRFB system was constructed to test the performance of these modified membranes, and efficiency and reliability of the system as a whole was measured.

(774b) Microporous Separators On Fe/V Redox Flow Battery: A Valuable Opportunity for Cost Reduction

Xiaoliang Wei[1], Qingtao Luo[1], Bin Li[1], Zimin Nie[1], Yuyan Shao[1], Feng Chen[1], Baowei Chen[1], Guanguang Xia[1], Liyu Li[2], Zhenguo Yang[2] and Wei Wang[1], (1)Pacific Northwest National Laboratory, Richland, WA, (2)UniEnergy Technologies, LLC, Mukilteo, WA

Microporous Separators on Fe/V Redox Flow Battery: A Valuable Opportunity for Cost Reduction

Wei, Xiaoliang; Luo, Qingtao; Li, Bin; Nie, Zimin; Shao, Yuyan; Chen, Feng; Chen, Baowei; Xia, Gordon; Li, Liyu; Yang, Z Gary; Wang, Wei

Pacific Northwest National Laboratory, 902 Battelle Boulevard, PO Box 999, Richland, WA 99354

Redox flow batteries are considered as one of the most promising medium-to-large scale energy storage technologies and have attracted much attention both academically and industrially.[1],[2] A number of redox chemistries have been proposed, and significant progress has been achieved by today.[3],[4] However, broad market penetration of redox flow batteries is still hindered by their intrinsic limitations such as high cost and low durability of components, narrow operational temperature range, limited solubility of active species, and/or low electrochemical activity.[5] The recently invented iron-vanadium (Fe/V) redox flow battery employing $Fe^{2+/3+}$ – $V^{2+/3+}$ redox couples have shown to be a promising option for stationary energy storage.[6],[7] The Fe/V flow battery demonstrated stable cycling performance with a nearly 100% utilization ratio over a broad temperature range of 0-50°C.

The positive charged species, Fe^{3+}, is a relatively weak oxidant. Therefore, hydrocarbon-based ion exchange membranes and/or separators are possible options for use in Fe/V flow battery system. This contribution investigated the cycling performance of a variety of polyethylene microporous separators on Fe/V flow cell. Among them, some separators exhibited energy efficiency of around 70% at temperatures ranging from 5-50°C and at current densities up to 80mA/cm^2. Because these separators are very inexpensive, their use significantly reduces the capital cost of Fe/V flow battery, delivering great potential for developing a low-cost energy storage system.

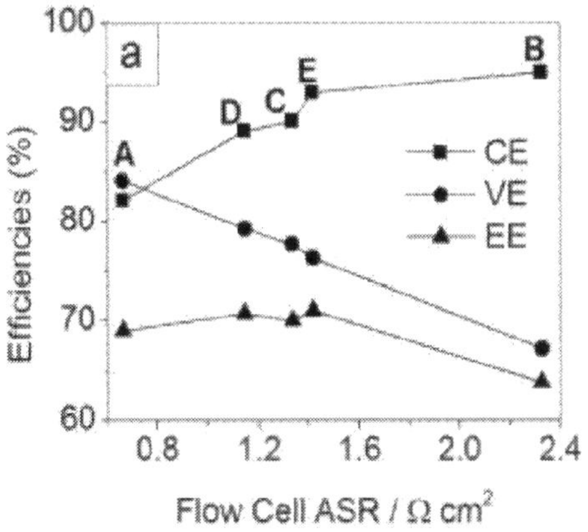

Figure 1 Flow cell cycling efficiencies (CE, VE, and EE) of microporous separators A, B, C, D, and E.

[1] Z. Yang, J. Zhang, M. C. W. Kintner-Meyer, X. Lu, D. Choi, J. P. Lemmon, and J. Liu, *Chem. Rev.* 2011, *111*, 3577-3613.

[2] J. Rugolo, and M. J. Aziz *Energy Environ. Sci.* 2012, **DOI:** 10.1039/C2EE02542F.

[3] M. Skyllas-Kazacos, M. H. Chakrabarti, S. A. Hajimolana, F. S. Mjalli, and M. Saleem *J. Electrochem. Soc.* 2011, *158*, R55-R79.

[4] L. Li, S. Kim, W. Wang, M. Vijayakumar, Z. Nie, B. Chen, J. Zhang, G. Xia, J. Hu, G. Graff, J. Liu, and Z. Yang *Adv. Energy Mater.* 2011, *1*, 394–400.

[5] S. Eckroad, *Vanadium Redox Flow Batteries: An In-Depth Analysis.* EPRI, Palo Alto, CA: 2007. 1014836.

[6] W. Wang, S. Kim, B. Chen, Z. Nie, J. Zhang, G. Xia, L. Li, and Z. Yang *Energy Environ. Sci.* 2011, *4*, 4068-4073.

[7] W. Wang, Z. Nie, B. Chen, F. Chen, Q. Luo, X. Wei, G. Xia, M. Skyllas-Kazacos, L. Li, and Z. Yang *Adv. Energy Mater*. 2012, *2*, 487–493.

(774c) In Situ Investigation of Vanadium Ions Transport in Redox Flow Battery

Qingtao Luo[1], Liyu Li[2], Zimin Nie[3], Wei Wang[1], Xiaoliang Wei[1], Feng Chen[1], Bin Li[4], Guanguang Xia[1], Baowei Chen[5], Gary Yang[1] and Sprenkle Vincent[1], (1)Pacific Northwest National Laboratory, Richland, WA, (2)Hydrocarbon Processing Group, Pacific Northwest National Laboratory, Richland, WA, (3)Pacific Northwest National Labratory, Richland, WA, (4)Pacific Northwest National Laboratory, Richland , WA, (5)Biological Sciences, Pacific Northwest National Laboratory, Richland, WA

The all-vanadium redox flow battery (VRB) is probably the most promising and extensively researched flow battery system. In order to shed light on the VRB capacity fading mechanism, it is necessary to investigate the transport behaviors of vanadium ions across the membrane. In this work, flow batteries with vanadium and iron redox couples as the electroactive species were employed to investigate the transport behavior of vanadium ions in the presence of an electric field. It was shown that the transport of vanadium ions was affected by a combination of different factors, such as the charge/discharge process, current density, and SOC. The electric field accelerated the positive-to-negative and reduced the negative-to-positive transport of vanadium ions in the charging process and affected the vanadium ion transport in the opposite way during discharge. With the increase of current density, the vanadium ion transport rate increases during the charge process but decreases during discharge. At each given current density, the vanadium ion transport rate, in general, demonstrated a trend of decreasing with the increase of SOC.

(774d) Electrolyte Optimization for Fe/V Redox Flow Battery System

Bin Li[1], Wei Wang[1], Zimin Nie[2], Vincent Sprenkle[1], Baowei Chen[3], Xiaoliang Wei[1], Qingtao Luo[1], Guanguang Xia[1], Liyu Li[4] and Zhenguo Yang[4], (1)Pacific Northwest National Laboratory, Richland, WA, (2)Pacific Northwest National Labratory, Richland, WA, (3)Biological Sciences, Pacific Northwest National Laboratory, Richland, WA, (4)UniEnergy Technologies, LLC, Mukilteo, WA

Electrolyte optimization for Fe/V redox flow battery system

Bin Li, Wei Wang, Zimin Nie, Vince Sprenkle, Baowei Chen, Xiaoliang Wei, Qingtao Luo, Guanguang Xia, Liyu Li and Zhenguo Yang

Pacific Northwest National Laboratory

902 Battelle Boulevard, PO Box 999, Richland, WA 99354

Abstract

Redox flow batteries (RFBs) research and development have recently enjoyed a renaissance primarily due to their ability to store large amounts of power and energy, up to multi-MW and multi-MWh, respectively, for renewable energy integration or smart-grid deployment.[1] Broad market penetration, however, has not been achieved due to several limitations such as high cost, instability in long-term operation, and narrow operational temperature range, etc. Recently invented Fe/V redox flow battery (IVRFB) has attracted more and more attentions due to its long-term cycling stability and the freedom to utilize membranes other than expensive Nafion.[2] In this study, extensive matrix study was performed on the Fe/V electrolyte in order to determine the relationships of the composition, state of charge (SOC), and temperature stability. As we know, the temperature stability of the RFB systems is of paramount importance for practical use. A RFB system with wider temperature stability will not only enable the RFB to be deployed with less geographical limitation, but also possibly eliminate the active temperature management system reducing parasitic loss. As previously reported, the current Fe/V electrolyte cannot be operated or stored at lower temperatures ($\leq 0^oC$) due to the precipitations ($FeCl_2$). In this research, factors influencing the stability of electrolytes in both positive and negative half-cells with different SOCs at different temperatures from -10^oC to 40^oC were systematically studied. Through the optimization of the electrolyte composition, an electrolyte stable in the temperature range from -5^oC to 40^oC was identified. Electrolyte stability data and IVRFB flow cell cycling performances using low-cost separators will be presented (Fig.1).

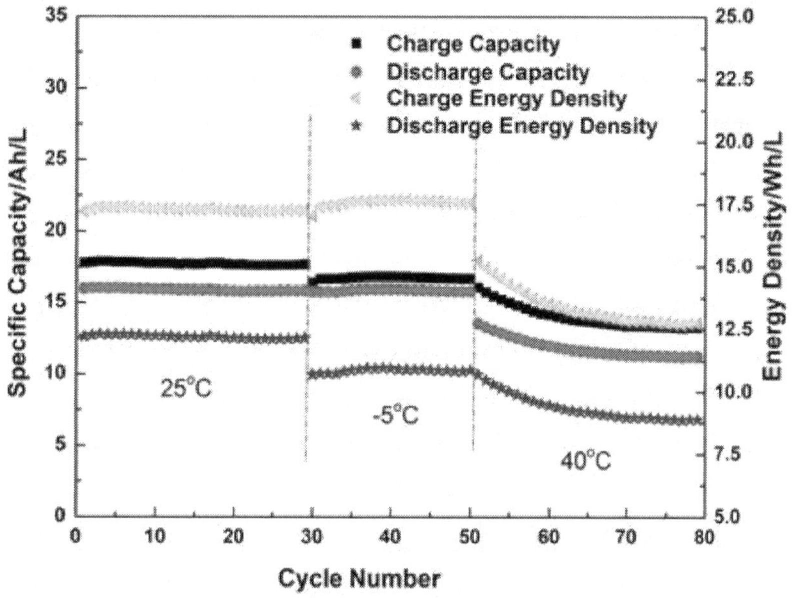

Fig.1 Cycling performances of optimized Fe/V electrolytes

References

(1) Yang, Z.; Zhang, J.; Kintner-Meyer, M. C. W.; Lu, X.; Choi, D.; Lemmon, J. P.; Liu, J. *Chemical Reviews* **2011**, *111*, 3577.

(2) Wang, W.; Kim, S.; Chen, B.; Nie, Z.; Zhang, J.; Xia, G.-G.; Li, L.; Yang, Z. *Energy & Environmental Science* **2011**, *4*, 4068.

(774e) Electrostatically Assembled Carbon Nanoparticle Electrodes for Vanadium Redox Flow Batteries

Abhinandh Sankar and **Anastasios Angelopoulos**, School of Energy, Environmental, Biological, and Medical Engineering, University of Cincinnati, Cincinnati, OH

Efficient charge and mass transfer at the surface of the electrode is critical to overall redox flow battery (RFB) performance. Fundamental structure-property correlations for RFB electrodes cannot presently be obtained using electrodes produced from carbon felts or carbon nanoparticle slurries and pastes due to the ill-defined distribution of active sites in these materials. For example, a recent study using high surface area multi-wall carbon nanotubes as electrode materials yielded only a two to three-fold increase in peak current density relative to glassy carbon. We here report on the first use of electrostatic layer by layer (LbL) assembly to fabricate RFB electrodes. Since its introduction by Iler over four decades ago, LbL assembly has been repeatedly shown capable of producing well-defined arrangements of nanoparticles and the method is now ubiquitous in scientific research. However, the approach has yet to be applied to produce well-defined electrode structures for RFB electrodes and has only rarely been explored in the general field of electrocatalysis. Key areas of concern include: 1) the impact of the polymer counter-ion layer that is typically required for assembly on ohmic resistance, 2) accessibility of active sites in tightly packed layered structures, and 3) durability. In this investigation, we demonstrate how LbL assembly of graphite nanoplatelets may be used to achieve unprecedented high current densities for the all-vanadium system through fine control over the thickness and morphology of the electrode. In particular, we show how nanoparticle packing density, size, and size distribution in addition to local polymer binder concentration and crystalline order all impact active site concentration and promote reagent accessibility to these sites in a manner that has not been previously recognized. Such basic understanding permits rational optimization of electrode chemistry and structure using carbon-based electrocatalysts as well as detailed interpretation of the promotional effects of inorganic electrode additives. Finally, we demonstrate through electrochemical cycling how these coatings are exceptionally stable relative to electrodes fabricated without LbL assembly.

(774f) Fractional Degradation Cost for Cycling a Zinc/Bromine Flow Cell Battery

Alex Bistrika[1], Alexandre F.T. Yokochi[1], Michael Antonishen[2], James Davidson[2], Douglas Halamay[2], David Naviaux[3], Jiajia Song[2], Ted Brekken[3] and Annette von Jouanne[3], (1)School of Chemical, Biological and Environmental Engineering, Oregon State University, Corvallis, OR, (2)Electrical Engineering and Computer Science, Oregon State University, Corvallis, OR, (3)School of Electrical Engineering and Computer Science, Oregon State University, Corvallis, OR

It is widely understood that intermittent resources such as solar and wind are dependent on environmental factors outside our ability to control. This allows for the possibility of either power output exceeding or falling short of forecast levels which leads to more frequent cycling of regional hydro generators and greater overall grid instability under the current infrastructure. Quantifying the cost associated with added maintenance attributed to an increased penetration of wind and the physical limitations of a growing dependence on Pumped Hydro Energy Storage (PHES), becomes essential in painting a more complete picture. However, alternative means of Large Scale Energy Storage (LSES), for storing excess power when the outputs exceeds demand and to supplement output power when it falls short of demand, is a widely accepted route for mitigating intermittent power generation.[1] Thus, it is crucial to establish an accurate and relative cost function for the commercially available and most economically viable storage technologies. The most promising electrical energy storage technology, to replace PHES, is flow cell batteries.[2] This begs the question what are the degradation mechanisms in these building size chemical systems and what do we do when they stop working. In this study we report on the key degradation mechanisms that could be expected for standard operation of a zinc/bromine flow cell battery using graphite electrodes and assign a cost function dependent of system operating parameters. In a previous study we reported the oxidation (or charging) reaction as the most damaging to the electrode material, see **Figure 1**. We discuss aging trends and electrode performance by considering surface chemistry as observed by Raman spectroscopy and scanning electron microscopy. Theoretical system performance based on steady state current density measurements is used to establish a normalized operating cost function for a zinc/bromine battery, as can be seen in **Figure 2**.

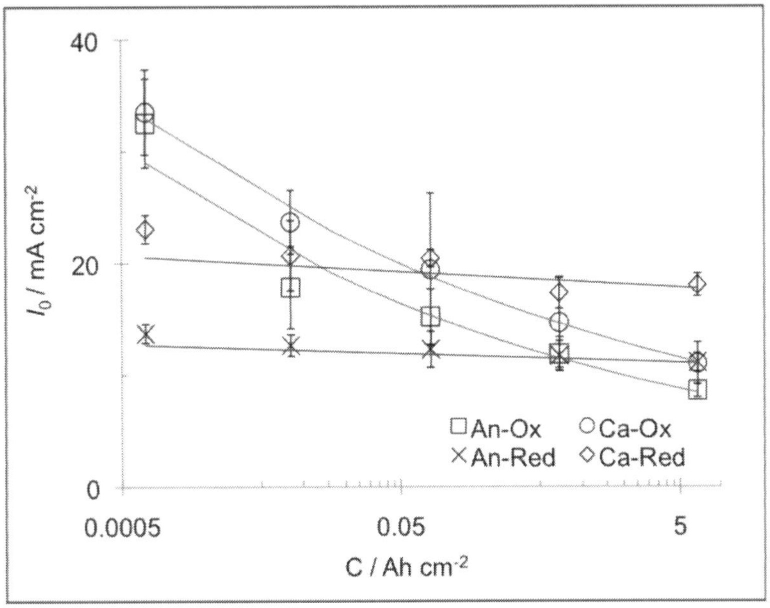

Figure 1 Oxidation and reduction exchange current; measured for both the cathode and anode in 6 [M] NH4Br. The x-axis represents the electrode aging, depicted by total coulombs passed through the electrode per unit surface.

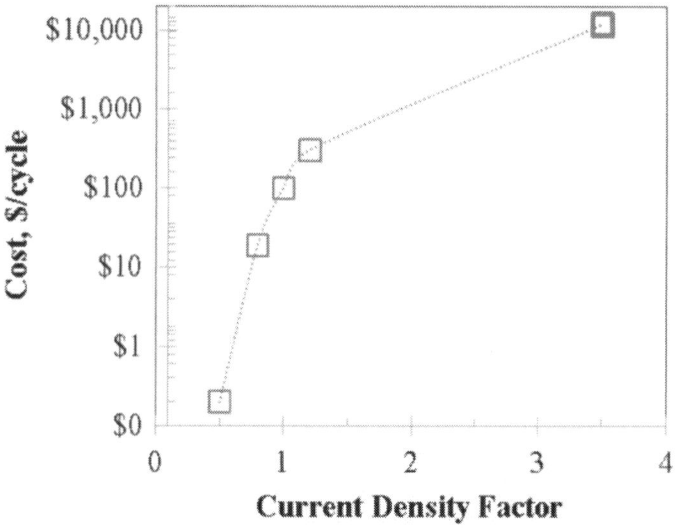

Figure 2 Cost of a 10 % state of charge cycle with respect to a representative operating parameter (current density factor).

1. Yang Z, Zhang J, Kintner-Meyer MCW, et al. Electrochemical Energy Storage for Green Grid. *Chemical Reviews.* 2012/04/07 **2011**;111(5):3577-3613.

2. Soloveichik GL. Battery Technologies for Large-Scale Stationary Energy Storage. *Annual Review of Chemical and Biomolecular Engineering.* 2012/04/24 **2011**;2(1):503-527.

(774g) Design Mesostructured LiMn2O4 As Low-Cost Cathode Materials for Lithium Ion Batteries

Jian Zhu, Chemical Engineering, Wayne State University, Detroit, MI, K. Y. Simon Ng, Department of Chemical Engineering and Materials Science, Wayne State University, Detroit, MI and Da Deng, Chemical Engineering and Materials Science, Wayne State University, Detroit, MI

Lithium ion batteries are the dominant power sources for mobile electronic devices and will play increasingly important role in expanding market of electric vehicles and green energy storage. Spinel $LiMn_2O_4$, alternative to expensive and toxic $LiCoO_2$, as cathode materials in lithium ion batteries are attracting much attention for advantages of low cost, environmental benignness, and abundance. However, the low lithium-ion diffusion coefficient and capacity fading caused by dismutation reaction of Mn^{3+} restrict the applications of $LiMn_2O_4$ at large/industrial scale where bulky batteries are required to have good cycling performance at high current rate with no significant capacity fading. Our strategy is to design and tailor the structure and composition of $LiMn_2O_4$ to improve its conductivity and cycling performance. In this presentation, our latest efforts on the design and facile fabrication of $LiMn_2O_4$ as high-power cathode materials will be reported. The unique structure is tailored and controlled to allow fast lithium ion diffusion while maintaining structure stability. The unique mesostructure is fully characterized by electron microscopy and x-ray diffraction and their electrochemical performance in lithium ion batteries is thoroughly evaluated.

SUBSTANTIAL INVOLVEMENT BY DOE TO ADDRESS SCALE-UP CHALLENGES FOR BIOREFINERIES

Gene Petersen and Glenn Doyle, US Department of Energy, Golden, CO; Art Wiselogel, CNJV, Golden, CO; Carol Babb, SAIC, Denver, CO; and Andras Marton, Independent Project Analysis, Inc., Ashburn, VA.

Extended Abstract

Since 2003, the Department of Energy (Department) has been substantially involved in cost-sharing the development of over thirty integrated biorefinery projects to process biomass feedstocks into transportation fuels and biobased products. These projects range in scale from developing engineering designs for pilot operations, to constructing integrated pilot- and demonstration-scale biorefineries, as well as commercial biorefinery demonstrations. Under direction of public law (Energy Policy Act of 2005), the Department was authorized to "carry out a program to demonstrate the commercial application of integrated biorefineries[1]." Under that mandate and subsequent law (e.g., the American Recovery and Reinvestment Act of 2009), the Department's Biomass Program conducted three separate competitive Funding Opportunity Announcements (FOAs) from 2006-2009 to solicit applications from the community to develop integrated biorefineries at these various scales. This presentation provides a brief overview of the careful selection process employed to meet the mandates of public law and describes how the Department utilizes project management best practices to address the challenges of scale-up and completion of these projects. An oft voiced criticism is that the federal government tries to pick technology winners. In fact, the Department's overall objective is to foster development of a robust, commercially-competitive biorefinery industry by selecting a broad suite of technology types that appear technically mature enough to be successful, not to advance any particular technology. A further objective is to reduce the risk associated with designing, constructing and operating first-of-a-kind biorefinery demonstrations such that the return on the investment can become attractive to private investment. In this vein, the overall benefit of the industry-federal partnerships will be discussed along with the practices and processes used to help both recipients of federal funds and the Department achieve successful development of Integrated Biorefineries.

Among the key best practices the Department has utilized to assist in the scale up of these new technologies and to manage the competitive process and oversight of these projects are the following:

Competitive Process and Selection

- Utilized competitive FOAs that identified the Department's minimum feedstock throughput and biofuels/bioproduct production threshold objectives and outlined the evaluation criteria for selecting integrated biorefinery projects. This allowed the industry the flexibility to propose projects intended to meet the criteria. In all three FOAs, the Department did not specify any particular technology or approach to meeting the objectives especially since the FOAs acknowledged the different levels

[1] Energy Policy Act of 2005, §932

of technology maturity by accepting applications for facilities ranging from pilot to demonstration scale.

- Conducted comprehensive merit reviews that compared responsive proposals against the published evaluation criteria using non-conflicted experts to determine the competitive range (i.e., technically meritorious applications) to be recommended to the Selection Official.
- Independent Project Analysis, Inc. prepared risk assessments on responsive proposals to compare them to its industry project management best practice benchmark database to determine each proposal's risk profile and state of readiness to begin construction.
- Program policy factors as outlined in each FOA were then used to assess the balance of the Biomass Program's integrated biorefinery investment portfolio and, along with the merit review results, recommendations for selection were submitted to the Selection Official.
- Based on the above factors and availability of Federal funds, the Selection Official made selections and notified the Contracting Officer to negotiate financial assistance awards.

Post-selection Award and Project Management Practices of DOE

- Upon selection, DOE chose to execute Federal financial assistance agreements instead of contracts to allow for greater flexibility in adjusting to technological, financial and project management challenges.
- The Biomass Program adapted and implemented relevant sections of Department of Energy Order 413.3B, *Program and Project Management for the Acquisition of Capital Assets* (DOE Order 413.3B) to the management of integrated biorefineries, which are process and capital intensive projects. Figure 1 compares the guidelines in DOE Order 413.3B with the development of a typical capital construction project and DOE's substantial involvement.
- DOE employed the industrial consulting firm, Independent Project Analysis, Inc. to comprehensively compare each selected proposal against its project management best practices database to assess the project cost and schedule risks and to provide the Department with an estimate of the amount of contingency needed to account for the identified risks.
- DOE's Golden Field Office contracted with RW Beck (now SAIC) to be the Department's Independent Engineer (IE) to assist in review of initial applications, and then evaluate and monitor progress on the selected projects. Also, the IE performs independent performance test analysis to assess actual biorefinery operations as compared to the target performance criteria. The IE's support to the Department is comparable to the role it would fulfill for a private financial institution involved in the project.
- Best project management practices were invoked, where applicable, to strengthen the Department's and the recipient's controls over the execution of the project. This includes the use of:
  - Separate budget periods to control the flow of federal cost share funds
    - Budget period 1: Engineering, Front-end loading design 2 and 3, National Environmental Policy Act analysis, permitting, secure intellectual property, secure site location and infrastructure needs,

secure feedstock agreements, identify and secure off-take agreements, and secure non-Federal cost share commitments
        - Budget period 2: Final design, construction, commissioning, startup
        - Budget period 3: Operations
    - DOE Critical Decision(CD) points
        - CD-2: Approve the performance baseline
        - CD-3: Approve start of construction
        - CD-4: Approve start of operations
    - Project management plans that invoke a Work Breakdown Structure framework and identify risks and barriers to be overcome
    - Resource-loaded Gantt charts and critical path schedules
    - Milestones
    - Go/No go decision points
    - Stage-gates to address technological maturity challenges
    - Earned value management
    - Responsible, Accountable, Consulted, Informed (RACI) charts
    - Risk Mitigation Plans and Risk registers
    - Contingency
    - Continual review of lessons learned to facilitate improving the process of federal and industry co-development of these biorefineries.

Examples of Good Practices and Lessons Learned

- The foremost challenge experienced by integrated biorefineries is successfully and continuously feeding solid biomass to the throat of the reactor. Adaptation of "commercial" technology to these first-of-a-kind technologies has been plagued by numerous challenges, resulting in changes in design, materials of construction, operating parameters, etc.
- The value of integrated pilot plant tests for scale up purposes cannot be overemphasized: For these first-of-a-kind demonstration- and commercial-scale IBRs, integrated pilot plant tests have proven to be invaluable in validating the design and establishing a credible performance baseline that controls scope creep, re-engineering, and minimizes change control actions during construction, all of which increase schedule delays and capital costs. Without the benefit of fully integrated pilot plant data to inform the demonstration or commercial designs, the risks to IBR cost and schedule are magnified. This reinforces key guidance provided to DOE by IPA, Inc.
- A major lesson learned is not to "assume" that a commercially available feed system will successfully integrate with a conversion technology. Similarly, unless "commercial" technology has been successfully integrated with new process technology at an appropriate scale before being commercialized, it cannot be assumed that it will work at the larger scales.
- The more new technology process steps, the higher the risk to successfully executing the project. A corollary to this would be, the more new technology process steps, the greater the need for fully integrated pilot plant tests of durations sufficient to validate and "lock" the demonstration- or commercial-scale design.
- Convening the appropriate team with the right skill and resource mix to successfully execute major capital projects depends on the stage of technical and organizational maturity: Oftentimes, due to the perceived need to protect intellectual property rights

and/or to keep costs at a minimum, R&D/entrepreneurial entities try to take on the challenges of scaling-up bench- or pilot-scale projects without having the necessary skills mix and resources to succeed. Delays in project schedules and cost increases often reflect the inability to quickly ramp up on the learning curve to execute major capital projects. R&D and entrepreneurial organizations need to educate themselves on the risks associated with scaling and plan to secure the appropriate skills and experience mix, either internally or through contracts, to improve their chances of success.

- Key R&D and/or pilot plant testing not complete: The urgency to be "first to market" sometimes results in R&D and/or integrated pilot plant testing not being performed or completed in order to inform a comprehensive ±10% design upon which to base EPC bids that are competitive and less likely to experience significant change actions.
- Adopting formal project management best practices improves the chances of success: Once the decision is made to build a demonstration or commercial biorefinery, the early adoption of formal project management practices and best practices will significantly strengthen the chances for a successful outcome. Assuring alignment with business objectives, establishing a well-defined performance baseline (scope, schedule, cost and risk), managing to this baseline, and controlling scope creep will greatly enhance the execution of the project.
- Pre-award meetings to clarify expectations for negotiating awards and meeting federal requirements were very valuable and provided companies with better roadmaps for proceeding with their project.
- Regular teleconferences (weekly or fortnightly) and face-to-face meetings were valuable tools to facilitate negotiations and monitor progress.
- Utilizing separate budget periods aligned with Critical Decisions (e.g., Budget Period 2 aligned with Critical Decision 3 – Approve Start of Construction) approach provided the requisite time needed to work out issues prior to actual start of construction. In one case, where a Budget Period 1 approach (see above for description of the activities typically involved in BP1) was not used, the project did not achieve successful operation.
- Annual comprehensive project reviews that involved DOE and the recipients allowed DOE to be better informed about actual progress and allowed DOE to provide some useful lessons from the collective experiences of other biorefinery efforts.

Benefits of Industry-Federal Partnerships

- Due to the reluctance of the private financing industry to invest in the "first" plant, the Department's overall objective in cost sharing in pilot-, demonstration- and commercial-scale integrated biorefineries is to "buy down" some of that inherent risk, advance the industry closer to being competitive with fossil-based fuels, and help the industry attract private investment for the 2nd through Nth plants.
- Through these financial assistance agreements, the Department and industry have demonstrated that the Federal government can play an effective role in seeding development of technologies that can help the nation move closer to energy independence while creating domestic employment opportunities.

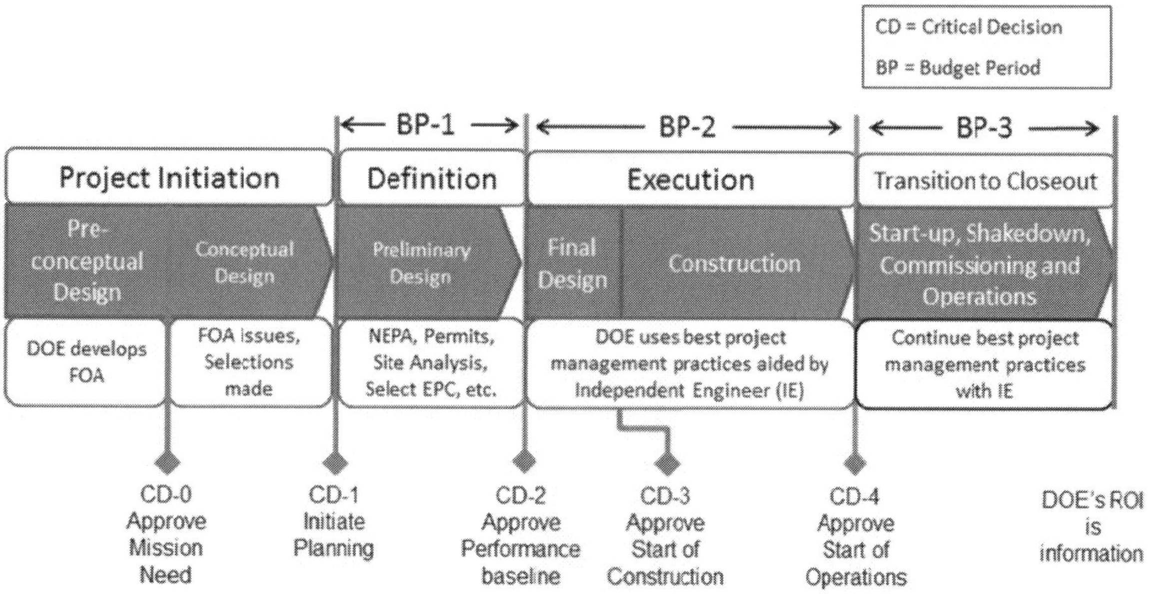

Figure 1 – Integration of Project Management for Biorefinery Development

(780b) Scale up of IH2 Process for Converting Biomass to Gasoline and Diesel

Terry Marker, Martin Linck, Michael Roberts, Pedro Ortiz-Toral, Jim Wangerow and Larry Felix, Gas Technology Institute, Des Plaines, IL

The IH^2 process is a new technology which directly converts biomass such as wood, cornstover, and lemna, into high quality liquid hydrocarbon transportation fuels using hydropyrolysis. Initial work was done on a small bench scale semi-continuous unit, which processes 0.45kg/hr of biomass. Technoeconomic and LCA studies of a commercial process based on these bench tests is very favorable, showing that gasoline and diesel can be produced from biomass at less than $2.00/gallon and have greater than 90% greenhouse gas reduction compared to fossel fuels.

Now the IH^2 process has been scaled up to a continuous unit which processes 50 kg/day of biomass. The new pilot plant unit helped us define key design parameters and also determine catalyst life and stability. GTI will discuss controlling process variables and show continuous data from the 50kg/day pilot plant. The 50kg/day tests lay the ground work for our design of a 50 t/day demo IH^2 unit which should lead directly to 2000 t/day commercial scale process in 2015-2016.

(780c) Piloting a Multiproduct Biorefinery

James J. Foster, Archer Daniels Midland Company, Decatur, IL and Michael Erpelding, Research, ADM, Decatur, IL

For some biorefineries, a single product is the goal of the project. Sometimes it is ethanol or an oil or perhaps a chemical compound, but it is focused on one product that will make or break the project. While it is good to be focused, it can lead to tunnel vision that can prematurely crash a project.

We believe a better method is run the biorefinery as having a number of co-products which allow a number of different outlets for the conversion of biomass. In our current biomass project, partially funded by the Department of Energy, we can produce ethanol, butyl acrylate, fiber and lignin. Butyl acrylate can be used as a drop in replacement for current petroleum based acrylates, fibers can be used for further conversions, and the lignin can be used as fuel source for power or steam generation while the fibers can be used as a feedstock for further conversation or sold as animal feed.

As part of the presentation we will list some of the challenges with the start-up of the pilot plant and the concerns of building a full scale plant including the technical issues and regulatory issues.

(780d) Scale-up of Ethanol Production From Sweet Sorghum Stems Using Advanced Solid-State Fermentation Technology

Shi-zhong Li, Institute of Nuclear and New Energy Technology, Tsinghua University, Beijing, China and Lei Zhang, Institute of New Energy Technology, Tsinghua University, Beijing, China

Compared with conventional submerged liquid fermentation (SLF), solid-state fermentation (SSF) can considerably reduce the process costs of substrate pretreatment and water usage, avoid the pollution problems resulted from the disposal of liquid residues. However, due to the inherent barriers of heat and mass transfer, SSF has not yet proved to be feasible in large-scale ethanol production. Accordingly, Tsinghua University had developed and operated a process of advanced solid-state fermentation (ASSF) at 127 m^3 scale using fermentable sugars from sweet sorghum stems in Inner Mongolia, China, in 2010. The field trial achieved high ethanol productivity that were comparable to conventional SLF, indicating that mass ethanol production in SSF with lower costs and less pollution problems has the great potential for its industrial application. In this study, to further optimize the process of ASSF, an improved rotary-drum fermenter at 555 m^3, the largest solid-state fermenter in the world so far, was constructed and deployed recently in Inner Mongolia, China, which was operated as continuous way for high process efficiency. During the process, the solid substrate and seed culture continue to be fed into the reactor and the fermented residues was conveyed immediately into distiller, making the whole process keep running without interval that improve the process efficiency to a great extent. Consequently, the continuous process achieved ~91% of theoretical ethanol yield that was parallel to 127 m^3 scale in the batch fermentation. The retention time of solid substrate (fermentation time) was only less than 30 h. In the meantime, tech-economic analysis (TEA) from National Renewable Energy Laboratory (NREL) indicated that 16 tons of sweet sorghum stems (with 13.5 % sugar content and 70% moisture) can be used to produce 1 ton ethanol; the fermented residues can be used to feed 1 cattle; the fuel ethanol production cost is $615.4/ton (49cent/litter) and the sweet sorghum stem price is at $30/t. In this study, the results of continuous fermentation and distillation at 555 m^3 scale are agree with the data from bench scale and pilot plant well. The TEA further demonstrates that ASSF is a cost-effective process that is highly feasible to be applied at industrial and commercial scale for fuel ethanol production, and eventually is able to improve the biofuel industry.

(780e) Evaluation of Bio-Oil Corrosion Characteristics

Patrick Johnston, Center for Sustainable Environmental Technologies, Iowa State University, Ames, IA and Robert C. Brown, Mechanical Engineering, Iowa State University, Ames, IA

Evaluation of Bio-oil Corrosion Characteristics

Patrick A. Johnston[1] and Robert C. Brown[1,2]

[1] *Center for Sustainable Environmental Technologies;* [2] *Department of Mechanical Engineering*

Iowa State University

Ames, IA 50011

Petroleum crude oils contain naphthenic acids that are one of the main causes of corrosion in the petroleum industry during storage, transport, and refining. "Naphthenic Acid Corrosion" NAC has been related to three parameters: Total Acid number (TAN), temperature, and flow rate [1, 2, 3, 4]. Bio-oils do not contain naphthenic acids but do however contain other organic acids such as acetic, formic, glycolic and propanoic acid that may have similar consequences to metal surfaces. A comprehensive corrosion study needs to be completed to determine the actual effects of the organic acids in bio-oil on select metal surfaces. As a result of this study, acid number and corrosion potential of pyrolysis bio-oils can be correlated and ranked to produce a fraction that is non-problematic with typical petroleum refining equipment.

The proposed project will evaluate and compare corrosion and acid number (AN) on fast pyrolysis bio-oil stage fractions produced at Iowa State University (ISU). The corrosion characteristics will be evaluated using a vacuum distillation apparatus that will simulate a distillation tower in a petroleum refinery. Metal specimens will be subjected to both the liquid and vapor phase of the bio-oil. Method D664 "Standard Test Method for Acid Number of Petroleum Products by Potentiometric Titration" will be used to determine acidy of the bio-oil samples. Ion Chromatography (IC) and High Performance Liquid Chromatography (HPLC) will be used to quantify organic acid concentration. The corrosiveness will be ranked gravimetrically according to mass loss during the study. Selected metal specimens will also be examined by Scanning Electron Microscopy (SEM) equipped with energy-dispersive spectroscopy (EDS) to determine morphological and composition changes on the metal surfaces.

References:

1. A. Jayaraman, H. Singh, and Y. Lefebvre, Rev. Inst. Fr.Pet.**41**, 265 (1986).

2. C. M. Cooper, Hydrocarbon Process.**51**, 75 (1972).

3. E. Babaian-Kibala, H. L. Craig, G. L. Rusk, et al., Mater.Perform. **32**, 50 (1993).

4. E.B. Zeinalov, V.M Abbasov, and L.I. Alieva, Petro. Chem.**49**, 185 (2009).

(790a) A Comparison of the Recoverability of Organic Solvent Mixtures in Waste Streams Using Process Simulation and Life Cycle Assessment Software

Eduardo J. Cavanagh[1], Mariano J. Savelski[2] and C. Stewart Slater[2], (1)Rowan University, Glassboro, NJ, (2)Chemical Engineering, Rowan University, Glassboro, NJ

Through the support of the U.S. Environmental Protection Agency and in partnership with Pfizer Inc., a flexible modeling tool that calculates avoided Life Cycle Assessment (LCA) emissions and Operative Cost Savings of solvent recovery systems for low volume pharmaceutical synthesis waste streams was developed. This new tool allows for a rapid assessment of the recoverability of solvent mixtures. ASPEN® Simulation Workbook was used for process modeling. Preliminary environmental impact was determined using SimaPro® and EcoSolvent® by comparing the proposed solvent recovery to the incineration of solvent waste and the manufacture of virgin solvent. Different homogeneous binary (azeotropic and non-azeotropic) solvent-solvent and water-solvent systems were analyzed. The proposed solvent mixtures are representative of those found in pharmaceutical manufacturing and include acetone, acetonitrile, toluene, tetrahydrofuran (THF), isopropyl alcohol, n-hexane, and methyl ethyl ketone (MEK).

The solvent recovery process model includes distillation, pervaporation (PV) and a combination of both. Sensitivity analysis for different system parameters such as reflux ratio and feed location was performed in several case studies, and points of minimum emissions and operating costs were also determined.

(790b) Sustainability of Battery Electric Vehicles Propelled by Lithium Ion Batteries : A Life Cycle Perspective

Qiang Dai and Christian M. Lastoskie, Civil & Environmental Engineering, University of Michigan, Ann Arbor, MI

With the foreseeable increasing penetration of battery electric vehicles (BEVs) into the U.S. market, the environmental impacts associated with the delivery of driven vehicle miles are worth investigating for this alternative mode of transportation as compared with conventional Internal Combustion Engine Vehicles (ICEVs). In this study, life cycle assessment has been conducted on Lithium Ion Battery (LIB)-powered BEVs representing the status-quo of current technology. The total life cycle impacts of BEVs, evaluated by Eco-indicator 99, Cumulative Energy Demand and Global Warming Potential, were compared with that of ICEVs burning petroleum to quantitatively demonstrate the environmental friendliness of BEVs. The importance of battery material chemistries, synthesis pathways and electricity mix was also investigated respectively. The results confirmed that BEVs equipped with LIBs offer substantial environmental benefits over ICEVs, including curtailing energy consumption, mitigating global warming, and reducing emissions capable of damaging human health and ecosystem quality. As use phase dominates the environmental burdens of BEVs, the switch to a greener electricity mix would be the key to a more sustainable transportation mode by BEV. Manufacturing and assembly phase of LIB, contributing to up to 20% of the total environmental footprints of BEVs, also provide windows of opportunity to alleviate environmental impacts through future BEV improvements, especially by extending cycle life and energy density of a battery pack.

(790c) Decision Making for Sustainable Processes

Raymond L. Smith, Gerardo J. Ruiz-Mercado and Michael A. Gonzalez, U.S. Environmental Protection Agency, Cincinnati, OH

Multi-objective decision support tools aim to represent the thought process of decision makers in the development and design of chemical processes. These tools often use calculations that mathematically describe preferences to combine various utility scores with weighting factors. Once utility scores are combined into a single objective, the alternative solutions for process development and design can be ranked. The creation of a single objective for a multi-attribute decision necessarily generates tradeoffs (except in the simplest win-win scenarios). While the intention in choosing (equal) weighting factors may be to make them value free, this contribution will show how combining utility scores is always a statement of a relationship between the weighted scores. The relationships between utility scores can be rewritten as a series of equations in terms of the weighting factors. Calculations can be used to determine the weighting factors for different sets of relationships, allowing for explicit representation of the implied tradeoffs for decision makers. In addition, a method will be described for using marginal rates of substitution between utility scores to calculate total utility. This method will simplify the relationships between the utility scores and present a powerful and intuitive way to calculate total utility.

The methods presented will be demonstrated in an example utilizing indicators derived from EPA's sustainability evaluation and design tool, GREENSCOPE (Gauging Reaction Effectiveness for the Environmental Sustainability of Chemistries with a multi-Objective Process Evaluator). GREENSCOPE has 140+ indicators in four basis areas: environment, economics, energy, and efficiency. Previous work has developed absolute sustainability limits for these indicators and provided data needs and solutions for calculations. The combination of GREENSCOPE with weighting / marginal rate of substitution methods and explicit representations of implied tradeoffs will provide users with tools to support multi-objective decisions.

(790d) Cradle-to-Gate Life-Cycle Assessment of Ionic Liquids

Amirhossein Mehrkesh and Arunprakash T. Karunanithi, Department of Civil Engineering, University of Colorado Denver, Denver, CO

Ionic liquids (ILs) are new class of "green" chemicals that are being developed as a replacement for volatile organic solvents for a range of industrial applications. The inherent greenness of ILs is often associated with negligible vapor pressures that they possess. Although, ILs lessens the risk of air pollution due to their insignificant vapor pressure, many are highly soluble in water and can become persistent pollutants. Existing studies show that certain ionic liquids have negative environmental and ecosystem impacts while others do not. Current strategies for designing novel ionic liquids tend to focus on tailoring their structures with an aim to impart specific functionalities (e.g. high CO2 solubility for carbon capture applications). However -in line with the assumption that all ionic liquids are green- they either do not consider the environmental and health impacts during the design or consider them after the fact. In this work, we present systems analysis view of certain ionic liquids and compare them with common organic solvents from a cradle-to gate life-cycle assessment (LCA) perspective. We will focus on the environment, ecosystem and health impacts associated with the production of different ionic liquids.

(790e) Life Cycle Assessment of Thin-Film CdTe Photovoltaics

Yuan Yao and Fengqi You, Department of Chemical and Biological Engineering, Northwestern University, Evanston, IL

Growing concern to adverse health effects and severe atmospheric pollution caused by the operation of fossil-fuel-burning power plants has fostered the research on more sustainable energy resources. Solar energy is one of the most promising alternatives for large-scale electricity production because of its abundance.[1] In the past 12 years, photovoltaics (PV) experienced an average growth of approximately 45% per year that was affected only slightly by the recent global financial crisis.[2] Among a variety of photovoltaic technologies, thin film is relatively new but developing rapidly for low cost, large flexibility[3] and small amount of materials used in modules. CdTe is one of the thin film solar cells that have the great potential to be produced in large scale with low cost.[4] However, any energy production, no matter how "green" they look like, will lead to environmental impacts in their entire life cycle.[5] Thus, it is necessary to apply Life Cycle Assessment (LCA) tools to CdTe photovoltaics to assess their environmental performance, including GHG and heavy metal emissions. Currently, most assessments are conducted based on the data from OECD countries (e.g., United States, Canada and Germany).[6, 7] However, most of solar cell companies prefer to move their manufacturing plants to non-OECD countries (e.g., China, India and Malaysia) because of low cost and abundant raw materials. Current LCA results of CdTe PV fail to provide decision makers with appropriate references on whether it is environmentally sustainable to have CdTe PV manufactured abroad, especially in none-OECD countries. It is of great importance to conduct comparative LCAs in two cases, where CdTe solar cells are manufactured domestically and oversea respectively.

There are several challenges to conduct this comparative study. The first one is to figure out all possible differences between two cases in each stage of CdTe PV life cycle. Since cadmium and tellurium are byproducts of zinc and copper production, LCAs should start with zinc and copper ore mining. Different grades of ores, efficiencies of metallurgical operation, electricity mixes, and some other factors may contribute to the difference of emission factors of life cycle stages between CdTe PV made in OECD and none-OECD countries. Another challenge is that the data for none-OECD countries are not as accessible as those for OECD members. A comprehensive understanding of material flows and emissions, and a deep data digging for non-OECD countries in all stages are required.

In this paper, a "Cradle to Gate" LCA, which includes raw material exploitation, material processing and PV manufacturing, of CdTe PV in two case studies is carried out. In the first case study, China is chosen as a representative of none-OECD countries. A CdTe PV manufacture plant is built in Sichuan Province where the first Chinese CdTe PV manufacturing line was located, [8] and CdTe solar panels are shipped to venders in the United States. In the second case, a local plant is built in Chicago, United States. The geographically representative data mainly come from industrial reports, literatures and Gabi.[9] GHG and heavy metal emissions in all life cycle stages are estimated in two case studies. Total life-cycle GHG emissions of CdTe PV made in China are 113 $kgCO_2$-eq/m^2 and heavy metal emissions are 1.17 mg Cd/m^2. While GHG emissions and heavy metal emissions of CdTe PV made in the United States are 74 $kgCO_2$-eq/m^2 and 0.37 mg Cd/m^2 respectively, which are much smaller than that in China. Emissions of other heavy metals, such as arsenic, lead, mercury and nickel are also estimated in two case studies. In order to integrate heavy metal and GHG emissions to the corresponding environmental impacts, Eco-indicator 99 [10] is applied to two case studies. The total indicator of CdTe PV in the United States is 4.32 points/m^2; while the indicator in China is 8.67 points/m^2, which indicates that the environmental damages caused by investing CdTe PV manufacturing in China is twice as severe as that in the United States. Thus CdTe PV made in the United States are more sustainable. Energy payback times (EPBT) are calculated based on 10.7% module efficiency, 80% performance ratio, 30 years life time and 1825 kWh/m^2/year that is forecast of radiation for Chicago area from National Solar Radiation Data Base by using time-series methods with Oracle Crystal ball.[11] EPBT for two case studies are very close, 0.7 years for CdTe PV in China and 0.62 years for that in the United States. This suggests that CdTe PV made in the United States spend relatively shorter time to generate the same amount of energy used in total life-cycle of CdTe PV. Based on LCAs results, CdTe PV made in the United States obtain a relatively smaller EPBT with much less environmental damages. Investing CdTe PV plants in non-OECD is not able to get a faster energy payback, while generates more emissions and causes severer environmental damages. Therefore, domestic manufacturing in OECD countries is a more environmentally sustainable approach suitable for CdTe PV.

References

[1] S. B. Darling, F. You, T. Veselka, and A. Velosa, "Assumptions and the levelized cost of energy for photovoltaics," *Energy & Environmental Science,* vol. 4, pp. 3133-3139, 2011.

[2] V. M. Fthenakis, "Sustainability metrics for extending thin-film photovoltaics to terawatt levels," presented at the MRS Bulletin, 2012.

[3] K. L. Chopra, P. D. Paulson, and V. Dutta, "Thin-film solar cells: An overview," *Progress in Photovoltaics,* vol. 12, pp. 69-92, 2004.

[4] K. Kato, T. Hibino, K. Komoto, S. Ihara, S. Yamamoto, and H. Fujihara, "A life-cycle analysis on thin-film CdS/CdTe PV modules," *Solar Energy Materials and Solar Cells,* vol. 67, pp. 279-287, 2001.

[5] V. M. Fthenakis and E. Alsema, "Photovoltaics energy payback times, greenhouse gas emissions and external costs: 2004–early 2005 status," *Progress in Photovoltaics: Research and Applications,* vol. 14, pp. 275-280, 2006.

[6] V. M. Fthenakis, W. M. Wang, and H. C. Kim, "Life cycle inventory analysis of the production of metals used in photovoltaics," *Renewable & Sustainable Energy Reviews,* vol. 13, pp. 493-517, 2009.

[7] V. M. Fthenakis and H. C. Kim, "Photovoltaics: Life-cycle analyses," *Solar Energy,* vol. 85, pp. 1609-1628, 2011.

[8] (Apr. 10th). *Chengdu COE Apollo Solar Co ., Ltd*. Available: http://www.coeapollo.com/en/about.asp

[9] Gabi, "Software and database contents for Life Cycle Engineering.," 2011 ed. Stuttgart: PE INTERNATIONAL AG.

[10] M. Goedkoop, R. Spriensma *et al*, "The Eco-indicator 99 a damage oriented method for life cycle impact assessment. Methodology report.," PRé-Consultants, Amersfoort2001.

[11] (24th, March). *Oracal Crystal Ball*. Available: http://www.oracle.com/us/products/applications/crystalball/index.html

(790f) Accounting for Ecosystem Services in Life Cycle Assessment by Eco-LCA: Advances in Methodology and Software

Bhavik R. Bakshi[1], Erin F. Landers[1], Shweta Singh[1], Laura A. Merugula[1], Oleg Mishchenko[2] and Joseph Fiksel[3], (1)William G. Lowrie Department of Chemical and Biomolecular Engineering, The Ohio State University, Columbus, OH, (2)Computer Science and Engineering, Ohio State University, Columbus, OH, (3)Center for Resilience, The Ohio State University, Columbus, OH

Ecologically-Based Life Cycle Assessment (Eco-LCA) is a framework to account for the role of ecosystem goods and services in supporting economic activities (Zhang et al., 2010). Ecosystem services are contributions from nature without which sustainability of any activity on earth would not be possible. They include provisioning services like food, biomass and water; regulating services like regulation of floods, climate and pests; supporting services like biogeochemical cycles and photosynthesis; and cultural services like tourism, aesthetics and spiritual development. Despite their importance, most sustainability methods have not accounted for their role.

In our previous research, we have developed ways of accounting for many ecosystem services, and have developed an environmentally extended input-output (EEIO) model of the United States economy to permit a quick Eco-LCA of economic sectors (Zhang et al., 2010). Hybrid Eco-LCA models have also been developed by combining detailed process level data with the IO model. Thermodynamic methods based on exergy and emergy analysis are used to aggregate various ecosystem services (Baral et al., 2012).

This presentation will describe some significant recent developments in the Eco-LCA data, method and software. A large number of provisioning services are now quantified in Eco-LCA as intermediate flows from relevant sectors by using the algebra of EEIO models. This relies on physical data about products such as grains, legumes, fish, biomass and medicinal plants. The Phosphorous cycle is also included in Eco-LCA by incorporating its flows into relevant sectors. This adds to the C and N cycles that are already included in Eco-LCA. Many ecosystem services, especially regulating services, are difficult to quantify. This challenge is overcome in this work by means of a novel approach of qualitative EEIO modeling and LCA. This approach relies on gathering information about the extent to which each sector of the economy depends on the selected ecosystem service. This dependence is represented in qualitative terms as "high," "medium," and "low," and this information is propagated through the IO model to get qualitative information about the extent to which the selected economic product depends on the ecosystem service.

These new developments have been implemented in the Eco-LCA software available at http://resilience.osu.edu/ecolca/. The software permits three types of analyses: thermodynamic, ecosystem services, and footprints. Thermodynamic analysis includes cumulative exergy consumption analysis with and without ecosystems. This analysis with ecosystems becomes analogous to emergy analysis. The tool enables calculation of metrics such as renewability index and thermodynamic return on investment. Analysis of ecosystem services provides information about these services by following classification systems like the one described in the first paragraph. These data are represented in multiple units, but may be aggregated by a hierarchical scheme that relies on thermodynamic methods. Footprint methods include univariate indicators like carbon, nitrogen, land use and water footprints. Use of this software for performing tiered hybrid LCA by combining process level data with the economy scale EEIO model will also be described and demonstrated, if time permits.

References

Baral, A., B. R. Bakshi, R. Smith, Assessing Resource Intensity and Renewability of Cellulosic Ethanol Technologies using Eco-LCA, *Environmental Science and Technology*, 46, 4, 2436–2444, 2012

Zhang, Y., A. Baral, B. R. Bakshi, Accounting for Ecosystem Services in Life Cycle Assessment, Part II: Toward an

Ecologically-Based LCA, *Environmental Science and Technology*, 44, 7, 2624-2631, 2010

(790g) Assessing the Critical Role of Natural Capital in Microalgal Biofuel Production

George Zaimes, Civil and Environmental Engineering, University of Pittsburgh, Pittsburgh, PA and Vikas Khanna, Department of Civil and Environmental Engineering, University of Pittsburgh, Pittsburgh, PA

Current global trends in resource consumption and fossil fuel depletion have driven research in biomass-to-energy systems. In recent years, microalgal biofuel have gained increased attention as a sustainable liquid-fuel alternative. Microalgal derived fuels do not suffer from many of the major drawbacks associated with first or second generation biofuels, and have the potential to act as "drop in" replacements for traditional petroleum fuels. As an emerging field, holistic evaluation of algae-to-fuel systems that considers the resource consumption, emissions, and their impact across the entire life cycle is critical to ensure the long-term sustainability of emerging algae-based energy systems. Current interest in microalgal biofuels has led to several microalgae based life cycle assessments (LCA). Most of these studies have focused on quantifying the greenhouse gas emissions and the net energy balance for microalgal biofuel production without sufficient attention to the spatial variability in algal biomass growth rates, climatological conditions, and stresses on ecological goods and services. As ecosystem goods and services (natural capital) provide the foundation for manmade capital, incorporating the broader role of ecosystem goods and services is critical to any accurate measure of sustainability. Proper quantification of ecosystem goods and services is also critical to assess the vulnerability of emerging microalgal biofuels to the depletion of natural capital and the role that nature plays in making ecological resources available for biofuel supply chains.

This work focuses on a comprehensive life cycle evaluation of 48 microalgal biomass production pathways. We build upon our prior research and examine multiple cultivation locations, culture conditions, cultivation and harvesting options, and industrial symbiosis within microalgal biomass production for the continental United States. Process flows and life cycle inventories are constructed based on first principles, peer reviewed literature, LCA databases, and best available engineering knowledge. The role of natural capital for emerging algae-to-energy systems is accounted for via the use of EcoLCA model. EcoLCA is a recent environmentally extended input-output life cycle oriented approach capable of accounting for the role of natural capital such as ecosystem goods and services in LCA. EcoLCA model considers a wide array of goods and services derived from nature and a hierarchical, thermodynamic aggregation scheme to permit meaningful interpretation. We quantify and aggregate the contribution of ecosystem goods and services in terms of mass, energy, and exergy (available energy) for the algae-to-energy systems. A variety of hierarchical metrics such as quality corrected thermodynamic return on investment and renewability index for the algae-to-energy systems are also developed. Tradeoffs between life cycle environmental impacts are analyzed via a series of two-dimensional pareto optimal surfaces. Synthesis of these results provides insights regarding the vulnerability of the emerging algae-to-energy systems to the depletion of specific resources.

(790h) Multi-Objective Synthesis of Biofuels' Processes by a Novel Concept of Eco-Profit, Extended to Eco-Net Present Value

Lidija Čuček and Zdravko Kravanja, University of Maribor, Faculty of Chemistry and Chemical Engineering, Maribor, Slovenia

Abstract

This paper's contribution introduces and discusses a novel concept of eco-net present value (Eco-NPV), also called eco-net present worth (Eco-NPW). Our recently proposed sustainability metric, based on the concept of eco-profit (Čuček et al., 2012; Kravanja and Čuček, 2012), which is on a yearly basis, has now been extended to Eco-NPV, which is based on the full lifetime of a particular project. Eco-profit is defined as the difference between unburdening (eco-benefit) and burdening (eco-cost) of the environment, where eco-benefit and eco-cost calculations are based on life cycle assessment (LCA). Since eco-profit is expressed in monetary units similarly to economic profit, they can be merged together into so-called 'total profit'. When the synthesis of a process, the complex of processes, or even a regional supply network is performed using e.g. the mixed-integer programming approach, the optimal solution is the one with the maximal total profit. The intention is not only to find economically the more attractive and environmentally least-burdening solutions, but also those that unburden the environment the most.

This concept has now been extended to the net present value (NPV). NPV is an economic measure which properly takes into account the complete economics of the project throughout the project's life-cycle; the initial investment expenditure, all the expected future annual cash flows, the time value of money, and the salvage value. Maximization of the NPV with a discount rate equal to the minimum acceptable rate of return (MARR), is probably the more appropriate method for making capital budgeting decisions, and for the optimization of process flow-sheets or their subsystems (Novak Pintarič and Kravanja, 2006). The Eco-NPV is an analogy of the economic NPV, where income in the NPV's yearly cash flow is now represented by unburdening (eco-benefit) and outcome by burdening (eco-cost) of the environment. Similarly as economic and eco-profit are merged into total profit, the economic NPV and Eco-NPV form total NPV, and the preferred solutions are those with maximal total NPV.

In this paper, this novel concept of eco-NPV is compared with the concept of eco-profit. This approach is illustrated through a couple of case studies of the supply chain synthesis for producing biofuels.

Keywords: Net present value, Eco-net present value, Eco-profit, LCA, Optimization, Biofuels

References

Čuček L., Drobež R., Pahor B., Kravanja Z., 2012, Sustainable synthesis of biogas processes using a novel concept of eco-profit, Computers & Chemical Engineering, doi: 10.1016/j.compchemeng.2012.01.010

Kravanja Z., Čuček L., 2012, Multi-objective optimization for generating sustainable solutions considering total effects on the environment, Applied Energy, doi: 10.1016/j.apenergy.2012.04.025

Novak Pintarič Z., Kravanja Z., 2006, Selection of the objective function for the optimization of process flow sheets, Industrial & Engineering Chemistry Research, 45, 4222-4232

(791a) Nanomaterials Enabled Lithium Chemistry for Advanced Rechargeable Batteries

Chao Wang, Materials Science Division, Argonne National Laboratory, Argonne, IL

Employment of nanomaterials in electrode assembly is promising for the development of advanced lithium-ion batteries. Nanostructured electrodes can potentially achiever higher power and better cyclability than their bulk counterparts, due to the facilitated charge transfer at electrode/electrolyte interface, electronic/ionic transport in the electrodes, and accommodation of strain during lithium insertion/extraction by reduction of material dimensions. Moreover, design of electrodes based on nanomaterials can enable new lithium insertion/extraction reactions that are not possible with bulk materials, and thereby expand our scope of material choices for energy storage. These advantages have lead to substantial literature effort on the synthesis of electrochemically active nanomaterials for battery applications.

Here we report the development of novel manganese oxide nanoparticles (10 - 100 nm) by organic solution synthesis and their solid chemical conversion into lithium manganese oxide ($Li_xMn_2O_4$) spinels for cathode application in lithium ion batteries. We managed to demonstrate that by reducing the material dimensions to nanoscale, charge/discharge up to $0<x<2$ can be achieved which however is not possible for conventional bulk or microscale materials (with only $0<x<1$). Our work shows the promise of expanding the scope of material choices for energy storage by size scaling and the great of potential of improving battery performnace by employing advanced nanomaterials.

(791b) Dynamics of Pumped Heat Energy Storage Systems

Hugo S. Caram and Fan Ni, Chemical Engineering, Lehigh University, Bethlehem, PA

The eventual success and widespread utilization of photovoltaic and wind energy generation will depend on the availability of efficient and economical storage capability. The current technology for long term (diurnal), high efficiency large scale energy storage of electrical energy appears to have been limited to Pumped Hydroelectric Storage (PHS) and Compressed Air Storage (CAES). Both of them are dependent of the availability of geological features such as elevated lakes or empty deep mines for PHS or caverns for the CAES. This features are not available everywhere and other technologies such flow batteries while capable of large scale storage are limited in their power output.

Pumped heat energy storage systems can provide a valuable alternative PHS or CAES. The proposed process uses a Brayton cycle where during the loading cycle heat is pumped to heat up a high pressure thermal regenerator while a low pressure thermal regenerator is cooled down to a low temperature. During the unloading period the gas circulation is reversed. Cold gas from the low pressure regenerator is compressed and heated in the high pressure regenerator to be then expanded to the low pressure vessel. Electrical energy storage capacities in the 30-50kWh/m^3 range appear possible and are comparable to lead acid batteries (25-50 kWh/m^3).

Analysis and optimization of these systems are complicated by the difficult dynamics of thermal regenerators. A novel matrix algorithm is proposed that allows us to directly calculate cyclic steady state as well as transients. The effects of different design parameters such as pressure ratios, compressor and expander efficiencies, type of gases, solid heat capacities and gas solid heat transfer coefficients are explored. Possible synergistic integration with natural gas combustion combined cycles will be discussed.

The effective utilization System for Renewable Energy Making use of Regional Characteristic

Masahiro Yamamoto, Tsuguhiko Nakagawa
Graduate school of *Okayama Prefectural University,*
Kuboki 111, Soja-shi, 719-1197, Okayama, Japan

ABSTRACT

The SMART factory project has been developing a process to reduce CO_2 emissions through researching and developing a process to make plastic from biomass at the biomass accumulated factory in Maniwa, Okayama.

At the biomass accumulated factory, wood chips have been made from shavings and unused wood parts from sawmills. The biomass factory consists of a building chips plant which makes the wood chips, a building silo which stores chips, and a barkyard which crushes bark. The energy used at the biomass accumulated factory consists of electric power and liquid fossil fuels. Liquid fossil fuels have been used by forklift trucks and others vehicles for wood chips. The CO_2 emissions by liquid fossil fuels are accounted for about 43% of the total from the biomass accumulated factory. In order to reduce 25% CO_2 emissions, the SMART factory is planned to be supplied by renewable energy from the wood chips plant and photovoltaic power generation.

In this study, in order to achieve this aim, the mathematical simulation model which calculates the energy demand and supply per hour in the SMART factory has been developed. Based on the mathematical simulation model, a system which is able to reduce consumption of liquid fossil fuels and electric power has been designed.

As a result, it has been proved that it is possible to reduce 25% or more CO_2 emissions using the system that combines photovoltaic power generation with electric forklift trucks.

1. INTRODUCTION

The SMART factory project in Maniwa has been developing a process to reduce CO_2 emissions using renewable energy through researching and developing a process to make plastic from biomass at the biomass accumulated factory.

In this study, we have examined the system which substituted fossil fuels to be used as fuel for mobile equipment such as forklift trucks for renewable energy such as photovoltaic power generation (hereafter refered to as "PV") which will be constructed at the biomass accumulated factory.

2. OUTLINE OF THE NEW ENERGY SYSTEM

Outline of the new energy system is shown in Fig.1. At the biomass accumulated factory, mobile equipment such as two forklift trucks, shovel loaders and two grapples have been used. These vehicles have previously used liquid fossil fuels. Among this mobile equipments we considered that forklift trucks be substituted for electric forklift trucks and the system that charges the PV electricity to the battery of electric forklift trucks.

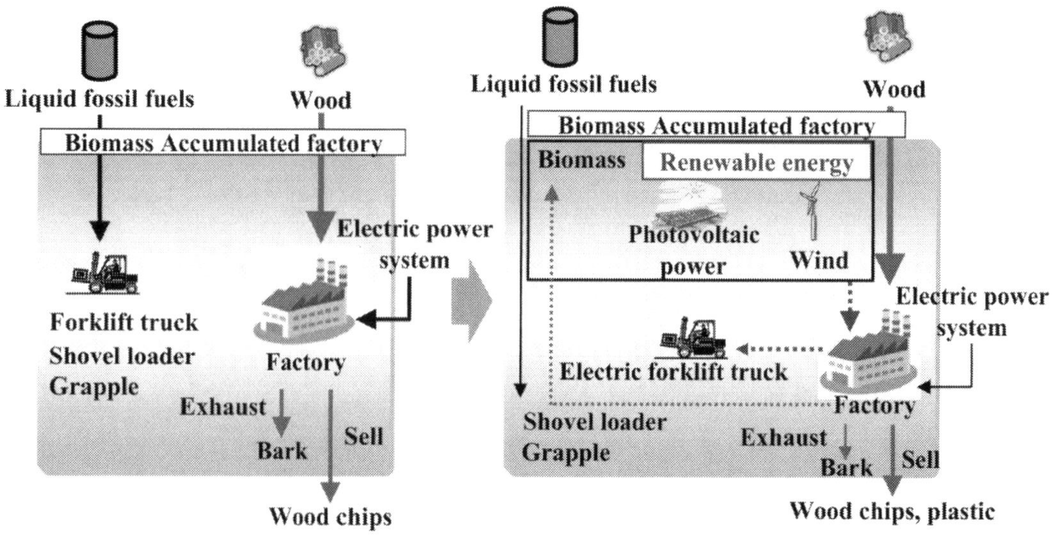

Fig.1. Images of the new energy system

3. THE EXAMINATION METHOD

3.1 Model Equation

For each production of wood chips, using equation (1), the mathematical simulation model which is able to calculate the energy demand and supply per hour in a year (8760hr) has been designed.*[1]

$$\sum_{i=1}^{n} \alpha_i = \frac{X_i}{\eta} + E_i - \beta_i \quad \cdots\cdots\cdots (1)$$

where α_i : electric power from electric power system [kWh]

E_i : electric power demand of factory [kWh]

X_i : electric power demand of electric forklift truck [kWh]

β_i : electric power generation of PV [kWh]

η : Charging&discharging efficiency (=0.85)

3.2 Production of wood chips and energy consumption

The relationship of monthly production of wood chips and electric power demand, liquid fossil fuels consumption of forklift truck at the biomass accumulated factory is shown in Fig.2 and Fig.3.

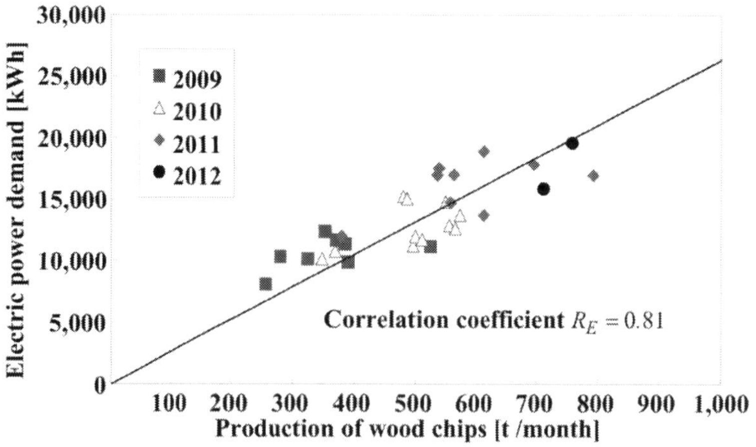

Fig.2. Relationship between electric power demand and production of wood chips

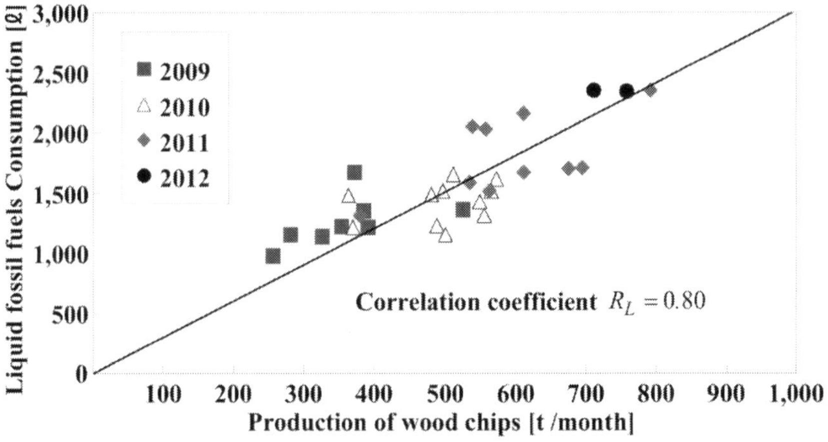

Fig.3. Relationship between liquid fossil fuels consumption of forklift trucks and production of wood chips

In Fig.2, it can be seen that there is a correlation between electric power demand and production of wood chips. For least-squares method,

$$y = 26.3x \quad \cdots\cdots\cdots\cdots (2)$$

From equation (2), it can be estimated electric power demand from production of wood chips. On the other hand, in Fig.3, similarly, it can be obtained by equation (3).

$$Z = 0.9x \quad \cdots\cdots\cdots\cdots (3)$$

where x : Production of wood chips [t/month]
y : electric power demand [kWh]
Z : liquid fossil fuels consumption of forklift trucks [ℓ]

Using equations (2) and (3) in this new energy system the optimal operation performance can be calculated depending on the production level of wood chips.

3.3 Result of examination

Regarding production of wood chips at 400[t/month] and 700[t/month], the relationship of CO_2 emissions and area of PV at the biomass accumulated factory is shown in Fig.4.

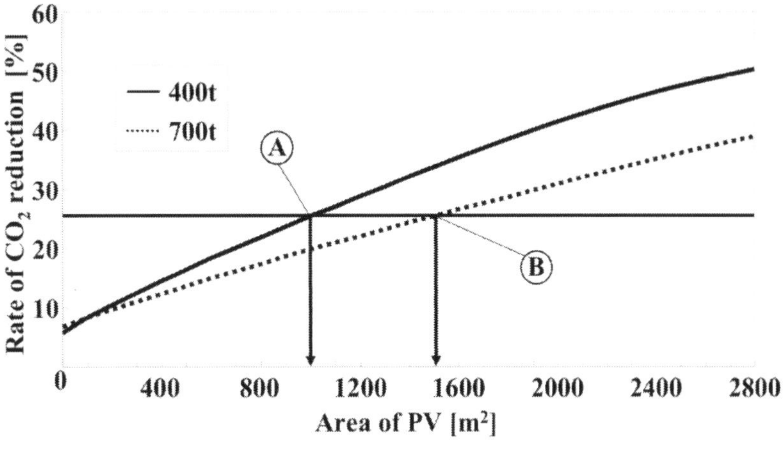

Fig.4.Relationship area of PV and rate of CO_2 reduction per production of wood chips

In Fig.4Ⓐ,Ⓑ, the breakdown in the case of reducing 25% CO_2 emissions is shown Fig.5 Ⓐ,Ⓑ. In Fig.4, PV is of low electric power generation efficiency. But, it is able to be installed on a sloping site e.g. the factory roofs. Also, it is amorphous that installation costs are low. Maximum output per unit area is 69.1[W/m^2].

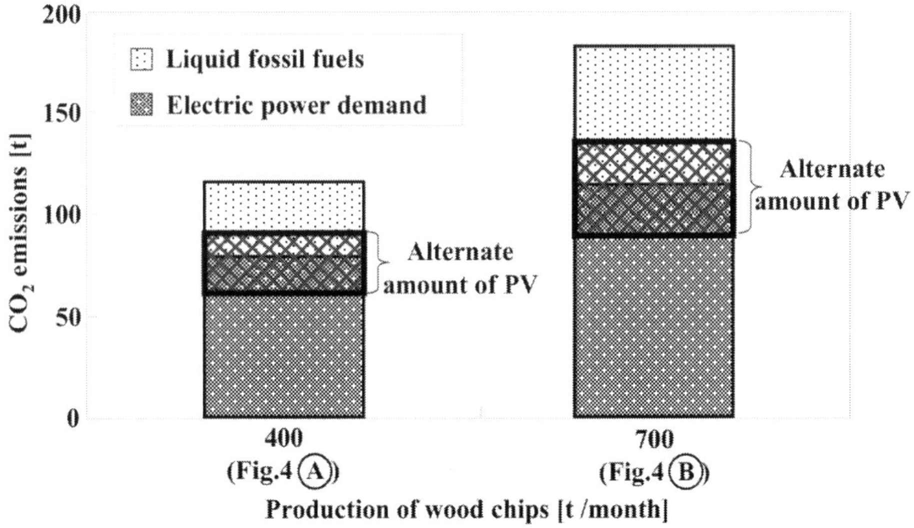

Fig.5. The breakdown in the case of reducing 25% CO_2 emissions

As a result, when production of wood chips is 700[t/month], it has been proved that the PV installation area should be greater than 1500m^2 to reduce 25% or more CO_2 emissions at the biomass accumulated factory.

4. CONCLUSION

Depending on the rate of Production of wood chips, it has been proved that it is possible to estimate liquid fossil fuels and electric power demand and calculate the design area of PV depending on the target of CO_2 emissions reduction.

In this study, we are subsidized by the funds form the Overall Promotion of Ministry of Education, Culture, Sports, Science and Technology. I am deeply grateful to them for their supports.

Reference

1. Shimohara, K., Y. Notoji, T. Nakagawa "Distributed Energy System Combination with Energy Conservation, PV and EV", *Kagaku kogaku Ronbunshu* Vol.38 No.4, pp.255−262, Japan (2012)

(791d) Thermochemical Cycle of a Mixed Metal Oxide for Augmentation of Thermal Energy Storage in Solid Particles

Brian D. Ehrhart[1,2], Nathan P. Siegel[3], Eric N. Coker[2] and Alan W. Weimer[1], (1)Chemical and Biological Engineering, University of Colorado at Boulder, Boulder, CO, (2)Sandia National Laboratories, Albuquerque, NM, (3)Mechanical Engineering, Bucknell University, Lewisburg, PA

Solid particle heat transfer and storage media have been shown to be able to operate at temperatures >1000°C in concentrated solar power (CSP) applications, much higher than the operational limit of 600°C for currently state-of-the-art molten nitrate salt heat transfer fluid. Solid particles can be endothermically reduced by direct exposure to concentrated solar energy, thus absorbing and storing thermal energy beyond that possible with sensible heating alone. The particles can then be oxidized exothermically at a later time, releasing the stored chemical heat and effectively augmenting the thermal energy storage capacity of the solid particles. A mixed metal oxide spinel material that reduces in the temperature range of interest (1000°C-1200°C) has been examined for applicability to this thermochemical energy storage concept. A description of this application, prospective materials, and details of the thermochemical cycle are presented. The heats of reduction and oxidation for the thermochemical cycle have been determined for various operating conditions to evaluate the amount of thermal energy that may be stored. Additionally, the extent of reduction possible at various operating conditions has been measured.

(791e) Carbohydrate Could Be the Best Electricity Storage Compound and a High-Density Hydrogen Carrier

Y.-H. Percival Zhang, Biological Systems Engineering, Virginia Polytechnic Institute and State University, Blacksburg, VA and Chun You, Biological Systems Engineering, Virginia Polytechnic Institute and State University

Electricity is a high-quality secondary energy, which can be generated from renewable primary energies and fossil fuels. Carbon-neutral electricity generation is becoming more and more important, but most of them, such as wind and solar electricity, suffer from large output variations. Therefore, it is vital to store electricity in the form of chemical compounds. Among numerous potential chemical compounds, such as hydrogen, isobutanol, methanol and hydrocarbons, we suggest that carbohydrates (e.g., starch, cellulose, glucose) could be the best electricity storage compound with a combustion energy density of 17 MJ/kg and an electricity output density of >3,000 Wh/kg [1]. We designed a novel biological CO2 fixation pathway that can fix CO2 by using hydrogen or electricity for producing carbohydrate mediated by cell-free cascade enzymes [1]. In this talk, we will present our efforts in developing building blocks and building modules for this ambitious project and conduct proof-of-concept experiments. Through the above bioconversion plus our previous achievements such as theoretical conversion of carbohydrate to hydrogen or electricity [2] and low-cost enzymatic conversion of cellulose to starch (unpublished), the electricity-carbohydrate-hydrogen cycle would not only supplement current and future primary energy utilization systems for facilitating electricity and hydrogen storage and enhancing secondary energy conversion efficiencies, but also address such sustainability challenges as transportation fuel production, CO_2 utilization, fresh water conservation, food and feed production, and maintenance of a small closed ecosystem in emergency situations [1]. In a word, carbohydrate would become a new oil in the future.

1. Zhang Y-HP, Huang W-D: **Constructing the electricity-carbohydrate-hydrogen cycle for a sustainability revolution**. *Trends Biotechnol.* 2012, **Epub. DOI: 10.1016/j.tibtech.2012.02.006**.

2. Zhang Y-HP, Evans BR, Mielenz JR, Hopkins RC, Adams MWW: **High-yield hydrogen production from starch and water by a synthetic enzymatic pathway**. *PLoS One* 2007, **2**:e456.

(791f) Energy Storage Design for Electric Transmission Systems

Donald J. Chmielewski, Illinois Institute of Technology, Chicago, IL

It is widely recognized that a major concern with the installation of renewable energy systems is the fact that wind and solar sources are non-dispatchable. That is, the power produced from renewable systems is dependent on environmental conditions and is likely uncorrelated with the power demand from load centers. In contrast, fossil based sources typically have the ability to track demand, but depending on the technology possess limits with respect to response time. Thus, many have advocated the use of massive energy storage systems to capture energy produced from renewable sources and then dispatch as load centers demand. An additional concern is the fact that renewable sources tend to be located far from the load centers (again in contrast to most fossil sources which are typically placed near demand). This combination of spatial and temporal mismatch with load center demand has caused many to worry about the stability of the electric power grid. Clearly, wide swings in power throughput over great distances will cause the weakest parts of the grid to fail.

To address both the spatial and temporal issues surrounding renewable power generation we have developed a scheme aimed at analyzing the capabilities of a regional power generation and distribution system. In particular, the scheme proposes a supervisory control law aimed at utilizing a mix of renewable and fossil sources along with energy storage systems to meet power load demands. This proposed control law takes into consideration not only the power output capacity of each source but also the rate at which power output can change as well as the storage capacity and efficiency losses associated with the storage units. In addition, the proposed scheme incorporates the maximum and minimum throughput limits on the distribution lines. If time allows, we will present an extension of the scheme aimed at sizing and placing energy storage systems as well as locating new power distribution lines.

AUTHOR INDEX

AUTHOR INDEX

AUTHOR INDEX

AUTHOR INDEX

AUTHOR INDEX

AUTHOR INDEX

AUTHOR INDEX

AUTHOR INDEX

AUTHOR INDEX

American Institute of Chemical Engineers
3 Park Avenue
New York, New York 10016-5991

proceedings.com

Education Division 2017

Core Programming Area at the 2017 AIChE Annual Meeting

Minneapolis, Minnesota, USA
29 October - 3 November 2017

AIChE
The Global Home of Chemical Engineers